D0437729

Biochemical and Cellular Mechanisms of Stress Tolerance in Plants

NATO ASI Series

Advanced Science Institutes Series

A series presenting the results of activities sponsored by the NATO Science Committee, which aims at the dissemination of advanced scientific and technological knowledge, with a view to strengthening links between scientific communities.

The Series is published by an international board of publishers in conjunction with the NATO Scientific Affairs Division

A	Life Sciences	Plenum Publishing Corporation
B	Physics	London and New York
C	Mathematical and Physical Sciences	Kluwer Academic Publishers Dordrecht, Boston and London
D	Behavioural and Social Sciences	
E	Applied Sciences	
F	Computer and Systems Sciences	Springer-Verlag Berlin Heidelberg New York
G	Ecological Sciences	London Paris Tokyo Hong Kong
H	Cell Biology	Barcelona Budapest
I	Global Environmental Change	

NATO-PCO DATABASE

The electronic index to the NATO ASI Series provides full bibliographical references (with keywords and/or abstracts) to more than 30 000 contributions from international scientists published in all sections of the NATO ASI Series. Access to the NATO-PCO DATABASE compiled by the NATO Publication Coordination Office is possible in two ways:

- via online FILE 128 (NATO-PCO DATABASE) hosted by ESRIN, Via Galileo Galilei, I-00044 Frascati, Italy.

- via CD-ROM "NATO Science & Technology Disk" with user-friendly retrieval software in English, French and German (© WTV GmbH and DATAWARE Technologies Inc. 1992).

The CD-ROM can be ordered through any member of the Board of Publishers or through NATO-PCO, Overijse, Belgium.

Series H: Cell Biology, Vol. 86

Biochemical and Cellular Mechanisms of Stress Tolerance in Plants

Edited by

Joe H. Cherry

Department of Botany and Microbiology
101 Life Science Bldg.
Auburn University, Auburn, AL 36849, USA

Springer-Verlag
Berlin Heidelberg New York London Paris Tokyo
Hong Kong Barcelona Budapest
Published in cooperation with NATO Scientific Affairs Division

Proceedings of the NATO Advanced Research Workshop on Biochemical and Cellular Mechanisms of Stress Tolerance in Plants, held at Maratea, Italy, June 20-24, 1993

ISBN 3-540-58215-0 Springer-Verlag Berlin Heidelberg New York
ISBN 0-387-58215-0 Springer-Verlag New York Berlin Heidelberg

CIP data applied for

© Springer-Verlag Berlin Heidelberg 1994
Printed in Germany

Typesetting: Camera ready by authors
SPIN 10089666   31/3130 - 5 4 3 2 1 0 - Printed on acid-free paper

PREFACE

Biological and environmental stresses represent the most limiting factors to agricultural productivity worldwide. These stresses impact not only crops that are presently being cultivated, but also are significant barriers to the introduction of crop plants into areas that are not at this time being used for agriculture. The rising trend in atmospheric CO_2 levels will increase climate variability and is likely to increase the severity of stresses (particularly those related to temperature, salinity and drought) to which plants will be exposed in the coming decades. Intensive irrigation in agriculture production has resulted in severe salinity. In Israel and in several developing countries, major problems related to drought, high temperature and salinity are already limiting agriculture productivity. Low temperature stress is responsible for a decline in productivity throughout many countries.

As part of a long range program to address the problems of abiotic stresses on plant productivity, a number of scientists have made a concerted effort to concentrate on fundamental research in plant stress biology. Programs established by these scientists include research on membrane changes that may be involved in salinity tolerance, the role of abscisic acid in adaptation of cells to NaCl, biochemical basis of low temperature tolerance, heat tolerance in suspension cells and whole plants, physiological mechanisms of inhibiting plant cell enlargement by mechanical stress, molecular cloning of genes for salt tolerance and thermoadaptation, molecular basis of adaptation to osmotic stress, regulation of gene expression and protein synthesis by abiotic stress factors.

When we began to organize the NATO/ARW, we believed it to be very timely to discuss the cellular, biochemical and molecular mechanisms that regulate tolerance for a variety of environmental stresses. We were particularly interested in ways that biotechnology may be applied to these problems in the modification of germplasm for tolerance that will impact agricultural productivity.

The NATO/ARW held in Italy in June, 1993 brought together a large number of scientists from 20 countries to discuss and compare results on molecular and cellular basis of abiotic stress tolerance.

The lectures published here reflect the diversity of current research in plant molecular biology and stress biology. Each lecture gives us a glimpse of the depth of questions being asked in various segments of this field of investigation. This research is directed at fundamental biological problems, but answers to these questions will provide knowledge essential for bringing about major changes in the way the world's agricultural enterprise can be improved.

It is unfortunate that this volume cannot bring to the reader the sense of discovery that came to participants as chapters of modern molecular plant biology were described, explained, and discussed enthusiastically. As organizers of this conference as well as participants we are grateful to the lecturers for clear expositions both in their formal presentations and in response to questions.

Speakers and students in this ARW represented a substantial fraction of all of the laboratories of the world working on molecular mechanisms of stress biology in plants. The generous financial support from the NATO International Scientific Exchange Programmes permitted participation not only by Europeans but also by speakers and students from the United States and elsewhere in the world.

In addition to the North Atlantic Treaty Organization, we are grateful for support from the following in the USA:

•U.S. Department of Agriculture, Washington, DC

•American Cyanamid Company, Princeton, NJ

•Ciba Plant Protection, Ciba-Geigy Corporation, Greensboro, NC

•Extended Product Life, 200 Four Falls Corporate Center, Conshohocken, PA

•Pioneer Hi-Bred International, Inc. Plant Breeding Division, Johnston, IA

and from the following donors from Italy:

•Microglass, Napoli

•Delchimica, Napoli

•Bioclinical Forniture, Napoli

•Gamma-Sud, Napoli

•Farmen Seed Comapny, Torre del Greco

•M-Medical, Firenze

•Ministero dell' Agricoltura e delle Foreste, Roma

•Centro Miglioramento Genetico degli Ortaggi-CRN, Research Center for Vegetable Breeding Portici

•Luigi Esposito, Ina-Assitalia

BIOCHEMICAL AND CELLULAR MECHANISMS OF STRESS TOLERANCE IN PLANTS

NATO Advanced Research Workshop

Maratea, Italy

June 20-24, 1993

_____ *Organized by* _____

Joe H. Cherry
Department of Botany & Microbiology
101 Life Sciences Building
Auburn University, AL 36849
USA

Luigi Monti
University of Naples
Department of Agronomy
    and Plant Sciences
Via Universitá 100
80055, Portici
ITALY

_____ *Organizing Committee* _____

Antonella Leone
National Research Council
Research Center for
  Vegetable Breeding
Via Universitá 133
80055, Portici
ITALY

Ray Bressan
Horticulture Department
Purdue University
West Lafayette, IN   47907
USA

Stefania Grillo
National Research Council
Research Center for
  Vegetable Breeding
Via Universitá 133
80055, Portici
ITALY

Narendra Singh
Department of Botany &
  Microbiology
101 Life Sciences Building
Auburn University, AL 36849
USA

TABLE OF CONTENTS

Chapter 1

HIGH TEMPERATURE STRESS

Chapter 2

DROUGHT STRESS

Chapter 3

SALINITY STRESS

Chapter 4

LOW TEMPERATURE STRESS

Chapter 5

ENGINEERING PLANTS FOR STRESS TOLERANCE

Chapter 1

HIGH TEMPERATURE STRESS

The heat stress response as part of the plant stress network:
An overview with six tables

Lutz Nover

Dept. Cell Biology, Biocenter of the Goethe-University, Marie-Curie-Str. 9, D-60439 Frankfurt and Dept. Stress Research, Institute of Plant Biochemistry, Weinberg 3, D-06120 Halle

1. Introduction

The molecular cell biology of the heat stress response starts with the report by Ritossa (1962) on heat stress (hs)-induced changes of gene activity in Drosophila salivary glands. Twelve years later, this was followed by the detection of the corresponding heat stress proteins (HSP) (Tissieres et al. 1974). With the developing new techniques rapid progress was made toward cloning the Drosophila hs genes (for summaries see Ashburner and Bonner 1979; Schlesinger et al. 1982). The explosive development of molecular stress biology in the following decade extended the investigations to all types of organisms. In all cases the heat stress (hs) response was found to comprise a highly complex but transient reprogramming of cellular activities necessary to protect cells from extensive damage and to provide optimum conditions for recovery after the stress period (Nover et al. 1990, Nover 1991).

As was noticed already by Ritossa, the rapid activation of hs genes is not restricted to hyperthermic conditions. Similar effects are observed by treatment with chemical stressors, including dinitrophenol and salicylate (Ritossa 1962) but also heavy metals, amino acid analogs and other drugs affecting protein synthesis, structure and function. Formation of denatured proteins or abortion of protein fragments by the translation apparatus may be central to the signal transduction mechanism (see recent summaries by Morimoto 1993 and Scharf et al. 1994). At the end of the signal transduction chain is the specific receptor element (HSE) in the promoter of all eukaryotic hs genes (Pelham 1982; Pelham and Bienz 1982; Nover 1987)

NATO ASI Series, Vol. H 86
Biochemical and Cellular Mechanisms
of Stress Tolerance in Plants
Edited by J. H. Cherry
© Springer-Verlag Berlin Heidelberg 1994

Table 1: : SURVEY OF EUKARYOTIC HSP FAMILIES

Table enlarged from Scharf et al. 1994: Members of each stress protein family are characterized by their size (apparent molecular weight in kDa), sequence homology and in many cases functional similarity (see last column). The E. coli proteins corresponding to the particular stress protein family are indicated in paranthesis. For summarizing information see Georgopoulos, 1992; Nover, 1991; Nover et al., 1990; Vierling, 1991. Additional references are given in footnotes and Tables 3 and 4. Proteins printed bold face are hs-inducible.

Organism (No. of genes)	Cellular localization Cytoplasm/ nucleus	ER	Chloropl. (c) / mitoch. (m)	Functional characteristics, remarks[a]
HSP110 family (Ec-ClpA/B)				
Mammals (2)	**HSP110**	-	-	In E. coli ATP-binding subunits of high m.w. protease complex; function in eukaryotes unclear; yeast HSP104 essential for general stress resistance
Plants (>2)	ClpB	-	ClpC	
Yeast (1)	**HSP104**	-	-	
HSP90 family (Ec-HtpG)				
Mammals (3)	**HSP89α, B**	GRP94	-	ATP-binding, autophosphorylating proteins; interact with Ca-calmodulin; in mammals characterized as inhibitory subunit of steroid hormone receptors complexed with the HSP56 immunophilin
Plants (1-6)	**HSP80**	?	-	
Yeast (2)	**HSP90**, HSC90	?	-	
HSP70 family (Ec-DnaK)				
Mammals (7-8)	**HSP70 A,B**, HSC72	GRP78	HSC70m	ATP-binding, autophosphorylating proteins with a potential calmodulin-binding site; ATPase activity stimulated by binding to protein substrates or distinct peptides; chaperones essential for folding, intracellular distribution, assembly/disassembly and degradation of proteins (see Table 4); biological activity requires interaction with DnaJ (HSP40) proteins
Plants (5-10)	**HSP70**	GRP78	HSC70c,**HSP68m**	
Yeast (8)	SSA 1-4, SSB 1,2	KAR2	SSC1m	
HSP40 family (Ec-DnaJ)				
Mammals (>1)	HDJ-1	?	?	In E. coli and yeast characterized as substrate-binding subunit of the DnaK/J complex; in yeast several DnaJ mutants found to be defective in protein sorting; Sec63 is a larger ER membrane protein with a DnaJ-domain exposed to the lumen
Plants (>2)	**ANJ1**	+	+	
Yeast (4)	YDJ1, SIS1	SEC63	**SCJ1**	

5

Family / Organism				Description
HSP60 family (Ec-GroEL)				ATP-binding, heptameric chaperone essential for assembly of protein complexes; two heptamers of GroEL interact with one heptamer of GroES to bind one molecule of substrate protein to the inner cavity; GroEL phosphorylated during hs; the cytoplasmic homologs (TCP1) form large complexes of 900 kDa including other proteins
Mammals (>2)	TCP1	—	**Cpn60m**	
Plants (>3)	TCP1	—	HSP60c, **HSP60m**	
Yeast (>2)	TCP1	—	HSP60m	
HSP10 family (Ec-GroES)				Heptameric complex interacts with corresponding proteins of complexes of the GroEL /HSP60 family; essential for normal function of GroEL
Mammals (1)	-	-	Cpn10m	
Plants (1-2)	-	-	Cpn10m, Cpn12c	
Yeast	?	-	?	
HSP20 family (?)				Form oligomeric complexes of 200-500 kDa, which further aggregate under hs conditions (hs granules); HSP25 and α,B-crystallin exhibit chaperone activity in vitro; phosphorylated HSP25/27 involved in smooth muscle contraction (actin binding); prominent protein in mammalian breast cancer cells
Mammals (2)	**HSP25/27, α,B-crystallin**	?	?	
Plants (15-20)	HSPs 15-18	**HSP18/19**	**HSP21/22c**	
Yeast (1)	HSP26	-	-	
Hsp8.5 family (lacking in E. coli)				Tag for protein degradation by ATP-dependent protease complex; polyubiquitin genes are stress-inducible in all eukaryotes; interacts with the proteasome system
Mammals (several)	Ubiquitin	-	-	
Plants (several)	Ubiquitin	-	-	
Yeast (several)	Ubiquitin	-	-	
Immunophilins (cyclophilin/FKBP)				Immunophilins (peptidylprolyl cis-trans isomerases) are ubiquitous chaperones with Mr 12-59 kDa found in all organisms and in all parts of the cell; some are stress-inducible; they are distinguished by binding of immune suppressive drugs cyclosporin A (CYP) or FK506 (FKBP); HSP56 is an FKBP associated with steroid hormone receptors
Mammals (~20)	FKBP12,25, CYP 18,22,40 HSP56	FKBP13, CYP22	?	
Plants	CYP	?	CYP, FKBP	
Yeast	CYP 1,2	FKBP13	?	

Footnotes to Table 1:

a. Family-specific characteristics are indicated in short. They frequently include results with the E. coli homologs. In many cases the allocation of a given stress gene product to the family is based on sequence homology only and not on its functional analysis. Proteins with a pronounced heat stress induction of synthesis are printed bold face, others are constitutively expressed (e.g. HSC72) or their synthesis increases under conditions of glucose deficiency (e.g. GRP78).

Additional references:

HSP110: Parsell et al. 1991; Sanchez et al. 1992, Squires and Squires 1992, Woo et al. 1992

HSP90: Callebaut et al. 1992; Conner et al. 1990; Csermely and Kahn 1991; Csermely et al. 1993; Felsheim and Das 1992; Lees-Miller and Anderson 1991; Legagneux et al. 1991; Pratt et al. 1992; Tai et al. 1992; Wiech et al. 1992

HSP70/DnaK: Flaherty et al. 1991; Landry et al. 1992; Leustek et al. 1992; McCarthy and Walker 1991; Neumann et al. 1993; Palleros et al. 1991; Sadis and Hightower 1992; Wang and Lin 1993; Watts et al. 1992. The number of genes /proteins of the HSP70 family may be much larger. Thus, there are 2-3 sperm-specific isoforms in mammals (Allen et al. 1988; Matsumoto and Fujimoto 1990; Rosario et al 1992;Wolgemuth and Gruppi 1991), a cold-stress induced CSP79 (BiP) in spinach (Neven et al. 1992), up to six different GRP78/BiP in tobacco (Denecke et al. 1991) and a specific SCE70 of the spinach chloroplast outer membrane (Ko et al. 1992)

HSP40/DnaJ: Atencio and Yaffe 1992; Blumberg and Silver 1991; Caplan et al. 1992; Cyr et al. 1992; Feldheim et al. 1992; Hattori et al. 1992; Raabe and Manley 1991; Zhong and Arndt 1993; Zhu et al. 1993; see review by Silver and Way 1993

HSP60/GroEL: Ahmad and Gupta 1990; Braig et al. 1993; Frydman et al. 1992; Gao et al. 1992; Hallberg et al. 1993; Langer et al. 1993; Lewis et al. 1992; Mori et al. 1992; Mummert et al. 1993; Prasad and Stewart 1992; Sherman and Goldberg 1992a; Waldinger et al. 1989; Yaffe et al. 1992

HSP10/GroES: Bertsch et al. 1992; Hartman et al. 1992a, 1992b

HSP20: Adamska and Kloppstech 1991; Behlke et al. 1991; Bitar et al. 1991; Chen and Vierling 1991; DeRocher et al. 1991; Gaestel et al. 1991; Inaguma et al. 1992; Jakob et al. 1993; Klemenz et al. 1991; Landry et al. 1992; Merck et al. 1993; Miron et al. 1991; Neumann et al. 1987; Nover et al. 1983, 1989; Rao et al. 1993; Tetu et al. 1992; see reviews by Bloemendal and deJong 1991 and Vierling 1991

Ubiquitin: Jentsch et al. 1991; Jentsch 1992; Seufert and Jentsch 1992

Immunophilins (Cyclophilin/FKBP): Braman et al. 1992; Callebaut et al. 1992; Gasser et al. 1990; Jin et al. 1992; Kieffer et al. 1992; Marivet et al. 1992; Massol et al. 1992; Partaledis et al. 1993; Peattie et al. 1992; Ratajczak et al. 1993; Sanchez 1990; Sykes et al. 1993; Tai et al. 1992; Thalhammer et al. 1992; see reviews by Heitman et al. 1992; Schmid et al. 1992; Stamnes et al. 1992

and the corresponding transcription factor (HSF) binding to it. Cloning and functional analysis of three HSF of tomato (Scharf et al. 1990; Treuter et al. 1993) showed that they are built of Leucine-zipper type hydrophobic repeats separating a highly conserved 90-100 amino acid DNA binding domain in the N-terminal part and a variable C-terminal activation domain. Short peptide motifs centered around tryptophan residues are essential for the activator function of the tomato HSF. HSF from other eukaryotes are basically similar with respect to their building modules and function (see summaries by Morimoto 1993; Scharf et al. 1994).

In addition to the intriguing mechanism of stress-induced transcription control, the interest of many research groups is focused on the role of stress proteins in general and for induced stress tolerance in particular. Following many discussions during the NATO meeting, I will try to summarize essential informations on stress proteins in five tables based on experimental material from all types of organisms. Finally, in correspondence with the overall topic of the workshop (Mechanisms of Stress Tolerance in Plants) I will conclude this overview with a sixth table compiling the constantly growing number of general stress proteins and stress metabolites. Table 6 reminds us of the highly sophisticated systems developed by plants to survive and reproduce in a hostile environment almost permanently characterized by a multi-stress situation.

2. Heat stress proteins

By sequence homology and fuctional analysis an increasing number of abundant cell proteins in pro- and eukaryotes were shown to belong to one of the conserved HSP families summarized in Table 1. Corresponding multigene families code for proteins found in different parts of the eukaryotic cell. References to Table 1 are restricted to recent contributions especially with respect to the functional characteristics of these proteins. A number of books and reviews can be used for access to earlier literature (Georgopoulos 1992; Kruse and Kloppstech 1992; Lindquist and Craig 1988; Nover 1991; Nover et al. 1990; Vierling 1991).

The essential role of these proteins as molecular chaperones not only under heat stress conditions (see Sect. 4) is reflected in the multiplicity of genes and regulatory patterns of their expression. Some are bona fide heat stress proteins (HSP), others are constitutively expressed

(HSC), or their synthesis increases under glucose deficiency (GRP) and other stress conditions respectively (Table 6). In addition, many examples for developmentally controlled expression were reported (Table 5 and summaries in Nover 1991 and Hightower and Nover 1991).

Two types of recent additions to this survey (Table 1) are worth being emphasized. (i) The close interaction of a small and a large subunits for both ATP-dependent chaperone systems of E. coli, i.e. DnaK /DnaJ and GroEL /GroES (see Table 3) is evidently also true for the eukaryotic homologs. An increasing number of DnaJ- and GroES-type proteins were characterized. (ii) A ubiquitous group of protein folding catalysts was independently detected by their interaction with immunosuppresive drugs and their activity as prolylpeptidyl cis-trans isomerases (see Table 3). According to the type of inhibitor these immunophilins can be separated into the group of cyclosporin A binding (cyclophilin, CYP) or FK506 binding proteins (FKBP).

3. The complexity of the heat stress response and thermotolerance

Though clearly in the center of interest, induced HSP synthesis is only a small part of a complex network of transient changes observed under heat stress conditions. Practically all aspects of cellular activities are involved. This concerns different parts of the gene expression apparatus (Table 2), but also DNA synthesis and cell division, the cytoskeletal systems and membrane structure and function (signal transduction, substrate transport, photosynthesis). For experimental details see Nover 1991.

Whenever tested in sufficient detail, the extent of stress damage can be diminished and the velocity of recovery can be improved by a prior conditioning treatment (induced thermotolerance). Synthesis of stress proteins but also of other stress metabolites contribute to the overall effect (see summary in Table 6). Though the picture is far from complete, the complexity of changes and the putative role of stress proteins for defined aspects of thermotolerance in a given cell compartment are evident from Table 2. Clearly, studies of cell survival are of limited value to elaborate the molecular details of such a system. This is vividly illustrated by the seemingly controversial interpretation of results reported for yeast mutants defective in the accumulation of individual HSPs or of all HSPs (De Virgilio et al. 1991; Hottiger et al. 1992; Sanchez et al 1992; Smith and Yaffe 1991a, b) and of other reports "proving" that a single HSP or HSP family is responsible for thermotolerance as judged by cell survival (Chretien and Landry 1988; Johnston and Kucey 1988; Landry et al. 1993; Loomis and Wheeler 1982). The investigations with mutants and/or transgenic systems with altered stress response must include the intracellular localization of stress proteins and stress metabolites as well as the ultrastructural and biochemical characterization of the compartment/process affected by the stress treatment (Neumann et al. 1987; Nover et al. 1983, 1986, 1989; Scharf and Nover 1987).

4. Stress proteins as molecular chaperones

The ongoing identification of more and more stress proteins led to an increasing search for their biological function. Activities converging from different directions (Table 3) were ultimately focused in the definition of GroEL-type proteins as molecular chaperones (Hemmingsen et al. 1988). This term was originally used by Laskey et al. (1978) for an acidic nuclear protein from Xenopus oocytes catalyzing the interaction of histones and DNA in nucleosome assembly. Meanwhile, the concept was extented to several other stress protein families (Table 3), whose members are active in folding, assembly and/or intracellular routing of newly synthesized proteins, but also in the disassembly and reactivation or degradation of stress-damaged proteins (see summaries by Ellis and Hemmingsen 1989; Ellis and van der Vies 1991; Gething and Sambrook 1992; Goldfarb 1992; Hartl et al. 1992).

In this rapidly expanding field several points deserve special attendance:
1. Usually cooperation between different chaperones is observed. Following the remarks to Table 1, this goes without saying for the interaction of GroEL / HSP60-type proteins with GroES /HSP10 and of DnaK /HSP70 with DnaJ /HSP40. But in addition to this, there is ample evidence that different chaperones are involved sequentially in the processing of a

Table 2: EFFECTS OF HEAT STRESS ON GENE EXPRESSION. ROLE OF HSPs FOR INDUCED THERMOTOLERANCE

For references to experimental literature see Table 3, 4 and summaries by Gething and Sambrook 1992; Morimoto 1993; Nover et al. 1990; Nover 1991; Scharf et al. 1994

Primary effect	Mediator	Response[a]	Thermotolerance (HSP involved)[b]
Transcriptional level			
Protein unfolding and aggregation, binding of HSP70	HSF activation (release of HSP70 ?)	Activation of hs genes, repression of house-keeping genes, induced HSP synthesis (G)	+ (HSP70 ?)
RNP processing			
?	Inactivation of U4/U6/U5 'assembly factor'	Stop of pre-mRNP splicing (M, I, P)	+ (HSP70)
?	?	Block of preribosomal processing, change of nucleolar ultrastructure (M, I, P)	+ (HSP70)
Translational level			
?	Inactivation of cap-binding factor (eIF-4F), collapse of cytoskeleton	Decay of polysomes, shift of translation specificity, segregation of house keeping mRNP to heat stress granules (M, I, P)	+ (HSP70, HSP20)
?	Inhibition of ribosomal PK	Dephosphorylation of small ribosomal subunit (protein S6) (G)	+

Folding, assembly, sorting and degradation of proteins

Unfolding or mis-folding of proteins	Deficiency of chape-rones	Formation of insoluble protein aggregates resistent to high salt and non-ionic detergents (G)	+ (HSP70, HSP20, ubiquitin)
		Defective or inefficient protein sorting and assembly (G)	? (HSP70, HSP60)

Signal transduction systems

Influx and intra-cellular redistribution of Ca^{2+}	Ca^{2+}- calmodulin	Collapse of cytoskeleton, release of proteins (G)	+ (HSP70, HSP90)
Release of inositol triphosphate, Ca^{2+} or other second messengers	Activation of protein kinases	Phosphorylation of HSP27, phosphorylation of HSF (M)	?

a) Similar results were reported for different types of organisms (G) or only in mammalian (M), plant (P) or insect (I) cells.

b) Thermotolerance is usually observed after a preconditioning treatment of cells or systems followed by an expression period. Except for protein folding and assembly the role of stress proteins for the indicated aspects of thermotolerance is circumstantial (see text).

Table 3: KEY CONTRIBUTIONS TO THE CHAPERONE CONCEPT AND THE ROLE OF HSPs

(For summaries see DeBoer and Weisboek 1991; Ellis and Hemmingsen 1989; Ellis and van der Vies 1991; Gatenby 1992; Glick and Schatz 1991; Gething and Sambrook 1992; Georgopoulos 1992; Hartl et al. 1992; Goldfarb 1992; Jaenicke 1991; Pfanner et al. 1991; Zeilstra-Ryalls et al. 1991)

References	Main results
A. Heat shock response	
Ritossa (1962)	Detection of hs-induced gene activity in Drosophila
Tissieres et al. (1974)	Induced synthesis of HSP in Drosophila
Ashburner and Bonner (1979)	Summary of results on induction and cloning of hsp genes
Pelham (1982); Pelham and Bienz (1982)	Definition and first experimental testing of the hs promoter element: (CTnGAAnnTTCnAG)
B. GroEL /HSP60 family	
Georgopoulos et al. (1972)	E. coli groE mutants are defective in phage head assembly
Hendrix (1979)	GroE forms complexes with ATPase activity
Barraclough and Ellis (1980)	Report on a 60 kDa protein (p60) of pea chloroplasts complexed with ribulose biphosphate carboxylase
Hemmingsen et al. (1988); McMullin and Hallberg (1988)	Sequences of HSP60, p60 and GroEL are closely related; they function as molecular chaperonins
Bochkareva et al. (1988; Hemmingsen et al. (1988); Goloubinoff et al. (1989)	Role of GroEL /HSP60 as molecular chaperones in protein folding and assembly
Ahmad and Gupta (1990); Gao et al. (1992); Frydman et al. (1992)	Cytoplasmic HSP60 (TCP1) and its function as chaperone
Langer et al. (1992a); Braig et al. (1993)	Fine structure of GroEL/S complex with protein substrate bound to the central cavity

(Table 3, continued)

    Mummert et al. (1993) — Role of plant TCP1 complex for photo-conversion of phytochrome

C. DnaK /HSP70 family

    Haas and Wabl (1983) — Identification of the Ig H-chain binding protein in the mammalian rER (BiP)

    Bardwell and Craig (1984) — Sequences of E. coli DnaK and Drosophila HSP70 are related

    Schlossman et al. (1984); Chapel et al. (1986) — Characterization of the 70 kDa clathrin uncoating ATPase and its identification as HSC70

    Welch and Feramisco (1985) — Method for rapid purification of mammalian HSP70 by ATP agarose columns

    Lewis and Pelham (1985) — ATP-dependent release of HSP70 from nuclear binding sites; HSP70 as protein shuttle

    Munro and Pelham (1986 /1987) — Grp78 (BiP) belongs to the HSP70 family; definition of the ER retention signal -KDEL

    Deshaies et al. (1988); Chirico et al. (1988) — Intracellular protein translocation in yeast is dependent on cytoplasmic HSP70

    Flaherty et al. (1990, 1991) — 3D crystal structure of the N-terminal ATPase domain of bovine HSC72

D. HtpG/ HSP90 family

    Hacket and Lis (1983); Farrelly and Finkelstein (1984); Bardwell and Craig (1987) — Sequences of the Drosophila, yeast and E. coli HSP90 proteins are similar

    Kulomaa et al. (1986); Sorger and Pelham (1987) — GRP94 in the ER belongs to the HSP90 family

    Brugge et al. (1983); Hutchison et al. (1992a) — HSP90 is a phosphoprotein associated with pp60src

    Dougherty et al. (1984); Sanchez et al. (1985); Picard et al. (1990) — HSP90 is part of the inactive hormon receptor complex; it facilitates activation after ligand binding

    Wiech et al. (1992) — HSP90 acts as chaperone (protein folding in vitro)

(Table 3, continued)

E. HSP20 family

Ingolia and Craig (1982)	Structural relation of the four Drosophila HSP20 and mammalian α,B-crystallin
Nover et al. (1983, 1989)	Hs-dependent aggregation of plant HSPs; formation of mRNP - containing hs granules
Benndorf et al. (1988); Fuqua et al. (1989)	Identification of a mammalian tumor-specific phosphoprotein as HSP25/27
Horwitz (1992); Jakob et al. (1993)	α,B-crystallin and mammalian HSP25/27 exhibit chaperone activity (in vitro unfolding and refolding of proteins)

F. Immunophilins (cyclophilin, FKBP)
(Peptidyl-prolyl cis-trans isomerases, PPI)

Fischer et al. (1984)	Report on PPI activity in pig kidney cells
Handschumacher et al. (1984)	First description of a cyclosporin A - binding protein (cyclophilin)
Fischer et al. (1989)	PPI and cyclophilin are identical
Lodish and Kong (1991)	Cyclophilin-type of PPI involved in transferrin folding in the ER of HepG2 cells
Ondek et al. (1992)	Drosophila ER membrane - bound cyclophilin (ninaA product) involved in rhodopsin biosynthesis
Callebaut et al. (1992); Tai et al. (1992)	HSP56, associated with steroid hormone receptors, has PPI activity (FK506-binding protein)

freshly synthesized proteins to their biologically active forms in the proper compartment (Brodsky et al. 1993; Gragerov et al. 1992; Langer et al. 1992; Liberek et al. 1991, see summaries by Georgopoulos 1992; Koll et al. 1992; Gething and Sambrook 1992; Pfanner et al. 1992). The whole complexity of the process is just emerging.

2. The existence of only one dnaK gene in E. coli led to the observation that defective mutants have a pleiotropic phenotype (Bukau and Walker 1989, see summary by Nover 1991). This is reflected by the remarkable multiplicity of effects reported for members of the HSP70 family (Table 4).The biochemical basis is probably common to all processes: ATP-dependent interaction with substrate proteins mediated by the DnaJ /HSP40 proteins. The diversity results from the type of substrate proteins affected and/or from the in-tracellular localization of different isoforms of the chaperone in cytoplasm, nucleus, mito-chondria, chloroplasts and ER.

3. Despite the high extent of sequence conservation, different members of the HSP70 family are functionally not identical even if localized in the same compartment. Some data illu-strating this were compiled for yeast (Boorstein and Craig 1990; Craig and Jakobsen 1985; Gao et al. 1991 Stone and Craig 1990; Werner-Weshburne et al. 1987). This concerns also the important question, why the generally abundant cytoplasmic HSC70 proteins are not sufficient to confer thermotolerance? What makes the difference between constitutive and stress-induced isoforms, which are beneficial for stress survival and recovery but may be toxic for normal growth and development (Feder et al. 1992)?

5. Developmental control of HSP synthesis

Our increasing knowledge about heat stress proteins and their role as molecular chaperones reinforced the interest in earlier results on stress-independent expression in animals, e.g. in defined developmental stages controlled by hormones (see summaries in Hightower and Nover 1991 and Nover 1991). Results with plants are compiled in Table 5. Because of the ubiquitous expression of members of the HSP70 family and the frequently lacking discrimi-

Table 4: MULTIPLICITY OF HSP70/DnaK EFFECTS

Process	References
Autoregulation (repression) of the hs response; interaction with HSF (eukaryotes), control of half life of sigma 32 (E. coli)	Baler et al. 1992; Beckmann et al. 1992; Gamer et al. 1992; Grossman et al. 1987; Liberek et al. 1992; Skelly et al. 1988; Stone and Craig 1990; Tilly et al. 1983; Yost and Lindquist 1991; see reviews by Sorger 1991; Morimoto 1993; Scharf et al. 1994
Autophosphorylation; may be involved in control of activity and hs induction	Amir-Shapira et al. 1990; Cegielska and Georgopoulos 1989; Dalie et al. 1990; Leustek et al. 1989; McCarthy and Walker 1991; Zylicz et al. 1983; see reviews by Craig and Gross 1991 and Leustek et al. 1992
Transport of proteins into the nucleus; protein shuttle	Imamoto et al. 1992; Lewis and Pelham 1985; Moore et al. 1992; Shi and Thomas 1992; see reviews by Dingwall and Laskey 1992; Nigg et al. 1991
In mammalian cells complexed with p53 tumor suppressor protein, steroid hormone receptors and DNA tumor virus T-antigens	Edwards et al. 1992; Hainaut and Milner 1992; Hutchison et al. 1992b; Lam and Calderwood 1992; Pallas et al. 1989; Pratt et al. 1992a,b; Sanchez et al. 1990; Sawai and Butel 1989; Walter et al. 1987
Stabilization of newly synthesized proteins along the pathway of folding and topogenesis	Beckmann et al. 1992; Gragerov et al. 1992; Langer et al. 1992b
Protein degradation and repair of damaged protein (reactivation); in E. coli DnaK needs interaction with DnaJ and GrpE	Beckman et al. 1992; Chiang et al. 1989; Dubois et al. 1991; Gaitanaris et al. 1990; Kroh and Simon 1991; Pinto et al. 1991; Sadis and Hightower 1992; Schröder et al. 1993; Sherman and Goldberg 1992b; Straus et al. 1988; Terlecky et al. 1992
Intracellular protein sorting across membranes from cytoplasm to the ER, mitochondria and chloroplasts needs interaction of HSP/C70 isoforms on both sites of the membrane	Brodsky et al. 1993; Chirico et al. 1988; Deshaies et al. 1988; Kang et al. 1990; Ko et al. 1992; Miernyk et al. 1992; Sanders et al. 1992; Soll and Waegemann 1992; Waegemann and Soll 1991

(Table 4, continued)

Integration of proteins into thylakoid membranes (in vitro)	Yalovsky et al. 1992
Reorganization of cytoskeleton (tubulin, intermediate filament system); function as clathrin uncoating ATPase	Chapel et al. 1986; Gao et al. 1991; Green and Liem 1989; Heuser and Steer 1989; Napolitano et al. 1985, 1987; Ohtsuka et al. 1986b; Schlossman et al. 1984; Wang et al. 1980; Weller 1988; Whatley et al. 1986
Assembly of translation initiation complex (in yeast)	Nelson et al. 1992b; Zhong and Arndt 1993
Interaction with RNP material in nucleolus (preribosomal RNP) and cytoplasm (mRNP in hs granules), function for rapid removal during recovery	Munro and Pelham 1984; Neumann et al. 1987, Nover et al. 1983, 1989; Ohtsuka et al. 1986a; Pelham 1984; Welch and Mizzen 1988; Welch and Suhan 1985
In bacteria DNA replication needs DnaK together with DnaJ and GrpE; required for primosome assembly	Alfano and McMacken 1989; Bukau and Walker 1989; Kawasaki et al. 1990; Sell et al. 1990; Wickner 1990; Wickner et al. 1991a, b, 1992

Table 5: DEVELOPMENTAL EXPRESSION OF PLANT HEAT STRESS GENES
(see summary by Nover 1991 and Winter and Sinibaldi 1991; na, not analyzed)

Organism; developmental process	HSPs[a]	hs mRNAs	Remarks	References
1. Lily, maize; pollen development	na	HSP18	Detected as meiosis-related multigene family; expressed during meiotic prophase	Bouchard 1990, Dietrich et al. 1991
2. Tobacco, tomato; developing anther	HSP17	na	Expression of hsp17 genes during meiotic stages of pollen development	Neumann et al., unpublished
3. Pea; ripening pods, seed germination	na	HSPs 70,20, 17	mRNA levels decreasing during germination	Vierling & Sun 1989
4. Tomato; fruit ripening, leaf senescence	na	pTOM66=HSP17	mRNA levels increasing in response to ripening or senescence; not induced by ethylene	Davies & Grierson 1989; Fray et al. 1990
5. Tomato; fruit ripening	na	HSC70	Highest levels in tapetum cells (ovary), seed integument and developing embryo	Duck et al. 1989
	na	HSP17, HSF30	Both mRNAs expressed together in young seeds	Materna et al., unpublished
6. Tobacco, tomato, pea, bean, maize; seed development and germination	HSP17	na	HSP17 accumulates in ripening seeds; degraded during germination	Neumann et al., unpublished

7. Wheat; germinating seeds	HSPs 94,70, 60,40,17,14	HSPs 70,60,40, 17, 14	Except for HSP70, levels of hs mRNAs rapidly decreasing	Helm & Abernethy 1990
8. Sunflower; seed ripening	na	HSP17.6	In seedlings also induced by heat, osmotic stress or ABA	Almoguera & Jordano 1992
9. Alfalfa; somatic embryogenesis	na	HSP18	Highest levels detected in early developmental stages	Györgyey et al. 1991
10. Arabidopsis; flower development	na	na	HSP18 promoter x gus construct expressed in sepals, filaments, styles	Tsukaya et al. 1992
11. Maize; developing seeds	GRP78 (BiP)		Storage protein synthesis consides with increase levels of BiP in rER	Boston et al. 1991; Denecke et al. 1991; Fontes et al. 1991; Jones and Bush 1991

nation between different isoforms reports on the developmental expression of HSP17/18 (HSP20 family) have a higher 'diagnostic' value. Two major developmental stages were characterized: (i) meiotic stages of pollen development (examples 1, 2) and (ii) ripening seeds (examples 3 - 10). The special requirements for HSP17/18 in these stages remain to be elaborated. Only the finding of increased levels of GRP78 (BiP) in the rER of ripening seeds (examples 11) can be clearly associated with the mass production and processing of seed proteins.

6. The network of plant stress response systems

When dicussing 'Biochemical and Cellular Mechanisms of Stress Tolerance in Plants' a number of problems are immediately apparent (see books by Levitt 1980; Nover et al. 1990 and other chapters in this volume): Due to their local fixation, plants are generally more affected by unfavorable environmental changes than animals. The usual situation for plant growth and development is best characterized by a daily multistress challenge. As a result of this, a multiplicity of partially overlapping stress response systems has evolved. The tight stress network is characterized by multivalent or even general stress metabolites and stress proteins respectively (Table 6). Hormones, in particular ethylene (Eth), abscicic acid (ABA) and jasmonic acid (JA) are not only frequently found as stress metabolites, but they form part of systemic signal systems (Kende 1993; Farmer and Ryan 1990; Pena-Cortes et al. 1989). Moreover, they serve the integration of stress gene expression into developmental processes (flowering, fruit ripening, senescence, seed maturation; see examples given in Tables 5 and 6).

The reference to developmental stages emphasizes important reservations with the data compiled in Table 6. It is not intended to suggest that the indicated compound is found in all plants and/or in all parts of a plant under a given stress condition. On the contrary, there are numerous examples of plant- and cell-specific modulations to the general theme. Moreover

Table 6: PLANT STRESS METABOLITES (from Nover et al. 1990, enlarged; + cellular level increased by the indicated stress treatment)

Stress-induced compounds	Wound-ing	Pathogen infect.[a], elicitor	Osmot. stress	Salt stress	Water defi-ciency	Heat stress	Cold stress	Anaero-biosis	Heavy metal stress	Radia-tion	Other inducing agents or conditions[e]
A. Non proteinaceous stress metabolites											
1. Polyamines, γ-aminobutyrate		+	+	+				+			Growth and cell division; cellular acidosis caused by SO_2 stress; NH_4^+ nutrition; K^+ deficiency
2. Glutathione (GSH)		+				+	+		+	+	Oxygen stress
3. Proline, glycine betaine, asparagine[b]		+	+	+	+	+	+			+	Ozone; SO_2 stress; cell division
4. Sucrose, polyols, trehalose					+	+	+				-
5. Cell wall material (suberin, cutin, + waxes, phenolics, lignin, callose)	+	+	+		+	+	+			+	Mineral deficiency; ABA; poly-L-Lys
6. Abscisic acid (ABA)	+	+	+	+	+	+	+			+	Nitrogen deficiency; cytokinins
7. Ethylene (Eth)	+	+	+	+	+	+	+	+	+	+	Oxygen stress (O_3); auxin; ripening; senescence
8. Jasmonic acid (JA)	+	+	+		+						Oxygen stress; senescence; glucose; reduced under cold stress
9. Oligosaccharide cell wall fragments	+	+									Agents causing local necrosis
10. Phytoalexins and other secondary metabolites	+	+			+				+	+	Local necroses caused by $HgCl_2$, chloroform, Triton, freezing; reduced growth; JA

B. Stress proteins

Protein	Inducers / Notes
1. Hydroxyproline-rich and glycine-rich cell wall proteins	ABA; JA; Eth; boron deficiency; high nitrate nutrition; GSH; flowering
2. Phenylalanine-ammonia lyase[c]	JA; GSH; vanadate,
3. Heat stress proteins	ABA
4. Stress protein 26-A[d]	ABA; 2,4-D
5. Proteinase inhibitors	ABA; JA; Eth; salicylate; sucrose; starvation; fruit ripening; poly-L-Lys; systemin
6. Chitinase, glucanase (PRP-types 2,3)	Flowering; SA; ethylene; induction inhibited by IAA or cytokinin
7. Pathogenesis-related proteins (PRP-types 1, 4, 5)	Ageing; flowering; 2,4-D; Eth or increased Eth levels by polyacrylic acid, salicylate or 2-chloroethylphosphonic acid
8. Ubiquitin	Starvation
9. Aldose reductase	-
10. Osmotin, thaumatin	Flower formation; protoplast culture; ABA; Eth
11. Late embryo abundant proteins (LEA)	ABA; developing seeds
12. Peroxidase, superoxide dismutase	Elicitors; salicylate; paraquat; oxygen stress
13. Thionins	JA
14. Lipoxygenase	Oxygen stress; senescence; JA; ABA
15. Vegetative storage proteins (soybean)	JA; seedling development
16. Cyclophilin/FKBP	-

Footnotes to Table 6:

a Pathogen induction includes fungi, bacteria and/or viruses as well as pathogen-derived elicitors.

b The amino acid or aa derivatives vary with the plants investigated (Jones 1984).

c In addition to phenylalanine-ammonia lyase, several other enzymes of general phenylpropanoid metabolism are induced as well (see summaries by Dangl 1992 and Hahlbrock and Scheel 1989).

d SP-26A is a general stress protein of soybean. The existence of similar proteins in other plants remains to be established.

e Depending on the cell type investigated, many stress proteins and metabolites mentioned in this table may also be synthesized stress-independently, under developmental control.

References to Table 6:

A1: Ayres 1984; Basso and Smith 1974; Flores et al. 1985; Murphy et al. 1983; Rhodes 1987; Slocum et al. 1984; Smith 1984, 1985; Young and Galston 1983

A2: Anderson and Gronwald 1991; Dean et al. 1990; Chiment et al. 1986; Dhindsa 1990; Edwards et al. 1991; Dohmen et al. 1990; Farago and Brunold 1990; Grill et al. 1982; Grill et al. 1989; Guy et al. 1984; Kunkel et al. 1991; Nieto-Sotelo and Ho 1986; Rüegsegger and Brunold 1992; Rüegsegger et al. 1990; Shaaltiel et al. 1988; Tukendorf and Rauser 1990; Yeune and Hatzios 1990; see review by Smith et al. 1990

A3: Ayres 1984; Dashek and Erickson 1981; Erickson and Dashek 1982; Jones 1984; Rhodes 1987; see also A. Hanson, this volume

A4: Rhodes 1987

A5: Asada and Matsumoto 1987; Baker 1974, Blum 1975; Bonhoff et al. 1987; Dixon et al. 1986; Hahlbrock and Scheel 1989; Jaffe and Telewski 1984; Köhle et al. 1984; Kolatukudy and Soliday 1985

A6: Daie and Campbell 1981; Degani and Itai 1978; Lalk and Dörfling 1985; Pena-Cortes et al. 1991; Ryan 1987; Thomas et al. 1992; Xu et al. 1993

A7: Fuhrer 1982; Harrison 1991; Hogsett et al. 1981; Meyer et al. 1987; Nakagawa et al. 1988; Stermer and Hammerschmidt 1987; see reviews by Yang and Hoffman 1984; Kende 1993

A8: Farmer and Ryan 1990, 1992; Gundlach et al. 1992; Ryan 1992; Mason and Mullet 1990; Mueller et al. 1993; Xu et al. 1993; see summaries by Parthier 1991, Staswick 1992, Sembdner and Parthier 1993

A9: Darvill and Albersheim 1984; Gross 1987; Ryan 1987, 1988; Ryan and Farmer 1991

24

A10: Darvill and Albersheim 1984; DiCosmo and Towers 1984; Dittrich et al 1992; Ebel 1986; Eilert 1987; Gershenzon 1984, Gross 1987; Hahlbrock and Scheel 1989; Hargreaves 1981; Kodama et al. 1988

B1: Creelman et al. 1992; Esquerre-Tugaye et al 1985; Neale et al. 1990; Roby et al. 1985; Showalter 1993; Showalter et al. 1991; Stermer and Hammerschmidt 1987; Wingate et al. 1988

B2: Dangl 1992; Eilert 1987; Gundlach et al. 1992; Hahlbrock and Scheel 1989; Joos and Hahlbrock 1992; Penharkar and Nair 1975; Rhodes and Wooltorton 1977; Steffens et al. 1989; Vayda and Schaeffer 1988; Wingate et al. 1988

B3: Almoguera and Jordano 1992; Borkhird et al. 1991; Edelman et al. 1988; Györgyey et al. 1991; Heikkila et al. 1984; Neven et al. 1992; Nover et al. 1990; Winter et al. 1988; Zhu et al. 1993

B4: Czarnecka et al. 1988; Farmer and Ryan 1992; Hagen et al. 1988

B5: An et al. 1989; Hildman et al. 1992; Johnson and Ryan 1990; Lincoln et al. 1987; McGure et al. 1992, Pearce et al. 1988; Ryan 1987, 1992; Walker-Simons and Ryan 1977, 1986

B6: Boller 1985, 1989; Gross et al. 1990; Hahlbrock and Scheel 1989; Hedrick et al. 1988, Kombrink et al. 1988; Linthorst 1991; Metraux et al. 1988; Neale et al. 1990;

B7 Granell et al. 1987, Hahlbrock and Scheel 1989; Linthorst 1991; Mühlbach 1987; Ohashi and Matsuoka 1985; Rigden and Coutts 1988; Traylor et al. 1987; van Loon 1983, 1985; Wagih and Coutts 1981

B8: Borkhird et al. 1991

B9: Bartels et al. 1991

B10: LaRosa et al. 1992; Nelson et al. 1992a; Singh et al. 1987, 1989; Stintzi et al. 1991

B11: Espelund et al. 1992; Hong et al. 1992; Skriver and Mundy 1990

B12: Tsang et al. 1991; Uknes et al. 1992; Ward et al. 1991

B13. Andresen et al. 1992, Bohlmann and Apel 1991

B14: Bell and Mullet 1991; Grimes et al. 1992; Maccarone et al. 1992; Melan et al. 1993

B15: Mason and Mullet 1990; Staswick et al. 1991

B16: Marivet et al. 1992

the pattern of stress metabolites and proteins detected in whole plants exposed to a long-term, multistress situation may considerably deviate from that found under experimental conditions with a selective, short-term challenge of seedlings, parts of plants or cell cultures. For sake of clarity Table 6 was subdivided into parts A with non-proteinaceous stress metabolites and B with stress proteins. However, in many cases increased synthesis of stress metabolites depends on the prior new-formation of the corresponding enzymes, which are stress-induced proteins. This is amply documented for A1, 3, 5, 7, 8, and 10. ACC synthase, the key enzyme of ethylene biosynthesis (A7), is probably the best studied example of a general stress protein.

With respect to the cellular mechanisms of stress tolerance in general there is no doubt that many proteins and stress metabolites may be involved acting alone or together in different parts of the cell. Survival, growth and productivity of plants under the normal multistress situation is an additive result of all protective effects (see Section 3). Unfortunately, we are far from a detailed understanding especially with respect the whole plant situation in its natural habitat.

The positive interaction of stress response systems is illustrated by common stress proteins and metabolites and by the frequently observed phenomena of mutual enhancement of the stress response and cross-tolerance (Nover 1991). However, inhibitory interactions were reported as well. Though extensive investigations are lacking, heat stress seems to be dominant in inhibiting expression of starvation stress proteins and at least part of the pathogen response (see examples summarized in Nover et al. 1990 and Nover 1991).

Table 6 helps to put the heat stress response into the frame of other stress response systems of plants. In addition to the well known HSP (Table 1), the increasing level of several common stress metabolites underlines the integration. Considering the stress-induced formation of members of the HSP families, only hs, heavy metals and few other well characterized chemical stressors (see Introduction), induce transcription of the whole set of hs genes. The induction depends on the activation of the heat stress transcription factor(s) (see summaries by Nover et al. 1990, Scharf et al. 1994). In contrast to this, other stress treatments are more selective. Thus, only HSP70 (which isoform?) was induced by wounding or water-stress adaptation (Borkhird et al. 1991; Heikkila et al. 1984) and few members of the HSP20 family are formed during osmotic stress or salt stress (Almoguera and Jordano 1992), whereas the cold stress induced HSC79 is probably the ER-localized BiP (Neven et al. 1992).

7. Perspectives

Lectures and fruitful discussions at the meeting emphasized major trends and fields of future experimental activities:

(i)   Filling the considerable gaps in our knowledge on stress response systems in their remarkable complexity;

(ii) Describing the interaction between plant stress response systems in terms of molecular cell biology;

(iii) Studies on signal systems involved in the environmental integration of plants and the triggering of the cellular stress response;

(iv) Attempts to connect the increasing amount of data on cellular mechanisms with the whole plant stress physiology.

8. References

Adamska I, Kloppstech K (1991) Evidence for the localization of the nuclear-coded 22-kDa heat-shock protein in a subfraction of thylakoid membranes. Europ J Biochem 198: 375-381

Ahmad S, Gupta RS (1990) Cloning of a Chinese hamster protein homologous to the mouse t-complex TCP-1: structural similarity to the ubiquitous 'chaperonin' family of heat shock proteins. Biochim Biophys Acta 1087: 253-255

Alfano C, McMacken R (1989) Heat shock protein-mediated disassembly of nucleoprotein structures is required for the initiation of bacteriophage lambda DNA replication. J Biol Chem 264: 10709-10718

Allen RL, O'Brien DA, Jones CC, Rockett DL, Eddy EM (1988) Expression of heat shock proteins by isolated mouse spermatogenic cells. Mol Cell Biol 8: 3260-3266

Almoguera C, Jordano J (1992) Developmental and environmental concurrent expression of sunflower dry-seed-stored low-molecular-weight heat-shock protein and Lea mRNAs. Plant Mol Biol 19: 781-792

Amir-Shapira D, Leustek T, Dalie B, Weissbach H, Brot N (1990) Hsp70 proteins, similar to Escherichia coli DnaK, in chloroplasts and mitochondria of Euglena gracilis. Proc Natl Acad Sci USA 87: 1749-1752

Anderson MP, Gronwald JP (1991) Atrazine resistance in a velvetleaf (Abutilon theophrasti) biotype due to enhanced glutathione S-transferase activity. Plant Physiol 96: 104-109

Andresen I, Becker W, Schlüter K, Burges J, Parthier B, Apel K (1992) The identification of leaf thionin as one of the main jasmonate-induced proteins of barley (Hordeum vulgare). Plant Mol Biol 19: 193-204

Asada Y, Matsumoto I (1987) Induction of desease resistance in plants by a lignification-inducing factor. In: Nishimura S, Vance CP, Dohe N (eds) Molecular Determinants of Plant Diseases. Springer Berlin, pp 223-233

Ashburner M, Bonner JJ (1979) The induction of gene activity in Drosophila by heat shock. Cell 17: 241-254

Atencio DP, Yaffe MP (1992) MAS5, a yeast homolog of DnaJ involved in mitochondrial protein import. Mol Cell Biol 12: 283-291

Ayres PG (1984) The interaction between environmental stress injury and biotic disease physiology. Annu Rev Phytopathol. 22: 53-75

Baker EA (1974) The influence of environment on leaf wax development in Brassica oleracea var. gemmifers. New Phytol 73: 955-966

Baler R, Welch WJ, Voellmy R (1992) Heat shock gene regulation by nascent polypeptides and denatured proteins - hsp70 as a potential autoregulatory factor. J Cell Biol 117: 1151-1159

Bardwell JCA, Craig EA (1984) Major heat shock gene of Drosophila and the Escherichia coli heat-inducible dnaK gene are homologous. Proc Natl Acad Sci USA 81: 848-852

Bardwell JCA, Craig EA (1987) Eukaryotic Mr 83.000 heat shock protein has a homologue in Escherichia coli. Proc Natl Acad Sci USA 84: 5177-5181

Barraclough R, Ellis RJ (1980) Protein synthesis in chloroplasts IX. Assembly of newly-synthesized large subunits into ribulose bisphosphate carboxylase in isolated intact pea chloroplasts. Biochim Biophys Acta 608: 19-31

Bartels D, Engelhardt K, Roncarati R, Schneider K, Rotter M, Salamini F (1991) An ABA and GA modulated gene expressed in the barley embryo encodes an aldose reductase related protein. EMBO J 10: 1037-1043

Basso LC, Smith TA (1974) Effect of mineral deficiency on amine formation in higher plants. Phytochemistry 13: 875-883

Beckmann RP, Lovett M, Welch WJ (1992) Examining the function and regulation of hsp70 in cells subjected to metabolic stress. J Cell Biol 117: 1137-1150

Behlke J, Lutsch G, Gaestel M, Bielka H (1991) Supramolecular structure of the recombinant murine small heat shock protein hsp25. FEBS L 288: 119-122

Bell E, Mullet JE (1991) Lipoxygenase gene expression is modulated in plants by water deficit, wounding and methyl jasmonate. Mol Gen Genetics 230: 456-462

Bertsch U, Soll J, Seetharam R, Viitanen PV (1992) Identification, characterization, and DNA sequence of a functional "double" groES-like chaperonin from chloroplasts of higher plants. Proc Natl Acad Sci USA 89: 8696-8700

Bienz M, Pelham HRB (1982) Expression of a Drosophila heat shock protein in Xenopus oocytes: conserved and divergent regulatory signals. EMBO J 1: 1583-1588

Bitar KN, Kaminski MS, Hailat N, Cease KB, Strahler JR (1991) HSP27 is a mediator of sustained smooth muscle contraction in response to bombesin. Biochem Biophys Res Commun 181: 1192-1200

Bloemendal H, DeJong WW (1991) Lens proteins and their genes. Progress Nucl Acid Res Mol Biol 41: 259-281

Blum A (1975) Effect of the Bm gene on epicuticular wax and the water relations of Sorghum bicolor L. (Moench). Israel J Bot 24: 50-51

Blumberg H, Silver PA (1991) A homoloue of the bacterial heat-shock gene DnaJ that alters protein sorting in yeast. Nature 349: 627-630

Bochkareva ES, Lissin NM, Girshovich AS (1988) Transient association of newly synthesized unfolded proteins with the heat-shock GroEL protein. Nature 336: 254-257

Bohlmann H, Apel K (1991) Thionins. Annu Rev Plant Physiol 42: 227-240

Boller T (1985) Induction of hydrolases as a defense reaction against pathogens. In: Key JL, Kosuge T (eds), Cellular and Molecular Biology of Plant Stress, Alan R. Liss Inc, New York, pp 247-262

Boller T (1989) Primary signals and second messenger in the reaction of plants to pathogenes. In: Boss WF, Morre IJ (eds) Second Messengers In Plant Growth and Development. Alan R. Liss Inc, New York, pp 227-255

Bonhoff A, Rieth B, Golecki J, Grisebach H (1987) Race cultivar-specific differences in callose deposition in soybean roots following infection with Phytophthora megasperma f. sp. glycinea. Planta 172: 101-105

Boorstein WR, Craig EA (1990) Transcriptional regulation of SSA3, an HSP70 gene from Saccharomyces cerevisiae. Mol Cell Biol 10: 3262-3267

Borkird C, Simoens C, Villarroel R, van Montagu M (1991) Gene expression associated with water-stress adaptation of rice cells and identification of two genes as hsp70 and ubiquitin. Physiol Plant 82: 449-457

Boston RS, Fontes EBP, Shank BB, Wrobel RL (1991) Increased expression of the maize immunoglobulin binding protein homolog b-70 in three zein regulatory mutants. Plant Cell 3: 497-505

Bouchard RA (1990) Characterization of expressed meiotic prophase repeat transcript clones of Lilium: meiosis-specific expression, relatedness, and affinities to small heat shock protein genes. Genome 33: 68-79

Braig K, Simon M, Furuya F, Hainfeld JF, Horwich AL (1993) A polypeptide bound by the chaperonin groEL is localized within a central cavity. Proc Natl Acad Sci USA 90: 3978-3982

Breiman A, Fawcett TW, Ghirardi ML, Mattoo AK (1992) Plant organelles contain distinct peptidylprolyl cis, trans-isomerases. J Biol Chem 267: 21293-21296

Brodsky JL, Hamamoto S, Feldheim D, Schekman R (1993) Reconstitution of protein translocation from solubilized yeast membranes reveals topologically distinct roles for BiP and cytosolic Hsc70. J Cell Biol 120: 95-102

Brugge JS, Erikson E, Erikson RL (1981) The specific interaction of the Rous sarcoma virus transforming protein, pp60^{src}, with two cellular proteins. Cell 25: 363-372

Bukau B, Walker GC (1989) Cellular defects caused by deletion of the Escherichia coli dnaK gene indicate roles for heat shock protein in normal metabolism. J Bacteriol 171: 2337-2346

Callebaut I, Renoir JM, Lebeau MC, Massol N, Burney A, Baulieu EE, Mornon JP (1992) An immunophilin that binds M_r 90,000 heat shock protein: Main structural features of a mammalian p59 protein. Proc Natl Acad Sci USA 89: 6270-6274

Caplan AJ, Cyr DM, Douglas MG (1992) YDJ1p facilitates polypeptide translocation across different intracellular membranes by a conserved mechanism. Cell 71: 1143-1155

Cegielska A, Georgopoulos C (1989) Functional domains of the Escherichia coli DnaK heat shock protein as revealed by mutational analysis. J Biol Chem 264: 21122-21130

Chappell TG, Welch WJ, Schlossman DM, Palter KB, Schlesinger MJ, Rothman JE (1986) Uncoating ATPase is a member of the 70 Kilodalton family of stress proteins. Cell 45: 3-13

Chen Q, Vierling E (1991) Analysis of conserved domains identifies a unique structural feature of a chloroplast heat shock protein. Mol Gen Genet 226: 425-431

Chiang HL, Terlecky SR, Plant CP, Dice JF (1989) A role for a 70-Kilodalton heat shock protein in lysosomal degradation of intracellular proteins. Science 246: 382-385

Chiment JJ, Alscher R, Hughes PR (1986) Glutathione as an indicator of SO_2-induced stress in soybean. Environ Exp Bot 26: 147-152

Chirico WJ, Waters MG, Blobel G (1988) 70 K heat shock related proteins stimulate protein translocation into microsomes. Nature 332: 805-810

Chretien P, Landry J (1988) Enhanced constitutive expression of the 27-kDa heat shock proteins in heat resistant variants from Chinese hamster cells. J Cell Physiol 137: 157-166

Christensen AH, Sharrock RA, Quail PH (1992) Maize polyubiquitin genes - Structure, thermal perturbation of expression and transcript splicing, and promoter activity following transfer to protoplasts by electroporation. Plant Mol Biol 18: 675-689

Conner TW, Lafayette PR, Nagao RT, Key JL (1990) Sequence and expression of a HSP83 from Arabidopsis thaliana. Plant Physiol 94: 1689-1695

Craig EA, Jakobsen K (1985) Mutations in cognate genes of Saccharomyces cerevisiae hsp70 result in reduced growth rates at low temperatures. Mol Cell Biol 5: 3517-3524

Craig EA, Gross CA (1991) Is hsp70 the cellular thermometer? Trends Biochem Sci 16: 135-140

Creelman RA, Tierney ML, Mullet JE (1992) Jasmonic acid/methyl jasmonate accumulate in wounded soybean hypocotyls and modulate wound gene expression. Proc Natl Acad Sci USA 89: 4938-4941

Csermely P, Kahn CR (1991) The 90-kDa heat shock protein (hsp-90) possesses an ATP binding site and autophosphorylating activity. J Biol Chem 266: 4943-4950

Csermely P, Kajtar J, Hollosi M, Jalsovszky G, Holly S, Kahn CR, Gergely P, Soti C, Mihaly K, Somogyi J (1993) ATP induces a conformational change of the 90-kDa heat shock protein (hsp90). J Biol Chem 268: 1901-1907

Cyr DM, Lux X, Douglas MG (1992) Regulation of Hsp70 function by a eukaryotic DnaJ homolog. J Biol Chem 267: 20927-20931

Czarnecka E, Nagao RT, Key JL, Gurley WB (1988) Characterization of Gm hsp26-A, a stress gene encoding a divergent heat shock protein of soybean: Heavy-metal-induced inhibition of intron processing. Mol Cell Biol 8: 1113-1122

Daie J, Campbell WF (1981) Response of tomato plants to stressful temperatures. Increase in abscisic acid concentrations. Plant Physiol 67: 26-29

Dalie BL, Skaleris DA, Kohle K, Weissbach H, Brot N (1990) Interaction of DnaK with ATP: Binding, hydrolysis and Ca^{2+}-stimulated autophosphorylation. Biochem Biophys Res Commun 166: 1284-1292

Dang JL (1992) Regulatory elements controlling developmental and stress-induced expression of phenylpropanoid genes. In: Boller T, Meins F (eds) Genes Involved in Plant Defense. Springer Wien, pp 303-326

Darvill AG, Albersheim P (1984) Phytoalexins and their elicitors - A defense against microbial infection in plants. Annu Rev Plant Physiol 35: 243-275

Dashek WV, Erickson SS (1981) Isolation, assay, biosynthesis, metabolism, uptake and translocation, and function of proline in plant cells and tissues. Bot Rev 47: 349-385

Davies KM, Grierson D (1989) Identification of cDNA clones for tomato (Lycopersicon esculentum Mill.) mRNAs that accumulate during fruit ripening and leaf senescence in response to ethylene. Planta 179: 73-80

Dean JV, Gronwald CV, Eberlein CV (1990) Induction of glutathione S-transferase isozymes in sorghum by herbicide antidotes. Plant Physiol 92: 467-473

DeBoer AD, Weisbeek PJ (1991) Chloroplast protein topogenesis - import, sorting and assembly. Biochim Biophys Acta 1071: 221-253

Degani N, Itai C (1978) The effect of radiation on growth and abscisic acid in wheat seedlings. Env Exp Bot 18: 113-115

Denecke J, Goldman MHS, Demolder J, Seurinck J, Botterman J (1991) The tobacco luminal binding protein is encoded by a multigene family. Plant Cell 3: 1025-1035

DeRocher AE, Helm KW, Lauzon LM, Vierling E (1991) Expression of a conserved family of cytoplasmic low molecular weight heat shock proteins during heat stress and recovery. Plant Physiol 96: 1038-1047

Deshaies RJ, Koch BD, Werner-Washburne M, Craig EA, Shekman R (1988) A subfamily of stress proteins facilitates translocation of secretory and mitochondrial precursor polypeptides. Nature 332: 800-805

DeVirgilio C, Piper P, Boller T, Wiemken A (1991) Aquisition of thermotolerance in Saccharomyces cerevisiae without heat shock protein hsp104 and in the absence of protein synthesis. FEBS Letters 288: 86-90

Dhindsa RS (1990) Drought stress, enzymes of glutathione metabolism, oxidation injury, and protein synthesis in Tortula ruralis. Plant Physiol 95: 648-651

DiCosmo F, Towers GHN (1984) Stress and secondary metabolism in cultured plant cells. Rec Advanc Phytochem 18: 97-175

Dietrich PS, Bouchard RA, Casey ES, Sinibaldi RM (1991) Isolation and characterization of a small heat shock protein gene from maize. Plant Physiol 96: 1268-1276

Dingwall C, Laskey R (1992) The nuclear membrane. Science 258: 942-947

Dittrich H, Kutchan TM, Zenk MH (1992) The jasmonate precursor 12-oxo-phytodienoic acid induces phytoalexin synthesis in Petroselinum crispum cell cultures. FEBS L 309: 33-36

Dixon RA, Bailey JA, Bell JN, Bolwell GP, Cramer CL, Edwards K, Hamdan MAM, Lamb CJ, Robbins MP, Ryder TB, Schuch W (1986) Rapid changes in gene expression in response to microbial elicitation. Phil Trans R Soc Lnd B 314: 411-426

Dohmen GP, Koppers A, Langebartels C (1990) Biochemical response of Norway spruce (Picea abies (L) Karst) towards 14-month exposure to ozone and acid mist: Effects on amino acid, glutathione and polyamine titers. Environ Pollut 64: 375-383

Dougherty JJ, Puri RK, Toft DO (1984) Polypeptide components of two 8S forms of chicken oviduct progesteron receptor. J Biol Chem 259: 8004-8009

Dubois MF, Hovanessian AG, Bensaude O (1991) Heat-shock-induced denaturation of proteins. J Biol Chem 266: 9707-9711

Duck N, McCormick S, Winter J (1989) Heat shock protein hsp70 cognate gene expression in vegetative and reproductive organs of Lycopersicon esculentum. Proc Natl Acad Sci USA 86: 3674-3678

Ebel J (1986) Phytoalexin synthesis: The biochemical analysis of the induction process. Annu Rev Phytopathol 24: 235-264

Edwards R, Blount JW, Dixon RA (1991) Glutathione and elicitation of the phytoalexin response in legume cell cultures. Planta 184: 403-409

Edwards DP, Estes PA, Fadok VA, Bona BJ, Onate S, Nordeen SK, Welch WJ (1992) Heat shock alters the composition of heteromeric steroid receptor complexes and enhances receptor activity in vivo. Biochemistry 31: 2482-2491

Ellis RJ, Hemmingsen SM (1989) Molecular chaperones: proteins essential for the biogenesis of some macromolecular structures. Trends Biochem Sci 14: 339-342

Ellis RJ, van der Vies SM (1991) Molecular chaperones. Annu Rev Biochem 60: 321-347

Eilert U (1987) Elicitation: Methodology and aspects of application. In: Constabel F, Vasil IK, (eds), 1987, Cell Culture and Somatic Cell Genetics of Plants 4, Acad Press, San Diego, pp 153-196

Erickson SS, Dashek WV (1982) Accumulation of soluble proline in sulphur dioxide-stressed Glycine max cvs. "Proctor" and "Excelsior" seedlings. Environ Pollut Ser A 28: 89-108

Espelund M, Saeboe-Larssen S, Hughes DW, Galan GA, Larsen F, Jakobsen KS (1992) Late embryogenesis-abundant genes encoding proteins with different numbers of hydrophilic repeats are regulated differentially by abscisic acid and osmotic stress. Plant J 2: 241-252

Esquerre-Tugaye MT, Mazau D, Pelissier B, Roby D, Rumean D, Toppan A (1985) Induction by elicitors and ethylene of proteins associated to the defense of plants. In: KeyJL, Kosuge T (eds), Cellular and Molecular Biology of Plant Stress, Alan R. Liss Inc, New York, pp 459-466

Farago S, Brunold C (1990) Regulation of assimilatory sulfate reduction by herbicide safeners in Zea mays. Plant Physiol 94: 1808-1812

Farmer EE, Ryan CA (1990) Interplant communication: Airborne methyl jasmonate induces synthesis of proteinase inhibitors in plant leaves. Proc Natl Acad Sci USA 87: 7713-7716

Farmer EE, Ryan CA (1992) Octadecanoid precursors of jasmonic acid activate the synthesis of wound-inducible proteinase inhibitors. Plant Cell 4:129-134

Farrelly FW, Finkelstein DB (1984) Complete sequence of the heat shock-inducible HSP90 gene of Saccharomyces cerevisiae. J Biol Chem 259: 5745-5751

Feder JH, Rossi JM, Solomon J, Solomon N, Lindquist S (1992) The consequences of expressing hsp70 in Drosophila cells at normal temperatures. Genes Devel 6: 1402-1413

Feldheim D, Rothblatt J, Shekman R (1992) Topology and functional domains of Sec63p, an endoplasmic reticulum membrane protein required for secretory protein translocation. Mol Cell Biol 12: 3288-3296

Felsheim RF, Das A (1992) Structure and expression of a heat-shock protein 83 gene of Pharbitis nil. Plant Physiol 100: 1764-1771

Flaherty KM, DeLuca-Flaherty C, McKay DB (1990) Three - dimensional structure of the ATPase fragment of a 70 K heat-shock cognate protein. Nature 346: 623-628

Flaherty KM, McKay DB, Kabsch W, Holmes KC (1991) Similarity of the three - dimensional structures of actin and the ATPase fragment of a 70 kDa heat shock cognate protein. Proc Natl Acad Sci USA 88: 5041-5045

Flores HE, Young ND, Galston AW (1985) Polyamine metabolism and plant stress. In: Key JL, Kosuge T (eds), Cellular and Molecular Biology of Plant Stress, Alan R. Liss Inc, New York, pp 93-114

Fontes EBP, Shank BB, Wrobel RL, Moose SP, O'Brian GR, Wurtzel ET, Boston RS (1991) Characterization of an immunoglobulin binding protein in the maize floury-2 endosperm mutant. Plant Cell 3: 483-496

Fray RG, Lycett GW, Grierson D (1990) Nucleotide sequence of a heat-shock and ripening-related cDNA from tomato. Nucl Acids Res 18: 7148

Frydman J, Nimmesgern E, Erdgument-Bromage H, Wall JS, Tempst P, Hartl FU (1992) Function in protein folding of TRiC, a cytosolic ring complex containing TCP1 and structurally related subunits. EMBO J 11: 4767-4778

Fuhrer J (1982) Ethylene biosynthesis and cadmium toxicity in leaf tissue of beans (Phaseolus vulgaris L). Plant Physiol 70: 162-167

Fuqua SAW, Blumsalingaros M, McGuire WL (1989) Induction of the estrogen-regulated 24K protein by heat shock. Cancer Res 49: 4126-4129

Gaestel M, Schröder W, Benndorf R, Lippmann C, Buchner K, Hucho F, Erdmann VA, Bielka H (1991) Identification of the phosphorylation sites of the murine small heat shock protein hsp25. J Biol Chem 266: 14721-14724

Gaitanaris GA, Papavassiliou AG, Rubock P, Silverstein SJ, Gottesman ME (1990) Renaturation of denatured λ repressor requires heat shock proteins. Cell 61: 1013-1020

Gamer J, Bujard H, Bukau B (1992) Physical interaction between heat shock proteins DnaK, DnaJ, GrpE and the bacterial heat shock transcription factor σ32. Cell 69: 833-842

Gao BC, Biosca J, Craig EA, Greene LE, Eisenberg E (1991) Uncoating of coated vesicles by yeast hsp70 proteins. J Biol Chem 266: 19565-19571

Gao Y, Thomas JO, Chow RL, Lee G-H, Cowan NJ (1992) A cytoplasmic chaperonin that catalyzes ß-actin folding. Cell 69: 1043-1050

Gatenby AA, (1992) Protein folding and chaperonins. Plant Mol Biol 19: 677-687

Georgopoulos C (1992) The emergence of the chaperone machines. Trends Biochem Sci 17: 295-299

Georgopoulos CP, Hendrix RW, Kaiser AD, Wood WB (1972) Role of the host cell in bacteriophage morphogenesis: Effects of a bacterial mutation on T4 head assembly. Nature NB 239: 38-42

Gething M-J, Sambrook J (1992) Protein folding in the cell. Nature 355: 33-45

Gershenzon J (1984) Changes in the levels of plant secondary metabolites under water and nutrient stress. Rec Advanc Phytochem 18: 273-320

Glick B, Schatz G (1991) Import of proteins into mitochondria. Annu Rev Gen 25: 21-44

Goldfarb DS (1992) Are the cytosolic components of the nuclear, ER, and mitochondrial import apparatus functionally related? Cell 70: 185-188

Goloubinoff P, Gatenby AA, Lorimer GH (1989) GroE heat shock proteins promote assembly of foreign prokaryotic ribulose bisphosphate carboxylase oligomers in Escherichia coli. Nature 337: 44-47

Gragerov A, Nudler E, Komissarova N, Gaitanaris GA, Gottesman ME, Nikiforov V (1992) Cooperation of GroEL / GroES and DnaK / DnaJ heat shock proteins in preventing protein misfolding in Escherichia coli. Proc Natl Acad Sci USA 89: 10341-10344

Granell A, Belles JM, Conejero V (1987) Induction of pathogenesis-related proteins in tomato by Citrus exocortis viroid, silver ion, and etephon. Physiol Mol Plant Pathol 31: 83-89

Green LAD, Liem RKH (1989) Beta-internexin is a microtubule-associated protein identical to the 70-kDa heat-shock cognate protein and the clathrin uncoating ATPase. J Biol Chem 264: 15210-15215

Grill D, Esterbauer H, Hellig K (1982) Further studies on the effect of SO_2-pollution on the sulfhydryl-system of plants. Phytopath Z 104: 264-271

Grill E, Löffler S, Winnacker E-L, Zenk MH (1989) Phytochelatins, the heavy-metal-binding peptides of plants, are synthesized from glutathione by a specific γ-glutamylcysteine dipeptidyl transpeptidase (phytochelatin synthase). Proc Natl Acad Sci USA 86: 6838-6842

Grimes HD, Koetje DS, Fransceshi VR (1992) Expression, activity, and cellular accumulation of methyl jasmonate - resposive lipoxygenase in soybean seedlings. Plant Physiol 100: 433-443

Gross D (1987) Chemische Abwehrstoffe der Pflanze. Biol Rundschau 25: 225-237

Grosset J, Meyer Y, Chartier Y, Kauffmann S, Legrand M, Fritig B (1990) Tobacco mesophyll protoplasts synthesize 1,3-ß-glucanase, chitinases and 'osmotins' during in vitro culture. Plant Physiol 92: 520-527

Grossman AD, Straus DB, Walter NA, Gross CA (1987) Sigma 32 synthesis can regulate the synthesis of heat shock proteins in Escherichia coli. Genes Devel 1: 179-184

Gundlach H, Müller MJ, Kutchan TM, Zenk MH (1992) Jasmonic acid is a signal transducer in elicitor-induced plant cell cultures. Proc Natl Acad Sci USA 89: 2389-2393

Guy CL, Carter JV, Yelenosky G; Guy CT (1984) Changes in glutathione content during cold acclimation in Cornus sericea and Citrus sinensis. Cryobiol 21: 443-453

Györgyey J, Gartner A, Nemeth K, Magyar Z, Hirt H, Heberlebors E, Dudits D (1991) Alfalfa heat shock genes are differentially expressed during somatic embryogenesis. Plant Mol Biol 16: 999-1007

Haas IG, Wabl M (1983) Immunoglobulin heavy chain binding protein. Nature 306: 387-389

Hackett RW, Lis JT (1983) Localization of the hsp83 transcript within a 3292 nucleotide-sequence from the 63B heat shock locus of Drosophila melanogaster. Nucl Acids Res 11: 7011-7030

Hagen G, Uhrhammer N, Guilfoyle TJ (1988) Regulation of expression of an auxin-induced soybean sequence by cadmium. J Biol Chem 263: 6442-6446

Hahlbrock K, Scheel D (1989) Physiology and molecular biology of phenylpropanoid metabolism. Annu Rev Plant Physiol 40: 347-369

Hainaut P, Milner J (1992) Interaction of heat shock protein-70 with p53 translated in vitro - Evidence for interaction with dimeric p53 and for a role in the regulation of p53 conformation. EMBO J 11: 3513-3520

Handschumacher R, Harding M, Rice J, Drugge R (1984) Cyclophilin: A specific cytosolic binding protein for cyclosporin A. Science 226: 544-547

Hargreaves JA (1981) Accumulation of phytoalexins in cotyledons of French bean (Phaseolus vulgaris L) following treatment with triton (T-octylphenol polyethoxyethanol) surfact-ants. New Phytol 87: 733-741

Hartl FU, Martin J, Neupert W (1992) Protein folding in the cell - the role of molecular chaperones Hsp70 and Hsp60. Annu Rev Biophys Biomol Structure 21: 293-322

Hartman DJ, Dougan D, Hoogenraad NJ, Hoj PB (1992a) Heat shock proteins of barley mitochondria and chloroplasts - Identification of organellar hsp10 and hsp12 - Putative chaperonin 10 homologues. FEBS L 305: 147-150

Hartman DJ, Hoogenraad NJ, Condron R, Hoj PB (1992b) Identification of a mammalian heat shock protein, a mitochondrial chaperonin 10 homologue essential for assisted folding of trimeric ornithine transcarbamoylase in vitro. Proc Natl Acad Sci USA 89: 3394-3398

Hattori H, Liu YC, Tohnai I, Ueda M, Kaneda T, Kobayashi T, Tanaba K, Ohtsuka K (1992) Intracellular localization and partial amino acid sequence of a stress-inducible 40-kDa protein in HeLa cells. Cell Struct Function 17: 77-86

Heitman J, Movva NR, Hall MN (1992) Proline isomerases at the crossroads of protein folding, signal transduction, and immunosuppression. New Biol 4: 448-460

Helm KW, Abernethy RH (1990) Heat shock protein and their mRNAs in dry and early inbibing embryos of wheat. Plant Physiol 93: 1626-1633

Hemmingsen SM, Woolford C, Van der Vies SM, Tilly K, Dennis DT, Georgopoulos CP, Hendrix RW, Ellis RJ (1988) Homologous plant and bacterial proteins chaperone oligomeric protein assembly. Nature 333: 330-335

Hendrix RW (1979) Purification and properties of GroE, a host protein involved in bacteriophage assembly. J Mol Biol 129: 375-392

Heuser J, Steer CJ (1989) Trimeric binding of the 70-kD uncoating ATPase to the vertices of
clathrin triskelia: A candidate intermediate in the vesicle uncoating reaction. J Cell Biol
109: 1457-1466

Hightower L, Nover L (eds) (1991) Heat Shock and Development. Springer Berlin

Hildman T, Ebneth M, Pena-Cortes H, Sànchez-Sèrrano JJ, Willmitzer L, Prat S (1992)
General role of abscisic and jasmonic acids in gene activation as a result of mechanical
wounding. Plant Cell 4: 1157-1170

Hogsett WE, Raba RM, Tingey D (1981) Biosynthesis of stress ethylene in soybean seedlings: similarities to endogenous ethylene biosynthesis. Physiol Plant 53: 307-314

Hong BM, Barg R, Ho THD (1992) Developmental and organ-specific expression of an
ABA-induced and stress-induced protein in barley. Plant Mol Biol 18: 663-674

Horwitz J (1992) α-Crystallin can function as a molecular chaperone. Proc Natl Acad Sci
USA 89: 10449-10453

Hottiger T, deVirgilio C, Bell W, Boller T, Wiemken A (1992) The 70-Kilodalton heat shock
protein of the SSA subfamily negatively modulate heat shock-induced accumulation of
trehalose and promote recovery from heat stress in the yeast, Saccharomyces cerevisiae.
Europ J Biochem 210: 125-132

Hutchison KA, Stancato LF, Jove R, Pratt WB (1992a) The protein-protein complex between
pp60 (v-src) and hsp90 is stabilized by molybdate, vanadate, tungstate, and an
endogenous cytosolic metal. J Biol Chem 267: 13952-13957

Hutchison KA, Czar MJ, Scherrer LC, Pratt WB (1992b) Monovalent cation selectivity for
ATP-dependent association of the glucocorticoid receptor with hsp70 and hsp90. J Biol
Chem 267: 14047-14053

Imamoto N, Matsuoka Y, Kurihara T, Kohno K, Miyagi M, Sakiyama F, Okada Y,
Tsunasawa S, Yoneda Y (1992) Antibodies against 70-kD heat shock cognate protein
inhibit mediated nuclear import of karyophilic proteins. J Cell Biol 119: 1047-1061

Inaguma Y, Shinohara H, Goto S, Kato K (1992) Translocation and induction of α,B-
crystallin by heat shock in rat glioma (GA-1) cells. Biochem Biophys Res Commun 182:
844-850

Ingolia TD, Craig EA (1982) Four small Drosophila heat shock proteins are related to each
other and to mammalian α-crystallin. Proc Natl Acad Sci USA 79: 2360-2364

Jaenicke R (1991) Protein folding - local structures, domains, subunits, and assemblies.
Biochemistry 30: 3147-3161

Jaffe MJ, Telewski FW (1984) Thigmomorphogenesis: Callose and ethylene in the hardening
of mechanically stressed plants. Rec Advanc Phytochem 18: 79-96

Jakob U, Gaestel M, Engel K, Buchner J (1993) Small heat shock proteins are molecular
chaperones. J Biol Chem 268: 1517-1520

Jentsch S (1992) The ubiquitin-conjugation system. Annu Rev Genet 26: 177-205

Jentsch S, Seufert W, Hauser H-P (1991) Genetic analysis of the ubiquitin system. Biochim
Biophys Acta 1089: 127-139

Jin YJ, Burahoff SJ, Bierer BE (1992) Molecular cloning of a 25-kDa high affinity rapamycin
binding protein, FKBP25. J Biol Chem 267: 10942-10945

Johnson R, Ryan CA (1990) Wound-inducible potato inhibitor II genes: Enhancement of
expression by sucrose. Plant Mol Biol 14: 527-536

Johnston RN, Kucey BL (1988) Competitive inhibition of hsp70 gene expression causes
thermosensitivity. Science 242: 1551-1554

Jones RGW (1984) Phytochemical aspects of osmotic adaptation. Rec Advanc Phytochem 18:
55-78

Jones RL, Bush DS (1991) Gibberellic acid regulates the level of a BiP cognate in the endoplasmic reticulum of barley aleurone cells Plant Physiol 97: 456-459

Joos J, Hahlbrock K (1992) Phenylalanine ammonia-lyase in potato (Solanum tuberosum L.). Genomic complexity, structural comparison of two selected genes and modes of expression. Europ J Biochem 204: 621-629

Kawasaki Y, Wada C, Yura T (1990) Roles of Escherichia coli heat shock proteins DnaK, DnaJ and GrpE in mini-F plasmid replication. Mol Gen Genetics 220: 277-282

Kende H (1993) Ethylene biosynthesis. Annu Rev Plant Physiol Plant Mol Biol 44: 283-307

Kieffer LJ, Thalhammer T, Handschumacher RE (1992) Isolation and characterization of a 40-kDa cyclophilin-related protein. J Biol Chem 267: 5503-5507

Klemenz R, Fröhli E, Steiger RH, Schäfer R, Aoyama A (1991) α,B-Crystallin is a small heat shock protein. Proc Natl Acad. Sci USA 88: 3652-3656

Ko K, Bornemisca O, Kourtz L, Ko ZW, Plaxton WC, Cashmore AR (1992) Isolation and characterization of a cDNA clone encoding a cognate 70-kDa heat shock protein of the chloroplast envelope. J Biol Chem 267: 2986-2993

Kodama O, Suzuki T, Miyakawa J, Akatsuka T (1988) Ultraviolet-induced accumulation of phytoalexins in rice leaves. Agric Biol Chem 52: 2469-2473

Köhle H, Young DH, Kauss H (1984) Physiological changes in suspension-cultured soybean cells elicited by treatment with chitosan. Plant Sci Letters 33: 221-230

Kolattukudy PE, Soliday CL (1985) Effects of stress on the defense barriers of plants. In: Key JL, Kosuge T (eds), Cellular and Molecular Biology of Plants Stress, Alan R. Liss Inc, New York, pp 381-400

Koll H, Guiard B, Rassow J, Ostermann J, Horwich AL, Neupert W, Hartl F-U (1992) Antifolding activity of hsp60 couples protein import into the mitochondrial matrix with export to the intermembrane space. Cell 68: 1163-1175

Kroh HE, Simon LD (1991) Increasing ATP-dependent proteolytic activity in Lon-deficient Escherichia coli strains lacking the DnaK protein. J Bacteriol 179: 2691-2695

Kruse E, Kloppstech K (1992) Heat shock proteins in plants: An approach to understanding the function of plastid heat shock proteins. In Barber J (ed) The Photosystems: Structure, Function and Molecular Biology. Elsevier Sci Publ Amsterdam, pp 409-442

Kulomaa MS, Weigel NL, Kleinsek OA, Beattie WG, Conneely OM, March C, Zarucki-Schulz T, Schrader WT, O'Malley W (1986) Amino acid sequence of a chicken heat shock protein derived from the complementary DNA nucleotide sequence. Biochemistry 25: 6244-6252

Kunkel DL, Steffens JC, Bellinder RR (1991) Effect of temperature and safeners on glutathione levels and glutathione S-transferase activity in maize. Z Naturforsch 46c: 856-860

Lalk J, Dörffling K (1985) Hardening, abscisic acid, proline and freezing resistance in two winter wheat varieties. Physiol Plant 63: 287-292

Lam KT, Calderwood SK (1992) HSP70 binds specifically to a peptide derived from the highly conserved domain (I) region of p53. Biochem Biophys Res Commun 184: 167-174

Landry SJ, Jordan R, McMacken R, Gierasch LM (1992a) Different conformations for the same polypeptide bound to chaperones DnaK and GroEL. Nature 355: 455-457

Landry J, Lambert H, Zhou M, Lavoie JN, Hickey E, Weber LA, Anderson CW (1992b) Human HSP27 is phosphorylated at serines-78 and serines-82 by heat shock and mitogen-activated kinases that recognize the same amino acid motif as S6 kinase-II. J Biol Chem 267: 794-803

Langer T, Pfeifer G, Martin J, Baumeister W, Hartl F-U (1992a) Chaperonin-mediated protein folding: GroES bind to one end of the GroEL cylinder, which accomodates the protein substrate within its central cavity. EMBO J 11: 4757-4765

Langer T, Lu C, Echols H, Flanagan J, Hayer MK, Hartl F-U (1992b) Successive action of DnaK, DnaJ and GroEL along the pathway of chaperone-mediated protein folding. Nature 356: 683-689

Laskey RA, Honda BM, Finch JT (1978) Nucleosomes are assembled by an acidic protein which binds histones and transfers them to DNA. Nature 275: 416-420

Lavoie JN, Gingrasbreton G, Tanguay RM, Landry J (1993) Induction of Chinese hamster HSP27 gene expression in mouse cells confers resistance to heat shock - HSP27 stabilization of the microfilament organization. J Biol Chem 268: 3420-3429

Lee YJ, Hou ZZ, Curetty L, Borrelli MJ (1992) Development of acute thermotolerance in L929 cells - Lack of HSP28 synthesis and phosphorylation. J Cell Physiol 152: 118-125

Lees-Miller SP, Anderson CW (1991) The DNA-activated protein kinase, DNA-PK: A potential coordinator of molecular events. Cancer Cells 3: 341-346

Legagneux V, Morange M, Bensaude O (1991) Heat shock increases turnover of 90 kDa heat shock protein phosphate groups in HeLa cells. FEBS L 291: 359-362

Leustek T, Dalie B, Amir-Shapira D, Brot N, Weissbach H (1989) A member of the Hsp70 family is localized in mitochondria and resembles Escherichia coli DnaK. Proc Natl Acad Sci USA 86: 7805-7808

Leustek T, Amir-Shapira D, Toledo H, Brot N, Weissbach H (1992) Autophosphorylation of 70 kDa heat shock proteins. Cell Mol Biol 38: 1-10

Levitt J (1980) Responses of Plants to Environmental Stresses. Academic Press, New York

Lewis MJ, Pelham HRB (1985) Involvement of ATP in the nuclear and nucleolar functions of the 70 kd heat shock protein. EMBO J 4: 3137-3143

Lewis VA, Hynes GM, Zheng D, Saibil H, Willison K (1992) T-complex polypeptide-1 is a subunit of a heteromeric particle in the eukaryotic cytosol. Nature 358: 249-252

Liberek K, Marszalek J, Ang D, Georgopoulos C, Zylicz M (1991) Escherichia coli DnaJ and GrpE heat shock proteins jointly stimulate ATPase activity of DnaK. Proc Natl Acad Sci USA 88: 2874-2878

Liberek K, Galitski TP, Zylicz M, Georgopoulos C (1992) The DnaK chaperone modulates the heat shock response of Escherichia coli by binding to the σ^{32} transcription factor. Proc Natl Acad Sci USA 89: 3516-3520

Lincoln JE, Cordes S, Read E, Fischer RL (1987) Regulation of gene expression by ethylene during Lycopersicon esculentum (tomato) fruit development. Proc Natl Acad Sci USA 84: 2793-2797

Lindquist S, Craig EA (1988) The heat - shock proteins. Annu Rev Genet 22: 631-677

Linthorst HJM (1991) Pathogenesis-related proteins in plants. Crit Rev Plant Sci 10: 123-150

Liu RY, Li XC, Li LG, Li GC (1992) Expression of human hsp70 in rat fibroblasts enhances cell survival and facilitates recovery from translational and transcriptional inhibition following heat shock. Cancer Res 52: 3667-3673

Loomis WF, Wheeler SA (1982) Chromatin-associated heat shock proteins in Dictyostelium. Devel Biol 90: 412-418

Maccarone M, Veldink GA, Vliegenthart JFG (1992) Thermal injury and ozone stress affect soybean lipoxygenases expression. FEBS L 309: 225-230

Marivet J, Frendo P, Burkard G (1992) Effects of abiotic stresses on cyclophilin gene expression in maize and bean and sequence analysis of bean cyclophilin cDNA. Plant Sci 84: 171-178

Mason HS, Mullet JE (1990) Expression of two soybean vegetative storage protein genes during development and in response to water deficit, wounding, and jasmonic acid. Plant Cell 2: 569-579

Massol M, Lebeau MC, Renoir JM, Faber LE, Baulieu EE (1992) Rabbit FKBP59-heat shock protein binding immunophilin (HBI) is a calmodulin binding protein. Biochem Biophys Res Commun 187: 1330-1335

Matsumoto M, Fujimoto H (1990) Cloning of a HSP70-related gene expressed in mouse spermatids. Biochem Biophys Res Commun 166: 43-49

McCarty JS, Walker GC (1991) DnaK as a thermometer: Threonine-199 is the site of autophosphorylation and is critical for ATPase activity. Proc Natl Acad Sci USA 88: 9513-9517

McMullin TW, Hallberg RL (1988) A highly evolutionarily conserved mitochondrial protein is structurally related to the protein encoded by the E. coli gene. Mol Cell Biol 8: 371-380

Melan MA, Dong X, Endara ME, Davis KR, Ausubel FM, Peterman TK (1993) An Arabidopsis thaliana lipoxygenase gene can be induced by pathogens, abscisic acid, and methyl jasmonate. Plant Physiol 101: 441-450

Merck KB, Groenen PJTA, Voorter CEM, DeHaard-Hoekman WA, Horwitz J, Bloemendal H, DeJong WW (1993) Structural and functional similarities of bovine α-crystallin and mouse small heat-shock protein - a family of chaperones. J Biol Chem 268: 1046-1052

Meyer A, Müller P, Sembdner G (1987) Air pollution and plant hormones. Biochem Physiol Pflanzen 182: 1-21

Miernyk JA, Duck NB, Shatters RG, Folk WR (1992) The 70-Kilodalton heat shock cognate can act as a molecular chaperone during the membrane translocation of a plant secretory protein precursor. Plant Cell 4: 821-829

Miron T, Van Compernolle K, Van de Kerckhove J, Wilchek M, Geiger B (1991) A 25-kD inhibitor of actin polymerization is a low molecular mass heat shock protein. J Cell Biol 114: 255-261

Moore MS, Blobel G (1992) The two steps of nuclear import, targeting to the nuclear envelope and translocation through the nuclear pore, require different cytosolic factors. Cell 69: 939-950

Mori M, Murata K, Kubota H, Yamamoto A, Matsushiro A, Morita T (1992) Cloning of a cDNA encoding the Tcp-1 (t complex polypeptide-1) homologue of Arabidopsis thaliana. Gene 122: 381-382

Morimoto RI (1993) Cells in stress - transcriptional activation of heat shock genes. Science 259: 1409-1410

Mühlbach HP (1987) Viroide: Freie infektiöse RNA-Moleküle als Erreger von Pflanzen-krankheiten. Biol uns Zeit 27: 65-78

Mueller MJ, Brodschelm W, Spannagel E, Zenk MH (1993) Signaling in the elicitation process is mediated through the octadecanoid pathway leading to jasmonic acid. Proc Natl Acad Sci USA 90: 7490-7494

Mummert E, Grimm R, Speth V, Eckershorn C, Schiltz E, Gatenby AA, Schäfer E (1993) A TCP1-related molecular chaperone from plants refolds phytochrome to its photoreversible form. Nature 363: 644-648

Munro S, Pelham HRB (1984) Use of peptide tagging to detect proteins expressed from cloned genes: deletion mapping functional domains of Drosophila hsp 70. EMBO J 3: 3087-3093

Munro S, Pelham HRB (1986) An hsp70 - like protein in the ER: Identity with the 78 kd glucose-regulated protein and immunoglobulin heavy chain binding protein. Cell 46: 291-300

Munro S, Pelham HRB (1987) A C-terminal signal prevents secretion of luminal ER proteins. Cell 48: 899-907

Murphy TM, Matson GB, Morrison SL (1983) Ultraviolet-stimulated $KHCO_3$ efflux from rose cells. Regulation of cytoplasmic pH. Plant Physiol 73: 20-24

Nakagawa N, Nakajima N, Imaseki H (1988) Immunochemical difference of wound-induced 1-aminocyclopropane-1-carboxylate synthase from the auxin-induced enzyme. Plant Cell Physiol 29: 1255-1259

Napolitano EW, Pachter JS, Chin SSM, Liem RKH (1985) ß-Internexin, a ubiquitous intermediate filament - associated protein. J Cell Biol 101: 1323-1331

Napolitano EW, Pachter JS, Liem RKH (1987) Intracellular distribution of mammalian stress protein. Effects of cytoskeletal-specific agents. J Biol Chem 262: 1493-1504

Neale AD, Wahleithner JA, Lund M, Bonnett HT, Kelly A, Meeks-Wagner DR, Peacock WJ, Dennis ES (1990) Chitinase, ß-1,3-glucanase, osmotin, and extensin are expressed in tobacco explants during flower formation. Plant Cell 2: 673-684

Nelson DE, Raghothama KG, Singh NK, Hasegawa PM, Bressan RA (1992a) Analysis of structure and transcriptional activation of an osmotin gene. Plant Mol Biol 19: 577-588

Nelson RJ, Ziegelhoffer T, Nicolet C, Werner-Washburne M, Craig EA (1992b) The translation machinery and 70 kd heat shock protein cooperate in protein synthesis. Cell 71: 97-105

Neumann D, Emmermann M, Thierfelder J-M, zur Nieden U, Clericus M, Braun H-P, Nover L, Schmitz UK (1993) HSP68 - a DnaK-like heat-stress protein of plant mitochondria. Planta 190: 32-43

Neumann D, zur Nieden U, Manteuffel R, Walter G, Scharf K-D, Nover L (1987) Intracellular localization of heat shock proteins in tomato cell cultures. Europ J Cell Biol 43: 71-81

Neupert W, Hartl F-U, Craig EA, Pfanner N (1990) How do polypeptides cross the mitochondrial membranes? Cell 63: 447-450

Neven LG, Haskell DW, Guy CL, Denslow N, Klein PA, Green LG, Silverman A (1992) Association of 70-kilodalton heat-shock cognate proteins with acclimation to cold. Plant Physiol 99: 1362-1369

Nieto-Sotelo J, Ho THD (1986) Effect of heat shock on the metabolism of glutathione in maize roots. Plant Physiol 82: 1031-1035

Nigg EA, Baeuerle PA, Lührmann R (1991) Nuclear import - export: In search of signals and mechanisms. Cell 66: 15-22

Nover L (1987) Expression of heat shock genes in homologous and heterologous systems. Enz Microb Technol 9: 130-144

Nover L (ed) (1991) Heat Shock Response. CRC Press, Boca Raton (Fd)

Nover L, Scharf K-D, Neumann D (1983) Formation of cytoplasmic heat shock granules in tomato cell cultures and leaves. Mol Cell Biol 3: 1648-1655

Nover L, Munsche D, Ohme K, Scharf K-D (1986) Ribosome biosynthesis in heat shocked tomato cell cultures I. Ribosomal RNA. Europ J Biochem 160: 297-304

Nover L, Scharf K-D, Neumann D (1989) Cytoplasmic heat shock granules are formed from precursor particles and are associated with a specific set of mRNAs. Mol Cell Biol 9: 1298-1308

Nover L, Neumann D, Scharf K-D (eds) (1990) Heat Shock and Other Stress Response Systems of Plants. Springer Berlin

Ohashi Y, Matsuoka M (1985) Synthesis of stress proteins in tobacco leaves. Plant Cell Physiol 26: 473-480

Ohtsuka K, Nakamura H, Sato C (1986) Intracellular distribution of 73000 and 72000 dalton heat shock proteins in HeLa cells. Int J Hyperthermia 2: 267-276

Ohtsuka K, Tanabe K, Nakamura H, Sato C (1986) Possible cytoskeletal association of 69.000 and 68.000 dalton heat shock proteins and structural relations among heat shock proteins in murine mastocytoma cells. Radiat Res 108: 34-42

Pallas DC, Morgan W, Roberts TM (1989) The cellular proteins which can associate specifically with polyomavirus middle T-antigen in human 293 cells include the major human 70-kilodalton heat shock proteins. J Virol 63: 4533-4539

Palleros DR, Welch WJ, Fink AL (1991) Interaction of hsp70 with unfolded proteins: Effects of temperature and nucleotides on the kinetics of binding. Proc Natl Acad Sci USA 88: 5719-5723

Parsell DA, Sanchez Y, Stitzel JD, Lindquist S (1991) Hsp104 is a highly conserved protein with 2 essential nucleotide-binding sites. Nature 353: 270-273

Partaledis JA, Berlin V (1993) The FKB2 gene of Saccharomyces cerevisiae, encoding the immuno-suppressant-binding protein FKBP-13, is regulated in response to accumulation of unfolded proteins in the endoplasmic reticulum. Proc Natl Acad Sci USA 90: 5450-5454

Parthier B (1991) Jasmonates, new regulators of plant growth and development: Many facts and few hypotheses on their action. Bot Acta 104: 446-454

Peattie DA, Harding MW, Fleming MA, DeCenzo MT, Lippke JA, Livingston DJ, Benasutti M (1992) Expression and characterization of human FKBP52, an immunophilin that associates with the 90-kDa heat shock protein and is a component of steroid receptor complexes. Proc Natl Acad Sci USA 89: 10974-10978

Pearce G, Ryan CA, Liljegren D (1988) Proteinase inhibitors I and II in fruit of wild tomato species: Transient components of a mechanism for defense and seed dispersal. Planta 175: 527-531

Pelham HRB (1982) A regulatory upstream promoter element in the Drosophila Hsp 70 heat-shock gene. Cell 30: 517-528

Pelham HRB (1984) Hsp 70 accelerates the recovery of nucleolar morphology after heat shock. EMBO J 3: 3095-3100

Pelham HRB, Bienz M (1982) A synthetic heat shock promoter element confers heat-inducibility on the herpes simplex virus thymidine kinase gene. EMBO J 1: 1473-1477

Pena-Cortes H, Sanchez-Serrano JJ, Mertens R, Willmitzer L, Prat S (1989) Abscisic acid is involved in the wound-induced expression of the proteinase inhibitor II gene in potato and tomato. Proc Natl Acad Sci USA 86: 9851-9855

Pena-Cortes H, Willmitzer L, Sanchez-Serrano JJ (1991) Abscisic acid mediates wound induction but not developmental-specific expression of the proteinase inhibitor II gene family. Plant Cell 3: 963-972

Pendharkar MB, Nair PM (1975) Induction of phenylalanine ammonia lyase (PAL) in gamma irradiated potatoes. Radiat Bot 15: 191-197

Pfanner N, Söllner T, Neupert W (1991) Mitochondrial import receptors for precursor proteins. Trends Biochem Sci 16: 63-67

Picard D, Khursheed B, Garabedian MJ, Fortin MG, Lindquist S, Yamamoto KR (1990) Reduced levels of hsp90 compromise steroid receptor action in vivo. Nature 348: 166-168

Pinto M, Morange M, Bensaude O (1991) Denaturation of proteins during heat shock - in vivo recovery of solubility and activity of reporter enzymes. J Biol Chem 266: 13941-13946

Prasad TK, Stewart CR (1992) cDNA clones encoding Arabidopsis thaliana and Zea mays mitochondrial chaperonin-HSP60 and gene expression during seed germination and heat shock. Plant Mol Biol 18: 873-885

Pratt WB, Hutchison KA, Scherrer LC (1992a) Steroid receptor folding by heat-shock proteins and composition of the receptor heterocomplex. Trends Endocrinol Metabol 3: 326-333

Pratt WB, Scherrer LC, Hutchison KA, Dalman FC (1992b) A model of glucocorticoid receptor unfolding and stabilization by a heat shock protein complex. J Steroid Biochem Mol Biol 41: 223-229

Privalov PL (1990) Cold denaturation of proteins. Crit Rev Biochem Mol Biol 25: 281-305

Raabe T, Manley JL (1991) A human homologue of the Escherichia coli DnaJ heat shock protein. Nucl Acids Res 19: 6645

Rao PV, Horwitz J, Zigler JS (1993) α-Crystallin, a molecular chaperone, forms a stable complex with carbonic anhydrase upon heat denaturation. Biochem Biophys Res Commun 190: 786-793

Ratajczak T, Carrello A, Mark PJ, Warner BJ, Simpson RJ, Moritz RL, House AK (1993) The cyclophilin component of the unactivated estrogen receptor contains a tetra-tricopeptide repeat domain and shares identity with p59 (FKBP59). J Biol Chem 268: 13187-13192

Rhodes MJC, Wooltorton LSC (1977) Changes in the activity of enzymes of phenylpropanoid metabolism in tomatoes stored at low temperatures.Phytochem 16: 655-659

Rhodes D (1987) Metabolic responses to stress: In: Davies E (ed), The Biochemistry of Plants 12, Acad Press, San Diego, pp 201-241

Rigden J, Coutis R (1988) Pathogenesis-related proteins in plants. Trends Genet 4: 87-89

Ritossa F (1962) A new puffing pattern induced by heat shock and DNP in Drosophila. Experientia 18: 571-573

Roby D, Toppan A, Esquerrè-Tugayè MT (1985) Cell surfaces in plant-microorganism interactions. V. Elicitors of fungal and plant origin trigger the synthesis of ethylene and of cell wall hydroxyproline-rich glycoprotein in plants. Plant Physiol 77: 700-704

Rosario MO, Perkins SL, O'Brien DA, Allen RL, Eddy EM (1992) Identification of the gene for the developmentally expressed 70 kDa heat-shock protein (P70) of mouse spermatogenic cells. Devel Biol 150: 1-11

Rüegsegger A, Brunold C (1992) Effect of cadmium on γ–glutamylcysteine synthesis in maize seedlings. Plant Physiol 99: 428-433

Rüegsegger A, Schmutz D, Brunold C (1990) Regulation of glutathione synthesis by cadmium in Pisum sativum. Plant Physiol 93: 1579-1584

Ryan CA (1987) Oligosaccharide signalling in plants. Annu Rev Cell Biol 3: 295-317

Ryan CA (1988) Oligosaccharines as recognition signals for the expression of defensive genes in plants. Biochemistry 27: 8879-8883

Ryan CA (1992) The search for the proteinase inhibitor-inducing factor, PIF. Plant Mol Biol 19: 123-133

Ryan CA, Farmer EE (1991) Oligosaccharide signals in plants: A current assessment. Annu Rev Plant Physiol Mol Biol 42: 651-674

Sadis S, Hightower LE (1992) Unfolded proteins stimulate molecular chaperone Hsc70 ATPase by accelerating ADP/ATP exchange. Biochemistry 31: 9406-9412

Sanchez ER (1990) Hsp56 - a novel heat shock protein associated with untransformed steroid receptor complexes. J Biol Chem 265: 22067-22070

Sanchez, ER, Toft DO, Schlesinger MJ, Pratt WB (1985) Evidence that the 90-kDa phosphoprotein associated with the untransformed L-cell glucocorticoid receptor is a murine heat shock protein. J Biol Chem 260: 12398-12401

Sanchez ER, Faber LE, Henzel WJ, Pratt WB (1990) The 56-59-kilodalton protein identified in untransformed steroid receptor complexes is a unique protein that exists in cytosol in a complex with both the 70- and 90-kilodalton heat shock proteins . Biochemistry 29: 5145-5152

Sanchez Y, Lindquist SL (1990) Hsp104 required for induced thermotolerance. Science 248: 1112-1115

Sanchez Y, Taulien J, Borkovich KA, Lindquist S (1992) Hsp104 is required for tolerance to many forms of stress. EMBO J 11: 2357-2364

Sanders SL, Whitfield KM, Vogel JP, Rose MD, Schekman RW (1992) Sec61p and BiP directly facilitate polypeptide translocation into the ER. Cell 69: 353-365

Sawai ET, Butel JS (1989) Association of a cellular heat shock protein with simian virus 40 large T-antigen in transformed cells. J Virol 63: 3961-3973

Scharf K-D, Nover L (1987) Control of ribosome biosynthesis in plant cell cultures under heat shock conditions. II Ribosomal proteins. Biochim Biophys Acta 909: 44-57

Scharf K-D, Rose S, Zott W, Schöffl F, Nover L, (1990) Three tomato genes code for heat stress transcription factors with a region of remarkable homology to the DNA-binding domain of the yeast HSF. EMBO J 9: 4495-4501

Scharf K-D, Materna T, Treuter E, Nover L (1994) Heat stress promoters and transcription factors. In L Nover ed, Plant Promoters and Transcription Factors. Springer, Berlin, in press

Schlesinger MJ, Ashburner M, Tissieres A (1982) (eds) Heat Shock: From Bacteria to Man. Cold Spring Harbor Lab., Cold Spring Harbor, New York

Schlossman DM, Schmid SL, Braell WA, Rothman JE (1984) An enzyme that removes clathrin coats: Purification of an uncoating ATPase. J Cell Biol 99: 723-733

Schmid FX (1993) Prolyl isomerase - Enzymatic catalysis of slow protein-folding reactions. Annu Rev Biophys Biomol Structure 22: 123-143

Schröder H, Langer T, Hartl F-U, Bukau B (1993) DnaK, DnaJ, GrpE form a cellular chaperone machinery capable of repairing heat-induced protein damage. EMBO J. in press

Sell SM, Eisen C, Ang D, Zylicz M, Georgopoulos C (1990) Isolation and characterization of dnaJ null mutants of Escherichia coli. J Bacteriol 172: 4827-4835

Sembdner G, Parthier B (1993) The biochemistry and the physiological and molecular actions of jasmonates. Annu Rev Plant Physiol Mol Biol 44: 569-589

Seufert W, Jentsch S (1992) In vivo function of the proteasome in the ubiquitin pathway. EMBO J 11: 3077-3080

Shaaltiel Y, Glazer A, Bocion PF, Gressel J (1988) Cross tolerance to herbicidal and environmental oxidants of plant biotypes tolerant to paraquat, sulfur dioxide, and ozone. Pestic Biochem Physiol 31: 13-23

Sherman MY, Goldberg AL (1992a) Heat shock in Escherichia coli alters the protein binding properties of the chaperonin groEL by inducing its phosphorylation. Nature 357: 167-169

Sherman MY, Goldberg AL (1992b) Involvement of the chaperonin DnaK in the rapid degradation of a mutant protein in Escherichia coli. EMBO J 11: 71-77

Shi Y, Thomas JO (1992) The transport of proteins into the nucleus requires the 70-kilodalton heat shock protein or its cytosolic cognate. Mol Cell Biol 12: 2186-2192

Shinshi H, Mohnen D, Meins F (1987) Regulation of a plant pathogenesis-related enzyme: Inhibition of chitinase and chitinase mRNA accumulation in cultured tobacco tissues by auxin and cytokinin. Proc Natl Acad Sci USA 84: 89-93

Showalter AM (1993) Structure and function of plant cell wall proteins. Plant Cell 5: 9-23

Showalter AM, Zhou J, Rumeau D, Worst SG, Varner JE (1991) Tomato extensin and extensin-like cDNAs: Structure and expression in response to wounding. Plant Mol Biol 16: 547-565

Silver PA, Way JC (1993) Eukaryotic DnaJ homologs and the specificity of Hsp70 activity. Cell 74: 5-6

Singh NK, Bracker CA, Hasegawa PM, Handa AK, Buckel S, Hermodson MA, Pfankoch E, Regnier FE, Bressan RA (1987) Characterization of osmotin: a thaumatin-like protein associated with osmotic adaptation in plant cells. Plant Physiol 85: 529-536

Singh NK, Nelson DE, Kuhn D, Hasegawa PM, Bressan RA (1989) Molecular cloning of osmotin and regulation of its expression by ABA and adaptation to low water potential. Plant Physiol 90: 1096-1101

Skelly S, Fu C-F, Dalie B, Redfield B, Coleman T, Brot N, Weissbach H (1988) Antibody to σ^{32} cross-reacts with DnaK: Association of DnaK protein with Escherichia coli RNA polymerase. Proc Natl Acad Sci USA 85: 5497-5501

Skriver K, Mundy J (1990) Gene expression in response to abscisic acid and osmotic stress. Plant Cell 2: 503-512

Slocum RD, Kaur-Sawhney R, Galston AW (1984) The physiology and biochemistry of polyamines in plants. Arch Biochem Biophys 235: 283-303

Smith TA (1984) Putrescine and inorganic icons. Rec Advanc Phytochem 18: 7-54

Smith BJ, Yaffe MP (1991a) A mutation in the yeast heat-shock factor gene causes temperature-sensitive defects in both mitochondrial protein import and the cell cycle. Mol Cell Biol 11: 2647-2655

Smith BJ, Yaffe MP (1991b) Uncoupling thermotolerance from the induction of heat shock proteins. Proc Natl Acad Sci USA 88: 11091-11094

Smith IK, Polle A, Rennenberg H (1990) Glutathione. In: Alscher RG, Cumming JR (eds), Stress Responses in Plants: Adaptation and Acclimation Mechanisms. Wiley-Liss, New York, pp 201-215

Soll J, Waegemann K (1992) A functionally active protein import complex from chloroplasts. Plant J 2: 253-256

Sorger PK, Pelham HRB (1987) The glucose-regulated protein grp94 is related to heat shock protein hsp90. J Mol Biol 194: 341-344

Stamnes MA, Rutherford SL, Zuker CS (1992) Cyclophilins: a new family of proteins involved in intracellular folding. Trends Cell Biol 2: 272-276

Staswick PE (1992) Jasmonate, genes, and fragrant signals. Plant Physiol 99: 804-807

Staswick PE, Huang J-F, Rhee I (1991) Nitrogen and methyl jasmonate induction of soybean vegetative storage protein genes. Plant Physiol 96: 130-136

Stermer BA, Hammerschmidt R (1987) Association of heat shock induced resistance to disease with increased accumulation of insoluble extensin and ethylene synthesis. Physiol Mol Plant Pathol 31: 453-462

Stintzi A, Heitz T, Kauffmann S, Legrand M, Fritig B (1991) Identification of a basic pathogenesis-related, thaumatin-like protein of virus-infected tobacco as osmotin. Physiol Mol Plant Pathol 37: 137-146

Stone DE, Craig EA (1990) Self-regulation of 70-kilodalton heat shock proteins in Saccharomyces cerevisiae. Mol Cell Biol 10: 1622-1632

Straus DB, Walter WA, Gross CA (1988) Escherichia coli heat shock gene mutants are defective in proteolysis. Genes Devel 2: 1851-1858

Squires C, Squires CL (1992) The Clp proteins: Proteolysis regulators or molecular chaperones? J Bacteriol 174: 1081-1085

Sykes K, Gething M-J, Sambrook J (1993) Proline isomerases function during heat shock. Proc Natl Acad Sci USA 90: 5853-5857

Tai P-K, Albers MW, Chang H, Faber LE, Schreiber SL (1992) Association of a 59-kilodalton immunophilin with the glucocorticoid receptor complex. Science 256: 1315-1318

Terlecky SR, Chiang HL, Olson TS, Dice JF (1992) Protein and peptide binding and stimulation of in vitro lysosomal proteolysis by the 73-kDa heat shock cognate protein. J Biol Chem 267: 9202-9209

Tetu B, Lacasse B, Bouchard HL, Lagace R, Huot J, Landry J (1992) Prognostic influence of HSP27 expression in malignant fibrous histiocytoma - A clinicopathological and immunohistochemical study. Cancer Res 52: 2325-2328

Thalhammer T, Kieffer LJ, Jiang T, Handschumacher RE (1992) Isolation and partial characterization of membrane - associated cyclophilin and a related 22-kDa glycoprotein. Europ J Biochem 206: 31-37

Thomas JC, McElwain EMF, Bohnert NJ (1992) Convergent induction of osmotic stress-responses. Abscisic acid, cytokinin and the effect of NaCl. Plant Physiol 100: 416-423

Tilly K, McKittrick N, Zylicz M, Georgopoulos C (1983) The dnaK protein modulates the heat-shock response of Escherichia coli. Cell 34: 641-646

Tissieres A, Mitchell HK, Tracy UM (1974) Protein synthesis in salivary glands of D. melanogaster. Relation to chromosome puffs. J Mol Biol 84: 389-398

Traylor EA, Shore SH, Ransom RF, Dunkle LD (1987) Pathotoxin effects in Sorghum are also produced by mercuric chloride treatment. Plant Physiol 84: 975-978

Tsang EWT, Bowler C, Heronart D, Van Camp W, Villarroel R, Genetello C, van Montageu M, Inze D (1991) Differential regulation of superoxide dismutases in plants exposed to environmental stress. Plant Cell 3: 783-792

Tsukaya H, Takahashi T, Naito S, Komeda Y (1993) Floral organ-specific and constitutive expression of an Arabidopsis thaliana heat shock HSP18.2 = Gus fusion gene is retained even after homeotic conversion of flowers by mutation. Mol Gen Genetics 237: 26-32

Tukendorf A, Rauser WE (1990) Changes in glutathione and phytochelatins in roots of maize seedlings exposed to cadmium. Plant Sci 70: 155-166

Uknes S, Mauch-Mani B, Moyer M, Potter S, Williams S, Dincher S, Chandler D, Slusarenko A, Ward E, Ryals J (1992) Acquired resistance in Arabidopsis. Plant Cell 4: 645-656

Van Loon LC (1983) Mechanisms of resistance in virus-infected plants. In: Bailey JA, Deverall BJ (eds), The Dynamic of Host Defence. Acad Press, Sydney, pp 123-190

Van Loon LC (1985) Pathogenesis-related proteins. Plant Mol Biol 4: 111-116

Vayda ME, Schaeffer HJ (1988) Hypoxic stress inhibits the appearance of wound-response proteins in potato tubers. Plant Physiol 88: 805-809

Vierling E (1991) The roles of heat shock proteins in plants. Annu Rev Plant Physiol Plant Mol Biol 42: 579-620

Vierling E, Sun A (1989) Developmental expression of heat shock proteins in higher plants. In: Cherry JH (ed), Environmental Stress in Plants, NATO ASI Series, Vol G19, Springer Berlin, pp 343-354

Vierstra RD, Burke TJ, Callis J, Hatfield PM, Jabben M, Shanklin J, Sullivan ML (1988) Characterization of the ubiquitin - dependent proteolytic pathway in higher plants. In: Schlesinger MJ, Hershko A (eds), The Ubiquitin System. Cold Spring Harbor Press, pp 119-125

Waegemann K, Soll J (1991) Characterization of the protein import apparatus in isolated outer envelopes of chloroplasts. Plant J 1: 149-158

Wagih EE, Coutts RHA (1981) Similarities in the soluble protein profiles of leaf tissue following either a hypersensitive reaction to virus infection or plasmolysis. Plant Sci Letters 21: 61-69

Waldinger D, Subramanian AR, Cleve H (1989) The polymorphic human chaperonin protein HuCha60 is a mitochondrial protein sensitive to heat shock and cell transformation. Europ J Cell Biol 50: 435-441

Walker-Simmons M, Ryan CA (1977) Wound -induced accumulation of trypsin inhibitor activities in plant leaves. Survey of several plant genera. Plant Physiol 59: 437-439

Walker-Simmons M, Ryan CA (1986) Proteinase inhibitor I accumulation in tomato suspension cultures. Induction by plant and fungal cell wall fragments and an extracellular polysaccharide secreted into the medium. Plant Physiol 80: 68-71

Walter G, Carbone A, Welch WJ (1987) Medium tumor antigen of polyomavirus transformation - defective mutant NG59 is associated with 73-kilodalton heat shock protein. J Virol 61: 405-410

Wang C, Asai DJ, Lazarides E (1980) The 68,000-dalton neurofilament-associated polypeptide is a component of nonneuronal cells and of skeletal myofibrils. Proc Natl Acad Sci USA 77: 1541-1545

Wang C, Lin BL (1993) The disappearance of an Hsc70 species in mung bean seed during germination - purification and characterization of the protein. Plant Mol Biol 21: 317-329

Ward ER, Uknes SJ, Williams SC, Dincher SS, Wiederhold DL, Alexander DC, Ahl-Goy P, Metraux JP, Ryals JA (1991) Coordinate gene activity in response to agents that induce systemic acquired resistance. Plant Cell 3: 1085-1094

Watts FZ, Walters AJ, Moore AL (1992) Characterization of PHSP1, a cDNA encoding a mitochondrial HSP70 from Pisum sativum. Plant Mol Biol 18: 23-32

Welch WJ, Feramisco JR (1985) Rapid purification of mammalian 70,000-dalton stress proteins: Affinity of the proteins for nucleotides. Mol Cell Biol 5: 1229-1237

Welch WJ, Mizzen LA (1988) Characterization of the thermotolerant cell: II Effects on the intracellular distribution of HSP70, intermediate filaments, and snRNP's. J Cell Biol 106: 1117-1130

Welch WJ, Suhan JP (1985) Morphological study of the mammalian stress response: characterization of changes in cytoplasmic organelles, cytoskeleton, and nucleoli, and appearance of intranuclear actin filaments in rat fibroblasts after heat-shock treatment. J Cell Biol 101: 1198-1211

Weller NK (1988) A 70-kDa microtubule-associated protein in NIL8 cells comigrates with the 70 kDa heat shock protein. Biol Cell 63: 307-318

Werner-Washburne M, Stone DE, Craig EA (1987) Complex interactions among members of an essential subfamily of hsp70 genes in Saccharomyces cerevisiae. Mol Cell Biol 7: 2568-2577

Whatley SA, Leung T, Hall C, Lim L (1986) The brain 68-kilodalton microtubule-associated protein is cognate form of the 70-kilodalton mammalian heat-shock protein and is present as a specific isoform in synaptosomal membranes. J Neurochem 47: 1576-1583

Wickner SH (1990) Three Escherichia coli heat shock proteins are required for P1 plasmid DNA replication: Formation of an active complex between E. coli DnaJ protein and the P1 initiator protein. Proc Natl Acad Sci USA 87: 2690-2694

Wickner S, Hoskins J, McKenney K (1991a) Function of DnaJ and DnaK as chaperones in origin - specific DNA binding by RepA. Nature 350: 165-167

Wickner S, Hoskins J, McKenney K (1991b) Monomerization of RepA dimers by heat shock proteins activates binding to DNA replication origin. Proc Natl Acad Sci USA 88: 7903-7907

Wickner S, Skowyra D, Hoskins J, McKenney K (1992) DnaJ, DnaK and GrpE heat shock proteins are required in ori P1 DNA replication solely at the RepA monomerization step. Proc Natl Acad Sci USA 89: 10345-10349

Wiech H, Buchner J, Zimmermann R, Jakob U (1992) Hsp90 chaperones protein folding in vitro. Nature 358: 169-170

Wingate VPM, Lawton MA, Lamb CJ (1988) Glutathione causes a massive and selective induction of plant defense genes. Plant Physiol 87: 206-210

Winter J, Sinibaldi R (1991) The expression of heat shock protein and cognate genes during plant development. In: Hightower L, Nover L (eds), Heat Shock and Development. Springer Berlin , pp 85-105

Wolgemuth DJ, Gruppi CM (1991) Heat shock gene expression during mammalian gametogenesis and early embryogenesis. In: Hightower L, Nover L (eds), Heat Shock and Development. Springer Berlin, pp 138-152

Woo KM, Kim KI, Goldberg AL, Ha DB, Chung CH (1992) The heat shock protein ClpB in Escherichia coli is a protein-activated ATPase. J Biol Chem 267: 20429-20434

Xu D, McElroy D, Thornburg RW, Wu R (1993) Systemic induction of a potato pin2 promoter by wounding, methyljasmonate and abscisic acid in transgenic rice plants. Plant Mol Biol 22: 573-588

Yaffe MB, Farr GW, Miklos D, Horvich AL, Sternlicht ML, Sternlicht H (1992) TCP1 complex is a molecular chaperone in tubulin biogenesis. Nature 358: 245-248

Yalovsky S, Paulsen H, Michaeli D, Chitnis PR, Nechushtai R (1992) Involvement of a chloroplast HSP70 heat shock protein in the integration of a protein (light - harvesting complex protein precursor) into the thylakoid membrane. Proc Natl Acad Sci USA 89: 5616-5619

Yang SF, Hoffman NE (1984) Ethylene biosynthesis and its regulation in higher plants. Annu Rev Plant Physiol 35: 155-189

Yenne SP, Hatzios KK (1990) Influence of oxime ether safeners on glutathione content and glutathione-related enzyme activity in seeds and seedlings of grain sorghum. Z Naturforsch 45c: 96-106

Young ND, Galston AW (1983) Putrescine and acid stress. Induction of arginine decarboxylase activity and putrescine accumulation by low pH. Plant Physiol 71: 767-771

Yost HJ, Lindquist S (1991) Heat shock proteins affect RNA processing during the heat shock response of Saccharomyces cerevisiae. Mol Cell Biol 11: 1062-1068

Zeilstra-Ryalls J, Fayet O, Georgopoulos C (1991) The universally conserved GroE (HSP60) chaperonins. Annu Rev Microbiol 45: 301-325

Zhong T, Arndt KT (1993) The yeast SIS1 - protein, a DnaJ homolog, is required for the initiation of translation. Cell 73: 1175-1186

Zhu JK, Shi J, Bressan RA, Hasegawa PM (1993) Expression of an Atriplex nummularia gene encoding a protein homologous to the bacterial molecular chaperone DnaJ. Plant Cell 5: 341-349

Zylicz M, LeBowitz JH, McMacken R, Georgopoulos C (1983) The DnaK protein of Escherichia coli processes an ATPase and autophosphorylating activity and is essential in an in vitro DNA replication system. Proc Natl Acad Sci USA 80: 6431-6435

Acknowledgements

I gratefully acknowledge the continuous assistance and critical advice of Ruth Laue, Justin Lee and Dr. Claus Wasternack during the preparation of this manuscript. Experimental work in our laboratories was supported by the Volkswagenstiftung Hannover, the Körberstiftung Hamburg, the Fonds der Chemischen Industrie and the Deutsche Forschungsgemeinschaft Bonn.

Developmental regulation and enhancement of heat shock gene expression

F. Schöffl, E. Kloske, A. Wagner, K. Severin, G. Schröder
Universität Tübingen
Lehrstuhl für Allgemeine Genetik
Auf der Morgenstelle 28
D-72076 Tübingen
Germany

Introduction

The heat shock (hs) response is one of the best-characterized gene regulatory systems in plants that is controlled by environmental stress. It affects a number of genes encoding hs proteins (HSPs), which become transcriptionally activated by the interaction between cis-regulatory promoter elements (HSE) and the trans-active hs activator protein HSF. In current research the molecular mechanism of transcriptional regulation is much emphasized. Soybean hs genes have played a key role in the identification of cis-active hs promoter elements. Their faithful regulation and use in chimeric constructions in transgenic tobacco and Arabidopsis suggested that a highly conserved and general regulatory mechanism of the hs response exists in plants. The isolation and characterization of genes encoding HSF of tomato confirmed this model by demonstrating the binding of HSF to synthetic HSEs (Scharf et al., 1990), and by transient activation of heterologous hs promoters, including also the soybean Gmhsp17.3-B promoter, in tobacco protoplasts (Treuter et al., 1993). The functional analysis of HSF from tomato and Arabidopsis and the manipulation of HSF expression in transgenic plants will have a major impact on our understanding of the molecular basis of the signal transfer induced by environmental factors.

However, besides the environmental regulation there are other aspects of the hs gene expression to which not much attention has been paid yet. These phenomena include the developmental regulation of hs protein (HSP) synthesis during embryogenesis and germination and the enhancement and stabilization of hs gene expression by scaffold attachment regions (SAR) in transgenic plants. Obviously, the two aspects seem to be unrelated but they both may limit or improve the use and efficiency of transgenes in plants. The investigation of the developmental control of HSP synthesis may ultimately reveal a new role of HSPs, perhaps important for the reproduction of plants.

NATO ASI Series, Vol. H 86
Biochemical and Cellular Mechanisms
of Stress Tolerance in Plants
Edited by J. H. Cherry
© Springer-Verlag Berlin Heidelberg 1994

Enhancement and stability of hs gene expression

In plants the amplitude of heat-inducible transcription of the
soybean hs genes Gmhsp17.3-B (Baumann et al., 1987) and
Gmhsp17.5-E (Czarnecka et al., 1989) is enhanced by upstream
sequences by a factor of five to ten. Such sequences are devoid
of HSE elements but require the downstream HSE-containing
sequences for the heat -inducible expression of genes. The
structural characteristics related to the enhancer-like activity
are short A- and T-rich sequences that proceed the generally
A-/T-rich intergenic regions which delimit the soybean hs genes.
Synergistic effects between upstream CCAAT box, A-/T-rich simple
sequences and HSEs of the Gmhsp17.3-B gene account for the
enhanced expression of the chimeric hs reporter gene expression
in transgenic tobacco (Rieping and Schöffl, 1992).

It is well established that the genes located at their authentic
chromosomal position are faithfully regulated and stably
expressed. On the other hand transfected genes, integrated
randomly into the genome, are frequently inactive, probably by
insertion into inactive chromatin (position effect) or become
inactivated in homozygous progeny. The phenomena of position
effects and allelic trans-interaction or co-suppression have been
described for the expression of different trans-genes in plants
(Horsch et al., 1984; 1985; Matzke et al., 1989; 1990; Napoli et
al., 1990; Krol et al., 1990; Joergensen, 1990; Meyer et al.,
1993). Position effects are also known for both authentic and
chimeric soybean hs -genes in transgenic tobacco (Baumann et al.,
1987; Schöffl et al., 1989; 1992; 1993); trans-inactivation
phenomena have not yet been critically investigated for chimeric
hs genes.

The level of expression of randomly integrated trans-genes is
generally unpredictable, varying with the chromosomal site of
insertion. This may be a consequence of the organization of the
chromosomes into topological constrained loops as functional
domains (Gasser and Laemmli, 1987; Jackson, 1986). A- and T-rich
consensus sequences termed MAR (matrix attachment region:
Cockerill and Garrard, 1986) or SAR (scaffold attachment regions:
Gasser and Laemmli, 1986) were mapped to chromosomal regions that
are attached to the nuclear matrix or scaffold in animal cells.
In plants, SAR sequences were identified for a number of genes or
chromosomal regions (Mielke et al., 1990; Hall et., 1991; Slatter
et al., 1991; Breyne et al., 1992) and also for soybean hs genes
(Schöffl et al., 1992). Some of these were tested for their
binding in vitro to isolated nuclear scaffolds; the highest
binding affinity was found for a 395 bp-fragment from the
3'-flanking region of Gmhsp17.3-B (Schöffl et al., 1993).

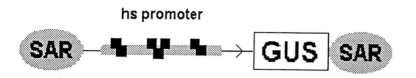

Fig. 1: SAR containing chimeric hs-GUS gene constructs
The construction of Gmhsp17.3-B soybean hs promoter fused to the
GUS reporter gene has been described by Schöffl et al., (1993).
SAR: scaffold attachment region sequences; boxes: HSE promoter
sequences

The effect of this SARL fragment on the expression of a hs
promoter-driven glucuronidase (GUS) reporter gene (Fig. 1) has
been examined by transgenic expression of a series of
constructions in tobacco (Schöffl et al., 1993). The mean level
of heat-inducible GUS activity increased significantly up to a
factor of ten when SARL was present at both sites or at least at
a position 5' to the chimeric gene. The orientation of SARL in
these constructs was not critical for the enhancing effect,
background levels (without hs) were very low, the induction
factor of GUS activity was about 30 for most constructs following
heat treatment. Despite position effects and a large range of the
inducible expression levels in these plants, there was a positive
correlation between expression levels and the copy number of the
trans-genes (Fig. 2). In transgenic plants containing constructs
without or only one 3' SARL, heat-inducible GUS activity was not
correlated with the gene copy number (Schöffl et al., 1993).
Thus, the enhancing effect of SAR may be due to a more stable and
gene dosage-correlated expression of the respective gene.

The molecular mechanism of SAR-mediated enhancement and/or
stability of gene expression is unknown and it has to be noted
that position effects are not eliminated. SARs might be involved
in the opening and closing of chromatin by the interaction with
other proteins, according to the chromatin switch model (Laemmli
et al., 1992), or in shielding an open chromatin structure from
the intrusion by chromatin closing activity. Alternatively, as
suggested by Jackson (1986), SAR-mediated elevation and
position-independent expression might apply for genes that are
reversibly linked via SAR to the transcriptionally active nuclear
scaffold. In contrast to developmentally regulated genes, whose
activity is temporarily and spatially confined to certain stages
in development, hs genes should be always accessible to

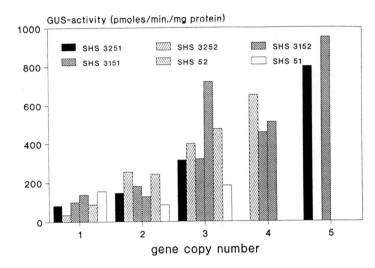

Fig. 2: Gene dosage correlated expression of SAR-containing gene
constructs. Individual plants containing the constructs SHS3251,
-3151, -3252, -3152 flanked at both sites by SAR or SHS51 and -51
only 5' of the hs-GUS gene were analysed (changed after Schöffl
et al., 1993). These constructs are known to mediate SAR-enhanced
expression of the reporter gene. The average heat-induced GUS
levels were plotted versus the transgene copy numbers of the
respective plants.

activation by heat stress and therefore located in constitutively
open chromatin domains. The availability of SAR sequences in the
flanking regions of hs genes may be one structural component of
the chromatin that warrants its determination for gene
expression.

The 395 bp SARL-fragment is by far the smallest SAR-containing
sequence that has been functionally identified for both, scaffold
binding and enhanced gene expression. It has yet to be tested
whether the stability of gene expression of the SARL-containing
constructs is maintained in the progeny of subsequent
generations; the enhanced expression levels were reproducibly
measured in individual plants of the F1-F3 progeny. For a general
use of SARL in heterologous constructions it has still to be
tested whether it exerts its effect also on the expression on
other promoters and genes linked to it.

Developmental regulation of the hs response

It is well established that upon temperature elevation HSPs are synthesized in essentially all vegetative tissues of the plant. Exceptions are generative and embryonic tissues. While it is known that germinating pollen and very early embryos do not exhibit full activation of hs genes upon heat stress (Zimmerman et al., 1989; van Herpen et al., 1989; Xiao and Mascarenhas, 1985) there are also some reports that at least several hs proteins are found without heat induction in late developing embryos and fully developed seeds of pea and wheat (Vierling and Sun, 1987; Helm and Abernethy, 1990) and also in several other legumes (Hernandez and Vierling, 1993). Developmental regulation of small HSP mRNAs has been also detected during meiotic prophase of pollen development in lily and maize (Bouchard, 1990; Dietrich et al., 1991) in certain stages of carrot embryogenesis (Zimmerman et al., 1989) and HSP90 mRNA during embryogenesis and pollen development in maize (Marrs et al., 1993).

Transgenic plants containing hs promoter-driven reporter genes are used in our laboratory to address the question about developmental regulation and developmental effects of hs gene expression in tobacco and Arabidopsis. Tobacco seeds and plants containing the GUS reporter gene controlled by a soybean hs promoter (see Fig. 1) were primarily used in these investigations (Kloske et al., 1993). The heat-induced levels of GUS activity varied considerably in the different stages of plant growth and development (Fig. 3). High constitutive levels of GUS activity (without hs) were only detected in dry seeds and during early germination. The constitutive expression of the GUS gene starts after flowering and increases during embryogenesis. The levels in dry seeds and early germination are as high as the inducible levels in leaves, but hs is unable to increase the GUS activity in seeds significantly. GUS expression seems to be restricted to the embryo only, other tissues of the imbibed seeds are not stained in histochemical assays of GUS activity (not shown). Three days after imbibition the GUS activity suddenly declines and cannot be induced by hs for several days until about that developmental stage when the primary leaves appear. In summary, the developmental regulation of the hs response in tobacco has two different aspects, the constitutive induction by the developmental program in embryogenesis and the lack of heat inducibility during germination.

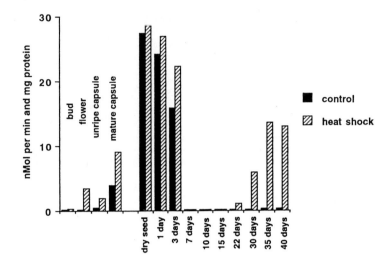

Fig. 3. Developmental regulation of a heat-inducible GUS gene in transgenic tobacco. Seeds from a transgenic plant containing a SAR flanked hs-GUS gene construct (see Fig. 1 and Schöffl et al., 1993) were subjected to germination and GUS activity was assayed before (control) and after a heat stress (heat shock) in samples of seeds, plantlets, leaves or respectively flower material taken after different times of germination and development (Kloske et al., 1993). The GUS activity was measured by the fluorescence method and is given in nmol methyl umbelliferone.

While testing the same chimeric hs-GUS gene construct in transgenic Arabidopsis, one of these two phenomena was not observed. There was no constitutive activity detectable in seeds and early germination, but in accordance with tobacco, there was also a lag-phase of heat-inducible GUS expression during germination until the first pair of leaves appeared (Kloske et al., 1993). We do not know whether this difference in developmental regulation between tobacco and Arabidopsis is fundamental or only gradual. To answer this question it has to be determined whether the synthesis of HSPs in both species follows the same pattern as exemplified by the heterologous reporter gene constructs. However, the lack of developmental induction in Arabidopsis is not due to a peculiarity of the used Gmhsp17.3-B promoter or the SAR sequences. Using the promoter of the Gmhsp17.6-L gene of soybean in a construct with the Adh (alcohol dehydrogenase) gene from Arabidopsis in an Adh- background (Severin et al., 1993), a very similar pattern of ADH activity respectively its inducibility can be observed (Fig. 4) as with the GUS construct described above.

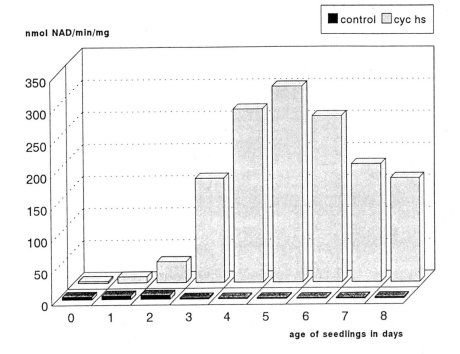

Fig. 4: Heat-inducibility of ADH expression during early plant development. The ADH activity was determined prior and after three cycles of heat stress and recovery in germinating seedlings of transgenic Arabidopsis plants containing the heat inducible Adh gene construct L+24adh (changed after Severin et al., 1993). Imbibed and vernalized seeds were germinated at 25°C and at the given times, samples of seeds or seedlings were taken and subjected to either hs or were kept at 25°C prior to protein extraction and determination of ADH activity.

The maximal levels of ADH activity were reached four to six days after imbibition. The peak region of ADH activity at this stage may reflect a temporary over-representation of ADH due to the mobilization and breakdown of storage proteins. The soluble protein content in mg per fresh weight is approximately three times higher during early than during later stages of germination.

The wt ADH enzyme in control plants (data not shown) has a much higher level before and a much lower level after that point in development when heat inducibility was maximal for the hs-Adh gene in Arabidopsis. Hence, developmental regulation of the hs-Adh gene is the major cause for the delay of its heat inducibility in development. The inducibility of HSP synthesis in embryos of early imbibed wheat (Helm and Abernethy, 1990) and sorghum (Howarth, 1989) seeds seems to contradict our data obtained for Arabidopsis, but there may be differences in sensitivity of the different methods, and perhaps also a less efficient recognition of heterologous hs promoters in the transgenic plants. It may be also possible that during embryogenesis and early germination the stability of HSPs is much higher than the stability of most other proteins including also reporter enzymes. The sudden breakdown of GUS activity after the third day of germination seems to indicate this for tobacco. In the case of the CAT (chloramphenicol acetyl transferase) reporter gene expressed as an heat-inducible construct driven also by the Gmhsp17.3-B promoter in tobacco, there may be an interference with the deposition in seeds because no developmental induction of CAT activity has been found (Schöffl et al., 1992).

The biological role of developmental regulation of the hs response is still not known. The examination of thermotolerance properties in the different stages of development indicates that early imbibed seeds are much more tolerant to heat treatment than later in germination. The acquisition of heat-induced tolerance is also impaired during germination and full competence is achieved after several days (Kloske et al., 1993). This is only a first clue to a protective function of HSPs during development and it is as well possible that the HSPs have an as yet unknown but specific role in development. A developmental function of HSPs is also implied for somatic embryogenesis, indicated by the appearance of hs-mRNA in Medicago sativa (Györgyey et al., 1991), tobacco (Heberle-Bors, personal communication), and HSPs in Brassica (Pechan, personal communication).

General conclusions

The developmental regulation of HSPs during embryogenesis and germination may be important for the reproduction of plants. There is some evidence that HSPs are required during zygotic and somatic embryogenesis in tobacco and perhaps also in other species. Limited by the relatively scarce data about this phenomenon, it is not yet possible to make a sound prediction on whether this is a general phenomenon or important to only some species. Our investigations of tobacco and Arabidopsis suggest that the requirement for HSP synthesis during embryogenesis is different in the two species. Developmental induction of a hs-GUS gene was only observed in tobacco. It will be very important to know whether endogenous hs genes are developmentally regulated in Arabidopsis. If there is a discrepancy to the heterologous soybean hs promoter it would be of great interest to identify the relevant promoter structures which should be different for at least tobacco and Arabidopsis but perhaps similar in soybean and tobacco.

The lack of heat inducibility of the hs-reporter gene construction during germination has some implications for the use of chimeric hs genes, e.g. as selectable markers for the isolation of mutants affecting signal transfer in plants (Severin et al., 1993). It will be also interesting to find out whether this negative regulation applies also to the native HSPs of the organism and at what level it occurs, transcription, RNA turnover, translation, or protein stability.
The main questions to be addressed in the investigation of developmental regulation are:

- Are hs genes always committed to expression (e. g. activation by hs and via HSF)?
- What is the molecular mechanism of developmental regulation of the hs response and is the interaction between HSE and HSF required?
- Is developmental regulation, e.g. the constitutive expression of HSP in embryogenesis, a response to or an anticipation of developmental/environmental stress?
- Are the different classes of HSPs differentially regulated during development?
- What is the cellular target for developmentally expressed HSPs?

The quantitative aspects (enhancement of hs gene expression), addressed by the functional investigation of SAR sequences have no direct relation to development. It seems that the enhancement by SAR sequences is perhaps secondary to a stabilizing effect on hs gene expression (e.g. inhibition of trans-inactivation or paramutation phenomena) and this phenomenon may have an impact on the use of transgenic plants in research and biotechnology.

Acknowledgements

The research was supported by grants from the Deutsche Forschungsgemein-schaft (Scho 242/5) and the Stiftung Volkswagenwerk.

References

Baumann G, Raschke E, Bevan M and Schöffl F (1987) Functional analysis of sequences required for transcriptional activation of a soybean heat shock gene in transgenic tobacco plants. EMBO J 6: 1161-1166

Bouchard RA (1990) Characterization of expressed meiotic prophase repeat transcript clones of Lilium: meiosis-specific expression, relatedness, and affinities to small heat shock protein genes. Genome 33: 68-79

Breyne P, Van Montagu M, Depicker A and Gheysen G (1992) Characterization of a plant scaffold attachment region in a DNA fragment that normalizes transgene expression in tobacco. Plant Cell 4: 463-471

Cockerill PN and Garrard WT (1986) Chromosomal loop anchorage of the kappa immunoglobulin gene occurs next to the enhancer in a region containing topoisomerase II sites. Cell 44: 273-282

Czarnecka E, Key JL, and Gurley WB (1989) Regulatory domains of the Gmhsp 17.5-E heat shock promoter of soybean: a mutational analysis. Mol Cell Biol 9: 3457-3463

Dietrich PS, Bouchard RA, Casey ES and Sinibaldi RM (1991) Isolation and characterization of a small heat shock protein gene from maize. Plant Physiol 96: 1268-1276

Gasser SM and Laemmli UK (1986) Cohabitation of scaffold binding regions with upstream/enhancer elements of three developmentally regulated genes of D. melanogaster. Cell 46: 521-530

Gasser SM and Laemmli UK (1987) A glimpse at chromosomal order. Trends in Genet 3: 16-22

Györgyey J, Gartner A, Nemeth K, Magyar Z, Hirt H, Heberle-Bors E and Dutits D (1991) Alfalfa heat shock genes are differentially expressed during somatic embryogenesis. Pl Mol Biol 6: 999-1007

Hall JG, Allen GC, Loer DS, Thompson WF and Spiker S (1991) Nuclear scaffolds and scaffold-attachment regions in higher plants. Proc Natl Acad Sci USA 88: 9320-9324

Helm KW and Abernethy RH (1990) Heat shock proteins and their mRNAs in dry and early imbibing embryos of wheat. Plant Physiol 93: 1626-1633

Hernandez LD and Vierling E (1993) Expression of low molecular weight heat shock proteins under field conditions. Plant Phys 101: 1209-1216

Horsch RB, Fraley RT, Rogers SG, Sanders PR, Lloyd A and Hoffmann N (1984) Inheritance of functional foreign genes in plants. Science 223: 496-498

Horsch RB, Fry JE, Hoffmann NL, Eichholtz D, Rogers SG and Fraley RT (1985) A simple and general method for transferring genes into plants. Science 227: 1229-1231

Howarth C (1989) Heat shock proteins in Sorghum bicolor and Pennisetum americanum. 1. Genotypic and developmental variation during seed germination. Plant Cell Environ 12: 471-477

Jackson DA (1986) Organization beyond the gene. Trends Biol Sci 6: 249-252 Jacobs M, Dolferus R and Van Den Bossche D (1988) Isolation and biochemical analysis of ethyl methanesulfonate-induced alcohol dehydrogenase null mutants of Arabidopsis thaliana (L.) Heynh. Biochem Genet 26: 105-122

Joergenson R (1990) Altered gene expression in plants due to trans interactions between homologous genes. TIBTECH 8: 340-344

Kloske E et al. (1993) Developmental regulation and tissue specific differences in the expression of chimeric hs-GUS genes in transgenic tobacco and Arabidopsis plants. (submitted)

Krol AR van der, Mur LA, Beld M, Mol JNM and Stuitje AR (1990) Flavonoid genes in Petunia: Addition of a limited number of gene copies may lead to a suppression of gene expression. Plant Cell 2: 291-299

Laemmli UK, Käs E, Poljak L and Adachi Y (1992) Scaffold-associated regions: cis-acting determinants of chromatin structural loops and functional domains. Curr Opin Gent Dev 2: 275-285

Marrs KA, Casey ES, Capitant SA, Bouchard RA, Dietrich PS, Mettler IJ, and Sinibaldi R (1993) Characterization of two maize HSP90 heat shock protein genes and expression during heat shock, embryogenesis, and pollen development. Dev Genetics, 14: 27-41

Matzke MA and Matzke AJM (1990) Gene interactions and epigenetic variation in transgenic plants. Dev. Genet. 11: 214-223

Matzke MA, Priming M, Trnovsky J and Matzke AJM (1989) Reversible methylation and inactivation of marker genes in sequentially transformed tobacco plants. EMBO J 8: 643-649

Meyer P, Heidmann I and Niedenhof I (1993) Differences in DNA-methylation are associated with a paramutation phenomenon in transgenic petunia. Plant J 4: 89-100

Mielke C, Kohwi Y, Kohwi-Shigematsu T, Bode J (1990) Hierarchical binding of DNA fragments derived from scaffold-attachment regions: Correlation of properties in vitro and function in vivo. Biochem 29: 7475-7485

Napoli C, Lemieux C and Joergensen R (1990) Introduction of a chimeric chalcone synthase gene into Petunia results in reversible co-suppression of homologous genes in trans. Plant Cell 2: 279-289

Rieping M and Schöffl F (1992) Synergistic effect of upstream sequences, CCAAT box elements, and HSE sequences for enhanced expression of chimeric heat shock genes in transgenic tobacco. Mol Gen Genet 321: 226-232

Scharf K-D, Rose S, Zott W, Schöffl F and Nover L (1990) Three tomato genes code for heat stress transcription factors with a region of remarkable homology to the DNA-binding domain of the yeast HSF. EMBO J, 9: 4495-4501

Schöffl F, Rieping M, Baumann G, Bevan M and Angermüller S (1989) The function of plant heat shock promoter elements in the regulated expression of chimeric genes in transgenic tobacco. Mol Gen Genet 217: 246-253

Schöffl F, Diedring V, Kliem M, Rieping M, Schröder G and Severin K (1992) The heat shock response in transgenic plants: the use of chimeric heat shock genes. In Inducible plant proteins. Their biochemistry and molecular biology, (Wray JL, ed). Cambridge: Cambridge University Press, pp 267-288

Schöffl F, Schröder G, Kliem M and Rieping M (1993) A SAR sequence containing 395 bp fragment mediates enhanced, gene dosage-correlated expression of a chimeric heat shock gene in transgenic tobacco plants. Transgenic Research 2: 20-27

Severin K, Wagner A and Schöffl F (1993) Heat-inducible Adh transgenes in Arabidopsis: reporter of developmental control of the heat shock response and selectable marker for the isolation of mutants. (submitted)

Slatter RE, Dupree P and Gray JC (1991) A scaffold-associated DNA region is located downstream of the pea plastocyanin gene. Plant Cell 3: 1239-1250

Treuter E, Nover L, Ohme K and Scharf K-D (1993) Promoter specificity and deletion analysis of three tomato heat stress transcription factors. Mol Gen Genet (in press)

van Herpen MMA, Reijnen WH, Schrauwen JAM, de Groot PF, Jager JWH and Wullems GJ (1989) Heat shock proteins and survival of germinating pollen of Lilium longiflorum and Nicotiana tabacum. J Plant Physiol 134: 345-351

Vierling E. and Sun A (1987) Developmental expression of heat
    shock proteins in higher plants. In: Environmental Stress in
    Plants (Cherry J, ed). Berlin: Springer, pp 343-354

Xiao CM and Mascarenhas JP (1985) High-temperature induced
    thermotolerance in pollen tubes of Tradescantia and
    heat-shock proteins. Plant Physiol 78: 887-890

Zimmerman JL, Apuya N, Darwish K and O'Carrol C (1989) Novel
    regulation of heat shock genes in carrot somatic embryo
    development. Plant Cell 1:1137-1146

NORMAL CELLULAR PROTEIN SYNTHESIS AND HEAT SHOCK

Mark R. Brodl, Jacqueline D. Campbell[1] , Kent K. Grindstaff[2] , and Lora Fielding
Department of Biology
Knox College
Galesburg, IL 61401
USA

Plant cells respond, as do the cells of all living organisms, to transient increases in temperature by inducing or enhancing the synthesis of a characteristic set of heat shock proteins (hsps). At the same time, there is in most organisms at least a partial suppression of normal cellular protein synthesis. This is done, presumably, to focus the cell's activities on addressing the physiological and biochemical changes which ensue with such stress; however, the precise function that these hsps perform in heat-shocked cells is not yet fully understood (for a recent review see Vierling 1991). In heat shock each cell responds to heat shock as a single cell; there is no system-wide, coordinated response.

The degree to which the synthesis of normal cellular proteins is effected by high temperature varies with regard to the organism under consideration. Heat shock causes a complete shut-down of normal cellular protein synthesis in *Drosophila* (Lindquist and DiDomenico 1985; Scott and Pardue 1981). The mechanism by which this is accomplished is by means of translational discrimination that favors the expression of hsps (Lindquist and DiDomenico 1985). The mRNAs encoding normal cellular proteins are present in these heat-shocked cells; however, they are not translated (Scott and Pardue 1981). In yeast, the discontinuation of normal cellular protein synthesis involves the normal, rapid turn over of normal cellular protein mRNAs in the absence of new transcription (Lindquist 1981). Therefore, the yeast mechanism appears to be transcriptional. On the other hand, heat shock does not appear to affect the synthesis of normal cellular proteins in root tips of maize seedlings (Cooper and Ho 1983).

The response of barley aleurone layers to heat shock differs from that of *Drosophila*, yeast, and maize in that the synthesis of only a subset of the normal cellular proteins is suppressed during heat shock (Brodl 1989). The expression of secretory proteins as a general class is affected by exposure to elevated temperature (Brodl and Ho 1991). The mRNAs encoding α-amylase (the predominant secretory protein synthesized by barley aleurone layers incubated in the presence of the phytohormone gibberellic acid, GA_3) are destabilized within 3 h of exposure to 40°C , whereas in the absence of heat shock these mRNAs have half-lives of at

---

[1] Current address: Department of Pathology, Washington University School of Medicine, St. Louis, MO 63110.

[2] Current address: Washington University, Division of Biology and Biomedical Sciences, Molecular Biology Program, St. Louis, MO 63110, USA.

NATO ASI Series, Vol. H 86
Biochemical and Cellular Mechanisms
of Stress Tolerance in Plants
Edited by J. H. Cherry
© Springer-Verlag Berlin Heidelberg 1994

least 100 h (Belanger et al. 1986). Concomitant with the destabilization of secretory protein mRNAs is the loss of lamellar structures of the endoplasmic reticulum (ER) upon which these proteins are synthesized (Belanger et al. 1986). Belanger et al. (1986) suggested that the heat shock-catalyzed loss of ER lamellar structures is responsible for the selectivity of the suppression of normal cellular protein synthesis. The observation that actin and tubulin mRNAs, which are nonsecretory protein mRNAs translated on free ribosomes, remain stable during heat shock supports this hypothesis (Brodl and Ho 1991).

In an effort to better understand the changes in ER membranes during heat shock and their relation to changes in the synthesis of normal cellular proteins, we have examined the effect of heat shock on membrane phospholipids in barley aleurone layers. The lipid component of membranes is thought to be an important control point for the homeostatic regulation of membrane fluidity and function under conditions of changing temperature, a property called "homeoviscous adaptation" (Sinensky 1974). In our studies, two types of heat shock were employed, one short-term (3 h) and the other long term (18 h). We have observed that heat shock causes a marked increase in the incorporation of [^{14}C]glycerol into phosphatidylcholine (PC). In samples that have been heat shocked for both short- and long-term exposures, the fatty acids extracted from ER membranes isolated on continuous sucrose gradients contain a higher degree of saturation as compared to samples incubated continuously at control temperatures. Furthermore, aleurone layers exposed to 18 h of heat shock regain the ability to synthesize and secrete the secretory proteins whose expression was initially suppressed by heat shock.

MATERIALS AND METHODS

Plant Material

Aleurone layers from barley grains (*Hordeum vulgare* L. cv. Himalaya, 1985 harvest, Washington State University, Pullman, WA) were prepared under sterile conditions as described by Belanger et al. (1986). The embryos were removed from the barley grains with a razor blade and were imbibed with incubation buffer (20 mM sodium succinate, pH 5.0, 20 mM CaCl$_2$) for 4 d prior to dissection of the endosperm from the aleurone layers. For aeration, isolated aleurone layers were incubated in a reciprocating shaking water bath (120 oscillations/min) at normal (25°C) or heat-shock (40°C) temperature.

In Vivo Labeling and Analysis of Proteins

Samples of secreted proteins were prepared by incubating 20 aleurone layers in 4 mL of incubation buffer in either the presence or absence of 1 μM GA$_3$ for 10 h at 25°C. Proteins were labeled by incubating the layers in 2 mL of fresh buffer, with or without GA_3, containing 200 μCi [^{35}S]methionine (specific activity >650 Ci/mmol, New England Nuclear, Boston, MA) for 6 h. At the end of the labeling period, the label was removed with two changes of buffer containing 1 mM nonradioactive methionine; then 2 mL of fresh, nonradioactive buffer was added. The medium was collected and replaced at 30-min intervals. During this time the temperature was either maintained at 25°C or shifted to 40°C. Proteins from collected media were precipitated on ice in 7 volumes of acetone for 30 min. The samples were centrifuged at 12,000-g for 20 min, and the pellets were resuspended in 300 μL of Laemmli (1970) gel-loading buffer.

Samples of intracellular proteins were prepared by incubating 10 aleurone layers in 2 mL of buffer, with or without 1 μM GA_3, for 16 h at 25°C. The aleurone layers were then rinsed with fresh buffer and shifted to 40°C or maintained at 25°C for a period of 3 h, the last 1 h of which was in the presence of 50 μCi of [^{35}S]methionine. After the labeling period, the aleurone layers were rinsed with buffer and 1 mM methionine, then homogenized in 5 μL of 100 μM leupeptin (a thioprotease inhibitor) (Sigma Chemical Co., St. Louis, MO, USA) in a mortar and pestle with acid-washed silica sand. The homogenates were mixed with 300 μL of SDS gel-loading buffer.

Proteins were analyzed by one-dimensional 11.5% SDS-PAGE as described by Laemmli (1970). Secreted proteins were quantified by the method of Bradford (1976) and radioactivity was assessed using a Beckman LS 6000IC liquid scintillation counter (Beckman Instruments, Palo Alto, CA).

RNA Isolations and Northern Blot Analysis

One-hundred aleurone layers per sample were incubated in sterile 250-mL flasks in 20 mL incubation buffer. RNA was isolated using the guanidine-HCl method of Chirgwin et al. (1979) as modified by Belanger et al. (1986). Fifteen μg of total RNA were prepared and electrophoresed into formaldehyde-containing 1.5% agarose gels (Maniatis et al. 1982). The electrophoresed RNA was capillary blotted onto "Gene Screen Plus" hybridization membranes (New England Nuclear) and prehybridized, hybridized and washed according to the procedures of Church and Gilbert (1984). Cloned cDNA for the high pI (pM/C) (Muthukrishnan et al., 1983) and low pI (clone E) (Rogers and Milliman, 1983), isozymes of α-amylase were nick-translated using a commercial nick-translation kit (Bethesda Research Laboratories, Bethesda, MD) in the presence of α-[^{32}P]dCTP (specific activity >500 Ci/mmol, New England Nuclear). The intensity of bands was digitized using NIH Image in a Macintosh Quadra 900 (Apple Computer, Cupertino, CA). The image was captured off of x-ray film using a high-resolution video camera (Javelin Electronics, model JE3462RGB, Tokyo, Japan).

Electron Microscopy

Tissue samples approximately 0.5 mm square were fixed in 2% formaldehyde, 3% glutaraldehyde buffered with 0.05 M Millonig's buffer, pH 7.0 (Hayat, 1984), for 4 h at 4°C. The tissue was rinsed twice in chilled, plain Millonig's buffer (15 min each) and then postfixed in Millonig's buffered 2% OsO_4 solution for 4 h at 4°C. After postfixation the samples were rinsed again, twice in plain buffer and twice in water (15 min each, at room temperature). The rinsed samples were dehydrated in a graded series of acetone baths (50%, 70%, 95%, 100%, 100%, 60 min each, at room temperature), infiltrated with "modified Ladd's" resin (Ringo et al., 1979), and polymerized under vacuum at 45°C overnight. Sections approximately 60 nm thick were cut on a diamond knife, stained with 1.5% aqueous uranyl acetate and Reynold's lead citrate (Hayat, 1984), and viewed in a JEOL JEM 100SK (JEOL Ltd., Tokyo, Japan) transmission electron microscope at original magnifications of 3,000 to 15,000 X.

Lipid Extraction, Phospholipid Isolation and Fatty acid Analysis

Twenty aleurone layers per sample were labeled with 10 μCi [^{14}C]glycerol (specific activity >10 mCi/mmol, New England Nuclear) during the final 3 h of their respective temperature regimens. The samples were homogenized in a mortar and pestle with acid-washed silica sand. Lipids were extracted from the samples using chloroform-methanol (Folch et al., 1957). The phospholipids in each sample were separated by one-dimensional thin layer chromatography (TLC) (20X20 K6 silica gel; Whatman, Maidstone, UK) according to the procedures of Gilfillan et al. (1983). The positions of individual phospholipids were determined by co-migration with standards (Sigma). The phospholipids were visualized by iodine vapor, scraped from the plates and radioactivity was assessed by liquid scintillation counting. Total phospholipid was quantified by the procedure of Bartlett (1959).

Thin-layer-chromatography-purified phosphatidylcholine, eluted from the silica with chloroform-methanol (2:1, v/v), was hydrolyzed and methylated according to the procedures of Geer et al. (1986). The resulting fatty acid methyl esters were analyzed by gas chromatography (Model 5890 Series II, Hewlett-Packard. Palo Alto, CA) using a column of 10% SP-2330 on 100/120 Supelcoport (Supelco, Bellefonte, PA). Peaks were identified by co-migration with fatty acid standards (Sigma). Samples of phosphatidylcholine with known fatty acid compositions (Sigma) which were TLC-purified, hydrolyzed, methylated and analyzed by gas chromatography (as above) indicated that sample oxidation during the extraction and analysis was negligible. In addition, butylated hydroxytoluene was added and samples were stored under nitrogen gas to minimize oxidation.

Cell Homogenization and Fractionation

One-hundred aleurone layers per sample were homogenized with razor blades using a modified electric carving knife retrofitted with single-edged razor-blade holders as described by Jones (1985). Homogenizations were done in 7 mL of a buffer containing 25 mM Hepes-BTP (pH, 7.4), 3 mM EDTA, 1 mM DTT, and 0.5% bovine serum albumin. The homogenate was filtered through Miracloth (Calbiochem, La Jolla, CA) and cell debris was pelleted at 1,000-g at 4°C for 10 min in a Sorvall SS-34 rotor in a Sorvall RC-5C refrigerated, high-speed centrifuge (Du Pont Instruments, Newton, CT). The 1000-g supernatant was centrifuged through a discontinuous sucrose (20 and 50%, w/w, sucrose, buffered with Hepes-BTP [pH 7.4], 1 mM DTT) gradient for 2 h at 70,000-g at 4°C in a Beckman Optima L-80 ultracentrifuge using SW27.1 buckets on an SW27 rotor (Beckman). The turbid band at the interface was collected and centrifuged through a continuous sucrose (18 to 45%, w/w, sucrose, buffered with Hepes-BTP [pH 7.4], 1 mM DTT) gradient at 70,000-g for 14 h using the same ultracentrifuge, rotor and buckets. One-mL fractions were collected using a Buchler Auto Densi-flow gradient fraction collector (Buchler Instruments, Fort Lee, NJ) and a Gilson Microfractionator (Gilson, Middletown, WI).

The position of the microsomal band was identified by cytochrome c reductase (ER-marker enzyme) activity according to the procedure of Jones (1980). Vanadate-sensitive ATPase (pH 6.5) activity was assayed according to the procedure of Sze (1984). IDPase activity was assayed according to the procedure of Nagahashi and Kane (1982). For both assays the release of inorganic phosphate was measured by the Fiske-SubbaRow assay (Fiske and SubbaRow, 1925). Fumarase and catalase were assayed spectrophotometrically (Lord et al., 1973) at room temperature. Assays used 2 to 14 μg of protein as estimated by Bradford assay (Bradford, 1975) and A_{280} measurements. Sucrose concentrations were measured by refractometry.

RESULTS

Heat Shock Increases the Incorporation of [¹⁴C]Glycerol into Phosphatidylcholine

The first step in our investigations was to assess whether the phospholipid component of barley aleurone layers was affected by heat shock. To this end, we labeled aleurone layers with [¹⁴C]glycerol for 3 h prior to homogenization and separated the phospholipids obtained from total lipid extracts by thin layer chromatography. For aleurone layers incubated in either the absence or presence of GA₃, the amount of [¹⁴C]glycerol incorporated into phosphatidylcholine (PC), phosphatidylethanolamine (PE), phosphatidylinositol (PI), phospha-tidylserine (PS), and phosphatidic acid (PA) were approximately equivalent in the two samples (Table 1). If during the last 3 h of incubation the aleurone layers were shifted to 40°C, the amount of [^{14}C]glycerol incorporated into PC increased from approximately 50% in control samples to 70% of total incorporation. This was true for aleurone layers incubated in either the absence or presence of GA$_3$. The proportion of [^{14}C]glycerol incorporated into PE, PI, PS and PA decreased compensatorily, but the relative proportions between these phospholipids remained similar, regardless of whether GA$_3$ was present in the incubation medium (PE from approximately 12% to 10 %, PI from approximately 20% to 11%, PS from approximately 8% to 2%, and PA from approximately 9% to 4%). If aleurone layers incubated in the presence of GA$_3$ were brought to 40°C and held at that temperature for 18 h, the increased incorporation of [^{14}C]glycerol into PC remained at approximately 70% of total incorporation, and the [^{14}C]glycerol incorporated into the other phospholipids remained little changed from those observed in samples heat shocked for 3 h. Aleurone layers incubated at 40°C during labeling incorporated between 1.75- and 2.5-fold more [^{14}C]glycerol into phospholipids than did their counterparts incubated at 25°C.

Table I. *Incorporation of [^{14}C]glycerol into phospholipids of barley aleurone layers incubated in the absence or presence of GA$_3$ and/or heat shock*
Aleurone layers were incubated in either the presence (+GA) or absence (−GA) of GA$_3$ for 19 h. The final 3 h of each sample's incubation were either at 25°C or 40°C (+HS) in the presence of [^{14}C]glycerol. Lipids were extracted from cell homogenates, separated by TLC, and quantified by liquid scintillation counting. Values represent the percentage of total [^{14}C]glycerol incorporation into the five listed phospholipid classes. Results are given as the mean ± SE of six independent experiments, three replicates each.

Phospholipid	[^{14}C]glycerol incorporation into phospholipid (% of total)[a]				
	−GA	+GA	−GA, +3 h HS	+GA, +3 h HS	+GA, 18 h HS
PA	9.58±1.26	8.22±1.88	4.37±0.38*	4.38±0.42*	7.01±1.08
PC	46.85±2.55	53.64±2.56	72.27±2.49**	71.43±1.62**	67.87±2.53**
PE	14.35±1.79	11.54±1.14	10.60±1.44	10.55±0.58	8.71±0.98
PI	20.53±1.70	19.99±0.73	10.63±0.38**	12.25±0.04*	12.41±0.07*
PS	8.70±0.93	7.22±0.92	2.10±0.46*	2.48±0.18**	4.00±0.87*
total cpm incorporated	1,354±263	1,199±220	3,515±187	2,633±161	2,118±103

[a]Means ± SE of heat-shocked samples that differ from their corresponding control group are indicated by *P<0.005, **P<0.01 (t-test).

In order to begin to understand the significance of such a change in phospholipid biosynthesis, we reconstituted PC out of the thin layer silica gel, hydrolyzed its fatty acids and analyzed the fatty acid composition by gas chromatography. For aleurone layers incubated in the absence of GA$_3$, the ratio of saturated to unsaturated fatty acids was 0.78 (Table II). For aleurone layers incubated in the presence of GA_3, this same ratio was 0.52. The principal difference between these two samples was the amount of mono- and poly-unsaturated 18-carbon fatty acids. Samples incubated in the presence of GA_3 had a higher mole percentage of these fatty acids and a correspondingly lower mole percentage of unsaturated 18-carbon fatty acid than that observed in samples incubated in the absence of the hormone. In addition, samples incubated in the absence of the hormone had a higher levels of 16:0 and 22:0 unsaturated fatty acids. In heat-shocked samples incubated in the presence of GA_3, the ratio of saturated to unsaturated fatty acids increased to 0.80. This increase resulted from increased amounts of 18:0 and 16:0 fatty acids and a corresponding decrease in 18-carbon mono- and poly-unsaturated fatty acids. In addition, there were measurable amounts of 14:0 and 22:0 fatty acids detectable. Heat shock had only a modest effect on the ratio of saturated to unsaturated fatty acids in aleurone layers incubated in the absence of GA_3.

Table II. *Fatty acid composition of TLC-isolated phosphatidylcholine from GA_3-treated, heat-shocked barley aleurone layers*
        Aleurone layers were incubated in either the presence (+GA) or absence (–GA) of GA_3 for 19 h. The final 3 h of each sample's incubation was either at 25°C or 40°C (+HS). Total lipids were extracted from each sample and separated by TLC. Spots co-migrating with PC were scraped from the TLC plates and reconstituted with chloroform-methanol. Fatty acids were hydrolyzed from the TLC-purified PC, methylated, and analyzed by gas chromatography. Results are given as mol % and are averages from three independent experiments, two replicates each. nd, not detected; t, trace. Errors of analysis were estimated to be within 5% of individual values.

	Treatment			
Fatty acid	-GA	+GA	-GA, +3 h HS	+GA, +3 h HS
14:0	4.6	nd	2.5	3.3
14:1	nd	t	nd	1.2
16:0	17.1	16.4	19.7	19.0
16:1	t	1.6	nd	nd
18:0	17.7	15.5	18.3	19.2
18:1	21.6	24.9	20.6	21.5
18:2	13.6	14.2	13.1	12.9
18:3	17.8	20.8	16.9	17.3
22:0	1.8	t	1.1	0.7
Sat./Unsat.	0.78	0.52	0.82	0.80

According to the tenets of the principles of homeoviscous adaptation, such a difference in fatty acid saturation might well result in a different level of tolerance for high temperature in the samples incubated in the absence of GA_3. To determine if this was the case, aleurone layers incubated in the presence or the absence of GA₃ were pulsed with [³⁵S]methionine for 6 h. Subsequently the samples were chased with cold methionine and either shifted to 40°C or retained at 25°C for 3 h. During this time, the incubation medium was collected from each sample at 30 min intervals and acetone precipitated. The amount of radioactivity released into the incubation medium during each interval was assessed by liquid scintillation counting. Figure 1 is a comparison of the amount of protein released from heat-shocked aleurone layers as a fraction nonheat-shocked controls for samples incubated in the presence and absence of GA₃. Heat shock reduced by approximately two-thirds the amount of radioactivity released into the incubation medium in samples of aleurone layers incubated in the presence of GA₃. In contrast,

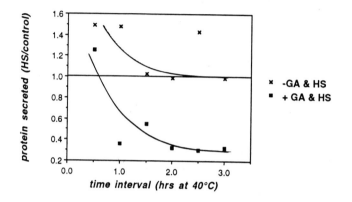

Figure 1. Secretion of TCA-precipitable radiolabeled proteins from heat-shocked aleurone layers. Aleurone layers were incubated in the presence (+GA) or absence (–GA) of 1 μM GA₃ for 16 h, the last 6 h of which was in the presence of [³⁵S]methionine. The layers were then chased with nonradioactive methionine and incubated for 3 h at either 25°C (–HS) or 40°C (+HS). During the incubation following the chase, the incubation medium was collected and replaced with fresh medium every 30 min. The samples of collected medium were acetone precipitated and radioactivity of each sample was analyzed by liquid scintillation counting. The fraction of TCA-precipitable radioactivity released from heat-shocked cells as compared to nonheat-shocked cells is given for each 30-min time interval for cells treated (♦) and not treated (x) with 1 μM GA₃.

heat shock had little effect on the release of proteins into the incubation medium in samples of aleurone layers incubated in the absence of the hormone. Compared on the basis of total protein secreted per aleurone layer during 16 h incubation, aleurone layers incubated in the presence of GA₃ secreted approximately 1.47-fold more protein into the medium than did layers incubated without GA₃ (0.212 μg protein/μL medium versus 0.144 μg protein/μL medium, respectively). During heat shock, the radioactivity released by aleurone layers incubated in the presence of GA₃ was reduced 4.5-fold. These data indicate that aleurone layers not incubated with GA₃ are actively secreting proteins and that the heat shock-induced reduction in secretion observed in layers incubated with GA_3 is not simply a return to some basal level of secretion. In both samples there was an burst in secretory activity during the first heat shock interval. This is most attributable to increased synthetic or secretory activity upon initial exposure to 40°C prior to equilibration to the elevated temperature.

Heat Shock Increases the Level of Fatty Acid Saturation in ER Membrane Fractions

The changes observed in PC biosynthesis and the changes in the saturation of PC-associated fatty acids are data obtained from whole cell extracts. In order to determine whether the lipid component of the ER membranes of barley aleurone cells were affected by heat shock in a similar manner, ER fractions were prepared on continuous sucrose gradients and the lipids were extracted from the fractions. Fatty acids were then hydrolyzed from the phospholipids in these extracts and analyzed by gas chromatography. The results of these analyses are presented in Table III. As had been observed with the fatty acids present is TLC-purified PC, samples from ER fractions prepared from aleurone layers incubated in the absence of GA_3 had a higher

Table III. *Fatty acid composition of microsomal membranes isolated by sucrose density centrifugation from heat-shocked barley aleurone layers*
Aleurone layers were incubated in either the presence (+GA) or absence (–GA) of GA_3 for 19 h. The final 3 h of each sample's incubation was either at 25°C or 40°C (+HS). Homogenized samples were layered onto sucrose step gradients. The turbid fraction at the 20%/50% interface was then layered on a continuous 20% to 50% sucrose gradient. Total lipid extractions were performed on fractions containing peak levels of cytochrome c reductase activity. Fatty acids in the samples were hydrolyzed from phospholipids (and any neutral fats), methylated and analyzed by gas chromatography. Results are given as mol % and are averages from three independent experiments, two replicates each. nd, not detected; t, trace. Errors of analysis were estimated to be within 5% of individual values.

			Treatment		
Fatty acid	–GA	+GA	–GA, +HS	+GA, +3 h HS	+GA, +18 h HS
12:0	1.87	nd	t	1.41	t
14:0	1.05	t	2.34	1.46	2.64
14:1	t	nd	nd	nd	t
16:0	11.42	12.51	20.83	28.74	26.49
16:1	3.03	t	1.17	t	nd
18:0	21.00	20.90	25.89	16.84	20.91
18:1	20.10	14.98	17.58	16.96	15.57
18:2	12.39	23.74	13.32	16.39	19.54
18:3	17.19	23.34	12.03	11.17	10.31
22:0	11.95	4.57	5.85	2.03	4.54
Sat./Unsat.	0.90	0.61	1.24	1.13	1.20

degree of fatty acid saturation than did their GA₃-treated counterparts (saturated to unsaturated fatty acid ratios of 0.90 and 0.61, respectively). In samples obtained from heat-shocked layers, there was an increase in the degree of fatty acid saturation in both untreated and GA₃-treated layers (saturated to unsaturated fatty acid ratios of 1.24 and 1.13, respectively). If the period of heat shock exposure was extended to 18 h total, this increased level of fatty acid saturation was maintained (1.20 as a ratio of saturated to unsaturated fatty acids). Assays for organellar marker enzymes (cytochrome c reductase, IDPase, catalase, fumarase, and vanadate-sensitive ATPase) revealed that the fractions obtained for each of these ER membrane preparations were highly enriched for ER and relatively free of contaminating organelles.

Exposure to Extended Heat Shock Results in High Temperature Acclimation

The observation that exposure to extended periods of heat shock results in an increased level of fatty acid saturation suggests that such a change might reflect an adaptation or acclimation to heat shock temperature. During initial exposure to heat shock (within 3 h), the synthesis and secretion of α-amylase and other secreted proteins is suppressed, and the mRNA encoding these proteins is selectively destabilized (Brodl and Ho, 1991). In addition, heat shock causes the loss of ER lamellar structures which serve as the site of secretory protein synthesis (Belanger et al., 1986). We examined secretory protein synthesis, α-amylase mRNA levels and ER ultrastructure in aleurone layers exposed to 18 h of 40°C in order to determine whether such high temperature acclimation could be observed.

Aleurone layers incubated in the presence of GA₃ vigorously synthesized α-amylase (Fig. 2, compare lanes 1 and 2), yet when exposed to heat shock for 3 h, the synthesis of this protein was suppressed and the synthesis of hsps was induced (Fig. 2, lane 3). If this heat shock period was extended for 24 h, it can be observed that there was an apparent recovery of α-amylase synthesis and the synthesis of hsps became much less vigorous (Fig. 2, lanes 4 through 10). In addition, the synthesis of several other proteins, whose synthesis was initially suppressed after 3 h of heat shock, resumed after exposure to extended periods of heat shock.

Because in the preceding experiment equal amounts of acid-precipitable radioactivity were electrophoresed for each sample, it is possible that the observed recovery of α-amylase synthesis is an artifact from the basis for "loading" the gel. For this reason the levels of α-amylase mRNA in samples exposed to extended periods of heat shock were monitored. Following a heat shock of 3 h, the mRNA levels for both the high (detected by pM/C) and low (detected by clone E) pI isoforms of α-amylase were reduced to less than 2% of control levels (Fig. 3, compare lanes 2 and 3). If the exposure to heat shock was extended for periods up to 24 h in length, the mRNA levels for both the high and low pI isoforms of α-amylase recovered to 85% of control levels within 18 h (Fig. 3, lanes 4 through 10). If this exposure to extended periods of heat shock took place in the presence of the transcription inhibitor cordycepin, this in-

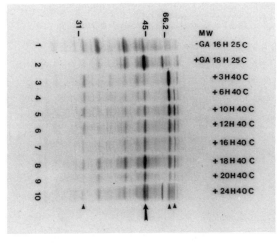

Figure 2. Protein synthesis in GA₃-stimulated aleurone cells exposed to extended periods of heat shock. Barley aleurone layers were incubated in the absence (lane 1) or presence (lane 2) of 1 μM GA₃ for 16 h at 25°C. GA₃-treated aleurone cells were also exposed to 40°C for 3 h to 24 h following an initial exposure at 25°C for 16 h (lanes 3-10). Each sample was labeled with 100 μCi of [³⁵S]methionine during the last 1 h of incubation. Equal amounts of TCA-precipitable radioactivity were electrophoresed into 11% SDS-containing polyacrylamide gels. The gel was dried and visualized fluorographically. MW and numbers indicate the positions and sizes of molecular weight standards in kD. Arrowheads indicate the positions of hsps. Arrow marks the position of α-amylase.

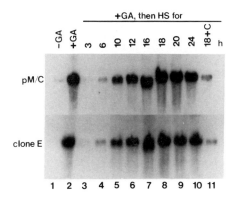

Figure 3. α-Amylase mRNA levels in GA₃-stimulated aleurone cells exposed to extended periods of heat shock. RNA was extracted from barley aleurone layers incubated as described in Fig. 2, except that for the sample in lane 11 the aleurone layers were exposed to both 40°C and 0.1 mM cordycepin for 18 h following initial incubation at 25°C for 16 h. An equal amount of total RNA from each sample was electrophoresed into formaldehyde-containing 1.5% agarose gels. The RNA was capillary blotted onto hybridization membranes and probed with nick-translated cDNA clones for the low (clone E) and high (pM/C) pI isozymes of α-amylase. Total RNA yields for each sample were equivalent, except after 24 h of heat shock where total RNA yield was approximately 85% of the other samples. The bands shown represent the only radioactivity detectable on the two fluorograms.

crease in α-amylase mRNA levels was not observed, indicating that active transcription is required for this recovery and the observed increase in α-amylase mRNA levels was not an artifact of electrophoresing equal amounts of total RNA in the northern analysis.

Heat shock causes the dissociation of ER lamellar structures in barley aleurone layers. We examined the ultrastructure of aleurone cells exposed to 18 h of heat shock by transmission electron microscopy in order to visualize whether the recovery in α-amylase synthesis and mRNA levels was accompanied by a reformation of ER lamellae. Aleurone layers incubated in the presence of GA₃ contained cells that possessed lamellar stacks of ER membranes (Fig. 4A). Ribosomes could be observed on the surface of the membranes, and the lumen of the ER was dilated, indicating active translation and translocation of secretory proteins. When these aleu-

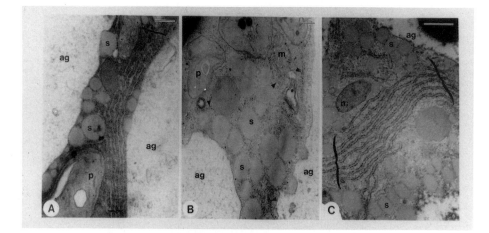

Figure 4A, B, C. Ultrastructure of aleurone cells exposed to extended periods of heat shock. **A** Electron micrograph of barley aleurone cells incubated in the presence of 1 μM GA₃ for 19 h at 25°C. **B** Electron micrograph of barley aleurone cells incubated in 1 μM GA₃ for 16 h at 25°C then shifted to 40°C for 3 h. **C** Electron micrograph of barley aleurone cells incubated in 1 μM GA₃ for 16 h at 25°C then shifted to 40°C for 18 h. ag, aleurone grain; arrowheads, ER fragments; brace ER lamellae; cw, cell wall; s, spherosome. Bar in each micrograph indicates 0.5 μm.

Table IV. *Ribosome density along ER membranes at normal and heat-shock temperatures*
The number of ribosomes along 1 μM of ER membrane were counted in a minimum of 7 regions per ER lamellar stack in a minimum of 4 cells for a minimum of three independent samples at original magnifications of 20,000 X. Results are given as the mean ± SE.

	Treatment		
	+GA	+GA, +3 h HS	+GA, +18 h HS
ribosome density	17.0±4.2	9.6±1.5	16.8±3.6

rone layers were exposed to a 3-h heat shock period, the lamellar stacks dissociated (Fig. 4B).Ribosomes were still apparent on the surface of the ER, but they were lower in density than in control samples (Table IV). In addition, the lumen of the ER was no longer so obviously dilated. If heat shock exposure continued for 18 h, lamellar stacks of ER membranes were again apparent (Fig. 4C). The ribosome density was comparable to that observed in control samples (Table IV), and the lumen of these membranes was noticeably more dilated that of the sample exposed to heat shock for 3 h.

DISCUSSION

The data presented in this paper provide evidence for a correlation between the degree of saturation in membrane phospholipids and the thermostability of ER membranes in barley aleurone cells. In aleurone samples where the degree of fatty acid saturation is greater, the heat-labile process of protein secretion is more resistant to high temperature effects. This is evidenced in aleurone layers incubated in the absence of GA_3. The degree of fatty acid saturation in these samples is greater than in aleurone layers incubated in the presence of GA_3, and these aleurone layers continue the release of proteins into the incubation medium during heat shock whereas their GA_3-treated counterparts release less than one-third the normal level of protein during heat shock. Further evidence for this correlation comes from data obtained in experiments examining the effects of extended heat shock. Aleurone layers exposed to 18 h of heat shock resumed the synthesis of α-amylase, and the levels of α-amylase mRNA approached levels comparable to samples incubated continuously at 25°C. Furthermore, the ER membranes present in samples heat shocked for 18 h were reorganized in lamellar stacks. In aleurone layers exposed to extended periods of heat shock, the degree of fatty acid saturation was comparable to that observed in aleurone layers incubated in the absence of GA_3.

Numerous studies have demonstrated a link between the degree of membrane phospholipid fatty acid saturation and membrane stability at a given temperature. Organisms exposed to chilling temperature contain membrane phospholipids with relatively unsaturated fatty acids (MacCarthy and Stumpf, 1980; Steponkus et al., 1988; Leathers and Scragg, 1989). This has been correlated with an increase in membrane fluidity at low temperature, which is believed to be important to the maintenance of proper membrane function (Quinn, 1984). Furthermore, increased levels of unsaturation in membrane fatty acids are correlated with the freezing tolerance (Yoshida, 1984). Steponkus, et al. (1988) demonstrated that freezing tolerance could be induced by fusing protoplasts with artificial liposomes made from mono- and diunsaturated species of PC or with liposomes derived from previously cold-acclimated plants. At the other extreme, plants exposed to high temperature have membrane phospholipids with reduced levels of unsaturated fatty acids (Pearcy, 1978; Raison et al., 1982). Mutants of *Arabidopsis thaliana* that are deficient in lipid desaturase activities have higher levels of saturation in the fatty acids of their membrane phospholipids, and these mutants are more thermotolerant (Hugly et al., 1989; Kunst et al., 1989).

The instability of biological membranes at extreme temperatures is believed to result from phase transitions in membrane phospholipids. The transition temperatures are lowered with increasing fatty acid unsaturation or are raised with increasing fatty acid saturation (for a review, see Quinn, 1984). We propose that the instability of the ER membranes in GA_3-treated, heat-shocked barley aleurone cells results from the comparatively high degree of fatty acid unsaturation in the membrane phospholipids of GA_3-treated cells. In contrast, the relatively greater levels of fatty acid saturation in the ER phospholipids of aleurone cells incubated without GA_3 enhance membrane thermostability. The difference in fatty acid saturation may cause a shift in phase transition temperature. We also propose that the heat shock-induced increase in the level of saturation in the fatty acids of membrane phospholipids is important to their eventual acclimation to heat shock temperature.

We have observed that heat shock enhances the incorporation of [^{14}C]glycerol into PC, and that heat shock increases the level of saturation in fatty acids isolated from TLC-purified PC. Work in other laboratories has documented that PC is synthesized by enzymes located primarily in the ER (Koehler and Varner, 1973; Moore, 1987), and PC is the major lipid constituent of barley aleurone layer's ER membranes (Johnson and Kende, 1971). These observations suggest that in the heat-shocked barley aleurone layer modification of the ER's PC component could be a route for regulating the fluidity of this membrane for homeoviscous adaptation. We did not, however, determine whether or not this change in phospholipid synthesis results in a change in total phospholipid composition. Furthermore although we know that heat shock increases the level of fatty acid saturation in total lipid extracts of ER membranes, we do not know whether the level of fatty acid saturation of specific membrane phospholipids, other than PC, is affected by heat shock.

The precise cause of membrane stability or instability during temperature stress may not reside completely in the degree of membrane fluidity and lipid phase transitions. Several workers have proposed that hsps may play an important role in conferring thermostability to biological membranes (Tanguay, 1985). Chen et al. (1988) reported that heat shock results in TEM-visible damage to ER membranes in soybean root cells and that the presence of hsps could thermoprotect the ER during subsequent heat shocks. Many hsps have been found to associate with membrane components. Maize hsp 72 and hsp 25 have been shown to localize to ER fractions in heat-shocked roots tips (Cooper and Ho, 1987). *In vitro* studies of hsp 22 in pea and *Chlamydomonas* demonstrate that this protein is located in the stroma of chloroplasts but is strongly associated with chloroplast membrane fractions after heat shock (Kloppstech et al.,

1985). Another potential determinant of membrane stability and instability concerns the cytoskeleton. The cytoskeleton plays an important role in the positioning and structuring of organelles and limiting membranes (Terasaki et al., 1986; Vogl et al., 1983). In animal cells, heat shock causes rapid and massive rearrangements of cytoskeletal components (Welch, 1990), which could contribute to membrane instability. Whether hsps of cytoskeletal changes affect the stability of ER lamellar structures in barley aleurone cells remains to be investigated.

ACKNOWLEDGEMENTS

The gift of cDNA clones (clone E and pM/C) from John Rogers (Washington University School of Medicine, St. Louis, Mo., USA) is greatly appreciated. We also wish to thank Bill Geer (Biology Department, Knox College), Tuan-hua David Ho (Biology Department, Washington University, St. Louis), and Russell Jones (Department of Plant Biology, University of California, Berkeley, USA) for their kind assistance and advice. This work was supported by National Science Foundation Grants DCB-8802026 and DCB-9105888 (to MRB) and Sigma Xi Grants-in-Aid of Research (to JDC and KKG).

LITERATURE CITED

Bartlett GR (1959) Phosphorous assay in column chromatography. J Biol Chem 234: 466-468

Belanger FC, Brodl MR, Ho T-hD (1986) Heat shock causes destabilization of specific mRNAs and destruction of endoplasmic reticulum in barley aleurone cells. Proc Natl Acad Sci USA 83: 1354-1358

Bradford MM (1976) A rapid and sensitive method for the quantification of protein utilizing the principle of protein dye-binding. Anal Biochem 86: 142-146

Brodl MR (1989) Expression of heat shock and normal cellular proteins in heat-shocked plant secretory cells. Physiol Plant 75:439-443

Brodl MR, Ho T-hD (1991) Heat shock causes selective destabilization of secretory protein mRNAs in barley aleurone cells. Plant Physiol 96:1048-1052

Chen Y-R, Chou M, Ren S-S, Chen Y-M, Lin C-Y (1988) Observations of soybean root meristematic cells in response to heat shock. Protoplasma 144: 1-9

Chirgwin JM, Pryzbyla AE, McDonald RJ, Rutter WJ (1979) Isolation of biologically active ribonucleic acid from sources enriched in ribonuclease. Biochem 18: 5294-5299

Church GM, Gilbert W (1984) Genomic sequencing. Proc Natl Acad Sci USA 81: 1991-1995

Cooper PS, Ho T-hD (1983) Heat shock proteins in maize. Plant Physiol 71: 215-222

Fiske CH, SubbaRow Y (1925) The colorimetric determination of phosphorous. J Biol Chem 66: 375-400

Folch J, Lees M, Sloane Stanley GH (1957) A simple method for the isolation and purification of total lipids from animal tissues. J Biol Chem 266: 497-505

Geer BW, McKechnie SW, Langevin ML (1986) The effect of dietary ethanol on the composition of lipids of *Drosophila melanogaster* larvae. Biochem Genet 24: 51-69

Gilfillan MA, Chu AJ, Smart DA, Rooney SA (1983) Single plate separation of lung phospholipids including disaturated phosphatidylcholine. J Lipid Res 24: 1651-1656

Hayat MA (1989) Principles and techniques of electron microscopy, 3rd ed. CRC Press, Boca Raton, FL

Hugly S, Kunst L, Browse J, Somerville C (1989) Enhanced thermotolerance and altered chloroplast ultrastructure in a mutant of *Arabidopsis* deficient in lipid unsaturation. Plant Physiol 90: 1134-1142

Johnson KD, Kende H (1971) Hormonal regulation of lecithin synthesis in barley aleurone cells: Regulation of the CDP-choline pathway by gibberellin. Proc Natl Acad Sci USA 68: 2674-2677

Jones R L (1980) The isolation of endoplasmic reticulum from barley aleurone layers. Planta 150: 58-69

Jones R L (1985) Endoplasmic reticulum. *In* HF Linskens, JF Jackson, eds, Modern Methods of Plant Analysis (N. S.). Springer-Verlag, New York, pp. 304-330

Kloppstech K, Meyer G, Schuster G, Ohad, I (1985) Synthesis, transport and localization of a nuclear coded 22-kd heat-shock protein in the chloroplast membranes of peas and *Chlamydomonas reinhardi*. EMBO J 4: 1901-1909

Koehler DE, Varner JE (1973) Hormonal control of orthophosphate incorporation into phospholipids of barley aleurone layers. Plant Physiol **52**: 208-214

Kunst L, Browse J, Somerville C (1989) Enhanced thermal tolerance in a mutant of *Arabidopsis* deficient in palmitic acid unsaturation. Plant Physiol **91**: 401-408

Laemmli UK (1970) Cleavage of structural proteins during the assembly of the head of bacteriophage T-4. Nature **227**: 680-685

Leathers RR, Scragg AH (1989) The effect of different temperatures on the growth, lipid content and fatty acid composition of *Theobroma cacao* cell suspension cultures. Plant Sci **62**: 217-227

Lindquist S (1981) Regulation of protein synthesis during heat shock. Nature **294**: 311-314

Lindquist S, DiDomenico BJ (1985) Coordinate and noncoordinate gene expression during heat shock: A model for regulation. In: Changes in eukaryotic gene expression in response to environmental stress, pp 71-90, Atkinson, B. G., Walden, D. B., eds. Academic Press, New York

Lord JM, Kagawa T, Moore TS, Beevers H (1973) Endoplasmic reticulum as the site of lecithin formation in castor bean endosperm. J Cell Biol **57**: 659-667

MacCarthy JJ, Stumpf PK (1980) The effect of different temperatures on fatty acid synthesis and polyunsaturation in cell suspension cultures. Planta **147**: 389-395

Maniatis T, Fritsch EF, Sambrook J (1982) Molecular cloning. Cold Spring Harbor Laboratory, Cold Spring Harbor, NY.

Moore TS (1987) Regulation of phospholipid head group composition in castor bean endosperm. *In* P Stumpf, JB Mudd, WD Nes, eds, The Metabolism, Structure and Function of Plant Lipids. Plenum Press, New York, pp 265-272

Muthukrishnan S, Chandra GR, Albaugh GP (1983) Modulation by abscisic acid and 5,2-amino-ethyl-cystiene of α-amylase mRNA in barley aleurone cells. Plant Mol Biol **2**: 249-258

Nagahashi J, Kane AP (1984) Triton-stimulated nucleoside diphosphatase activity: subcellular localization in corn root homogenates. Protoplasma **112**: 167-173

Pearcy RW (1978) Effect of growth temperature on the fatty acid composition of the leaf lipids in *Atriplex lentiformis* (Torr.) Wats. Plant Physiol **61**: 484-486

Quinn PJ (1984) Polar lipid structures and the stability of cell membranes. *In* P-A Siegenthaler, W Eichenberger, eds, Structure, Function and Metabolism of Plant Lipids. Elsevier Science Publishers, Amsterdam, pp. 331-338

Raison JK, Roberts JKM, Berry JA (1982) Correlations between the thermal stability of chloroplast (thylakoid) membranes and the composition and fluidity of their polar lipids upon acclimation of the higher plant, *Nerium oleander*, to growth temperature. Biochim Biophys Acta **688**: 218-228

Ringo DL, Cota-Robles EH, Humphrey BJ (1979) Low viscosity embedding resins for transmission electron microscopy. Annu Proc Electron Microsc Soc Amer **37**: 348-349

Rogers JC, Milliman C (1983) Isolation and sequence analysis of a barley alpha-amylase cDNA clone. J Biol Chem **258**: 8169-8174

Scott MP, Pardue ML (1981) Translational control in lysates of *Drosophila melanogaster* cells. Proc Natl Acad Sci USA **78**: 3353-3357

Sinensky M (1974) Homeoviscous adaptation - A homeostatic process that regulates the viscosity of membrane lipids in *Escherichia coli*. Proc Natl Acad Sci USA **71**: 522-525

Steponkus PL, Uemura M, Balsamo RA, Arvinte T, Lynch DV (1988) Transformation of the cryobehavior of rye protoplasts by modification of the plasma membrane lipid composition. Proc Natl Acad Sci USA **85**: 9026-9030

Sze H (1984) H+-translocating ATPases of the plasmamembrane and tonoplast of plant cells. Physiol Plant **61**: 683-691

Terasaki M, Chen LB, Fujiwara K (1986) Microtubules and the endoplasmic reticulum are highly interdependent structures. J Cell Biol **103**: 1557-1568

Tanguay RM (1985) Intracellular localizations and possible functions of heat shock proteins. *In* B G Atkinson, DB Walden DB, eds, Changes in Eukaryotic Gene Expression in Response to Environmental Stress. Academic Press, New York, pp 91-113

Vierling E (1991) The roles of heat shock proteins in plants. Annu Rev Plant Physiol Molec Biol **42**: 579-620

Vogl AW, Linck RW, Dym M (1983) Colchicine-induced changes in the cytoskeleton of the golden-mantled ground squirrel (Spermophilous lateralis) Sertoli cells. Am J Anat **168**: 99-108

Welch WJ (1990) The mammalian stress response: Cell physiology and biochemistry of stress proteins. *In* RI Morimoto, A Tissières, C Georgopoulos, eds, Stress Proteins in Biology and Medicine. Cold Spring Harbor Laboratory, Cold Spring Harbor, NY, pp 223-278

Yoshida S (1984) Studies on freezing injury of plant cells. Plant Physiol **75**: 38-42

THERMAL ACCLIMATION AND HEAT STRESS RESPONSE OF *SYNECHOCYSTIS* PCC6803: THE POSSIBLE ROLE OF THYLAKOID PHYSICAL STATE, LIPID SATURATION AND MOLECULAR CHAPERONES

László Vigh, Zsolt Török, Eszter Kovács, Attila Glatz, Nándor Balogh* and Ibolya Horváth
Institute of Biochemistry and *Biophysics, Biological Research Center of the Hungarian
Academy of Sciences, 6701 Szeged, POB 521., Hungary.

There is unequivocal evidence that in higher plant cells exposed to heat stress the photosynthetic apparatus is irreversibly damaged prior to impairment of other cellular functions (Berry and Bjorkman 1980). The general assembly of the photosynthetic membranes in cyanobacteria is similar to that of higher plants, whereas the genetic organization is simpler and can be manipulated more easily. Correspondingly, our attention has been focused on the molecular mechanism of adaptation and resistance to high temperatures of a cyanobacterial strain, *Synechocystis* PCC6803 (Vigh et al. 1990, Lehel et al. 1992,1993a,1993b).

Recently Murata et al. examined the effect of the fatty-acid unsaturation on the thermal stability of PSII activity, using isothermally grown cells of the mutant and transformant of *Synechocystis* PCC6803 which are defective in desaturationa at the Δ^{12} position of C_{18} fatty acids (Gombos et al.1991). Results gained from these studies suggested, that thermal stability is not affected directly by changes in the unsaturation level of membrane lipids. Whilst the authors predicted that factors others than lipids are responsible for the adaptation of photosynthesis to high temperature in cyanobacteria, other lines of evidences indicate that lipid unsaturation, being usually highly reduced upon the increase of growth temperature in most various organisms, is also a component of thylakoid thermal tolerance. Using *Arabidopsis*, the critical temperature at which chlorophyll a-b binding protein complex dissociated from the photosynthetic reaction centers in the fadB and fadC mutants indicated a significant enhancement of the thermal stability if compared to wild-type chloroplast membranes (Hughly et al. 1989, Kunst et al. 1989). A similar enhancement of

NATO ASI Series, Vol. H 86
Biochemical and Cellular Mechanisms
of Stress Tolerance in Plants
Edited by J. H. Cherry
© Springer-Verlag Berlin Heidelberg 1994

thermal stability was observed following catalytic hydrogenation of pea thylakoids (Thomas et al. 1986, Vigh et al. 1989). This latter method permitted the *in situ* evaluation of the role of lipid unsaturation without allowing the organism to change other components of the membrane in order to compensate for the primary effects of mutation on the membrane physical state (Vigh et al. 1985, Quinn et al. 1989). In the present work we will demonstrate, that whilst the mutant/transformant *Synechocystis* cells can withstand large decreases in the amount of polyunsaturated fatty acids (without any significant alteration of the level of *saturated* components) their thylakoid membranes are simultaneously reorganized, most probably in order to counterbalance the effects of altered lipid unsaturation on membrane fluidity and protein-lipid interactions, as well. Our data underline the operation of a homeostatic control mechanism and could account for the apparently identical thermotolerance of wild-type and mutant cyanobacteria deficient in lipid unsaturation.

Here we propose that the high-temperature acclimation phenomenon of *Synechocystis* thylakoids could be characterized by the operation of a two-phase mechanism. The actual level of thylakoid heat stability can be enhanced by the raise of growth temperature. It is known, that increasing the growth temperature causes an upward shift of the optimal temperature of photosynthesis in numerous species and renders the photosynthetic apparatus more tolerant to heat stress (Süss and Yordanov 1986, Havaux 1993). In the present study, this long-term heat-hardening of photosynthesis has been correlated with changes in the lipid unsaturation and ratio of protein-to-lipid of thylakoid membranes. Growth-temperature dependent adaptive alterations were in full accordance with the fluidity changes, detected. Heat-induced deactivation of PSII appears to involve denaturation of certain functional membrane proteins. The critical temperature at which denaturation events occur is presumably linked to destabilization of specific lipid-protein interactions, which in turn are associated with increased fluidity of the thylakoid membranes at elevated temperatures (Berry and Bjorkman 1980, Süss and Yordanov 1986, Vigh et al. 1990).

In the second part of our study, we intended to present evidences, that besides the long-term phase of acclimation there exists a rapid phase of photosynthetic adaptation to heat stress, triggered by a short, sublethal heat treatment. Santarius and Muller have reported for spinach and most recently Havaux for potato leaves, that a brief warming also causes a measurable increase in the resistance of the photosynthetic membranes to subsequent heat stress, an effect which occurs without concomitant changes in the lipid and fatty acid composition (Santarius and Muller 1979, Havaux 1993). Our study with the photosynthetic blue-green alga, *Synechocystis* PCC6803 fully agree with these findings. Whereas we found no measurable changes concerning lipids during rapid adaptation phase, the mild heat-stress which is able to induce PSII thermotolerance resulted in an elevated level of the ratio of protein-to-lipid in thylakoids and a reduction of membrane fluidity. As we will highlight, synthesis and cellular redistribution of specific HSPs including GroE-type molecular chaperones could provide the tool of thermoprotection during the rapid phase of acclimation.

MATERIALS AND METHODS

Growth of cells. *Synechocystis* PCC6803 cells were grown photoautrophically at different temperatures as in (Lehel et al. 1992). Cultures in exponential growth phase were used for experiments. Cells deficient in fatty acid desaturation (Fad6/*desA*::Km^r) were kindly provided by Dr. N. Murata, Okazaki, Japan.

Determination of photosynthetic activity. Heat inactivation of photosynthesis was measured by means of oxygen exchange with a Clark-type oxygen electrode. Evolution of oxygen was monitored in intact cells (total photosynthesis). Alternatively, PS II activity was measured by the transport of electrons from H_2O to 1,4-benzoquinone. For further details, see Lehel et al. 1993/b.

Extraction and analysis of lipids and proteins from thylakoid. After isolation of thylakoid as in (Vigh et al.1985), lipids were extracted and total lipids were subjected to methanolysis. The standard method was applied for total membrane protein determination. Protein-to-lipid ratio was calculated as in (Gombos et al.1987).

Fluidity measurements. To detect changes in microviscosity of the isolated thylakoids steady-state fluorescence anisotropy was determined as described earlier (Gombos et al.1987) by using DPH as a membrane probe. The fluidity of the photosynthetic membranes and lipids was measured by using ESR spectroscopy of 14-PGSL as described in (Vigh et al. 1985).

Determination of the fraction of motionally restricted stearic acid spin-label. Thylakoid membranes were labelled by SASL and ESR spectra were recorded as in (Li et al.1989).

Fraction of protein-restricted spin-label pool was determined according to (Schlame et al.1990).

Electrophoresis and fluorography. For protein separation 8-15 or 8 % SDS-PAGE was used. Gels were stained with Coomassie Brilliant Blue.

Northern analysis. RNA was extracted and separated from DNA as described in (Lehel et al.1993/a). Probes for cpn60 and groESL were synthesized from *Synechocystis* genomic DNA using polymerase chain reaction and purified by agarose electrophoresis.

RESULTS AND DISCUSSION

Long-term acclimation

<u>Effect of growth temperature on thylakoid heat stability</u>

Upward shift in growth temperature afforded significant protection to otherwise lethal temperatures in wild-type *Synechocystis* cells. According to Fork et al. (1987) a critical temperature can be identified, below which heat-stress had little or no effect and above which there is a dramatic reduction in O_2 evolution efficiency (T_1). A second characteristic temperature can also be defined (T_2) above which cyanobacteria show a complete loss of O_2 evolution capability. According to this criteria, the corresponding T_1 values for cells grown at 22 and 36°C were 42 and 46°C, respectively, whereas T_2 values were 47 and 49 °C. In addition to the above studies, thermal sensitivity of PSII was also determined (Fig.1). The actual T_1 values determined for PSII were 40 and 45°C for the low and high temperature adapted cells, respectively. The critical temperatures obtained for T_2 were identical with those measured on entire electron transport. It is noted, that threshold temperatures at which efficiency of PS II became impaired overlapped with characteristic temperatures above which heat-treatment caused a severe decrease in survival of *Synechocystis* cells tested by counting colonies of heat-pretreated and plated cells 8 days following treatment (for details see Lehel et al.1993/b).

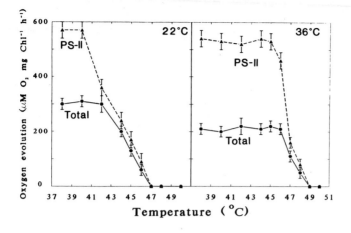

Figure 1. The profiles of heat inactivation of photosynthesis and PSII in intact cells of *Synechocystis* PCC6803 grown at 22 and 36°C, respectively. Photosynthetic oxygen evolution was measured with no exogeneously added electron acceptor. PSII activity measured as light induced evolution of oxygen with 1,4-benzoquinone added as electron acceptor. Cells were treated for 60 min at designated temperatures in darkness and oxygen evolution was measured at 22°C.

In accordance with findings of Gombos et al. (1991), we did not observe measurable difference between the heat-stability of PSII of 36°C-grown wild-type and mutant/transformant cells (data not shown). Similar results could be obtained if thylakoid thermal stability was tested by following chlorophyll fluorescence yield *in vivo* by heating cell suspension at a rate of 1°C/min from 25 to 54°C (Török et al., unpublished).

<u>Changes in fatty acid composition and the ratio of protein-to-lipid within thylakoids</u>

No significant differences between the thylakoid membranes of 22 and 36°C-adapted wild type and 36°C-grown mutant/transformant cells were observed concerning lipid classes (data not shown). The major fatty acids of thylakoid total lipids are shown in Table I. In accordance with previous investigations (Wada and Murata 1990), thylakoid of wild-type

Table I. Fatty acid composition of total thylakoid lipids derived from wild-type *Synechocystis* cells adapted to 22 and 36°C, respectively as well as 36 °C-grown Fad6/*desA*::Km[r] cells.

Fatty acids (mol %)	Wild-22	Wild-36	Fad6/*desA*-36
16:0	51	58	56
16:1	3	2	4
18:0	tr	2	3
18:1(9)	2	9	37
18:2(9,12)	6	15	tr
18:3(9,12,15)	8	tr	-
18:3(6,9,12)	22	14	-
18:4(6,9,12,15)	8	-	-

strains contained 16:0, $18:2(\Delta^{9,12})$ and $18:3(\Delta^{6,9,12})$ as the most abundant fatty acids. Increasing the growth temperature of wild-type cells resulted in a remarkable reduction in the level of fatty acyl unsaturation. Upon the elevation of growth temperature, the level of saturated fatty acids 16:0 and 18:0 increased by about 10%, parallel with complete elimination of $18:3(\Delta^{9,12,15})$ and $18:4(\Delta^{6,9,12,15})$. Simultaneously with a one-third drop in the level of $18:3(\Delta^{6,9,12})$ both $18:1(\Delta^{9})$ and $18:2(\Delta^{9,12})$ increased markedly. As a result of the combination of mutation and gene disruption, thylakoid lipids of 36°C-grown Fad6/*desA*::Km[r] strain contained only saturated and monounsaturated fatty acids, mostly 18:1. It was demonstrated in a separate study, that whereas Fad6 mutant was defective in desaturation at the Δ^{6} position of C_{18} fatty acids in MGDG and DGDG, *desA*::Km[r] transformants have lost their ability to introduce a double bond at the Δ^{12} position of C_{18} acids in all lipid classes (Wada et al.1992).

Acclimation of wild-type *Synechocystis* cells to high temperatures resulted in a 20 %

increase in ratio of protein-to-lipid of thylakoids (Fig.2.) In contrast, in membranes of Fad6/*desA*::Km^r cells the actual level of protein-to-lipid ratio was about 30% less than in wild-type grown at identical temperature. Besides fatty acid unsaturation, protein-to-lipid ratio is known to be an other determinant of the overall fluidity of biomembranes (Gombos et al. 1987, Vigh et al. 1990, Webb and Green 1991). Our presented result suggest that parallel with reduced lipid unsaturation, *Synechocystis* cells may adjust their thylakoid fluidity

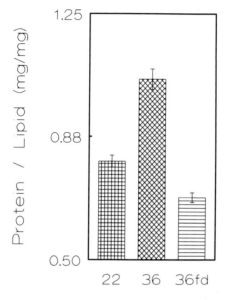

Figure 2. Ratio of protein-to-lipid in thylakoids isolated from 22 and 36°C-grown adapted wild-type and 36°C grown mutant/transformant cells.

in response to increased growth temperature by the elevation of the ratio of protein-to-lipid, consistently with the idea of homeoviscous membrane adaptation (Quinn et al.1989). Thus, by supposing that the gene-engineered additional depletion of cis double bonds would reduce fluidity to an abnormal level one can predict that cells are also able to compensate for the decreased fluidity by readjusting the protein-to-lipid ratio during the development of their thylakoid membranes.

Temperature dependent changes in membrane dynamics tested by measuring DPH fluorescence anisotropy and correlation time of 14-PGSL spin label.

Fluorescence anisotropy values were determined by DPH as fluorescent probe in the isolated thylakoids. Comparing results gained with intact thylakoids from low- and high-temperature preadapted wild-type cells led us to the conclusion that there is a significant

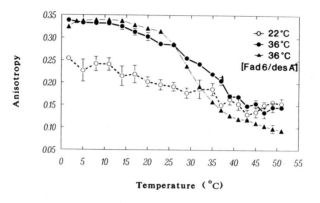

Figure 3. Fluidity measurements on various thylakoid samples by using DPH probe.

homeostatic adaptation of membrane fluidity, probably to maintain a particular value at the temperature of growth, similar to some other cyanobacteria (Gombos et al.1987). In addition, we failed to observe any significant effects of genetically-induced changes of fatty acid composition on overall level of membrane fluidity (Fig.3). Surprisingly, the genetic manipulation apparently resulted in some extra fluidization of whole-thylakoid above 30°C.

To further confirm, that the above described changes of lipid unsaturation and protein-to-lipid ratio truly exerts an unexpected influence on membrane properties, we made also ESR studies by using spin labelled phosphatidylglycerol (14-PGSL) as a probe. Microviscosity changes were detected both in the aqueous dispersions of total polar lipid extracts (Fig.4/B) derived from membranes and from the corresponding intact thylakoids (Fig.4/A), by measuring the mobility of the nitroxide spin label. It is well stated, that the level of overall microviscosity in whole thylakoids is always higher than that of their extracted and reconstituted lipids (Quinn et al.1989). Accordingly, we could measure in all thylakoid samples lower fluidities than in the corresponding lipid dispersions by screening temperature induced changes of the correlation times. Comparisons of fluidity of lipid dispersions or membranes of wild-type cells have concluded that samples derived from high-temperature grown cells are significantly more rigid than their low-temperature counterparts.

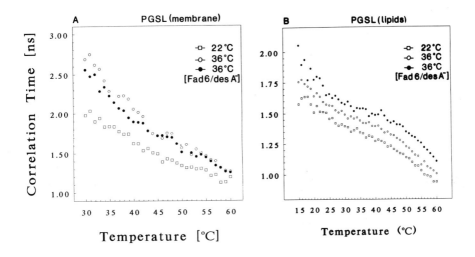

Figure 4. Temperature dependence of the correlation time for 14-PGSL in thylakoid membranes and their total lipid dispersions.

Unsaturated fatty acids have lower melting points than their corresponding saturated analogues and a decrease in the degree of unsaturation would be expected to increase microviscosity. According to this expectation, spin label mobility was substantially lower in lipid dispersion of mutant/transformant than in wild-type samples at all temperatures. By contrast, the investigation of the temperature dependence of the fluidity of the intact thylakoids derived from isothermally grown wild-type and gene-engineered samples revealed, that the genetic manipulation of desaturases resulted in either a small decrease in microviscosity (between 30-40°C) or rather had no any effect on the motional characteristics of membranes. This finding was consistent with the presented investigations conduced with fluorescence probes. Moreover, ESR study supported the notion, that even if deficiency in

Δ^6 and Δ^{12} desaturase activities brought about changes in overall physical state of thylakoid lipids, *Synechocystis* cells were able to compensate the primary alterations of fluidity by changing the ratio of protein-to-lipid within membranes.

Determination of the fraction of motionally restricted spin-labeled fatty acid component affected by growth temperature and changing level of lipid unsaturation

Besides predicting that the co-ordinated changes of lipid unsaturation and the ratio of protein-to-lipid of the membranes serve together to maintain a relatively constant microviscosity at the growth temperature, we presume that some alternative mechanism may also govern membrane homeostasis and necessitate the above described readjusment of protein-to-lipid ratio. There is gathering support for the idea that short range specific interactions associated with the formation of oligomeric complexes within the membrane involve the interaction of membrane proteins with particular membrane lipids. It was suggested, that the motionally restricted lipids surrounding the membrane proteins provide the interface between the fluid bilayer regions and functional protein units within membrane (Li et al.1989, Schlame et al.1990). Whereas a functional association of specific lipids with chlorophyll-protein complexes could exist, this population of lipids was supposed to stabilize particular protein complexes against changes in temperature or overall fluidity. On the other hand, a critical portion of free matrix-lipids would be an absolute requirement for lateral migration of membrane proteins, like those involved in stateI - stateII transitions (Li et al.1989).

As shown by Fig.5, using spin-labeled stearic acid as a probe, the fraction of motionally restricted lipids was found to be quantitatively in line with the actual values of protein-to-lipid ratios of different thylakoid membranes investigated. It was recently demonstrated, however, that the overall level of lipid unsaturation is also a critical factor which determines the size of lipid solvation shell of membrane proteins (Schlame et al.1990). By subjecting 22°C-grown *Synechocystis* thylakoids to catalytic hydrogenation, reduction of double bond content in a range compatible with that accompanied by high temperature adaptation resulted in a dramatic increase in size of protein-immobilized lipid pool (Fig.5.). Combining these facts, we suppose that in order to maintain the quantity of motionally restricted lipids on a functionally optimal level, Fad6/*desA*::Km^r *Synechocystis*

87

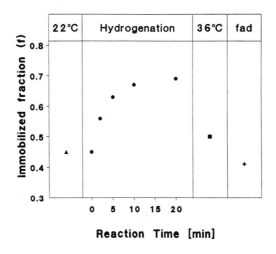

Figure 5. The fraction of protein immobilized lipids determined by using SASL spin probe in membranes of 22 and 36°C adapted wild-type and 36°C adapted mutant/transformant (fad) as well as 22°C adapted but hydrogenated thylakoids.

cells have an additional reason for decreasing the protein-to-lipid ratio in their thylakoid membranes.

It has been suggested that heat induced break in the protein boundary lipid layer could reduce thermal stability of the membrane in a direct manner by destroying native structure of photosynthetic complexes (Vigh et al.1990). It was supposed, that during thylakoid thermal adaptation an appearance of specific lipid molecular species with saturated fatty acyl chains lead to a decreased fluidity of membranes at supraoptimal temperatures bringing about a conformation-induced rise in thermal stability of PSII units (Yordanov et al.1986). On the other hand, the authors have failed to establish a correlation between gross fatty acid polyunsaturation and the thermal stability of the photosynthetic apparatus. Here we have shown, that acclimation of wild-type *Synechocystis* cells to 36°C was also accompanied by large decrease in overall level of lipid unsaturation parallel with increase in the thermal stability of photosynthesis. Since the magnitude of genetically-induced reduction of cis double bond content in Fad6/*desA*::Km^r cells was even greater, one can conclude that fatty acid unsaturation *per se* is not the determining factor which contributes to thylakoid heat resistance. But our data also imply that the critical role of **membrane physical state** still can not be excluded. By predicting that homeostatic mechanisms are operating both on the level of overall fluidity and the size of protein-immobilized lipid pool in mutant/transformant *Synechocystis*, their existence would serve to ensure an eventually unaltered heat stability of thylakoid. Evidence in favour of this hypothesis is, that conversion of monounsaturated fatty acids to their saturated counterparts by catalytic hydrogenation of isolated thylakoids, taking place in the **absence** of any compensatory readjustment, resulted in an elevation in thermal resistance of PS II mediated electron transport (Gombos et al., unpublished), likewise found with *Arabidopsis* mutant. Finally, our studies also imply, that a special care is required with use of desaturase mutants to analyze the precise physiological role of polyunsaturated fatty acids. Whereas the essential requirement of acyl chain unsaturation for fulfilling specific functions could clearly be documented by their application, additional factors which might be attributed to the operation of some regulatory mechanisms controlling membrane physical state and thus superimposed to the primary effect of manipulation should also be considered, simultaneously.

Rapid adaptation of photosynthesis to heat

Acquisition of PSII thermotolerance is paralleled by increased membrane microviscosity

As was discussed above, various observations suggest the existence of short-term heat-adaptation processes in photosynthetic organisms (Havaux 1993). The present study further confirms this suggestion: short-term, non-stressful heat-treatment (42°C, 3 hours) induced additional raise of PSII thermotolerance in wild-type, 30°C grown *Synechocystis* cells (Fig.6). The enhancement of PSII thermoresistance was about 4°C, thus appeared to be in a range of the increase of thermostability induced by long-therm acclimation. As discussed recently by Havaux (1993), the time constant of the rapid adaptation principally excludes the possibility of profound reorganization of thylakoid membranes by *de novo* lipid

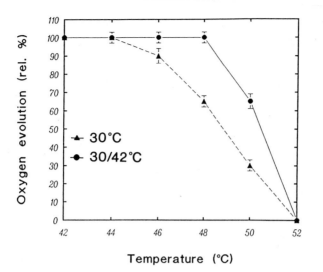

Figure 6. Effect of rapid heat-hardening on the heat-inactivation of PSII. Algal cells were grown at 30 and exposed to 42°C for 3 hours. Cells were treated for 30 minutes at designated temperatures in darkness and PSII was measured at 30°C.

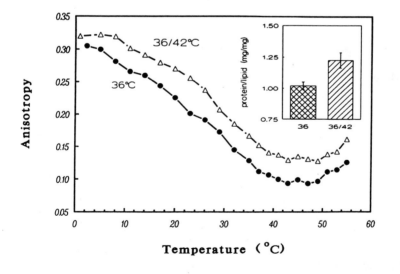

Figure 7. Changes in membrane fluidity and ratio of protein-to-lipid (insert) in thylakoids isolated from control (36°C grown) and heat-hardened (42°C, 3 hours) cells.

and protein synthesis. In fact, we found no measurable changes in glycerolipid and fatty acid composition during rapid acclimation (data not shown). By contrast, if thylakoids derived from 36°C grown cells exposed to mild heat-stress the ratio of protein-to-lipid increased by cca. 20% (Fig.7, insert). In accordance with the assumed documented effect of alteration in protein-to-lipid ratio on membrane dynamical properties, short-therm adaptation resulted in an increased microviscosity in thylakoids (Fig.7). Since the fluidity measurements were conduced on isolated membranes, we examined the thermal stability of isolated thylakoid vesicles. Enhancement of thermostability could also be observed at this level by suggesting, that the thermoprotector factor(s), which could possibly be involved in the rapid adjustment of PSII thermal stability remained in isolated thylakoid membranes.

Heat-induced expression of the two groE-genes

Recently we have presented detailed characterization of synthesis and regulation of heat-shock proteins (HSPs) in *Synechocystis* PCC6803 (Lehel et al.1992,1993/a/b). Proteins of four size classes, 70, 64, 15 and 14 kDa, were specifically induced when cells were

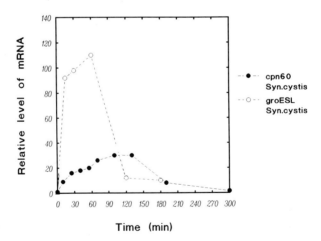

Figure 8. Transcription of cpn60 and groESL operon under heat stress. The transcript level was quantitated by dot blot analysis of total RNA obtained from cells at different times after shifting 30°C-grown cultures to 42°C using cpn60 and groES probes.

shifted from 30°C to 42°C. Western analysis revealed that HSP70 is the member of the 70 kDa molecular chaperone family. According to partial protein sequencing and immunological evidences (Lehel et al,1992) the 64 kDa HSP proved to be a GroEL-related cyanobacterial chaperonin. We postulated, that in contrast with prior findings described by Chitnis and Nelson (1991), at least two groEL-type gene should be present in the genome of *Synechocystis* PCC6803. Molecular cloning and sequencing of the groESL operon fully supported this contention (Lehel et al. 1993/b). It is noted, that similar to *Synechocystis* the presence of more than one copy of groEL genes has been reported for different prokaryotic organisms (de Wit et al.1992). The unexpected observation concerning the finding of multiple GroEL proteins implied that they might fulfil physiologically distinct roles. In fact, the level of the bicistronic 2.2-kb transcript of the groESL operon increased 100-fold upon exposing *Synechocystis* cells to heat stress, whilst the transcript level of cpn60 (possessing no groES in its neighbouring regions) accumulated about 30-fold under identical conditions (see Fig.8.). In accordance with the difference found on the level of transcription, the vast majority of chaperonin gene products formed during rapid heat treatment and assembled into tetradecameric species appeared to be the same HSP64 what we described previously (Lehel et al. 1992).

Heat-induced membrane association of the GroE-proteins

Although the exact role of HSPs is not known, acquisition of thermotolerance has been widely postulated as an essential function of these proteins. In some cases, their synthesis and decay have been correlated with the acquisition and decay of the thermotolerant state (Lindquist and Craig 1988). The GroES and GroEL proteins were shown to be essential for the growth of *E.coli* at all temperatures (Fayet et al.1989). The acquisition of thermotolerance appears to depend not only upon the synthesis of HSPs but also on their cellular localization. Several of the small, nuclear-encoded HSPs were shown to be transferred into the chloroplast of pea, soybean and *Chlamydomonas* cells. In pea, some of these HSPs have been demonstrated to become associated with the thylakoid membranes: heat induced alteration of the thylakoid was shown to be essential for the membrane binding of stress proteins (Glaczinski and Kloppstech,1988).

In line of this presumption, we proposed that if HSPs of *Synechocystis* PCC6803 and

92

especially the molecular chaperone proteins played any protective role against heat-damage to the photosynthetic apparatus, they would probably accumulate in thylakoid membrane during rapid acclimation phase. In fact, whereas the total level of the two GroEL-analogs (being unresolved by 1D-SDS-PAGE) was increasing in the cytosol, heat-treatment resulted in a strikingly elevated level of association of GroEL-like proteins to thylakoid membranes, simultaneously (Fig.9). Moreover, whilst the level of the cytosolic co-chaperonin, GroES

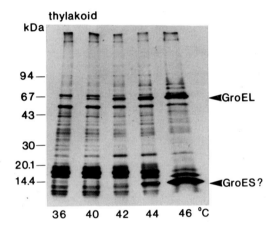

Figure 9. Heat-induced thylakoid association of GroE-proteins in cells grown at 30°C and treated for 3 hours at designated temperatures in light.

(identified previously by Lehel et al.1993/a) was also enhanced gradually upon heat-treatment, its membrane binding occured at a discrete threshold temperature, and corresponded to the temperature (44°C) at which the efficiency of Photosystem II was shown to decline (see Fig.6). Detailed characterization of the heat-induced membrane binding of chaperonin proteins has recently been accomplished and published elsewhere (Kovács et al, submitted). Based on the apparent correlation between increase in heat stability of PSII and the simultaneous association of GroE-proteins with thylakoids, we predict that these findings might be explained in terms of the production of partially denatured membrane proteins and the capability of molecular chaperones to prevent the formation of non-functional aggregates by binding to the partially denatured proteins (Georgopoulos 1992, Hartman et al.1993). A recent study on mitocondrial malate dehydrogenase (MDH) not only demonstrates the chaperonin mediated reconstitution of heat shock effect, but underlines also the need of the concerted action of GroES and GroEL proteins (Hartman et al.1993). Whereas GroEL alone did not protect MDH against thermal inactivation but prevented its aggregation, reactivation could be achieved by addition of GroES.

In conclusion, our presented data on heat-induced GroEL-, GroES-thylakoid interaction observed parallel with rapid heat-adaptation of PSII might provide the first evidence on the possible implication of chaperonin proteins in the development of thermal stability of thylakoids. In order to gain further insights into regulation and roles of cpn60 and groESL gene-products, construction of chaperonin mutants defected in either of the groE-type genes is in progress in our laboratory.

A main point of interest was to document, that both the long- and short-term phases of thermal acclimation resulted in a subsequent enhancement of membrane microviscosity. It is tempting to postulate, that the ability of cyaanobacterial cells to optimize the physical state (fluidity, lipid-protein interactions) of their photosynthetic membranes is part of a mechanism ensuring increased thermostability.

Membranes are not only site of the thermal damage but they might act also as thermal sensors. Recently Maresca and his co-workers suggested, that upon heat shock, perturbation of lipoprotein membrane complexes activates a putative heat shock sensor that monitors temperature changes and stimulates transcription of heat shock genes. In favour of their hypothesis, modification of membrane lipids by addition of exogeneous saturated fatty acids to the fungus *Histoplasma capsulatum* caused an increase of heat shock gene (HSP82, HSP70) expression at a specific heat shock temperature (personal communication). Recently we have reported, that the increased transcription of the desA gene can be achieved not only by a decrease in temperature but also by direct catalytic hydrogenation of membranes of living cyanobacterial cells (Vigh et al.1993). Identification and characterization of the hypothetic primary sensor, which detects the change in molecular motion of membrane lipids and would also influence HS gene expression needs further studies.

Acknowledgements

This research was supported by grants from Hungarian Research Foundation (OTKA) to L.V. and I.H.

REFERENCES

Berry JA, Björkman O (1980) Photosynthetic response and adaptation to temperature in higher plants. Annu Rev Plant Physiol 31:491-453

Chitnis PR, Nelson N (1991) Molecular cloning of the genes encoding two chaperone proteins of the cyanobacterium *Synechocystis* sp. PCC 6803. J Biol Chem 266:56-65

Fayet O, Ziegelhoffer T, Georgopoulos C (1989) The groES and groEL heat shock gene products of *Escherichia coli* are essential for bacterial growth at all temperatures. J Bacteriol 171:1379-1385

Fork DC, Sen A, Williams WP (1987) The relationship between heat-stress and photobleaching in green and blue-green algae. Photosynthesis Res 11:71-87

Georgopoulos C (1992) The emergence of the chaperone machines. TIBS 17:295-299

Glaczinski H, Kloppstech K (1988) Temperature-dependent binding to the thylakoid membranes of nuclear-coded chloroplast heat-shock proteins. Eur J Biochem 173:579-583

Gombos Z, Kiss M, Páli T, Vigh L (1987) Nitrate starvation induces homeoviscous regulation of lipids in the cell envelope of the blue-green alga, *Anacystis nidulans*. Eur J Biochem 165:461-465

Gombos Z, Wada H, Murata N (1991) Direct evaluation of effects of fatty-acid unsaturation on the thermal properties of photosynthetic activities, as studied by mutation and transformation of *Synechocystis* PCC6803. Plant Cell Physiol 32:205-211

Hartman DJ, Surin BP, Dixon NE, Hoogenraad NJ, Hoj PB (1993) Substochiometric amounts of the molecular chaperones GroEL and GroES prevent thermal denaturation and aggregation of mammalian mitochondrial malate dehydrogenase in vitro. Proc Natl Acad Sci 90:2276-2280

Havaux M (1993) Rapid photosynthetic adaptation to heat stress triggered in potato leaves by moderately elevated temperatures. Plant Cell Environm 16:461-467

Hugly S, Kunst L, Browse J, Somerwille C (1989) Enhanced thermal tolerance of photosynthesis and altered chloroplast ultrastructure in a mutant of *Arabidopsis* deficient in lipid desaturation. Plant Physiol 90:1134-1142

Kovács E, Lehel Cs, Gombos Z, Mustárdy L, Török Zs, Horváth I and Vigh L (1993) Heat stress induces association of the GroEL-analogue chaperonin with thylakoid membranes in cyanobacterium, *Synechocystis* PCC6803. (submitted to Plant Phys Biochem)

Kunst L, Browse J, Somerwille C (1989) Enhanced thermal tolerance in a mutant of *Arabidopsis* deficient in palmitic acid unsaturation Plant Physiol 91:401-408

Lehel CS, Wada H, Kovács E, Török Zs, Gombos Z, Horváth I, Murata N, Vigh L (1992) Heat shock protein synthesis of the cyanobacterium *Synechocystis* PCC6803: purification of the GroEL-related chaperonin. Plant Mol Biol 18:327-336

Lehel Cs, Los D, Wada H, Györgyei J, Horváth I, Kovács E, Murata N, Vigh L (1993/a) A second groEL-like gene, organized in a groESL operon is present in the genome of *Synechocystis* PCC6803. J Biol Chem 268:1799-1804

Lehel Cs, Gombos Z, Török Zs, Vigh L (1993/b) Growth temperature modulates thermotolerance and heat shock response of cyanobacterium, *Synechocystis* PCC 6803. Plant Phys. Biochem. 31:81-88

Li G, Knowles PF, Murphy DJ, Mars D (1990) Lipid-protein interactions in thylakoid membranes of chilling-resistant and -sensitive plants studied by spin label electron spin resonance spectroscopy. J Biol Chem 265:16867-16872

Lindquist S, EA Craig (1988) The heat shock proteins. Annu Rev Genet 22:631-677

Quinn P, Joo F, Vigh L (1989) The role of unsaturated lipids in membrane structure and stability. Progr Biophys Mol Biol 53:71-103

Schlame M, Horváth LI, Vigh L (1990) Relationship between lipid saturation and lipid-protein interaction in liver mitochondria modified by catalytic hydrogenation with reference to cardiolipin molecular species. Biochem J 265:79-85

Süss K-H, Yordanov I (1986) Biosynthetic cause of in vivo acquired thremotolerance of photosynthetic light reactions and metabolic responses of chloroplasts to heat stress. Plant Physiol 81:192-199

Thomas PG, Domini PJ, Vigh L, Mansourian AR, Quinn PJ, Williams WP (1986) Increased thermal stability of pigment-protein complexes of pea thylakoids following catalytic hydrogenation of membrane lipids. Biochim Biophys Acta 849:131-140

Vigh L, Gombos Z, Horváth I, Joo F (1989) Saturation of membrane lipids by hydrogenation induces thermal stability in chloroplast inhibiting the heat-dependent stimulation of Photosystem I-mediated electron transport. Biochim Biophys Acta 979:361

Vigh L, Joo F, Droppa M, Horváth LI, Horváth G (1985) Modulation of chloroplast membrane lipids by homogeneous catalytic hydrogenation. Eur J Biochem 147:477-481

Vigh L, Lehel CS, Török Zs, Gombos Z, Balogh N, Horváth I (1990) Factors affecting thylakoid thermal stability in cyanobacterium, Synechocystis sp.PCC6803. In: Qin PJ, Harwood JL, eds, Plant Lipid Biochemistry, Structure and Utilization. Portland Press Limited, London, pp 373-381

Vigh L, Los D, Horváth I, Murata N (1993) The primary signal in the biological perception of temperature: PD-catalyzed hydrogenation of membrane lipids stimulated the expression of the desA gene in Synechocystis PCC6803. (submitted to Proc Natl Acad Sci)

Wada H, Gombos Z, Sakamoto T, Murata N (1992) Genetic manipulation of the extent of desaturation of fatty acids in membrane lipids in the cyanobacterium Synechocystis PCC6803. Plant Cell Physiol 33:535-540

Wada H, Murata N (1990) Temperature-induced changes in the fatty acid composition of the cyanobacterium, Synechocystis PCC6803. Plant Physiol 92:1062-1069

Webb MS, Green BR (1991) Biochemical and biophysical properties of thylakoid acyl lipids. Biochim Biophys Acta 1060:133-158

de Wit TFR, Bekelie S, Osland A, Miko TL, Hermans PWM, van Solingen D, Drijfhout JW, Schöningh R, Janson AAM and Thole JER (1992) Mycobacteria contain two, groEL genes: the second Mycobacterium leprae groEL gene is arranged in an operon with groES. Molecular Microbiology 6:1995-2007

Yordanov I, Dilova S, Petkova R, Pangelova T, Goltsev V, Süss K-H (1986) Mechanisms of the temperature damage and acclimation of the photosynthetic apparatus. Photobiochem Photobiphys 12:147-155

STUDIES OF A CHLOROPLAST-LOCALIZED SMALL HEAT SHOCK PROTEIN IN ARABIDOPSIS

Katherine W. Osteryoung, Brian Pipes, Nadja Wehmeyer and Elizabeth Vierling
Department of Biochemistry
University of Arizona
Tucson, Arizona 85721
USA

INTRODUCTION

Thermotolerance, or the ability of organisms to withstand exposure to potentially lethal high temperatures, is thought to be conferred by the induction of heat shock proteins (HSPs) (Lindquist 1986). Several major families of HSPs classified according to molecular weight and amino acid similarity have been defined, although how they contribute to the development of thermotolerance is not well understood. Among the HSPs most highly induced by heat stress in eukaryotes is a class of nuclear-encoded proteins ranging in size from 15 to 30 kD referred to as the small HSPs (sHSPs). Though diverse in size and primary structure, the sHSPs can be identified based on their similar hydropathy profiles and homology to the α-crystallin proteins of the eye lens (Lindquist and Craig 1988). In a number of organisms the sHSPs are found in oligomeric complexes of 200 to 800 kD which can aggregate reversibly into even larger particles termed heat shock granules (Arrigo and Welch 1987; Collier et al. 1988; Nover et al. 1989). Although a number of functions have been postulated for the sHSPs, recent reports have shown that both sHSPs and α-crystallins can prevent thermal aggregation of other proteins and facilitate reactivation of denatured enzymes in vitro (Horwitz 1992; Jakob et al. 1993). These data suggest that sHSPs may be "molecular chaperones", or proteins that interact transiently with other polypeptides to facilitate attainment of a functional conformational state. Molecular chaperone activity has already been described for the HSP60 and HSP70 groups of HSPs (Gething

NATO ASI Series, Vol. H 86
Biochemical and Cellular Mechanisms
of Stress Tolerance in Plants
Edited by J. H. Cherry
© Springer-Verlag Berlin Heidelberg 1994

and Sambrook 1992). Further work is needed to clarify the mechanism of sHSP function and to determine the nature of sHSP substrates in vivo.

The sHSPs are most abundantly represented among higher plants, which synthesize up to 30 distinct sHSPs upon exposure to elevated temperatures (Vierling 1991). In contrast with the situation in other organisms in which sHSP localization is confined to the cytoplasm, the small HSPs in plants are defined by four distinct gene families whose products are localized to three subcellular compartments. Two of these gene families encode proteins localized to the cytoplasm (Vierling 1991) and the other two encode proteins localized to the endoplasmic reticulum (Helm et al. 1993) and chloroplasts (Kloppstech et al. 1985; Vierling et al. 1988), respectively. The multicompartmental distribution of the sHSPs in plants argues that they play a particularly important role in the ability of plants to withstand exposure to high temperatures.

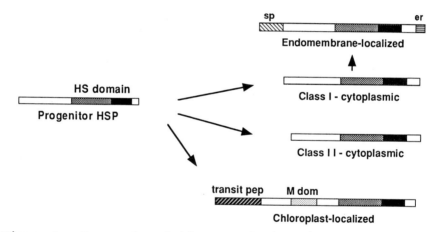

Figure 1. Proposed evolutionary relationships among the sHSPs in plants. A progenitor sHSP gene was duplicated to give rise to three of the four classes of sHSPs found in higher plants. A more recent duplication of the class I cytosolic sHSP gene and addition of sequences encoding a signal peptide (sp) and ER retention signal (er) established the ER-localized class. Sequence divergence and addition of a chloroplast transit peptide and Met-rich domain (M-dom) gave rise to the chloroplast sHSPs. The class II cytosolic sHSP retained the original cytosolic localization, but diverged to give rise to a distinct class of sHSPs.

Recent studies indicate that plant sHSPs are quite stable following heat stress (Chen et al. 1990; DeRocher et al. 1991), suggesting they may serve an important function not only during heat stress but also in recovery from heat stress. Although a definitive role for sHSPs in conferring thermotolerance in plants has not been established, data showing a correlation between the level of sHSP accumulation and acquired thermotolerance suggest that sHSPs exert a protective function in plants exposed to high temperatures (Vierling 1991).

We are investigating the major sHSP localized to the chloroplasts, HSP21. This protein has been characterized at the cDNA level in several divergent plant species including soybean, *Arabidopsis thaliana*, maize, petunia, pea, and wheat (Vierling et al. 1988; Nieto-Sotolo et al. 1990; Chen and Vierling 1991; Weng et al. 1991). Like other nuclear-encoded chloroplast proteins, HSP21 is synthesized in the cytoplasm as a large precursor bearing a cleavable N-terminal transit peptide that directs it post-translationally to the chloroplast (Vierling et al. 1986; Vierling et al. 1988). The protein is not detectable in the absence of stress, but is induced at least 200-fold at 38°C in pea (Chen et al. 1990). In heat-stressed pea plants, HSP21 is found primarily in the stromal fraction in a 200 kD complex (Chen 1992). The evolutionary relationship between the chloroplast sHSP and the other sHSPs of plants may be inferred from a comparison of their primary structures (Fig. 1). While HSP21 retains the "heat shock domain" that is conserved among all the plant sHSPs, it has also acquired two unique domains. One is the transit peptide, which targets the protein to the chloroplast, and the other is a Met-rich domain predicted to form an amphipathic a-helix (Chen and Vierling 1991). The other three classes of plant sHSPs also exhibit features that distinguish them as distinct gene families (Fig. 1).

The chloroplast-localized sHSP presents an attractive model for studying sHSP function in plants for several reasons: 1) It has a clear evolutionary relationship to the other sHSPs and likely performs a similar function within the chloroplast during heat stress; 2) it is localized in an easily isolated organelle and can therefore be readily manipulated in vitro; and 3) it is encoded by only one or a few genes in most plants and is therefore amenable to genetic manipulation. We are combining several approaches to understand the function of this protein during heat stress. Because HSP21 is encoded by a single gene in *Arabidopsis* (Osteryoung et al. 1993), and because of the ease of generating transformed *Arabidopsis* plants, we have chosen this species for investigating the role of HSP21 in the acquisition of thermotolerance in plants. To this end, we have generated transgenic *Arabidopsis* plants that either underexpress or constitutively overexpress HSP21. Here we report the initial characterization of these plants. Our studies represent the first experimental alteration of sHSP protein levels in plants.

MATERIALS AND METHODS

Vector construction, plant transformation and establishment of transgenic lines

A DNA fragment encoding nearly the entire transcribed region of the *Arabidopsis* HSP21 gene was obtained by PCR amplification from a genomic clone (Osteryoung et al. 1993) for use in vector construction. Constitutive overexpression and antisense constructs were prepared by ligating this fragment in either the sense or antisense orientation into the T-DNA vector pMON530 (Rogers et al. 1987). This vector incorporates the constitutive CaMV 35S promoter as well as a gene conferring plant kanamycin resistance (kan^r). A heat-inducible antisense gene was constructed in the same vector by subcloning a soybean heat-shock promoter (Ainley and Key 1990) between the 35S promoter and the HSP21 insert. The resulting vectors were transferred to *Agrobacterium tumefaciens* strain GV3111SE (Rogers et al. 1986), and subsequently introduced into *Arabidopsis thaliana* ecotypes "Norway", "RLD" and "Landsberg" using a modified root transformation procedure (Valvekens et al. 1988; and Márton et al. 1991).

Primary transformants were selected by regeneration and

rooting on kanamycin and were allowed to self-fertilize. The resulting seed were analyzed for segregation of the kan^r phenotype by germination on agarose plates containing 50-100 mg l^-1 kanamycin. Kan^r individuals were transferred to soil and grown for about three weeks prior to initial assay for HSP21 expression by immunoblot analysis (described below). The genotypes of these individuals were determined by scoring for kan^r in their progeny. This allowed the establishment both of homozygous transgenic lines and of sibling wild-type lines in which the transgene had segregated away for use as controls in subsequent experiments.

Plant growth and heat stress conditions

Seeds were sown in moist soil or on agarose plates containing kanamycin as described above and placed at 4°C for 3-5 d in the dark to increase uniformity of germination. Plants were germinated and grown under fluorescent light (100-150 mmol m^-2 sec^-1) at 22-25°C using a daylength of 16 hr. Kan^r seedlings started on agarose plates were transferred to soil 5-7 d following germination. For phenotypic analysis of transgenic lines that constitutively overexpress HSP21, plants were grown at 20°C at a density of two to three per pot.

Heat stress treatments were administered in a programmable growth chamber (Conviron) containing a humidifier to prevent evaporative cooling of the leaves. Whole plants were heat-stressed as described previously (Chen et al. 1990) by raising the temperature gradually from 22°C to the indicated maximum temperature, maintaining the plants at that temperature for the stated period of time, and, when plants were not harvested immediately, lowering the temperature gradually back to 22°C. For abrupt stress treatments, the growth chamber was prewarmed to the maximum temperature before the plants were introduced, and the temperature was lowered gradually to 22°C at the end of the stress period.

Thermotolerance assays

Arabidopsis plants were grown under the temperature and

lighting conditions described above in 2-in pots at a density of 25-50 plants per pot. Three- to four-week-old plants were heat-stressed either gradually or abruptly at the indicated temperature. Following heat stress the plants were returned to their original growth conditions and their appearance and survival were monitored for six days. Percent survival in most cases was estimated by visual inspection rather than by direct counting. Two pots were included for each treatment and the experiment was replicated five times.

Immunoblot analysis

Proteins were extracted from leaves, roots and flowers in SDS-PAGE sample buffer and analyzed by SDS-PAGE and immunoblotting as described previously (Osteryoung and Vierling 1992) using a chemiluminescent detection method (Amersham). For quantitative analysis of HSP21 expression in transgenic plants, immunoreactive protein was detected using ^{125}I-protein A as described (Chen et al. 1990) followed by image analysis on a phosphorimager (Molecular Dynamics).

Native gel electrophoresis

For electrophoretic analysis of native proteins, leaves were homogenized in 10 ml mg^{-1} cold native sample buffer (60 mM Tris-HCl pH 8.0, 15% sucrose, 5 mM ϵ-amino-N-caproic acid, 1 mM benzamidine, 0.01% bromophenol blue). The homogenate was centrifuged in a microcentrifuge for 15 min at 4°C and the supernatant applied to a 4-20% acrylamide pore exclusion gradient gel. This gel system allows a direct estimate of molecular size for nondenatured proteins (Anderson et al. 1972). Following electrophoresis at 65-75 V for 48-72 hr, HSP21 was detected by immunoblotting.

RESULTS

Thermotolerance in *Arabidopsis*

To be able to evaluate the effects of altered HSP21 expression in transgenic *Arabidopsis*, it is important to determine how wild-type plants respond to heat stress and to verify that *Arabidopsis* behaves similarly to other plant species. Accordingly, *Arabidopsis* plants (ecotype "Norway") were subjected to either a gradual or an abrupt heat stress over a range of temperatures and their survival rates recorded after six days (Fig 2). Gradual heat stress has been shown in other species to induce thermotolerance, which refers to the ability to survive an otherwise lethal heat treatment (Lindquist 1986).

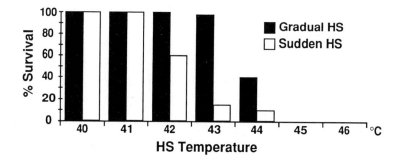

Figure 2. Analysis of thermotolerance in wild-type *Arabidopsis* plants. Three- to four-week-old plants (ecotype Norway) were subjected to either a gradual or an abrupt heat stress and their survival was recorded six days later. For the gradual stress, the temperature was raised from 22°C to the indicated temperature at a rate of 4°C h^{-1} and maintained at that temperature for 4 h. For the abrupt stress, plants were placed directly at the indicated temperature for 4 h. Following both treatments the temperature was decreased to 22°C at a rate of 4°C h^{-1}.

At 40° and 41°C all plants survived regardless of whether the stress was imposed gradually or abruptly. At higher temperatures both the condition of the plants and the survival after six days were dependent on how the stress was imposed. In plants subjected to abrupt stress, signs of injury, such as leaf browning, were evident at 41°C the day following the stress even though all the plants survived after six days. Survival decreased to less than 60% at 42°C, and at 44°C only 10% survived. Sensitivity to an abrupt 42°C heat stress has been previously documented in *Arabidopsis* (Binelli and Mascarenhas 1990) In contrast, plants subjected to a gradual stress exhibited no signs of injury and no mortality up to 42°C. Only a small decrease in survival occurred at 43°C. At 44°C, 40% of the plants remained alive after six days. No survival was observed at 45°C or above for either gradually or abruptly stressed plants. Preliminary results suggest that other ecotypes of *Arabidopsis* behave similarly to Norway in their responses to heat stress.

Decreased expression of HSP21 by antisense inhibition

In an effort to define the function of HSP21, we are working to generate transgenic plants that exhibit reduced expression of HSP21 in response to heat stress. Towards this end, antisense HSP21 genes incorporating either the constitutive CaMV 35S promoter or the heat-inducible GmHSP17.5-E promoter from soybean (Ainley and Key 1990) were transferred to *Arabidopsis* plants via *Agrobacterium*-mediated root transformation. Transfer of T-DNA to putative transformants was verified by germinating seed collected from selfed primary transformants on kanamycin. Kan^r progeny of primary transformants were transferred to soil and tested for induction of HSP21 following a gradual heat stress. HSP21 accumulation was assayed by Western blot analysis using similarly stressed wild-type plants as controls. In an earlier communication we reported on one transgenic line in the Landsberg ecotype that showed significantly reduced accumulation of HSP21 following heat stress (Osteryoung and Vierling 1992). This line carries the heat-inducible antisense gene construct. The reduced level of expression in this line has been stably maintained in subsequent generations. We have now tested nearly 70 antisense transformant lines. Although all exhibit resistance to kanamycin, we have not identified any additional transformants that show greatly reduced HSP21 accumulation upon heat stress. A few other RLD and Norway transformant lines carrying either the constitutive or the heat-inducible antisense gene have been identified that show slightly reduced HSP21 expression levels (not shown).

Overexpression of HSP21 in transgenic plants

Three independent lines of *Arabidopsis* that constitutively overexpress HSP21 were generated, two in ecotype RLD (designated 5M and 5Y) and one in ecotype Norway (4W). In 5M and 5Y, segregation ratios for kan^r indicate that the transgene is inserted at a single locus. 4W appears to have insertions at two or more loci. These conclusions are supported by the results of Southern analysis (not shown). Segregation of the HSP21 overexpression trait in the 5M and 5Y lines exhibits the expected ratio for a single transgene: heterozygous individuals produce about three constitutive overexpressing progeny for every one nonexpressing wild-type. In addition, for all three overexpression lines, all kan^r individuals tested to date constitutively overexpress HSP21, suggesting that HSP21 overexpression cosegregates with the T-DNA. Progeny of heterozygous individuals from the 5M and 5Y lines segregating for zero, one or two copies of the transgene were identified and those with two copies (homozygotes) or zero copies (sibling wild-types) were used to generate seed for subsequent experiments.

All three transgenic lines accumulate considerably more HSP21 in the absence of heat-stress than do heat-stressed wild-type plants (Fig. 3).

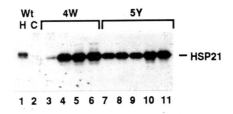

Figure 3 . Accumulation of HSP21 in transgenic *Arabidopsis* plants transformed with the constitutive HSP21 overexpression construct. Leaf proteins were extracted with SDS sample buffer and separated by SDS-PAGE. HSP21 was detected by immunoblotting. Lane 1: Wild-type plant heat-stressed at 40°C. Lane 2: Unstressed wild-type plant. Lanes 3-6: Unstressed kan^r individuals from transgenic line 4W. Lanes 7-11: Unstressed kan^r individuals from transgenic line 5Y.

The levels of expression shown in Fig. 3 for several kan^r individuals of the 4W and 5Y lines are typical of all three overexpression lines. In these lines, HSP21 can be detected in leaves, stems, roots and flowers (not shown), consistent with the known expression pattern of the CaMV 35S promoter (Odell et al. 1985). On SDS gels the overexpressed protein comigrates with mature, authentic HSP21 (Fig 3), suggesting that the transit peptide has been cleaved and the protein is properly localized in the chloroplast.

To quantify the relative levels of HSP21 protein in the transgenic plants, leaf extracts from unstressed homozygous individuals and from wild-type plants heat-stressed gradually to 40°C were analyzed by immunoblotting using ^{125}I-protein A to detect the immunoreactive protein. Based on this analysis, 5M and 5Y homozygotes accumulate at least 5- to 10-fold more HSP21 in the absence of stress than do heat-stressed wild-type plants.

To determine whether constitutive overexpression of HSP21 affects growth and development, 15 to 20 progeny of heterozygous individuals of the 5Y and 5M transgenic lines were grown at 20°C using a 16-hr day. The genotypes of these progeny should be one-quarter homozygous for the transgene, one-half heterozygous, and one-quarter nonexpressing sibling wild-types. Plants were tested for constitutive HSP21 expression by immunoblotting and their genotypes were identified by segregation analysis of kan^r in their progeny. Table 1 shows the results of one experiment. The data confirm the cosegregation of the kan^r and overexpression phenotypes. No obvious differences in any readily observable parameter of growth and development were noted between the sibling wild-type and overexpressing plants. Neither were any discernible differences noted between the multiple-insertion line 4W and wild-type (nonsibling) Norway plants. Preliminary observations had suggested the possibility of early flowering in one overexpression line (Osteryoung and Vierling 1992). This was not confirmed by the above analysis; however, we are repeating the experiment using a shorter daylength to maximize any potential differences in time-to-flowering or other aspects of vegetative growth.

Table 1. HSP21 expression in several individuals of the 5Y and 5M transgenic overexpression lines and segregation of kan[r] in their progeny. Parents of individuals shown were heterozygous for the transgene. Individuals shown are a subset of the the same plants used for analysis of the effect of HSP21 overexpression on growth and development.

Plant	Constitutive HSP21 expression	Segregation ratio Kan[r]: Kan[s]	Genotype
5M-4	+	654 : 0	Homozygous
5M-16	+	228 : 68	Heterozygous
5M-6	-	0 : 400	Wild-type
5Y-7	+	376 : 0	Homozygous
5Y-6	+	186 : 65	Heterozygous
5Y-14	-	0 : 320	Wild-type

Native structure of HSP21 in wild-type and transgenic *Arabidopsis*

An important consideration for interpreting the effects of HSP21 overexpression is whether the protein acquires the same native structure in transgenic plants as in wild-type plants. To examine the native structure of *Arabidopsis* HSP21, which has not been previously characterized, leaf extracts from heat-stressed wild-type *Arabidopsis* plants were analyzed by native gel electrophoresis followed by immunoblotting. Leaf extracts from unstressed overexpressing transgenic plants were analyzed in parallel. The major immunoreactive bands in both samples comigrate at about 300 kD (Fig. 4). These results indicate that HSP21 in *Arabidopsis* is found in a high molecular weight complex as has been documented in pea (Chen 1992), and more importantly, that the protein does acquire its native structure in the transgenic plants. Since the 300 kD complex is found in the absence of heat stress in these plants, we conclude that assembly of HSP21 into the native complex does not require any other heat-inducible factors or modifications.

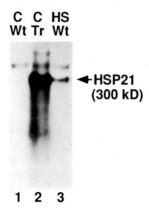

Figure 4. Native structure of HSP21 in wild-type and transgenic *Arabidopsis*. Leaf extracts were prepared as described in Materials and Methods and analyzed by electrophoresis on nondenaturing 4-20% acrylamide gels followed by immunoblotting. Lane 1: Extract from a control (unstressed) wild-type *Arabidopsis* plant. Lane 2: Extract from an unstressed transgenic plant that constitutively overexpresses HSP21. Lane 3: Extract from a wild-type *Arabidopsis* plant subjected to a gradual heat stress with a maximum temperature of 40°C.

DISCUSSION

Our goal in manipulating HSP21 expression in transgenic *Arabidopsis* is to determine the role of this protein in the acquisition of thermotolerance and ultimately to establish its biochemical function within the chloroplast. As a prelude to evaluating the effects of altered HSP21 expression in transgenic *Arabidopsis*, we have examined how this species responds to heat stress under conditions that either confer (gradual heat stress) or do not confer (abrupt heat stress) thermotolerance. The ability of a gradual increase in temperature to confer thermotolerance in plants has been documented by Altschuler and Mascarenhas (1982) who demonstrated that protein synthesis and survival in heat-stressed soybean seedlings continued at higher temperatures when the heat stress was imposed gradually than when imposed abruptly. Similar observations have also been made in other plant species (Howarth 1991). Our studies show that *Arabidopsis* plants subjected to a gradual increase in temperature also become thermotolerant and are thus able to withstand temperatures about 2°C higher than plants subjected to an abrupt temperature increase. Numerous investigators have proposed that thermotolerance is conferred by synthesis of heat shock proteins (Lindquist 1986; Lindquist and Craig 1988). However, only a single HSP, HSP104 from yeast, has been directly demonstrated to be required for the acquisition of thermotolerance (Sanchez and Lindquist 1990). Manipulation of HSP21 expression in transgenic *Arabidopsis* will allow us to test directly the role of this protein in the development of thermotolerance in higher plants.

Thus far, our attempts to reduce HSP21 expression during heat stress by antisense RNA inhibition have yielded limited success. Antisense techniques have proven remarkably effective as a method for analysis of protein function in plants (Watson and Grierson, 1993). The reason for our inability to significantly affect HSP21 levels in the antisense plants is not clear, but may be a consequence of the relatively high stability of the protein. Previous studies have shown that the half-life of HSP21 in pea is more than 50 hr (Chen et al. 1990); thus, it may require relatively little HSP21 mRNA to obtain a significant accumulation of protein. Another possibility is that the antisense construct itself may be ineffective in rendering inactive the endogenous HSP21 transcript. Since the mechanism of antisense inhibition is unclear, there are few rules available for designing effective antisense genes (Watson and Grierson, 1993); thus trial and error remains the standard method for determining the effectiveness of a specific construct. We have begun constructing several new antisense genes that utilize a different heat-inducible promoter and incorporate different portions of the HSP21 transcript for use in future experiments. We are also generating crosses of the antisense lines that do show reduced HSP21 expression as another approach to effecting a greater decrease in HSP21 accumulation during heat stress.

Transformation of *Arabidopsis* with the overexpression construct has yielded several transgenic plant lines that accumulate HSP21 in the absence of heat stress. The constitutive overexpression trait has been stably inherited for four generations to date. These transgenic lines accumulate HSP21 to levels well in excess of the levels typically induced by a 4-hour heat stress in wild-type plants. This is somewhat surprising because at least some sHSP mRNAs are rapidly degraded upon return to nonstress temperatures (Kimpel et al. 1990). The high level of accumulation may again be a consequence of the fairly long half-life of HSP21. Another possibility is that, coincidentally, all three transgenic lines have exceptionally high rates of transgene transcription leading to high steady-state levels of mRNA. We are currently examining transgene mRNA levels in the three overexpression lines.

An interesting aspect of HSP21 expression in the overexpression lines is that the protein assembles into its native oligomeric form in the absence of heat stress. Although HSP21 is normally synthesized only under heat-stress conditions, this observation implies that assembly of the protein in the chloroplast is not dependent on other heat-inducible factors or modifications. To our knowledge, this is the first demonstration that sHSPs do not require heat shock conditions for assembly in plants. Because HSP21 is normally not expressed in the absence of heat stress, it might be postulated that overexpression of the protein could be detrimental to the plant. However, our phenotypic analysis has so far provided no evidence for any effect on aspects of growth or development observable at the whole-plant level under optimum growth conditions. We are currently generating the essential controls that will allow us to conclude unambiguously that any phenotypes we observe in future analyses are a consequence of the transgene. These include obtaining multiple independent overexpression lines, establishing sibling wild-type lines derived from the same primary transformants as the transgenic plants, verifying that the transgenes are stably inherited with the T-DNA, backcrossing the transgenes to wild-type plants, and characterizing the properties of the overexpressed protein. Future studies will focus on the effects of HSP21 overexpression on chloroplast ultrastructure, photosynthetic characteristics, and thermotolerance.

ACKNOWLEDGMENTS

We are grateful to Teri Suzuki for helpful discussions and to Debra Claus-Walker for assistance with photography. This work is supported by National Institutes of Health grant R01 GM42762-04 to E.V. and by National Institutes of Health Postdoctoral Fellowship F32 GM14953-10A1 to K.W.O.

REFERENCES

Ainley WM, Key JL (1990) Development of a heat shock inducible expression cassette for plants: Characterization of parameters for its use in transient expression assays. Plant Mol Biol 14:949-967

Altschuler M, Mascarenhas JP (1982) Heat shock proteins and effects of heat shock in plants. Plant Mol Biol 1:103-115

Anderson LO, Borg H, Mikaelsson M (1972) Molecular weight estimation of proteins by electrophoresis in polyacrylamide gels of graded porosity. FEBS Lett 20:199-202

Arrigo A-P, Welch WJ (1987) Characterization and purification of the small 28,000-dalton mammalian heat shock protein. J Biol Chem 262:15359-15369

Binelli G, Mascarenhas JP (1990) Arabidopsis: Sensitivity of growth to high temperature. Dev Genet 11:294-298

Chen Q (1992) The major chloroplast low molecular weight heat shock protein. PhD Dissertation, University of Arizona, Tucson, AZ

Chen Q, Lauzon LM, DeRocher AE, Vierling E (1990) Accumulation, stability, and localization of a major chloroplast heat-shock protein. J Cell Biol 110:1873-1883

Chen Q, Vierling E (1991) Analysis of conserved domains identifies a unique structural feature of a chloroplast heat shock protein. Mol Gen Genet 226:425-431

Collier NC, Heuser J, Levy MA, Schlesinger MJ (1988) Ultrastructural and biochemical analysis of the stress granule in chicken embryo fibroblasts. J Cell Biol 106:1131-1139

DeRocher AE, Helm KW, Lauzon LM, Vierling E (1991) Expression of a conserved family of cytoplasmic low molecular weight heat shock proteins during heat stress and recovery. Plant Physiol 96:1038-1047

Gething M-J, Sambrook J (1992) Protein folding in the cell. Nature 355:33-45

Helm KW, LaFayette PR, Nagao RT, Key JL, Vierling E (1993) Localization of small HSPs to the higher plant endomembrane system. Mol Cell Biol 13:238-247

Horwitz J (1992) α-Crystallin can function as a molecular chaperone. Proc Natl Acad Sci USA 89:10449-10453

Howarth CJ (1991) Molecular reponses of plants to an increased incidence of heat shock. Plant, Cell Environ 14:831-841

Jakob U, Gaestel M, Engel K, Buchner J (1993) Small heat shock proteins are molecular chaperones. J Biol Chem 268:1517-1520

Kimpel JA, Nagao RT, Goekjian V, Key JL (1990) Regulation of the heat shock response in soybean seedlings. Plant Physiol 94:988-995

Kloppstech K, Meyer G, Schuster G, Ohad I (1985) Synthesis, transport and localization of a nuclear coded 22-kd heat-shock protein in the chloroplast membranes of peas and Chlamydomonas reinhardi. EMBO J 4:1902-1909

Lindquist S (1986) The heat shock response. Annu Rev Biochem 45:39-72

Lindquist S, Craig EA (1988) The heat shock proteins. Annu Rev Genet 22:631-677

Márton L, Browse J (1991) Facile transformation of Arabidopsis. Plant Cell Reports 10:235-239

Nieto-Sotolo J, Vierling E, Ho T-H D (1990) Cloning, sequence analysis and expression of a cDNA encoding a plastid localized heat shock protein in maize. Plant Physiol 93:1321-1328

Nover L, Scharf K-D, Neumann D (1989) Cytoplasmic heat shock granules are formed from precursor perticles and are associated with a specific set of mRNAs. Mol Cell Biol 9:1298-1308

Odell JT, Nagy F, Chua N-H (1985) Identification of DNA sequences required for activity of the cauliflower mosaic virus 35S promoter. Nature 313:810-812

Osteryoung KW, Sundberg H, Vierling E (1993) Poly(A) tail length of a heat shock protein RNA is increased by severe heat stress, but intron splicing is unaffected. Mol Gen Genet, 239:323-333.

Osteryoung KW, Vierling E (1992) Genetic approaches to the function of the chloroplast low molecular weight heat shock protein. In N Murata, ed, Research in Photosynthesis. Kluwer Academic Publishers, Dordrecht, Netherlands Vol IV:129-136

Rogers SG, Horsch RB, Fraley RT (1986) Gene transfer in plants: production of transformed plants using Ti plasmid vectors. Methods Enzymol. 118:627-641

Rogers SG, Klee HJ, Horsch RB, Fraley RT (1987) Improved vectors for plant transformation: Expression cassette vectors and new selectable markers. Meth Enzymol 153:253-277

Sanchez Y, Lindquist SL (1990) HSP104 is required for induced thermotolerance. Science 248:1112-1115

Valvekens D, Van Montagu M, Van Lijsebettens M (1988) *Agrobacterium tumefaciens*-mediated transformation of *Arabidopsis thaliana* root explants by using kanamycin selection. Proc Natl Acad Sci 85:5536-5540

Vierling E (1991) The roles of heat shock proteins in plants. Annu Rev Plant Physiol Plant Mol Biol 42:579-620

Vierling E, Mishkind ML, Schmidt GW, Key JL (1986) Specific heat shock proteins are transported into chloroplasts. Proc Natl Acad Sci USA 83:361-365

Vierling E, Nagao RT, DeRocher AE, Harris LM (1988) A heat shock protein localized to chloroplasts is a member of a eukaryotic superfamlily of heat shock proteins. EMBO J 7:575-581

Watson CF, Grierson D (1993) Antisense RNA in plants. *In* A Hiatt, ed, Transgenic Plants: Fundamentals and Applications. Marcel Dekker, New York, pp 255-281

Weng J, Wang Z, Nguyen HT (1991) Nucleotide sequence of a *Triticum aestivam* cDNA clone which is homologous to the 26 kDa chloroplast-localized heat shock protein gene of maize. Plant Mol Biol 17:255

THE LOW MOLECULAR WEIGHT HEAT SHOCK PROTEINS OF SOYBEAN SEEDLINGS

Joe L. Key, Yuh-Ru Julie Lee, Virginia Goekjian, and Ronald T. Nagao

Department of Botany
The University of Georgia
Athens, GA   30602

Introduction

The heat shock (HS) response has been studied extensively for about 30 years since the discovery of the phenomenon in *Drosophila* in the early 1960s; studies in plants have been restricted generally to less than 15 years. Heat shock is the subject of numerous reviews in recent years (e.g. Craig, 1985; Kimpel and Key, 1985a; Key et al., 1985a; 1985b; Lindquist, 1986; Lindquist and Craig, 1988; Nagao et al., 1986; Key et al., 1987a; 1987b; Nagao and Key, 1989; Nagao et al., 1990; Neumann et al., 1989; Vierling, 1991; Gurley and Key, 1991). Since various aspects of the HS response in a number of plant systems are covered in this volume, the comments of this contribution will focus on our work on the low molecular weight (LMW) heat shock proteins (hsps) of soybean seedlings, much of which is detailed in some of the reviews cited above, with emphasis on expression of the HS genes encoding hsps, and some properties of those proteins, in the 15 to 27 kD range. This relates to a rather unique feature of HS in plants in that most plants synthesize a large number of LMW hsps, and some of these are generally the most abundant hsps expressed in plants. Most other eukaryotes, in contrast, synthesize a limited number of LMW hsps, and these are generally much less abundant than the 70 kD and 83/90 kD hsps, with hsp70 proteins often representing the majority of hsp synthesis in many organisms. Plants synthesize a complement of hsps in the 60 kD, 70 kD, 83/90 kD, and 104/110 kD MW range which appear to be homologs of the corresponding hsps of other eukaryotes. The basis

[1]This work was supported by the U.S. Department of Energy under grant DE-FG09-86ER13602.

NATO ASI Series, Vol. H 86
Biochemical and Cellular Mechanisms
of Stress Tolerance in Plants
Edited by J. H. Cherry
© Springer-Verlag Berlin Heidelberg 1994

for and potential significance of the difference in complexity and in the relative abundance of the different hsps synthesized in different organisms is not understood.

The HS response is characterized by a significant slowing of most normal protein synthesis (the level of reduction varying substantially among the different proteins of a given organism and the relative intensity of the HS temperature) as a result of a dramatic decline both in transcription of genes which were active prior to HS and the translation of those encoded mRNAs. Most normal mRNAs persist to varying degrees during a typical HS treatment. Concomitant with the cessation or slowing of normal mRNA and protein synthesis at the onset of HS is a rapid activation of the transcription of HS genes and accumulation of HS mRNAs; these mRNAs are preferentially translated resulting in the substantial accumulation of hsps.

In discussing HS genes and hsps, it is essential to recognize that some HS genes (e.g. hsp70, hsp83/90) have cognates or homologs which are expressed at the normal growth temperature of the organism. Thus, within a given organism, the expression of some hsps, for example hsp70, results from the activation at the HS temperature of a previously silent gene(s) and the enhanced expression of the HS cognate gene(s); expression of other cognate members of the same family may not be measurably affected by HS. It is also notable that some HS genes including some encoding LMW hsps (i.e. expressed only at HS temperatures in most cell types/tissues) are expressed at a specific stage of embryogenesis/development in response(s) to some developmental que(s). Additionally, many (most) HS genes are expressed in response to other stress signals (e.g. arsenite, some amino acid analogs, cadmium). HS genes encoding LMW hsps are expressed (at a detectable level) only at a HS temperature in soybean seedlings, except for the pCE54 group (Table 1), and the pCE54 group is recognized to encode general stress proteins which differ substantially from hsps.

The Low Molecular Weight HS Gene Groups of Soybeans

In discussing the different cDNA clone groups of soybean, the term group may be recognized as synonymous with classes and/or subfamilies. The different groups of soybean LMW HS mRNAs/hsps that have been identified to date by my research group are presented in Table 1. The designation as different groups is based on nucleotide sequence analysis of cDNA and/or genomic clones, hybrid-select translation and IEF 2-D gel analysis of the translation products, and in some cases cellular localization of the hsps. This grouping most likely is incomplete since we do not believe that clones have been identified to all of the LMW hsps of soybean. While each group of Table 1 will be discussed below, it should be noted that the group represented by cDNA clone, pCE54 (Czarnecka et al., 1984) and genomic clone, Gmhsp26A (Czarnecka et al., 1988) probably should not be included as hsps, even though some members of this group are induced several fold by HS temperatures; rather, as discussed below, these are general stress proteins (Nagao and Key, 1989; Vierling, 1991), responding to essentially every stress insult to which soybean seedlings have been introduced (Czarnecka et al., 1984; Edelman et al., 1988)! Four of these groups fall into the Class I, Class II, chloroplast-localized (Class III), and endomembrane-localized (Class IV) groups defined by Vierling (1991), represented by cDNA/genomic clones pCE53/Gmhsp17.5E, pCE75/Gmhsp17.9D, cDNA clone pEV3, and pFS2033/Gmhsp22, respectively, of Table 1. The predicted amino acid sequences of a representative of the Class I to IV of the hsps of Table 1 are presented in Figure 1. Their sequence identities and similarities are presented in Table 2 along with the sequence relationships for pEV1, pEV2, and pCE54. The pEV1 and pEV2 cDNA clones probably do not fall into any of these classes. However, these probably are representative of other groups of the super-gene family of LMW hsps; these different multigene classes likely have arisen from ancient gene duplications that occurred before the divergence of monocots and dicots (Vierling, 1991).

Table 1. cDNA Clone Groups to Low Molecular Weight Heat Shock Proteins of Soybean

Representative cDNA Clone	Molecular Mass of Encoded hsp(s)	Estimated No. of Members	Cellular Localization
pCE75*	15 to 16 kD	8	Cytoplasm
pCE53	15 to 18 kD	13	Cytoplasm
pEV3	22 and 27 kD	2	Chloroplast
pFS2033	22 to 24 kD	4	Endomembrane(ER)
pEV2	23 to 24 kD	3 - 4	Endomembrane or?
pEV1	21 to 24 kD	4	Cytoplasm?
pCE54**	25 to 26 kD	4 - 5	"Soluble" Cytoplasm

* See text for possible translation products of pCE75 other than 15 to 16 kD hsps.
** Considered not to be a class of hsps, but rather a class of general stress proteins.

The cDNA Clone, pCE75 Group

This is one of the least studied of the identified LMW groups of hsps of soybean. In hybrid select translation analyses with pCE75, at least 8 mRNAs are hybridized which translate into 15 to 16 kD polypeptides having pIs ranging from approximately 5 to 7 on IEF 2-D gels (Key et al., 1985a). DNA sequence analysis of pCE75 indicates an open reading frame with at least five possible translation start sites predicting amino acid sequences of 15.4 to 18.0 kD. By using the consensus rules of a purine at -3 and a guanine at +1 next to the initiator methionine codon, polypeptides of 18.0, 17.3, and 15.7 would be favored. While one might predict preferential utilization of the first consensus initiator codon leading to translation of an 18.0 kD protein, as previously mentioned,

```
               1                                                    50
Gmhsp17.5E    ..........  ..........  ........MS  LIPGFFGGRR  SNVFDPFSLD
    pCE75     ..........  ..........  ..........  ....MDMSED
     pEV3     GDNKDNSVEV  QHVSKGDQGT  AVEKKPRRTA  MDISPFGILD  PWSPMRSMRQ
  Gmhsp22     ........M  RLQQLNLFFL  LLCVAKANGS  LLP.FMDPPI  TLLADLWSDR

               51                                                   100
Gmhsp17.5E    MWDPFKDF.H  VPTSSVSAEN  S.AFVSTRVD  WKETPEAHVF  KADIPGLKKE
    pCE75     AAGENKTYSA  PTRSYVRDAK  AMAATPADV.  .KEYPNSYVF  EIDMPGLKSG
     pEV3     ILDTMDRVFE  DTMTFPGRNI  GGGEIRAPWD  IKDEEHEIRM  RFDMPGLAKE
  Gmhsp22     FPDPFRVLEH  IPFGVDKDEA  SMAMSPARVD  WKETPEGHVI  MLDVPGLKRE

               101                                                  150
Gmhsp17.5E    EVKVQIEDDR  VLQISGERNV  EKEDKNDTWH  RVERSSGKFT  RRFRLPENAK
    pCE75     DIKVQVEDDN  VLLISGERKR  DEEKEGVKYL  RMERRVGKFM  RKFVLPENAN
     pEV3     DVKVSVEDD.  MLVIKGGHKS  EQEHGGDDS.  WSSRTYSSYD  TRLKLPDNCE
  Gmhsp22     EIKVEVEENR  VLRVSGERKK  EEEKKGDHWH  RVERSYGKFW  RQFRLPQNVD

               151                                                  200
Gmhsp17.5E    VNEVKASMEN  GVLTVTVPK.  ..EEVKKPDV  KAIEISG...  ..........
    pCE75     TDAISAVCQD  GVLSVTVQKL  PPPEPKKPRT  IEVKVF....  ..........
     pEV3     KDKVKAELKN  GVLYITIPKT  KVE....RKV  IDVQVQ....  ..........
  Gmhsp22     LDSVKAKLEN  GVLTLTLDKL  SPGKIKGPRV  VSIAGEDHQQ  GNLNNDGAKQ

               201
Gmhsp17.5E    ..
    pCE75     ..
     pEV3     ..
  Gmhsp22     EL
```

Figure 1. Comparative alignment of the deduced amino acid sequence of representatives of four classes of soybean low molecular weight heat shock proteins. Class I (Gmhsp17.5E), Class II (pCE75), Class III (pEV3), and Class IV (Gmhsp22) amino acid identity is bolded and shaded. The conserved LMW HS domain is underlined. The derived amino acid sequences are from Gmhsp17.5E (Czarnecka et al., 1985) and Gmhsp22.0 (Helm et al., 1993).

hybrid select translation analysis with pCE75 suggested synthesis of polypeptides in the 15 to 16 kD range (see Key et al., 1985a).

The predicted polypeptide(s) from the nucleotide sequence of pCE75 (see Fig. 1) has about 90% similarity to the dicot Class II hsps (Vierling, 1991); accordingly, the pCE75 group should be considered as Class II hsps, since much lower homologies are noted between the pCE75 polypeptide(s) and the Class I, III, or IV hsps. The soybean genomic clone, Gmhsp17.9D, which was characterized by Rashcke et al. (1988), is apparently the only Class II genomic clone sequenced

Table 2. Amino Acid Sequence Relationship of Low Molecular Weight Heat Shock Proteins of Soybean[a]

Percent amino acid identity (similarity) to:[b]

	Class I				Class II		Class III	Class IV		??	General Stress
	Gmhsp17.5E	Gmhsp17.5M	Gmhsp17.6L	Gmhsp18.5	Gmhsp17.9D	pCE75	pEV3	Gmhsp22	pEV6	pEV2	Gmhsp26A
Class I											
Gmhsp17.5E	—										
Gmhsp17.5M	92.2 (96.1)	—									
Gmhsp17.6L	91.0 (96.1)	90.9 (95.4)	—								
Gmhsp18.5	91.6 (93.5)	92.2 (93.5)	87.1 (91.6)	—							
Class II											
Gmhsp17.9D	38.4 (56.3)	33.1 (57.9)	34.9 (57.9)	35.9 (57.7)	—						
pCE75cDNA	43.8 (59.1)	44.8 (60.3)	40.9 (58.4)	44.5 (61.3)	92.1 (97.1)	—					
Class III - Chloroplast											
pEV3cDNA	30.7 (52.0)	30.9 (52.3)	34.5 (56.6)	30.7 (53.6)	29.0 (57.4)	29.4 (57.4)	—				
Class IV - Endomembrane											
Gmhsp22	50.0 (69.5)	46.7 (66.7)	48.4 (68.6)	43.2 (65.4)	34.0 (62.3)	40.0 (62.8)	27.9 (59.4)	—			
pEV6cDNA	43.2 (68.4)	43.3 (66.0)	43.6 (65.8)	41.4 (63.6)	33.3 (59.1)	36.4 (59.3)	29.6 (56.8)	72.9 (85.4)	—		
pEV2cDNA ??	47.7 (68.9)	44.2 (65.8)	47.0 (68.2)	44.9 (66.7)	35.8 (56.6)	40.7 (58.6)	34.4 (56.4)	44.1 (67.7)	42.4 (62.3)	—	
General Stress from Heat Shock											
Gmhsp26A	25.0 (41.9)	16.3 (41.8)	23.7 (46.6)	18.4 (42.8)	18.9 (49.7)	18.6 (42.8)	16.0 (37.4)	14.0 (42.2)	20.1 (45.0)	17.0 (46.3)	—

[a] Sequences are as follows: Gmhsp17.5E (Czarnecka et al., 1985); Gmhsp17.5M and Gmhsp17.6L (Nagao et al., 1985); Gmhsp18.5 and Gmhsp17.9D (Raschke et al., 1988); Gmhsp26A (Czarnecka et al., 1988) and pCE75, pEV2, pEV3 and pEV6 (unpublished data).

[b] Percentage identity and similarity were calculated using the GAP computer program (set with default parameters) of the Wisconsin GCG sequence analysis software.

to date from soybean. The identification of pCE75 to a Class II hsp raises some question about the identity and molecular weight of translation products from mRNA hybrid selected by pCE75. Comparison between the deduced amino acid sequence of Gmhsp17.9D and pCE75 (92.1% identity) show that they are clearly members of the same class; however, a significant question remains whether pCE75 mRNA may translate into polypeptides of 15.7 and 18.0 kD *in vivo* while apparently translating into only the LMW form *in vitro*. We do not know the size(s) of *in vivo* translation products of pCE75 and related sequences; however, the possibility of additional preferential translational or possible transcriptional controls is intriguing. The pCE75-related hsps appear to be cytoplasmic localized, and probably occur in higher molecular weight multimers as has been described for other LMW hsps (see Vierling, 1991; Neumann et al., 1989).

The cDNA Clone, pCE53 Group

This group represents the most complex and most abundant group of hsps of soybean, and also the most studied group. Based on predicted amino acid sequences encoded by the cDNA and genomic clones sequenced to date, the pCE53 group falls into the Class I hsps (Czarnecka et al., 1985; Nagao et al., 1985; Nagao and Key, 1989). One of these sequences (Gmhsp17.5E) is presented in Figure 1. Data presented on the induction and accumulation of HS mRNAs (Fig. 2) and of the encoded hsps (Fig. 3) of this group are representative of most groups/classes of soybean LMW hsps.

28°C 5 10 15 45 60 120 180
Time (min) at 40°C

Figure 2. Northern blot analysis of the time course of Class I low molecular weight heat shock mRNAs. RNA isolated from soybean seedlings incubated at 28°C for one hour and at 40°C for the time indicated was separated on an agarose gel, transferred to nitrocellulose and hybridized to a LMW Class I cDNA clone, pCE53 (from Roberts and Key, 1991).

The mRNAs accumulate very rapidly (Fig. 2) and to very high levels (e.g. up to 20,000 copies or more per cell per transcript, Schöffl and Key, 1982).

The rapid accumulation of HS mRNAs at the onset of the HS treatment relates to activation of transcription of the HS genes based on results of nuclear run-off experiments (Schöffl et al., 1987; Kimpel et al., 1990). As with most soybean HS mRNAs, maximum accumulation of this group of mRNAs occurs at 1 to 2 hr of continuous HS followed by a gradual decline over the next 10 to 12 hr, remaining at detectable levels even at 12 hr. Transcription of these HS genes slows after 1 to 2 hr and is not detectable by 4 hr (Kimpel et al., 1990). This phenomenon has been referred to as self-regulation or autoregulation (DiDomenico et al., 1982; see Gurley and Key, 1991). In contrast to the maintenance of substantial levels of the HS mRNAs and relatively slow decay rates during continuous HS, HS mRNAs decay very rapidly when the HS tissues are returned to non-HS temperatures and are hardly detectable (if at all) after 4 hr at the normal growth temperature (Nagao et al., 1986). Thus, steady-state HS mRNA levels relate to both transcriptional controls and to post-transcriptional stability differentials. These HS mRNAs are rapidly translated into the corresponding hsps (Key et al., 1981), and the hsps accumulate rapidly up to 4 to 6 hr (see Fig. 3; from Hsieh et al., 1992). Even though the HS mRNAs persist at HS temperatures out to 12 hr or more, *in vivo* translation is not detectable after 6 to 8 hr (see Kimpel et al., 1990), consistent with the kinetics of accumulation and decay of the HS mRNAs (Nagao et al., 1986). Normal protein synthesis seems to resume after 6 to 8 hr of continuous HS, in concert with the decline in hsp synthesis (see Kimpel et al., 1990). The hsps are stable for many hr after their synthesis (Lin et al., 1984; Chen et al., 1990; DeRocher et al., 1991; Hsieh et al., 1992), consistent with maintenance of the thermotolerant state (see below) for many hr following HS (Edelman, 1988). The mRNAs corresponding to cDNA clone pCE53 are

123

not detected prior to HS in young soybean seedlings or plants (see Schöffl and Key, 1982; Kimpel and Key, 1985b). Yet, a few members of this group/class appear to be synthesized at some stage(s) during embryogenesis since a small number of these hsps are detected in the dry seed and disappear during the early hours of post-imbibition germination (see Mansfield, 1986; Hernandez and Vierling, 1993).

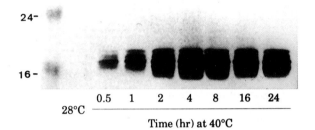

Figure 3.   Immunoblot analysis of the 15 to 18 kD hsps synthesized under continuous HS at 40°C. Protein samples (50 µg) from seedlings treated at 28°C or at 40°C for 0.5 hr, 1 hr, 2 hr, 4 hr, 8 hr, 16 hr, and 24 hr were analyzed by 1-D PAGE and immunodetected by antibody raised from Class I LMW hsp. The 16 and 24 kD MW markers are noted on the left. (from Hsieh et al., 1992).

While this phenomenon has not been studied in great detail in the soybean system, some interesting data are available on the occurrence, disappearance, and onset of synthesis during the early phases of germination of the soybean (Mansfield, 1986). The hsp corresponding to the cDNA clone pFS2019 (encoding probably the most abundant hsp of soybean, and a member of the pCE53 group, Key et al., 1985a) is present at low levels during early germination at 28°C, is not induced at 40°C until 12 to 16 hr post-imbibition, but is induced during the initial 12 hr of germination at 45°C. By 12 to 16 hr, a 45°C treatment is lethal while it was not during the initial 12 hr. Thus, the ability to induce effectively the HS response at the optimum (40°C) HS temperature for soybean seedlings corresponds to the time at which the germinating embryo loses its innate thermotolerant state (Mansfield, 1986). A more detailed analysis of hsp synthesis at non-HS temperatures has been made in alfalfa by Hernandez and Vierling (1993), during alfalfa somatic embryogenesis by Györgyey et al. (1991), and in carrot embryos by Zimmermann et al. (1989). Other examples of members of this group of hsps being synthesized at non-HS temperatures have been reported (e.g. Almoguera and Jordano, 1992; and Schöffl, this volume). The nature of the signal(s) or developmental que(s) responsible for expression of these HS genes at non-HS temperatures in embryos is not understood.

Chloroplast-Localized LMW HSPs

Vierling et al. (1986) demonstrated that certain nuclear-encoded HS mRNAs of a number of species including soybean were translated in the cytoplasm and imported *in vitro* into isolated pea chloroplast. Similar results were obtained by Kloppstech et al. (1985) using pea and *Chlamydomonas*, and the list of species exhibiting this phenomenon has grown (see Vierling, 1991). In the case of soybean, poly(A)RNAs were isolated from photosynthetically active soybean cells growing in liquid culture, and the mRNAs were translated *in vitro* in wheat germ extracts. The translation products were then incubated with pea chloroplasts under conditions that protein import and processing occurred (Vierling et al., 1986). In the case of translation products from control (25°C) cultured cells, the major imported protein was the small subunit of Rubisco. The import of this protein also was evident from the translation products of mRNAs from HS (39°C for 2 hr) cells; however, the major imported products from the HS mRNAs were 22 and 27 kD proteins which were not observed in the control samples. These proteins were processed from higher MW precursors (which migrated very differently on 2-D gels than the processed products) and internalized into the chloroplasts in a protease-resistant form. These hsps appeared to localize primarily in the soluble phase of the chloroplasts. Under HS conditions varying amounts of these LMW hsps associate *in vivo* with the thylakoid membranes depending upon the severity of the HS temperature (see Vierling, 1991). In addition to having the highly conserved domains I (the HS domain) and II, common to other LMW hsps, the chloroplast-localized hsps also have a highly conserved domain III which is unique to chloroplast-localized LMW hsps (Vierling, 1991). These hsps occur in higher molecular mass structures in their native state within the chloroplasts. The HS-induced mRNAs which encode the 22 and 27 kD chloroplast-localized proteins are also induced by HS in etiolated soybean tissues, and their *in vitro* translation products are imported into isolated pea chloroplasts. This is suggestive that at least some function(s) of these hsps in the chloroplast may be common to the proplastids/etioplasts of the nonphotosynthetic tissues. These LMW hsps, along with other high molecular weight hsps (e.g., 60 and 70 kD), undoubtedly provide protection to (i.e. thermoprotection or thermotolerance) various temperature-sensitive functions of the chloroplast during HS and enable recovery from HS after the temperature stress is alleviated. One example of protection from heat/high temperature stress comes from the work of Weis and Berry (1988) where chlorophyll fluorescence (Fo) was measured as a function of temperature in bean leaves that had received a nondamaging 40°C 3-hr pretreatment and those that had not. An increase in Fo is indicative of irreversible heat damage in photosystem II. Schuster et al. (1988) also have presented evidence that the chloroplast-localized hsps prevent damage to the photosystem II reaction center in light during HS. The tolerance limit observed by Weis and Berry (1988) was increased by about 4°C as a result of the 40°C, 3-hr preincubation. The kinetics of the acquisition of this increased tolerance relative to the increase in Fo at the damaging temperature of 46°C are very similar to the kinetics of acquisition of thermotolerance as measured by growth in soybean seedlings by a

40°C treatment prior to imposition of an otherwise lethal treatment of 45°C for 2 hr (Lin et al., 1984). This acquired thermotolerance has generally been attributed to the induction of synthesis and accumulation of hsps (see Lindquist, 1986; Nagao et al., 1986; Lindquist and Craig, 1988; Vierling, 1991), suggestive of a role(s) for the chloroplast-localized hsps in thermoprotection of processes such as that noted above.

Endomembrane/Endoplasmic Reticulum-Localized LMW HSPs

In addition to the localization of LMW hsps to the chloroplast, a feature of obvious uniqueness to plants, cell fractionation studies indicate that some of the LMW hsps may be localized in other plant organelles (e.g. mitochondria, Lin et al., 1984 and unpublished data; endoplasmic reticulum or ER, Cooper and Ho, 1987). While localization in the ER of specific HMW hsps (e.g. a hsp70 and hsp70-related proteins) has been demonstrated in a number of eukaryotic systems (e.g. yeast, human cells), no LMW hsps have been shown to be specifically associated with the ER in any organism until recently. Several features of one group/class (Class IV of Table 2) of LMW hsps were suggestive that they might in fact be ER/endomembrane-localized proteins. Sequence analysis of the genes/cDNAs of some Class IV hsps revealed the presence of an amino terminal sequence similar to known consensus signal peptides; this was true both for pea hsp22.7 and for soybean pFS2033/Gmhsp22. While not identical to the carboxy terminal consensus sequence for the ER retention signal, each of these proteins contains a sequence of sufficient similarity to predict such an activity (i.e. Gmhsp22 and pEV6 have a KQEL terminus whereas the consensus is KDEL). Additionally, the mRNAs for the group (four highly related members) of soybean hsps represented by pFS2033 are translated by ER-associated polyribosomes (unpublished data of LaFayette and Key; Helm et al., 1993). The *in vitro* translation product of the pea hsp22.7, a homolog of Gmhsp22 (sequence presented in Fig. 1), is processed by microsomes and transported into a protease-resistant form (Helm and Vierling, 1990). More recently Helm et al. (1993) demonstrated that the pea hsp22.7 is in fact ER-localized. Sucrose gradient fractionation was used to resolve organellar/membrane fractions and marker enzymes were used to identify the various fractions on the gradients. A pea hsp22.7 antibody was then used to localize this hsp to the ER fraction. This raises interesting questions both about the function(s) of LMW hsps in the ER/endomembrane system and about why these have been identified in plants but not in other eukaryotes.

A Family of General Stress (and HS-Induced) Proteins

As noted in the introduction, the pCE54/Gmhsp26A cDNA/gene is considered not to encode hsps. First, there is exceedingly low amino acid identity between the encoded proteins and those of hsp Classes I to IV (see Table 2); further, the pCE54 group of proteins lacks the typical HS domain (Czarnecka et al., 1988) characteristic of the LMW hsps (Classes I to IV and other related but different HS clones). It is noteworthy that the cDNA clone pCE54 was isolated by differential screening of HS cDNA libraries with cDNAs made to poly(A)RNAs isolated from tissues stressed in a number of ways (HS, water stress, salt stress, heavy metals, ABA-treated, auxin, etc.). A few cDNA clones were identified which showed a marked differential in signal intensity when the library was screened with control cDNA versus the "stress" cDNAs (Czarnecka et al., 1984). Results from Northern blot analyses demonstrated that a wide range of physical stresses such as noted above enhanced accumulation of the mRNA homologous to cDNA clone pCE54 above the constitutive level present in control (nonstressed) soybean hypocotyl. The pCE54 clone hybrid-selected mRNAs which translated into four or five 26 to 27 kD proteins.

In the course of assessing the influence of this wide range of stress agents on the expression of pCE54 mRNA, a number of the HS clones isolated by Schöffl and Key (1982) and by Czarnecka et al. (1984) were hybridized to Northern blots of these stress RNAs (Czarnecka et al., 1984; Edelman et al., 1988). As an example, the pCE53 clone hybridized at very low levels (100-fold or so less than to HS mRNA) to the wide range of stresses which induced pCE54 RNA to HS levels or higher. There were two or three notable exceptions to the generalization that these diverse stress agents did not induce significantly the HS mRNAs; these were arsenite and to a lesser extent cadmium and some amino acid analogs (azetidine, and canavanine, with the former being much more effective). Western blot analysis also demonstrates that the pCE53 hsps accumulate to substantial levels in response to Aze and arsenite, but to lower levels than induced by HS (Fig. 4). Some accumulation is observed in cadmium-treated tissue; canavanine may induce accumulation above the control background level. In a more detailed analysis (Lee, Lin, Nagao, and Key, unpublished), Aze was shown to induce most HS mRNAs to high levels, but generally lower than HS levels; however, Aze does not induce significantly the mRNAs for the pEV3 encoded chloroplast Class III protein or for the pFS2033 encoded ER Class IV protein. Aze induces pCE54 mRNA to levels well above the HS level.

Interestingly, 2,4-D at high concentrations was one of the treatments which induced pCE54 mRNA to high levels while not substantially inducing any of the HS mRNAs (Czarnecka et al., 1984). Since that time, a number of laboratories (Hagen et al, 1988; Droog et al., 1993; Itzhaki and Woodson, 1983; Dominov et al., 1992; Takahashi and Nagata, 1992) have isolated clones to ethylene/auxin/2,4-D-induced poly(A)RNAs which have very high sequence identity to the pCE54 family.

In fact the pGH2 and pGH4 clones isolated by Hagen et al. (1988) have identical sequences to a pCE54-related genomic clone, Gmhsp26A (Czarnecka et al., 1988). Other of these related clones have sequence identities in the range of 50% to Gmhsp26A. The results of these studies based on amino acid sequence identity of

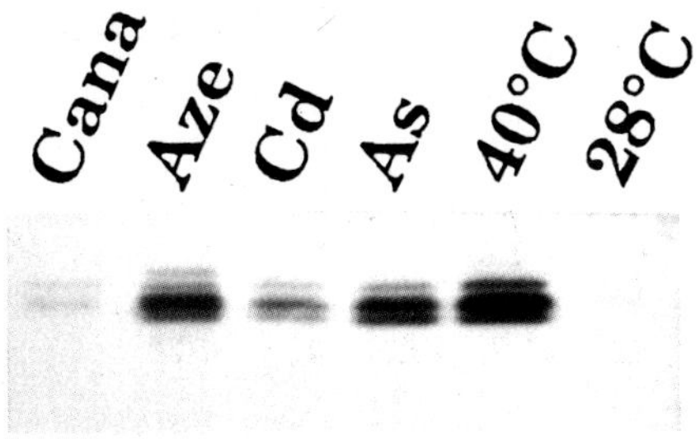

Figure 4. Immunoblot analysis of Class I low molecular weight heat shock proteins induced by various stress agents. Soybean seedlings were incubated for 6 hr at 40°C and 28°C as indicated and for 12 hr at 28°C containing Cana (canavanine, 5.7 mM), Aze (azetidine-2-carboxylic acid, 5 mM) Cd (cadmium chloride, 1 mM) or As (sodium arsenite, 250 μM). Total protein was isolated from the seedlings and 25 μg of protein/lane was separated by 1-D SDS PAGE and immunodetected with antibody raised against a Class I LMW hsp (kindly provided by Dr. C. Y. Lin).

50% or more show that this group or class of sequences likely encodes glutathione S-transferases. In addition to the significant amino acid identity, at least one of the auxin-regulated genes of tobacco encodes a protein which has glutathione S-transferase activity *in vitro* (Droog et al., 1993). Also of interest in this context is the fact that a 25 kD protein has been identified as an auxin-binding protein (ABP) with high homology to some glutathione S-transferases (Bilang et al., 1993).

Since a number of studies are suggestive that other stresses (e.g. water/salt stress) may induce some HS mRNAs/hsps, we have performed recently some experiments to evaluate further whether the complex and abundant group of pCE53-related hsps are induced by water stress using antibodies made to one of the 13 members (pCE53/Gmhsp17.5E) of this group of hsps; this antibody recognizes at least 13 hsps in the 15 to 18 kD group. As presented in Figure 5, water stress generated by mannitol or PEG or treatment with the water stress-related hormone, ABA, does not lead to the accumulation of the pCE53 family of hsps, consistent with the Northern blot data of Czarnecka et al. (1984).

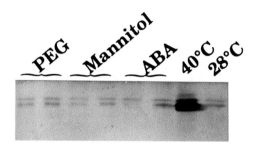

Figure 5. Immunoblot analysis of Class I low molecular weight heat shock proteins induced by various water stress related agents. Soybean seedlings were incubated at 28°C with 26 mM PEG, 0.3 M mannitol, and 1 mM ABA for 12 hr (left lane) and 6 hr (right lane), respectively for each treatment; 40°C and 28°C lanes were incubated for 6 hr. Total protein was isolated from the seedlings and 25 μg of protein/lane was separated by 1-D SDS PAGE and immunodetected with antibody raised against a Class I LMW hsp. Color development for the immunoblot was extended to allow visualization of the low level of reactivity in the non-HS samples.

Also a Western blot of a 2-D gel shows no immunoreactive proteins other than the control background proteins whereas HS induces very high levels of these pCE53 proteins. Other studies done with field-grown and growth chamber-grown soybean plants (Kimpel and Key, 1985b) show that water stress does not cause induction/accumulation of any of the HS mRNAs which were measured by Northern blot analysis using various cDNA probes in addition to the pCE53 group. We do not understand the basis for the failure to observe water stress-induced accumulation of any HS mRNAs in soybeans while some reports indicate water stress-induced accumulation of some HS mRNAs and/or hsps in other species.

Other Groups of LMW hsps

Two groups of cDNA clones, pEV1 and pEV2, listed in Table 1, cannot yet be assigned to a defined class of LMW hsps. The pEV2 clone is listed in Table 2 as a possible member of the endomembrane Class IV hsps. The pEV2 sequence shares in common with pEV6 and pFS2033 translation of the respective mRNAs on ER-associated ribosomes, a distinct leader sequence at the amino terminus, and processing by dog pancreas microsomes (unpublished data of LaFayette, Nagao, and Key). However, the amino acid identity/similarity of pEV2 is no more related to that of pEV6 and pFS2033/Gmhsp22 than to other classes of the soybean LMW hsp sequences (Table 2) or those of other species summarized by Vierling (1991). The pEV2 sequence also does not have a carboxy terminal sequence related to the consensus ER retention signal as do the pEV6- and pFS2033-encoded proteins. A distinct group of mRNAs are hybrid selected by pEV2 relative to those of other soybean cDNA clones identified to date. Thus, if pEV2 encodes an endomembrane protein, it likely is not ER-retained. While its cellular localization remains to be determined, the divergence in sequence from other defined classes is such that pEV2 may be representative of a different class of hsps with unique cellular localization.

The pEV1 cDNA clone hybrid selects mRNAs for a distinct group of 22 to 24 kD hsps (Table 1 and unpublished data). Sequence analysis of this cDNA clone has not been completed, so its relatedness to the other groups/classes has not been determined. Partial sequence analysis indicates that the pEV1 encoded protein may have about 50% amino acid identity to pEV3 (Class III) with much lower identity to the other classes (groups) noted in Table 1 and Figure 1. No information on cellular localization of pEV1 hsps is available. However, we speculate that this will represent a new class of hsps. We believe that a group of 22 to 24 kD hsps are mitochondrial-localized in soybean seedlings. During HS a large number of hsps associate with the mitochondrial fraction obtained by sucrose gradient fractionation of seedling homogenates (Lin et al., 1984); most of these hsps (Class I and Class II) "dissociate" during a chase at non-HS temperatures, and the nature of their association renders them sensitive to protease digestion. However, hsps in the 22 to 24 kD range which become mitochondrial-associated during HS do not chase from the mitochondrial fraction during subsequent incubation (Lin et al., 1984), and they are associated with the mitochondria in a protease-resistant manner (unpublished data of Ainley and Key). Taken together, these data are suggestive of internalization of this group of hsps within the mitochondria. No data on uptake of these hsps into mitochondria *in vitro* are available to confirm the suggestive evidence presented above for specific localization of a group of LMW hsps in the mitochondrion. The only clone group which has been isolated to date and which has low abundance and encodes proteins with size properties which would make the hsps candidates for mitochondrial localization would be the pEV1 group. No data are available to support or disprove this possibility. Sequence analyses currently underway should provide information on the potential for import of the pEV1 proteins into mitochondria (i.e. a mitochondrial targeting sequence at the amino terminus, see Schatz, 1993). If this proves negative, we must conclude that there is at least one (or more) group/class of LMW hsp cDNA clones yet to be isolated/identified.

Summary Comments

The information provided above serves to establish the complexity of the HS response generally and specifically in soybeans. A large number of HS genes encode LMW (15 to 27 kD) hsps in soybean. They are related at the sequence level sufficiently to assume that these many genes arose from some common primordial gene; yet they are sufficiently diverged to be logically categorized into different classes or subfamilies of a large supergene family (Vierling, 1991).

Specific cellular localization correlates at least in some cases with class identity.

There are many potentially important and interesting questions yet to be fully resolved about the HS response, e.g. perception and "transduction" of the high temperature signal including temperature sensing, self-regulation/autoregulation of the HS response, identifying other groups/classes of HS genes, the signals or ques which cause a few otherwise HS-responsive genes to be expressed at some specific embryogenic/developmental stage at non-HS temperatures, function(s) of the LMW hsps, and the list goes on. It is generally accepted that hsps provide thermotolerance or thermoprotection to otherwise lethal temperatures. Each organism has an "optimal" temperature to elicit a full HS response; this temperature does not significantly impair the organism long term but does lead to activation of transcription of the HS genes, translation of the HS mRNAs, and accumulation of the hsps. If on the other hand, the organism is exposed to an abrupt temperature change of about 5°C to 7°C above the "optimum" HS temperature, the treatment if for more than a few minutes typically is lethal; the HS system is not activated and sustained sufficiently to provide any protection to the organism at the supra-optimal or lethal HS temperature. A wide range of time/temperature regimes cause a full HS response and provide thermotolerance in soybean to a subsequent lethal HS treatment (e.g. 45°C for 1 to 2 hr); for example in soybean seedlings (Lin et al., 1984; Key et al., 1985; Nagao et al., 1986), (1) a 40°C, 2 hr HS treatment ("optimum" or near optimum for soybean), (2) a 40°C, 1 or 2 hr HS treatment followed by 4 hr or more at the normal growing temperature (28°C to 30°C), (3) a brief (5 to 10 min) 45°C treatment followed by 1 to 2 hr or more at 30°C, (4) a gradual increase in temperature (e.g. 3°C per hr) from 30°C up to about 48°C, or (5) induction of a near typical "HS response" by arsenite treatment also leads to a thermotolerant state, recognizing that there are deleterious effects caused by arsenite. All of these treatments, and many variations thereof, have in common the synthesis and accumulation of a full complement of HS mRNAs and hsps. The induction of a HS-like response by amino acid analogs, on the other hand, does not lead to the acquisition of thermotolerance as a result of accumulation of abnormal, analog-containing hsps. A question of importance in this conference was the relationship if any between short-term HS-induced acquired thermotolerance and long-term high temperature adaptation. The ability of plants to undergo the short-term thermoadaptation may well contribute to the ability of the plants to undergo the long-term adaptation. But it seems safe to speculate that there are many differences at the physiological and biochemical level. The work of Berry's group (e.g. Weiss and Berry, 1988) and others address in part this question.

As to function of the LMW hsps, recent evidence indicates that these hsps interact with normal cellular proteins to prevent their denaturation and subsequent aggregation as well as to renature denatured proteins (covered in more detail by Lin in this volume). In fact Lin's laboratory (Jinn et al., 1989) was the first to show that the addition of an enriched fraction of LMW hsps to an extract of normal control proteins provided protection to thermal denaturation and aggregation of about 50% of the normal proteins over the appropriate time/temperature range. While this work has been extended and reported in this volume, other workers have shown that purified LMW hsps and specific proteins (various enzymes such as elastase, citrate synthetase, and α-glucosidase) interact in such a way as to protect these proteins from heat inactivation or denaturation in an ATP-independent fashion (e.g. Jakob et al., 1993; Merck et al., 1993). It has been known for many years that the LMW hsps shared a region of high sequence relatedness and an overall similar secondary structure to α-crystalline. They also share a greater thermostability to high temperature than do most normal cellular proteins. α-crystalline and the LMW hsps appear to function as molecular chaperons, by suppressing heat-induced aggregation of other proteins. They function also in refolding of denatured (e.g. urea) proteins again in an apparent ATP-independent fashion. These observations along with many earlier reports on the HMW hsps (e.g. hsp60, 70, 90, see Welch, 1991) are suggestive that most or all hsps function in preventing denaturation/aggregation of normal proteins during HS and even in renaturing/refolding of denatured/partially denatured proteins during recovery from HS. This remains a most fertile area for further research.

LITERATURE CITED

Almoguera C, Jordano J (1992) Developmental and environmental concurrent expression of sunflower dry-seed-stored low-molecular-weight heat-shock protein and Lea mRNAs. Plant Mol Biol 19:781-792

Bilang J, Macdonald H, King PJ, Sturm A (1993) A soluble auxin-binding protein from *Hyoscyamus muticus* is a glutathione S-transferase. Plant Physiol 102:29-34

Chen Q, Lauzon L, DeRocher A, Vierling E (1990) Accumulation, stability, and localization of a major chloroplast heat shock protein. J Cell Biol 110:1873-1883

Cooper P, HO T-HD (1987) Intracellular localization of heat shock proteins in maize. Plant Physiol 84:1197-1203

Craig EA (1985) The heat shock response. CRC Crit Rev Biochem 18:239-280

Czarnecka E, Edelman L, Schöffl F, Key JL (1984) Comparative analysis of physical stress responses in soybean seedlings using cloned heat shock cDNAs. Plant Mol Biol 3:45-58

Czarnecka E, Gurley WB, Nagao RT, Mosquera LA, Key JL (1985) DNA sequence and transcript mapping of a soybean gene encoding a small heat shock protein. Proc Natl Acad Sci USA 82:3726-3730

Czarnecka E, Nagao RT, Key JL, Gurley WB (1988) Characterization of *Gmhsp26-A*, a stress gene encoding a divergent heat shock protein of soybean: heavy-metal-induced inhibition of intron processing. Mol Cell Biol 8:1113-1122

DeRocher AE, Helm KW, Lauzon LM, Vierling E (1991) Expression of a conserved family of cytoplasmic low molecular weight heat shock proteins during heat stress and recovery. Plant Physiol 96:1038-1047

DiDomenico BJ, Bugaisky GE, Lindquist W (1982) Heat shock and recovery are mediated by different translations mechanisms. Proc Natl Acad Sci USA 79:6181-6185

Dominov JA, Stenzler L, Lee S, Schwarz JJ, Leisner S, Howell SH (1992) Cytokinins and auxins control the expression of a gene in *Nicotiana plumbaginifolia* cells by feedback regulation. Plant Cell 4:451-461

Droog FNJ, Hooykaas PJJ, Libbenga KR, van der Zaal EJ (1993) Proteins encoded by an auxin-regulated gene family of tobacco share limited but significant homology with glutathione S-transferases and one member indeed shows *in vitro* GST activity. Plant Mol Biol 21:965-972

Edelman L (1988) The stress responses in soybean to heat shock, arsenite, and cadmium treatments. Ph.D. dissertation. University of Georgia, Athens, Georgia.

Edelman L, Czarnecka E, Key JL (1988) Induction and accumulation of heat shock-specific poly(A+)RNAs and proteins in soybean seedlings during arsenite and cadmium treatments. Plant Physiol 86:1048-1056

Györgyey J, Gartner A, Németh K, Magyar A, Hirt H, Heberle-Bors E, Dudits D (1991) Alfalfa heat shock genes are differentially expressed during somatic embryogenesis. Plant Mol Biol 16:999-1007

Gurley WB, Key JL (1991) Transcriptional regulation of the heat shock response: a plant perspective. Biochemistry 30:1-12

Hagen G, Uhrhammer N, Guilfoyle TJ (1988) Regulation of expression of an auxin-induced soybean sequence by cadmium. J Biol Chem 263:6443-6446

Helm KW, Vierling E (1990) A member of the eukaryotic superfamily of small heat shock proteins is located in the endomembrane system of *Pisum sativum*. J Cell Biol 111:69a

Helm KW, LaFayette PR, Nagao RT, Key JL, Vierling E (1993) Localization of small heat shock proteins to the higher plant endomembrane system. Mol Cell Biol 13:238-247

Hernandez LD, Vierling E (1993) Expression of low molecular weight heat-shock proteins under field conditions. Plant Physiol 101:1209-1216

Hsieh M-H, Chen J-T, Jinn T-L, Chen Y-M, Lin C-Y (1992) A class of soybean low molecular weight heat shock proteins. Plant Physiol 99:1279-1284

Itzhaki H, Woodson WR (1993) Characterization of an ethylene-responsive glutathione S-transferase gene cluster in carnation. Plant Mol Biol 22:43-58

Jakob U, Gaestel M, Engel K, Buchner J (1993) Small heat shock proteins are molecular chaperones. J Biol Chem 268:1517-1520

Jinn T-L, Yeh Y-C, Chen Y-M, Lin C-Y (1989) Stabilization of soluble proteins *in vitro* by heat shock proteins-enriched ammonium sulfate fraction from soybean seedlings. Plant Cell Physiol 30(4):463-469

Key JL, Lin C-Y, Chen Y-M (1981) Heat shock proteins of higher plants. Proc Natl Acad Sci USA 78:3526-3530

Key JL, Gurley WB, Nagao RT, Czarnecka E, Mansfield MA (1985a) Multigene families of soybean heat shock proteins. *In* L van Vloten-Doting, GSP Grout, TC Hall, eds, Molecular Form and Function of the Plant Genome, NATO ASI Series, Series A: Life Sciences, Vol 83, Plenum Press, pp 81-100

Key JL, Kimpel J, Vierling E, Lin C-Y, Nagao RT (1985b) Physiological and molecular analyses of the heat shock response in plants. *In* BG Atkinson, DB Walden, eds, Changes in Eukaryotic Gene Expression in Response to Environmental Stress, Academic Press, Inc., New York, pp 327-348

Key JL, Kimpel JA, Nagao RT (1987a) Heat shock gene families of soybean and the regulation of their expression. UCLA Symp Mol Cell Biol **62:**87-97

Key JL, Nagao RT, Czarnecka E, Gurley WB (1987b) Heat stress: expression and structure of heat shock protein genes. *In* D von Wettstein and NH Chua, eds, Plant Molecular Biology, Plenum Press, New York, pp 87-97

Kimpel JA, Key JL (1985a) Heat shock in plants. Trends Biochem. Sci. **10:**353-357

Kimpel JA, Key JL (1985b) Presence of heat shock mRNAs in field grown soybeans. Plant Physiol **79:**672-678

Kimpel JA, Nagao RT, Goekjian V, Key JL (1990) Regulation of the heat shock response in soybean seedlings. Plant Physiol **94:**988-995

Kloppstech K, Meyer G, Schuster G, Ohad I (1985) Synthesis, transport and localization of a nuclear coded 22-kd heat-shock protein in the chloroplast membranes of peas and *Chlamydomonas reinhardi*. EMBO J **4:**1902-1909

Lin C-Y, Roberts JK, Key JL (1984) Acquisition of thermotolerance in soybean seedlings: synthesis and accumulation of heat shock proteins and their cellular localization. Plant Phsyiol **74:**152-157

Lindquist S (1986) The heat shock response. Annu Rev Biochem **55:**1151-1191

Lindquist S, Craig EA (1988) The heat shock proteins. Annu Rev Genet **22:**631-677

Mansfield M (1986) The heat shock response of soybean: synthesis, accumulation and distribution of the low molecular weight heat shock proteins. Ph.D. dissertation. University of Georgia. Athens, Georgia.

Merck KB, Groenen PJTA, Voorter CEM, de Haard-Hoekman WA, Horwitz J, Bloemendal H, de Jong WW (1993) Structural and functional similarities of bovine α-crystallin and mouse small heat-shock protein. J Biol Chem **268**:1046-1052

Nagao RT, Czarnecka E, Gurley WB, Key JL (1985) Genes for low-molecular-weight heat shock proteins of soybeans: sequence analysis of a multi-gene family. Mol Cell Biol **5**:3417-3428

Nagao RT, Kimpel JA, Vierling E, Key JL (1986) The heat shock response: a comparative analysis. *In* BJ Miflin, ed, Oxford Surveys of Plant Molecular & Cell Biology, Vol 3, Oxford Univ. Press, Oxford, pp 384-438

Nagao RT, Key JL (1989) Heat shock protein genes of plants. *In* I Vasil, J Schell, eds, Cell Culture and Somatic Cell Genetics of Plants, Vol 6, Academic Press, Inc, New York, pp 297-328

Nagao RT, Kimpel JA, Key JL (1990) Molecular and cellular biology of the heat-shock response. *In* J Scandalious, ed, Genomic Responses to Environmental Stresses, Academic Press, Inc., New York, pp 235-274

Neumann D, Nover L, Parthier B., Reiger R, Scharf K-D (1989) Heat shock and other stress response systems of plants. Biol Zentralbl **108**: 1-156

Raschke E, Baumann G, Schöffl F (1988) Nucleotide sequence analysis of soybean small heat shock genes belonging to two different multigene families. J Mol Biol **199**:549-557

Roberts JK, Key JL (1991) Isolation and characterization of a soybean hsp70 gene. Plant Mol Biol **16**:671-683

Schatz G (1993) The protein import machinery of mitochondria. Protein Science **2**:141-146

Schöffl F, Key JL (1982) An analysis of mRNAs for a group of heat shock proteins of soybean using cloned cDNAs. J Mol Appl Genet 1:301-314

Schöffl F, Rossol I, Angermueller S (1987) Regulation of the transcription of heat shock genes in nuclei from soybean (*Glycine max*) seedlings. Plant Cell Envir 10:113-119

Schuster G, Even D, Kloppstech K, Ohad I (1988) Evidence for protection by heat-shock proteins against photoinhibition during heat-shock. EMBO J 7:1-6

Takahashi Y, Nagata T (1992) *parB*: An auxin-regulated gene encoding glutathione S-transferase. Proc Natl Acad Sci USA 89:56-59

Vierling E (1991) The roles of heat shock proteins in plants. Annu Rev Plant Physiol Plant Mol Biol 42:579-620

Vierling E, Mishkind ML, Schmidt GW, Key JL (1986) Specific heat shock proteins are transported into chloroplasts. Proc Natl Acad Sci USA 83:361-365

Weis E, Berry JA (1988) Plants and high temperature stress. *In* SP Long, FI Woodward, eds, Plants and Temperature, Symposia of the Society for Experimental Biology, Vol 42, The Company of Biologists Limited, Cambridge, pp 329-346

Welch WJ (1991) The role of heat-shock proteins as molecular chaperones. Current Opinion in Cell Biol 3:1033-1038

Zimmermann JL, Apuya N, Darwish K, O'Carroll C (1989) Novel regulation of heat shock genes during carrot somatic embryo development. Plant Cell 1:37-46

Class I low molecular weight heat shock proteins in plants: immunological study and thermoprotection against heat denaturation of soluble proteins. [1]

Chu-Yung Lin, Tsung-Luo Jinn, Ming-Hsiun Hsieh and Yih-Ming Chen

Department of Botany, National Taiwan University, Taipei, Taiwan, R.O.C.

ABSTRACT

Antibodies prepared against two major polypeptides of the 15- to 18-kD class of soybean heat shock proteins (HSPs) individually reacted with its antigen and cross-reacted with 12 other 15- to 18-kD HSPs. We also found that this antibody preparation cross-reacted with the same class low molecular weight (LMW) HSPs of mung bean, rice and other seven plant species tested based on western blot analysis. The 70 to 100% ammonium sulfate (AS) fraction from heat shocked seedlings of mung bean and rice, as in soybean, contained a high percentage of all the HSPs. The proteins in this fraction were resistant to heat denaturation, as judged by their unpelletability after heat treatment. Moreover, this fraction showed a significant ability to protect the soluble proteins from heat denaturation. The HSPs-enriched fractions prepared from mung bean and rice heat shocked seedlings were able to thermostabilize the homologous soluble proteins. Additionally, the HSPs-enriched fractions were exchangeable among these three plant species for thermostabilization.

In soybean, when the HSPs-enriched fraction depleted of the 15- to 18-kD class HSPs, the effectiveness in prevention of heat denaturation was lost. However, when the 15- to 18-kD HSPs recovered in the AS fraction, the thermostabilization was restored again. This suggests that the HSPs of 15- to 18-kD class are important for providing the protection from heat denaturation.

[1] Supported by National Science Council of Taiwan under grants NSC 78-0211-B002-02, NSC 79-0211-B002-02 and NSC 80-0211-B002-02.

NATO ASI Series, Vol. H 86
Biochemical and Cellular Mechanisms
of Stress Tolerance in Plants
Edited by J. H. Cherry
© Springer-Verlag Berlin Heidelberg 1994

INTRODUCTION

The induction of HSP synthesis in response to thermostress occurs in wide range of organisms from bacteria to humans (18). When soybean seedlings are shifted from a normal growth temperature of 28°C to 40°C, a new set of proteins, the HSPs, is rapidly synthesized (1, 5).

While the physiological function of HSPs has not been clearly documented, their accumulation is correlated well with the thermotolerance. Based on this correlation it has been hyposethized that accumulation of HSPs is an essential component of the protection process from heat damage (8, 9, 14, 17).The kinetics of localization of the HSPs during HS and recovery from HS in soybean seedlings (9) are also consistent with the notion that this dynamic localization process is important for thermoprotection.

Although studies of *Drosophila* have provided a basis for research in many systems, there are significant differences among organisms in response to HS. A major difference is that relative abundance of proteins synthesized during HS. Plant synthesize HSPs with molecular masses of 92, 84, 70,68, and 15 to 27 kD, the complex and abundant proteins (6, 22) compared to *Drosophila* (10) which synthesize the 70-kD HSP. The 15- to 18-kD class of HSPs synthesized during the shift to HS temperature of 40°C in soybean seedlings was shown to accumulate to a detectable level after SDS-PAGE separation followed by Coomassie blue staining (4, 13). They were shown to protect *in vitro* oxidative phosphorylation of mitochondria at HS temperature of 42.5°C (2) and the protection of soluble proteins from heat denaturation *in vitro* occurred with the addition of the HSP-enriched aliquot which was predominantly composed of the 15- to 18-kD class HSPs (4).

The experiments of thermostabilizing effect by HSPs conducted in the previous studies (4) was based on the model proposed by Minton et al. (15) for the biological role of HSPs. In this study we used the soybean antibody preparation to test the immunological kinship among the plant species and also to test the thermostabilization effect of HSPs and their exchangeability among the plant species using HSPs-enriched AS fraction. We also show in this study the concrete evidence in soybean system that the 15- to 18-kD class HSPs contribute most to provide thermostability.

MATERIALS AND METHODS

Plant materials:

The following plant species: soybean (*Glycine max* cv. Taita Kaohsiung No.8), mung bean (*Vigna radiata* L.) and rice (*Oryza sativa* L. cv. Tainong No.67) were used in this study. All seeds were surface sterilized in 10% Clorox for 10 min, rinsed thoroughly in water, and then germinated in a roll of moist paper towel at 28°C dark growth chamber (9).

Extraction of ³H-labeled proteins:

Two-day-old seedlings of soybean and mung bean without cotyledons and five-day-old rice seedlings without endosperm were incubated for 3 h in 10 ml incubation buffer [5 mM potassium phosphate buffer, pH 6.0, containing 1% sucrose, 50 μg/ml chloramphenicol and L-[4,5-³H] Leucine (New England Nuclear, Boston, MA, USA) at a concentration of 20 μCi/ml] with gentle shaking. Labeled seedlings were processed as described previously for preparation of soluble ³H-labeled proteins (9).

Preparation of postribosomal supernatant (PRS):

Seedlings were homogenized with a polytron, in 0.2 M Tris-HCl buffer, pH 8.8, containing 0.5 M sucrose, 0.1 M KCl, 30 mM MgCl₂, 1mM DTT, and 1mM PMSF. The homogenate was filtered through a layer of Miracloth and centrifuged for 15 min at 23.000 x g. Ribosomes were pelleted as described previously (9) but without the 1.7 M sucrose cushion.

AS fractionation:

PRS prepared from heat-shocked seedlings were fractionated in 0 to 50 %, 50 to 70 %, and 70 to 100 % saturation by AS. The precipitates from each fraction were pelleted, dissolved in 50 mM Tris-HCl buffer, pH 8.8, containing 1 mM EDTA and 0.1 % 2-mercaptoethanol, and dialyzed overnight against the same buffer as described previously (4). All experimental were carried out at 4°C.

Measurement of ¹H-labeled proteins and quantitative estimation of proteins:

For measurement of ¹H-labeled proteins, a sample aliquot was blotted on a 3 MM filter paper and processed as described by Mans and Noveli (12). Proteins were measured according to the method of Lowry et al. (11).

Gel electrophoresis and flurography of in vivo synthesized proteins:

Labeled proteins were extracted with SDS extraction buffer at room temperature as described previously (4). One dimensional gel electrophoresis was performed according to Laemmli (7) using 12.5% acrylamide gels. Fluorography of the gels was accomplished using ENHANCER (New England Nuclear, Boston, MA, USA) and Kodak film (XAR-5).

Antibodies:

Preparation of the antibody against a soybean LMW HSP has been described (3). The antibody against the 16.9 kD HSP from rice was prepared using protein synthesized from a pGEX-2T expression vector (23) carrying the cDNA pTS1 (21).

Immunoblotting:

For immunoblotting, proteins were transferred from SDS gels to Immobilon PVDF Transfer Membranes (Millipore) with glycine electrode buffer according to Towbin et al. (20). Protein bands cross-reacted with the LMW HSPs antibodies were identified by reaction with alkaline phosphatase conjugated to goat anti-rabbit IgG. Bound antibodies were visualized by reaction with BCIP and NBT according to the manufacturer's specification.

Assay for thermal denaturation of normal soluble proteins:

After addition of HSPs-enriched fraction (prepared from 70 to 100% AS saturation) to ¹H--labeled normal soluble proteins, the mixture was heated at 55°C for 30 min with shaking. The denatured proteins were pelleted at 16,000 x g for 15 min and the radioactivity in the pellet was measured after suspension in the Laemmli's sample buffer (7).

Localization and delocalization of HSPs during HS and recovery from HS:

Three samples of soybean seedlings were incubated at 40°C for 3 h following incubation at 28°C for 4 h. Two of them were subsequently shifted to 45°C for 30 min and one of which was transferred to 28°C for 4 h. After each treatment, seedlings were processed for preparation of postribosomal supernatant. Treatment of 40°C (3 h) → 28°C (4 h) → 45°C (30 min), HSPs associated with cellular components occurred but 40°C (3 h) → 28°C (4 h) → 45°C (30 min) → 28°C (4 h), HSPs delocalized from cellular components and return to cytosol (9).

RESULTS AND DISCUSSIONS

Antibodies preparation and immunoreaction:

The HSPs induced and accumulated by 40°C HS treatment of soybean seedlings were enriched in the 65 to 100% AS fraction prepared from the PRS (4). After 2-D PAGE, the resolved proteins were stained with Coomassie blue. The 15- to 18-kD HSPs were shown to accumulate to a stainable level (Fig. 1A). The most densely stained spots (shown with thick arrows plus asterisks) were cut out individually from the 2-D gels, collected from 30 gels, and used for preparation of polyclonal antibodies. This antibody preparation cross-reacted with at least 13 polypeptides of the 15- to 18-kD LMW HSPs (class I LMW HSPs) group (Fig. 1B), but neither high molecular weight HSPs nor 20- to 30- kD LMW HSPs were cross-reacted.

This soybean antibody showed a similar immunoreaction pattern to the same class of LMW HSPs from heat shocked seedlings of mung bean (Fig. 2, lane 4) and rice (Fig. 2, lane 6). Rice antibody prepared against the 16.9 kD HSP generated by fusion of cDNA pTS1 to pGEX-2T expression vector also give similar immunoreaction patterns as the soybean antibody (unpublished data). The cross-reactivity of the soybean antibody with the similar class of LMW HSPs of mung bean and rice is shown in Fig. 2. Tests with other plant species, i.e. pea, cucumber, tobacco, *Arabidopsis*, maize, wheat and barley (data not shown) all indicate that the 15- to 18- kD class HSPs among these plant species share the same

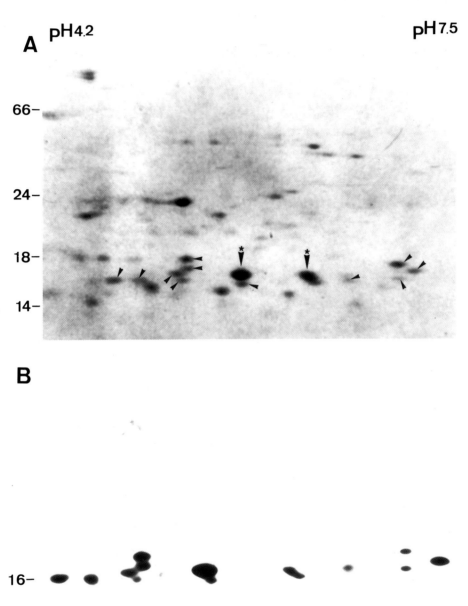

Figure 1. Coomassie-blue staining of proteins from 40°C heat shocked soybean seedlings (A). LMW HSPs indicated with a large thick arrow + asterisk were used individually for antiserum preparation (B). Western blot analysis of (A). Cross-reacted polypeptides in (B) are equivalent to the Coomassie-blue stained spots indicated by arrows in (A).

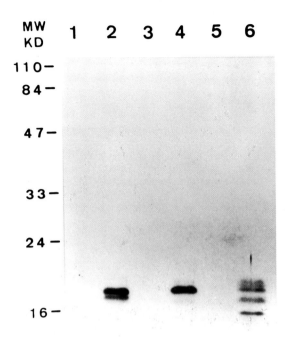

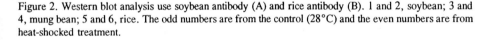

Figure 2. Western blot analysis use soybean antibody (A) and rice antibody (B). 1 and 2, soybean; 3 and 4, mung bean; 5 and 6, rice. The odd numbers are from the control (28°C) and the even numbers are from heat-shocked treatment.

antigenicity but not with other class of LMW HSPs of plants and also of *Drosophila* (data not shown). The antibody preparation against 70 kD from chick cells was shown to cross-react with the 70 kD HSPs from all organisms: yeast, *Dictyostelium*, *Drosophila*, human cell lines and higher plants (19). The same antigenicity of 70 kD HSPs but not in the class of 15- to 18- kD HSPs existed between *Drosophila* and plants indicate the uniqueness of 15- to 18- kD HSPs may play different physiological functions compared to 70 kD HSPs in plants.

AS fractionation of PRS:

Three different AS concentrations were used to separate soluble proteins of PRS from heat shocked seedlings of mung bean and rice into three fractions, i.e. 0 to 50%, 50 to 70% and 70 to 100% saturation. Fig.3A and Fig. 3B show fluorograms of each fraction after SDS-PAGE from mung bean and rice respectively. Separation on SDS-PAGE revealed that the HSPs were enriched in the 70 to 100% saturation, as we reported in soybean (4). In rice, however, some of the HMW HSPs were also found in other AS fractions (Fig. 3B, lanes 2 & 4). Among different plant species, HSPs were all enriched in 70 to 100% AS saturation. This fraction was used for the assay of protection effect from heat denaturation of soluble proteins.

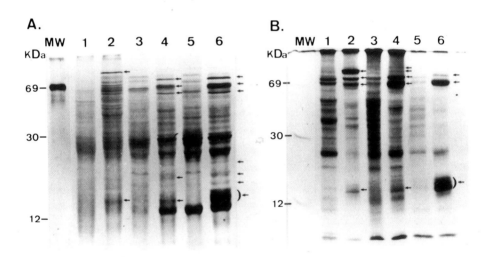

Figure 3. Fluorograms of SDS-PAGE analysis of proteins from mung bean (A) and rice (B). After AS fractionation. Lane 1 and 2, 0 to 50% AS saturation; 3 and 4, 50 to 70% AS saturation; 5 and 6, 70 to 100% AS saturation. 1, 3 and 5 are proteins from 28°C and 2, 4 and 6 are proteins from 40°C. Arrows indicate HSPs. The MW markers are shown in the left lane of lane 1.

Thermostabilization of soluble proteins against heat denaturation by addition of HSPs-enriched fraction from heat shocked seedlings of soybean, mung bean and rice:

Table 1 shows that HSPs-enriched fractions obtained from each plant species was able to provide protection to its own soluble proteins from heat denaturation at 55°C by nearly 50% in the homologous system. It is also noteworthy that the source of HSPs-enriched fraction is exchangeable for thermostabilizing soluble proteins among three plant species: soybean, mung bean and rice. In these experiments, the HSPs-enriched fraction from rice was very effective to protect its own soluble proteins from heat denaturation; however, it provided a relatively low level of protection (69% protein denatured) to the soluble proteins of soybean. On the other hand, HSPs-enriched fraction from soybean provided a much high protection to the rice soluble proteins (compared to the protection normally obtained by its own HSPs-enriched preparation). This probably reflect the actual amount of HSPs in each preparation of HSPs-enriched fraction due to a different level of accumulation of HSPs rather than the difference in functional activity of HSP molecules. According to our experience, the 15- to 18- kD HSPs in soybean could be easily accumulate to a detectable amount by Coomassie blue stain (4) but less in rice (data not shown).

Protection ability of HSPs without 15- to 18- kD class of HSPs in soybean:

Fig. 4A (fluorogram) and Fig. 4B (Western blot) show 70 to 100% AS saturation fraction from soybean contains HSPs. Lane 1 was from 28°C (control) treated seedlings and lane 3 was from the treated seedlings causing the association of 15- to 18- kD HSPs and small fraction of 68- to 70- kD with organelles fraction so that from which the AS fraction prepared was deficient in 15- to 18-kD HSPs but containing the HMW HSPs. Lane 2 and 4 show that 15- to 18-kD HSPs delocalized from cellular components and returned to cytosol after recovery at 28°C for 4h. Western blot analysis shown in Fig. 5A, 15- to 18-kD LMW HSPs were found in the lanes 2 and 4.

The protection ability with or without 15- to 18- kD of HSPs-enriched fraction were studied by using these four different 70 to 100% AS saturation fractions shown in Fig. 5. As shown in Table 2, treatment of 40°C (3 h) → 28°C (4 h) some 50% of PRS proteins were protected by HSPs, 40°C (3 h) → 45°C (30

Table 1. Stabilization of ^{3}H-proteins against heat denaturation by addition of HSPs-enriched fractions from soybean, mung bean, and rice.

Addition 70-100% AS fraction		Proteins denatured (cpm x 10^{-2})		
		soybean	mung bean	rice
Expt. 1				
soybean	C (28°C)	898 (100%)	2,368 (100%)	--
	HS (40°C)	477 (53.1%)	1,372 (57.9%)	--
mung bean	C	1,610 (100%)	1,414 (100%)	--
	HS	835 (51.8%)	735 (51.9%)	--
Expt. 2				
soybean	C (28°C)	1,750 (100%)	--	2,590 (100%)
	HS (40°C)	1,010 (57.7%)	--	1,000 (38.5%)
rice	C	330 (100%)	--	1,340 (100%)
	HS	230 (69.7%)	--	700 (52.2%)

To 2 mg of ^{3}H-proteins from the postribosomal supernatant, 1 mg of a 70 to 100% AS fraction (from either 28°C or 40°C treated seedlings) was added; the mixture was heated at 55°C for 30 min. After heat treatment, the samples were centrifuged at 16,000 x g for 15 min. The pellets were suspended in 250 μl of Laemmli's sample buffer, and 25 μl duplicate samples were assayed for radioactivity. The amount of ^{3}H-protein that pellets after heating by addition of C (28°C) fraction in this Table and other Tables were found always equal to 98% to 95% of pelletable proteins by heating when nothing was added.

Table 2. Thermostabilization of ^{3}H-proteins by addition of a HSPs-enriched fraction prepared from one of four different temperature treatments of soybean seedlings.

Addition 70-100% AS fraction	Proteins denatured (cpm)
1. no addition	186,880 ± 1,020 (100%)
2. 28°C (3 h)	182,360 ± 720 (97.6%)
3. 40°C (3 h)→28°C (3 h)[a]	105,100 ± 1,290 (56.3%)
4. 40°C (3 h)→45°C (30 min)[b]	180,650 ± 980 (96.7%)
5. 40°C (3 h)→45°C (30 min)→28°C (4 h)[c]	97,650 ± 1,120 (52.3%)

To 2 mg of ^{3}H-proteins from the postribosomal supernatant (675,400 cpm), 1 mg of the 70 to 100% AS fraction from seedlings of four different treatments was added; the mixtures were heated at 55°C for 30 min. After the heat treatment, the samples were centrifuged at 16,000 x g for 15 min. The pellets were suspended in 250 µl of Laemmli's sample buffer, and 25 µl duplicate samples were assayed for radioactivity.

[a] The subsequent treatment at 28°C for 3 h cause the partial HSPs associated with organelles to redistribution into the cytosol (9).

[b] Under this treatment regime, many LMW HSPs remain associated with organelles (9).

[c] Under this treatment regime, HSPs are redistributed into the cytosol (9).

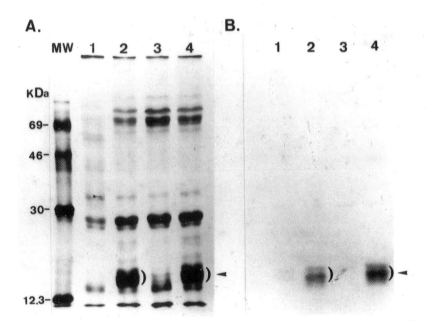

Figure 4. Fluorogram (A) and Western blot (B) analysis of 70 to 100% AS fractions from four different treatments of soybean seedlings separated into protein components by SDS-PAGE. Lane 1, 28°C (3 h); Lane 2, 40°C (3 h) → 28°C (3 h); Lane 3, 40°C (3 h) → 45°C (30 min); Lane 4, 40°C (3 h) → 45°C (30 min) → 28°C (3 h). In lane 2, 3 and 4 were loaded with equal cpm and equal amount at Fig. A and B respectively. Arrow head indicate the 15- to 18- kD HSPs.

in) depleted the 15- to 18- kD HSPs from HSPs-enriched fraction (Fig. 5, lane 3) and lost the effectiveness of heat protection entirely even with the presence of HMW HSPs. While 40°C (3 h) → 45°C (30 min) →28°C (4 h) treatment, the 15- to 18- kD HSPs were recovered in 70 to 100% AS saturation fraction (Fig. 5, lane 4); accordingly, the thermostabilization ability was restored; and some 50% of PRS proteins were protected by HSPs again. These data suggest that only the 15- to 18-kD HSPs, which were present in 70 to 100% AS saturation fractions, were required for thermoprotection of soluble proteins.

Table 3. Thermostabilization of ^{3}H-proteins in a low or high salt extract by addition of HSPs-enriched fraction from soybean.

Addition 70-100% AS fraction	Proteins denatured (cpm)	
	low salt extract[a]	high salt extract from residue of low salt extract[b]
1. no addition	17,130 ± 670 (100%)	12,130 ± 410 (100%)
2. 28°C	16,230 ± 600 (94.7%)	12,180 ± 380 (100%)
3. 40°C(3h)→28°C(3h)	15,890 ± 490 (92.7%)	9,110 ± 210 (75.1%)

To 750 µg of the postribosomal supernatant proteins from a low salt extract (62,000 cpm) or from a high salt extract of the residues from the low salt extract (55,890 cpm), 750 µg of the 70 to 100% AS fraction from seedlings of two different treatments were added; the mixtures were then heated at 55°C for 30 min. The salt concentration in the assay mixtures from the low salt extract was adjusted to the same concentration as in the high salt extract. After heat treatment, the samples were centrifuged at 16,000 x g for 15 min. The pellets were suspended in 250 µl of Laemmli's sample buffer, and 25 µl duplicate samples were assayed for radioactivity.

[a] The low salt extraction medium contained 50 mM Tris-HCl buffer, pH 8.8, 0.5 M sucrose, 50 mM KCl, 5 mM $MgCl_2$, 1 mM DTT, and 1mM PMSF.

[b] The residues were extracted with the medium as described in methods.

Proteins thermoprotected by the HSPs in soybean:

Table 3 shows the soluble proteins extracted with two different salt concentrations are protected from heat denaturation differently. Those high salt extractable proteins which are membrane system related proteins could be protected by HSPs. The cell sap proteins are much more stable to heat denaturation than the soluble proteins extracted from the residues of cell sap preparation (data not shown).

LITERATURE CITED

Barnett. T, Altssschuler M, Mcdaniel CN, Mascarenhas JP (1980). Heat shock induced proteins in plant cells. *Dev Genet* 1: 331-340.

Chou, M., Chen, Y.M. and Lin C.Y. (1989). Thermotolerance of isolated mitochondria associated with heat shock proteins. *Plant Physiol.* 89: 617-621.

Hsieh, M.H., Chen, J.T., Jinn, T.L., Chen, Y.M., and Lin, C.Y. (1992). A class of soybean low molecular weight heat shock proteins: immunological study and quantitation. *Plant Physiol.* 99: 1279-1284.

Jinn, T.L., Yeh, Y.C., Chen, Y.M., and Lin, C.Y. (1989). Stabilization of soluble proteins *in vitro* by heat shock proteins-enriched ammonium sulfate fraction from soybean seedlings. *Plant Cell Physiol.* 30: 463-469.

Key, J.L., Lin, C.Y., and Chen, Y.M. (1981). Heat shock proteins of higher plants. *Proc. Natl. Acad. Sci. USA* 78: 3526-3530.

Kimpel, J.A., and Key, J.L. (1985). Heat shock in plants. *Trends Biochem Sci* 10:353-357.

Laemmli, U.K. (1970). Cleavage of structural proteins during the assembly of the head of bacteriophage T4. *Nature* 227: 680-685.

Li, G.C. and Werb, Z. (1982). Correlation between synthesis of heat shock proteins and development of thermotolerance in Chinese hamster fibroblasts. *Proc. Natl. Acad. Sci. USA* 98:3218-3222.

Lin, C.Y., Roberts, J.K. and Key, J.L. (1984). Acquisition of thermotolerance in soybean seedlings: synthesis and accumulation of heat shock proteins and their cellular localization. *Plant Physiol.* 74: 152-160.

Lindquist, S. (1986). The heat-shock response. *Annu. Rev. Biochem.* 55: 1151-1191.

Lowry, O.H., Rosebrough, N.J., Farr, A.L. and Randoll R.J. (1951). Protein measurement with the Folin phenol reagent. *J. Biol. Chem.* 193: 265-275.

Mans, R.J. and Noveli, G.D. (1961). Measurement of the incorporation of radioactive amino acids into protein by a filter paper disk method. *Arch. Biochem. Biophys.* 4: 48-53.

Mansfield, M.A., and Key, J.L. (1987). Synthesis of the low molecular weight heat shock proteins in plants. *Plant Physiol.* 84: 1007-1017.

McAlister, L., and Finkelstein, D.B. (1980). Heat shock proteins and thermal resistance in yeast. *Biochem. Biophys. Res. Commun.* 93: 819-824.

Minton, K.W., Karmin, P., Hahn, G.H. and Minton, A.P. (1982). Nonspecific stabilization of stress-susceptible proteins by stress-resistant proteins: a model for the biological roles of heat shock proteins. *Proc. Natl. Acad. Sci. USA* 79: 7107-7111.

O'Farrell, P.H. (1975). High resolution two-dimensional electrophoresis of proteins. *J. Biol. Chem.* 250: 4007-4021.

Peterson, N.S. and Mitchell, H.K. (1981). Recovery of protein synthesis after heat shock: prior heat-treatment affects the ability of cells to translate mRNA. *Proc. Natl. Acad. Sci. USA* 78: 1078-1711.

Schlesinger, M.J., Ashburner, M. and Tissieres, A. (1982). Heat shock from bacteria to man. New York: Cold Spring Harbor Laboratory Press. 440 pp.

Schlesinger, M., Alliterate, G. and Kelley, P.M. (1982). The response of cells to heat shock. *Trends Biochem. Sci.* 1: 222-225.

Towbin, H., Staehelin, T. and Gordon, J. (1979). Electrophoretic transfer of protein from polyacrylamide gels to nitrocellulose sheet: procedure and some application. *Proc. Natl. Acad. Sci. USA* 76: 4350-4354.

Tseng, T.S., Yeh, K.W., Yeh, C.H., Chang, F.C., Chen, Y.M. and Lin, C.Y. (1992). Two-rice (*Oryza sativa*) full-length cDNA clones encoding low-molecular-weight heat-shock proteins. *Plant Mol. Biol.* 18: 963-965.

Vierling, E. (1991). The roles of heat shock proteins in plants. *Annu. Rev. Plant Physiol. Plant Mol. Biol.* 42: 579-620.

Wu, S.H. (1992). The biological function of low molecular weight heat shock proteins in plants. Master thesis. Department of Botany, National Taiwan University, Taipei, Taiwan, Republic of China

GENETIC AND MOLECULAR EVIDENCES OF THE REGULATION

OF GENE EXPRESSION DURING HEAT SHOCK IN PLANTS

N. Marmiroli[1], E. Maestri[1], V. Terzi[3], M. Gulli[1], A. Pavesi[1], G. Raho[4], E. Lupotto[4], G. Di Cola[1], R. Sinibaldi[5,] C. Perrotta[2].

[1] Department of Evolutionary Biology - University of Parma - Italy

[2] Department of Biology - University of Lecce - Italy

[3] Experimental Institute for Cereal Research - Fiorenzuola d'Arda (PC) - Italy

[4] Experimental Institute for Cereal Research - Bergamo - Italy

[5] Sandoz Agro Inc.- Palo Alto, California - U.S.A.

Abstract

In cultivated plants, as well as in wild species, there is a certain degree of variability in the pattern of Heat Shock Proteins (HSPs) induced after different types of temperature stress. In barley, this variability is too complex to correlate with the degrees of thermotolerance measured by physiological indexes such as osmotic potentials ($-\Psi$) or growth rates (CGR).

To make more consistent this approach we have analysed the progeny F_1 and F_2 from reciprocal crosses involving two genotypes (cv Onice and Georgie) which differed in i) the level of thermotolerance and ii) a limited number of HSPs induced after heat shock. The genetic analysis of the variations in HSPs synthesis showed: i) a uniparental heredity for some of the HSPs and ii) the existence of epistatic regulation. To establish the role of the various genetic elements involved in thermotolerance we have analysed: i) the structure of heat shock genes and proteins and their regulation, ii) their role in thermoprotection and iii) their involvement in determining thermoresistance and stress resistance through the isolation and analysis of specific mutants. The isolation of heat shock genes such as the Hvhsp18 and Hvhsp17 has allowed the comprehension that their heat-inducibility occurs through specific Heat Shock Elements (HSE). The utilization of a maize protoplast system for transient assay and of specific expression vectors has allowed the identification of the relative importance in heat shock response of the two HSEs that we have identified within the gene Hvhsp17. Transformation with similar genetic constructs has been used to obtain transgenic plants of tobacco to monitor their role and expression *in vivo*. Moreover heat shock promoters have been utilized to produce other expression vectors which allowed the comparison of mRNA synthesis and processing in normal and heat shock condition. Heat shock which effectively induced expression of a gene under the control of a heat shock promoter strongly inhibited the splicing of introns. In condition of induced thermotolerance, i.e. after induction of HSPs, the splicing of the intron was partially maintained. The isolation of stress-tolerant mutants has indicated that genetic factors other than the heat shock genes may be involved in thermoprotection.

Introduction

The deterioration of soil and atmosphere conditions, basically due to various types of pollution, has increased the concern for the possibility of weather changes on planet areas scale; this phenomenon has also been referred to as Global Change. Global Change rises a challenge to all living organisms, animals and plants and to several man made activities, including agriculture. In particular the possibility that the Green House Effect may increase the temperature on a planet scale of only few degrees is dangerous enough to threaten the growth of many common crops as well as several wild species.

The interest in the comprehension of the mechanisms of heat tolerance and of heat stress response is therefore not only academical, and justifies the many years of researches so far carried out.

One of the most striking evidences reported about temperature stress response was the induction of Heat Shock Proteins (HSPs) in almost all the plant species analysed. Indeed, if the induction of HSPs is a prominent figure of the temperature stress response, their role in conferring thermoprotection has been shown mostly on correlative basis (Lindquist, 1986; Key et al., 1987; Vierling, 1991).

Several genetic and molecular elements involved in the heat shock response have been described: some heat shock genes, their cis-activating elements or Heat Shock Elements (HSE) and the Heat Shock Trans Activating Factors (HSTF) are among these; moreover the involvement of enhancer and silencer elements have been ascertained (Nover 1991).

However none of the evidences so far obtained has been direct enough to establish more than correlations between HSPs induction and thermotolerance and the same has occurred also for their practical utilization in the genetic breeding.

For this purpose more genetic analysis on heat tolerance is needed such as: the search for genetic linkage with the presence of specific HSPs in cultivars or wild species, the linkage with mapped heat shock genes i.e. by RFLP, and ultimately the isolation and characterization of heat shock tolerant or heat stress tolerant mutants.

MATERIALS AND METHODS
Genetic materials

Arabidopsis thaliana ecotype Columbia has been utilized for the induction and isolation of stress-tolerant mutants.

Hordeum vulgare cvs Onice, Georgie, Huron, Tipper and Igri were utilized for one and two dimension analysis of HSPs induction and for the determination of physiological indexes under different temperature stress conditions. The cvs Onice and Georgie were also the parentals of two reciprocal crosses for the production of an F_1 and F_2 progeny utilized to analyse the segregation of several HSPs. Barley protoplasts were prepared from the cv Onice. Other barley materials utilized are reported in the legends to the Figures.

Nicotiana tabacum cv Petit Havana whole plants and protoplasts were utilized for transformation experiments with vectors containing the HSEs of the Hvhsp17 gene and the reporter gene GUS in transient assay (protoplasts) and in stable transformants (whole plants).

Zea mays cv Black Mexican Sweet cell derived protoplasts were utilized in the splicing experiments whereas leaf derived protoplasts of maize inbred line 20 904 were utilized in the Transient Gene Expression Assay (TGEA) experiments.

Triticum aestivum protoplasts were obtained from the cv Oderzo (bread wheat) and from the cv Farneto (durum wheat). The cultivar Chinese Spring and the nullisomic-tetrasomic series of its mutants utilized for RFLP mapping of Hvhsp17 (Sears, 1954)

Cultivation of *Arabidopsis thaliana* and mutagenic treatment.

Arabidopsis thaliana cv Columbia seeds were cultivated in sterile condition at 22°C with a 16 h photoperiod under 8000 lux illumination. After 8 weeks the plant flower showed the first appearance of siliques. The formation of the fruit was completed in about 8 weeks and every fruit contained from 30 to 60 seeds. The seeds were conserved at room temperature. The mutagenic agent EMS (Ethyl methanesulfonate) was applied to the seeds in liquid condition by treating 40 μg of seeds with 20 μg of the mutagen. The seeds were exposed to the mutagen for 7 hours, with continuos shaking and then rinsed with a continuos flow of water for a total of 1 hour and 30 min. The rinsed seeds were then collected and dried for conservation. The seeds were sterilized with NaClO 2% for 1 h and 30 min. The seasoning was obtained by incubating the seeds at 4°C overnight. The germination of the seeds was carried out *in vitro* with a Murashige and Skoog (1962) medium in glass containers, 20/30 seeds per vessel at the temperature of 22°C with photoperiod of 16 h at 8000 lux. After 5 days the germination medium must be replaced by the growth medium (Murashige and Skoog medium with sucrose 3% and vitamins).

Temperature stress and selection for stress tolerance

Preliminary experiments were carried out to determine the heat shock condition which had the strongest effect upon survival of the germinated seeds. For this reason the survival was determined after heat shock from 22°C to temperatures as high as 50°C for time intervals ranging from 30 min to 6 hours. The temperature treatment which gave the best selection evidence was 46°C for 2 hours with a survival of almost 0% in repeated experiments. For this reason we decided to utilize it as the selection condition for the mutagenized seeds.

Barley genetic materials and *in vivo* labelling

In vivo labelled proteins were obtained from two barley cultivars with different thermotolerance (Spring cv Georgie and Winter cv Onice). Analyses were also extended to the F1's derived from the crosses Onice x Georgie and Georgie x Onice, and to about 30 individuals of the F2 derived from selfing the F1 (Onice x Georgie). Seedling were incubated in 500 μl of sterile water supplemented with 20 μCi of ^{35}S methionine (specific activity > 1000 Ci/mmol) for 4 hours at the control temperature of 23°C and at the stress temperature of 34°C.

Protein extraction from roots of individual seedlings was carried out according to Hurkman and Tanaka (1986). Labelled proteins in heat shock conditions were also characterized in the cvs Huron, Tipper and Igri.

Gel electrophoresis of plant proteins

Labelled proteins were separated by both one- and two dimension polyacrylamide gel electrophoresis and visualized by direct autoradiography of gels with Amersham ß-max X-ray films with intensifying screen, at -70°C. The isoelectric focusing (IEF) forming gels in the pH range 3.5-7 were performed according to O'Farrell (1975). The SDS-separating gels were composed of 10% polyacrylamide. Silver staining was performed with the Silver Stain Kit according to the manufacturer instruction (BioRad, Richmond, CA).

DNA and RNA isolation

High molecular weight nuclear DNA was isolated from harvested sterile barley (cv Onice) seedlings according to Rogers and Bendich (1988). Poly(A)$^+$ RNA was isolated from normal (25°C) and heat shocked seedlings harvested immediately after heat treatment (40°C, 1 hour) (Nagy et al. 1988).

Construction and screening of barley cDNA library (*cv Onice*)

Poly(A)$^+$ RNAs isolated from the seedlings were qualitatively assessed by *in vitro* translation in a rabbit reticulocyte lysate (Amersham, U.K.). Complementary DNA (cDNA) was prepared from heat shock poly(A)$^+$ RNAs and inserted into pBR322 vector. Specific heat shock clones were selected by differential hybridization (Sambrook et al. 1989).

Northern hybridization analysis

Purified poly(A)$^+$ RNAs were size-fractionated by electrophoresis on 1% agarose in the presence of formaldehyde, according to Sambrook et al. (1989) and transferred to Zeta-Probe membranes (BioRad, Richmond, CA). The DNA fragment used as probe came from a *Pst* I digest of the p378 cDNA clone containing a portion of the Hvhsp18 gene, labelled by nick-translation with [a-^{32}P]dCTP (Amersham U.K., specific activity 3,000 Ci/mmol) and hybridized to RNA blots as described by Sambrook et al. (1989).

Construction and screening of barley genomic library

Total genomic DNA was partially digested and a genomic DNA library was constructed using the cloning vector lambda EMBL4. The screening of the phage library was performed with a purified Hvhsp18 cDNA insert. The plaques which gave a positive hybridization signal were isolated, the phages purified and mapped by restriction digestion and Southern blotting using standard techniques.

Sequencing of cDNA and genomic clones

The Hvhsp18 and Hvhsp17 *Apa*I/*Apa*I inserts were sequenced by the dideoxy chain termination procedure using the modified T7 polymerase chain reaction. DNA fragments were generated by restriction endonuclease digestion of the cloned DNA and the excised fragments were subcloned into the appropriate cloning site in pGEM plasmids (Promega). Single strand DNA was derived from R408 helper phage superinfection of plasmid bearing *E.coli* JM109. T3 and T7 primers were used (depending on the orientation of the DNA fragment) to dideoxy sequence the DNA fragments in the subclones.

Transformation of cell line derived maize protoplasts (BMS)

Black Mexican Sweet (BMS) cell lines were cultivated in MS medium (Murashige and Skoog, 1962). The protoplasts were prepared after digestion with chitinase (1%), cellulase (0,3%) and pectinase (0,02%). A concentration of 1×10^6 protoplasts were mixtured with 30μg of plasmid DNA and transformed according to the method of Fromm et al. (1985). An electric field of 200V was applied to the electroporation cell in order to promote the uptake of exogenous DNA by the protoplasts.

The plasmids utilized for the transformation experiments all derive from pUC18-19 with insertion in the polylinker cloning site of a GUS cassette including: the 35S promoter of CaMV or the promoter of the gene HSP82 or HSP18 of maize, the reporter gene GUS and the terminator NOS of Ti. Some constructs have introns (ADH6 or HSP82). Some of the plasmids have been mutagenized in the 3' or 5' splicing site or in both (see also Figure 10).

Determination of the expression of the various plasmids by a transient assay

The transformed protoplasts were maintained at 25°C before the various thermal treatments: 2h at 35°C, 37°C, 39°C, 41°C, 43°C, 45°C. The thermoprotection experiments were carried out with protoplasts previously exposed at 40°C and then stressed at 45°C for 2 hours. The extracts prepared according to Jefferson (1987) were assayed spectrophotometrically at $\lambda = 415$ nm with the substrate p-nitrophenyl-ß-glucuronide in order to detect the activity of the enzyme ß-glucuronidase (GUS). The GUS activity was expressed as relative activity (specific activity); protein concentration in the samples was measured according to Bradford (1976). The results were further elaborated by the aid of two computer programmes: Microexcel and Criket.

TGEA (transient gene expression assay) with cereal cell systems

Mesophyll cell derived protoplasts of cereal species were isolated according to published procedures for barley (Teeri et al. 1988), and personal protocols (Lupotto, unpublished) for the other species. As control, dicotiledonous plant *Nicotiana tabacum* protoplasts were isolated according to standard procedures (Saul et al. 1988). Transient gene expression assays (TGEA) were performed according to the protocol established for tobacco (Bilang and Schrott, 1991) except that osmotic pressure was 0.6 M instead of 0.4 M mannitol. PEG solution (40% PEG 4000 Merck) was maintained at 40°C before the addition to the protoplasts. Transformed protoplasts were incubated in K3 medium for tobacco, and MS-Z for cereal species (MS medium salts and vitamins, 0.6 M glucose, 0.6 g/l MES, 1 mg/l 2,4-D, pH 5.6, supplemented with 250 μg/ml cefotaxime), incubated 16-18h in the dark at 26°C. Heat stress was performed incubating the petri dishes on a grid over the water surface of a thermostatic bath at 40°C for 1h. Protoplasts were collected, washed, and extracted for GUS fluorimetric assay according to Jefferson (1987).

Computer analysis

The nucleotide and deduced amino acid sequences comparisons of the Hvhsp18 and Hvhsp17 inserts vs. DNA and protein data banks (EMBL data library release 25 and SWISSPROT release 17) were performed by a Lipman and Pearson (1985) procedure (PC-Gene package software release 6.5). Protein coding regions of both clones were recognized by selecting those ORFs which better correlated with barley codon usage, as estimated using protein coding genes available in GENBANK release 62. In order to recognize the metal-ion response element (MRE), the heat shock promoter element (HSE), and the putative polyadenylation signals, DNA sequence analysis of Hvhsp17 genomic sequence was performed making reference to algorithms of signal search (Staden, 1984). The multiple alignment of HSPs was obtained by a series of binary alignments (Lipman and Pearson, 1985). Hydrophobicity analysis of the aligned protein sequences was constructed using a window scanning of 19 residues to discriminate putative membrane spanning segments according to Kyte and Doolittle (1982).

RESULTS
Different level of thermotolerance in different barley cultivars
Figure 1 shows the behaviour of five barley cultivars when subjected to heat shock treatments to measure changes in the osmotic potential (-Ψ) or in the crop growth rate (CGR). The two physiological indexes have been widely used to describe stress tolerance in plants. The temperature stress from 25 to 40°C for three hours reduced CGR significantly in all the five cultivars, the most sensitive being the genotype Tipper. The other treatments 25->35°C and 35->40°C had minor effects. Evidences on the lasting of thermoprotection afforded by temperature pretreatments are shown by the CGR values in the plants pretreated at 35°C for three hours.

The osmotic potentials in Figure are expressed in bars; within the five cultivars the response to the different temperature treatments showed many similarities. Exceptions are the following thermal regimens: 35 (3h) -> 40°C and 25->35°C (3 days and 6 days), at which the five cultivars showed different values of osmotic potential. These two conditions corresponded to those effective also in inducing a degree of lasting thermotolerance.

166

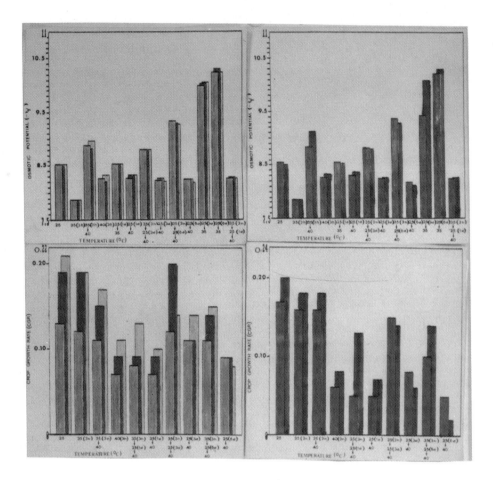

Figure 1

Variations observed in Osmotic Potential (-y) and Crop Growth Rate (CGR) in barley cultivars subjected to different thermal treatments.
Upper panels: the plants were grown at 25°C for 15 days in a greenhouse before the temperature treatments. Left: osmotic potentials for the cultivars Igri (red), Onice (yellow) and Huron (blue); right: osmotic potentials calculated for the cultivars: Georgie (green) and Tipper (red). The osmotic potentials were calculated according to the following equation: y = (RT/M) ln aw, where aw = water activity in the sealed chamber measured with a thermocouple psychrometer as described by Richards and Ogata (1958).
Lower panels: Crop Growth Rates (CGR) calculated 15 days after returning the stressed plants to the greenhouse at 25°C, CGR is calculated according to Kvet et al. (1971):

$$CGR = \frac{SS_2 - SS_1}{t_2 - t_1} \; g/m^2/d$$

Each histogram represents a temperature regimen as indicated, and the corresponding CGR calculated for: left, Igri (red), Onice (yellow), Huron (blue) and right, Georgie (green) and Tipper (red).

Variability in the HSPs patterns induced in different barley cultivars

Figure 2 shows a schematism of the HSP protein pattern induced in the five barley cultivars under the different temperature regimens. The five cultivars showed differences in the heat shock response: i) at the various heat shock temperatures and ii) after the pretreatment at 35°C in condition of induction of thermotolerance. The higher number of HSPs was induced during the treatment 25->40°C and 35->40°C. Again the cv Tipper showed the highest number of HSPs induced at each temperature. The 70 kDa HSP is induced in all the five genotypes in every stress condition utilized. The intensity of many high molecular weight HSPs is significantly increased at 35°C in Tipper, Huron and Georgie and then declined at 40°C or after a pretreatment at 35°C. The induction of the low molecular weight HSPs showed the largest variability between genotypes at 35°C and at 40°C with differences in the size of the HSPs preferentially induced.

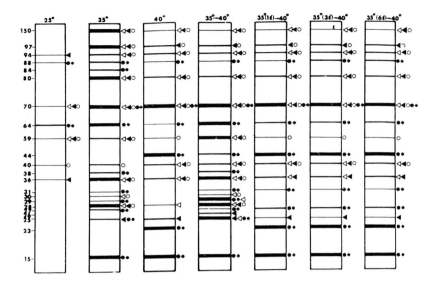

Figure 2

Drawn illustrating the major changes in protein synthesis observed in barley cultivars subjected to the indicated thermal treatments. The symbols indicate: ⌂, Tipper; ▲, Huron; ·, Georgie; ●, Onice and *, Igri; and are pointed to the protein bands which correspond to stress induced or stress increased proteins. The protein molecular weights (in kDa) are indicated on the left border.

Two-dimension analysis of HSPs in barley genotypes and crosses

A comparison of protein patterns from unstressed and stressed roots of Onice and Georgie, two genotypes sufficiently different in thermotolerance (Figure 3), was carried out through a two-dimension analysis. There are differences between the two genotypes: the spring variety (Georgie) showing the lowest number of HSPs induced at 34°C. The pattern of HSPs synthesis in F1 showed all the Onice parental spots (Figure 4). The two reciprocal crosses (Onice x Georgie and Georgie x Onice) gave identical results (Terzi et al. 1990). The analysis of the F2 seedlings gave segregation patterns of the major HSP as reported in Table 1. A schematic explanation of some of the results obtained is reported in Figure 5.

Table 1

Segregation patterns of some HSPs in the F_2 progeny obtained from selfing the F_1 Onice x Georgie.

Presence (+) / absence (-) of HSPs in Onice, Georgie, F_1 and F_2 were inferred from 2D-analysis as in Figure 4. The HSPs indicated with conventional numbers are referred to the spots showed in Figure 5

HSP	Onice	Georgie	F1	F2	Most probable Mendelian ratio (x^2 not significant)
1-2	+	-	+	13+:9-	(9:7)
3	+	-	+	13+:9-	(9:7)
4	+	+	+	+	
5-6-7-8	+	-	+	+	
9	+	+	+	+	
10	+	-	+	+	
11	+	-	+	+	
12	+	+	+	+	
13	+	+	+	+	
14	+	-	+	+	
16	+	+	+	+	
21	+	-	+	15+:9-	(9.7)

Absence of segregation is shown for the HSP 5-6-7-8, 10, 11, 14. From the segregation ratios obtained for the spots 1-2, 3 and 21 we can speculate on their inheritance mechanism: for example, these spots are present in Onice, in F_1 and segregate in F_2 as determined by duplicate recessive epistatic genes. The genetic consistence of this hypothesis has been partially proved through the analysis of the segregation patterns in the backcross.

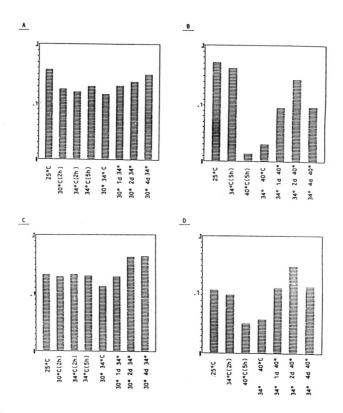

Figure 3

Induction and maintenance of a thermotolerance by non-injuring temperature treatments in the two barley cvs Onice and Georgie. Plants were grown at the stage of tillering at 25°C in a greenhouse. Mild (A-C) and high (B-D) temperature treatments in the cultivars Onice and Georgie respectively. All data are expressed as Crop Growth Rate.

Identification of low molecular weight HSPs in barley

The p378 cDNA clone was isolated out of a cDNA library in pBR322 by differential hybridization. The putative clone for a small HSP was confirmed by Northern analysis and by hybrid-arrested / hybrid-released translation (HART): Hvhsp18 sequence is able to select *in vitro* for the translation of a 18 kDa protein band and the *in vitro* translation of the same HSP band is arrested by

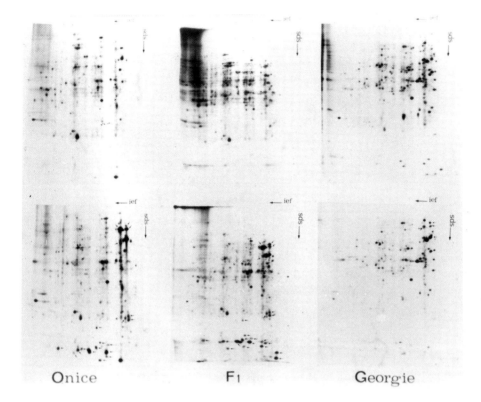

Figure 4

Two-dimensional separation (IEF/SDS-PAGE) of proteins extracted from *in vivo* labelled seedlings of the cvs. Onice, Georgie and one of their F1. The arrows indicate the proteins which are induced or enhanced after a heat shock from normal temperature 25°C (upper panels) to 34°C for 3 hours (lower panels).

addition of the same DNA insert (Marmiroli et al. 1993). The kinetic of induction of the hs mRNAs complementary to this probe reached its maximum at 40°C and after 10-15 min. from the stress and then remains at high level for

3-4 hours; after 1 hour the level of hs mRNAs was about 80% of its maximal level (Marmiroli et al. 1993).

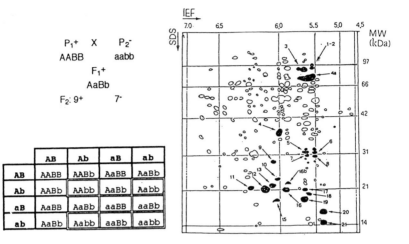

Figure 5

Schematism illustrating the two-dimension pattern of protein synthesis observed in F2 progeny of the hybrid barley (Onice x Georgie). The Figure shows also a possible genetic interpretation of some of the segregations observed in F_2 and reported in Table 1. In condition of heat shock only, a hybrid F_1 carrying two duplicate recessive epistatic genes in the heterozygous condition would show the relevant HSP spot but segregates an F_2 with 7 out of 16 plants which were totally or partially avoided of that HSP spot. Therefore the existence of double recessive epistasis between two hsp loci can change the typical Mendelian segregation 9:3:3:1 into the epistatic ratio 9:7 which is close to the 13:9 segregation observed for some of the HSP.

The Hvhsp18 insert revealed a maximum of similarity (80.5%) with wheat (*Triticum aestivum*) cDNA clone C5-8 which encodes a 16.9 kDa class I HSP (McElwain and Spiker 1989). The longest open reading frame starts at nucleotide position 37 and encodes a 18 kDa predicted gene product.

The coding region shows a prevalent content of GC nucleotides with 56% of nucleotides G or C, moreover 73.3% of third position are G or C. As far as amino acid coding is concerned, GC-rich codons are more frequently used with respect to the AT-rich codons, e.g. valine; GTC are used a total of 19 times vs. 5 for GTT and GTA combined. Amino acid sequence showed significant identities with wheat 16.9 kDa HSP (63%), soybean (*Glycine max*) 17.5 kDa HSP (54%), *Arabidopsis thaliana* 17.6 kDa HSP (49%), and pea (*Pisum sativum*) 17.7 kDa HSP (30%), all class I small cytoplasmic HSPs. A reduced identity, 30%, was found with the wheat class II 17.3 kDa HSP (Figure 6).

Isolation and analysis of the genomic clone

Among the phage plaques isolated from the lambda EMBL4 library that hybridized to the Hvhsp18 cDNA a heat shock gene designated Hvhsp17 was analysed. The clone Hvhsp17 shows a strong similarity (76.2%) with wheat cDNA encoding the 16.9 kDa HSP. The coding region predicts a polypeptide of 151 amino acids with a deduced molecular weight of 17.03 kDa (we will refer to it as the 17 kDa HSP). Highly different percent identities were found with wheat 16.9 kDa HSP (86.1%) and pea 17.7 kDa HSP (34%) a class I and a class II low molecular weight HSP respectively, *Neurospora crassa* HSP30 (28.5%), *Chlamydomonas reinhardii* HSP22 (30.5%), and *Drosophila melanogaster* HSP26 (21.9%) (Figure 6). Again, a lower identity was found with wheat 17.3 kDa HSP (34%). The presumptive site for initiation of transcription is at position -93; two heat shock elements (HSE) were identified at position -168 and -283.

A sequence showing similarity (8 out of 12) to the MRE (mammalian metal-ion response element) has been also localised at position -418. This sequence has been shown to be essential for transcriptional induction of the human metallothionein-II$_A$ gene by cadmium ions (Karin et al. 1984). These facts can give a rational explanation to the evidences that in barley induction of small HSPs is turned on also by treatment with cadmium ions (data not shown). The polyadenylation signal has been located from nucleotide position +484 to +509. The upstream and downstream context of the AATAAA box-like showed a fairly good similarity with the respective consensus sequences, as proposed by Joshi (1987) in higher plants.

```
     .  .   ...*   *.**.  ......*...*....*  ..***   ..  .. ...  *.*
 1.  GRRLEAHVFKADLPGVKKEEVKVEVEDGNVLIVSGERTKEKEDKNDKWHRVER
 2.  GRRAVAHVFKADLPGVKKEEVKVEVEDGNVLVVSGERTKEKEDKNDKWHRVER
 3.  ETP-QAHVFKADLPGLKKEEVKVELEEGKVLQISGERNKEKEEKNDKWHRVER
 4.  ETP-QAHVFKADLPGLKKEEVKVEVEEGKVLQISGERNKEKEEKNNKWHRVEF
 5.  EYP-NSYVFEIDMPGLKSGDIKVQVEDDNLLLICGERKRDEEKEGAKYLRMER
 6.  ETP-EAHVFKADIPGLKKEEVKVQIEDDKVLQISGERNVEKEDKNDTWHRVER
 7.  ELP-GAYAFVVDMPGLGTGDIRVQVEDERVLVVSGERRREEREDDAKYLRMER
 8.  ELP-GAYAFVVDMPGLGTGDIKVQVEDERVLVISGERRREERED-AKYLRMER
 9.  ETP-EAHVFKADLPGLRKEEVKVEVEDGNILQISGERSNENEEKNDKWHRVER
10.  ETP-EAHVFKVDLPGVKKEEVKVEVEDGNVLVVSGERSREKEDKNDKWHRVER
11.  ELP-GAYAFVVDMPGLGSGDIKVQVEDERVLVISGERRREEKED-AKYLRMER
```

```
     . **. * * **... . . .***.*** *  * *** .  ....
 1.  RSGKFVRPFRLPEDGKVDEVKAGLENGVLTVTVPK---AEVKKPEVKAIEISG
 2.  SSGKFVRPFRLPEDAKVEEVKAGLENGVLTVTVPK---TEVKKPEVKAIEISG
 3.  SSGKFLRRFRLPENAKVDEVKAAMANGVLTVTVPK---VEIKKPEVKAIDISG
 4.  SSGKFLRRFRLPENANVDEVKAGMENGVLTVTVPK---VEMKKPEVKSIDISG
 5.  RVGKLMRKFVLPENANTDAISAVCQDGVLSVTVQKLPPPEPKKPRTIQVKVA-
 6.  SSGKFMRRFRLPENAKVEQVKASMENGVLTVTVPK----EEKKPDVKAIEISG
 7.  RMGKFMRKFVLPDNADVDKVAAVCRDGVLTVTVEKLPPPEPKKPKTIEVKVA-
 8.  RMGKFMRKFVLPDNADMDKISAVCRDGVLTVTVEKLPPPEPKKPKTIEVKVA-
 9.  SSGKFTRRFRLPENAKMEEIKASMENGVLSVTVPK--VPE-KKPEVKSIDISG
10.  SSGKFVRRFRLPEDAKVEEVKAGLENGVLTVTVPK---AEVKKPEVKAIEISG
11.  RMGKLMRKFVLPENADMEKISP-CRDGVLTVTVDKLPPPEPKKPKTIQVQVA-
```

Figure 6

The amino acid sequences of the most conserved domain, located at or near the carboxy terminus of higher plants HSPs. 1. Barley HSP17 (class I), aa. 49-151. 2. Barley HSP18 (class I), aa. 3-105. 3. Carrot (*Daucus carota*) HSP17.7 (class I), aa. 56-157 (Darwish et al. 1991). 4. Carrot HSP17.9 (class I), aa. 58-159 (Darwish et al. 1991). 5. Soybean HSP17.9 (class I), aa. 56-159 (Raschke et al. 1988). 6. Soybean HSP18.5 (class II), aa. 56-156 (Raschke et al. 1988). 7. Maize HSP18.3 (class II), aa. 61-164 (Goping et al. 1991). 8. Maize HSP18.9 (class II), aa. 53-161 (Goping et al. 1991). 9. *Arabidopsis thaliana* HSP17.6 (class II), aa. 56-157 (Helm and Vierling 1989). 10. Wheat HSP17 (class I), aa. 50-151 (McElwain and Spiker 1989). 11. Wheat HSP17.3 (class II), aa 56-157 (Weng et al. 1991). Asterisks indicate invariant positions, dots indicate variant positions with neutral or conservative amino acid replacements.

Analysis of Hvhsp17 heat shock promoter region

In order to analyse the activity of the promoter region, functional analysis of the 1700 bp upstream 5' region was performed. Expression vectors were constructed linking complete or partially deleted promoter regions of Hvhsp17 to the reporter GUS gene (Figures 7 and 8). These constructs were analysed in transient assays for GUS activity in protoplasts derived from leaf mesophyll cells of cereal species and tobacco. The complete (1.7 kb) promoter region-GUS gene fusion (pHSGUS) resulted activated following treatment of 1 h at 40°C in the monocot cell systems whilst was silent in tobacco protoplasts. One deletion construct comprising a 0.6 kb promoter region in which the two HSEs were present (pHSGD1) behaved similarly. A minimal promoter region of 0.17 kb-GUS gene fused (pHSGD2) including the shortest HSEs only was, on the contrary, silent. TGEA of the chimeric constructs with tobacco and cereal protoplasts have evidenced characteristic features of the HS promoter (Table 2):

- the HS promoter is activated with 1h at 40°C treatment and therefore it behaves as a real heat inducible promoter;
- the 600 bp long deletion mutant promoter pHSGD1, which maintains the two HSE, is even more active than the 1700 bp long pHSGUS construct. This result is in accordance to data obtained in other species (e.g. soybean);
- the shortest promoter alone, including the 15 bp long HSE closest to TATA box, is not active; therefore both, or maybe even only the longest HSE is responsible for the thermal inducibility of the promoter;
- in monocot cell systems, the HS promoter is quantitatively expressed at higher levels as respect to the constitutive CaMV35S promoter, giving a 4 to 10 fold higher response.

Stably transformed tobacco plants were obtained by the transformation of protoplasts with plasmids carrying both the BAR gene under the constitutive 35S CaMV promoter and the GUS gene under the control of Hvhsp17 complete promoter. About 100 independent regenerated plants were selected on phosphinotricin containing medium. Out of these plants only 30 looked

Table 2

Transient gene expression in different protoplast systems determined by GUS fluorimetric assay

SPECIES	N.T.	pCaMVGUS		pHSGUS		pHSGD1		pHSGD2	
		26°C	1h 40°C	26°C	1h 40°C	26°C	1h 40°C	26°C	1h 40°C
MAIZE	-	+	+	-	+	-	++	-	-
BREAD WHEAT	-	+	+	-	+	-	++	-	-
DURUM WHEAT	-	+	+	-	+	-	+	-	-
BARLEY	-	+	+	-	+	-	+	-	-
TOBACCO	-	+	+	-	-	-	-	-	-

Protoplasts were isolated from leaf mesophyll of *Nicotiana t.* (cv Petit Havana), *Zea m.* (inbred line LO 904), *Hordeum v.* (cv Onice) and *Triticum ae* . (cvs Oderzo and Farneto). Transformation of the protoplasts was obtained through PEG mediated transfer of the various plasmid vector DNAs. The GUS activity was determined by measuring the fluorescence of 4-methylumbelliferone (4MU) produced by hydrolysis of 4-methylumbelliferyl glucuronide and expressed as nmol 4 MU min $^{-1}$ μg proteins $^{-1}$. The results reported indicate:

(-) activity between 50 and 100

(+) activity between 500 and 2000

(++) activity above 2000

N.T.= non transformed

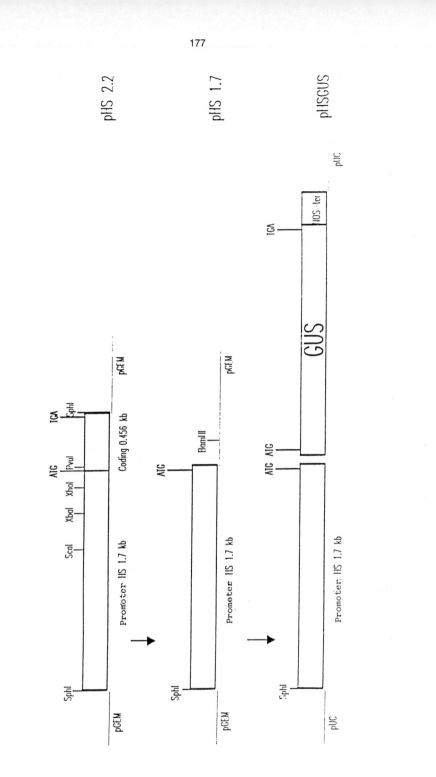

Figure 7

Strategy utilized for the construction of the expression vector carrying the Hvhsp17 heat shock promoter.
A genomic clone 2200 bp long containing both the coding region for the HSP17 and the 5' non-coding region of this gene was utilized. By cutting the DNA at the *Pvu* I site downstream the ATG start signal all the coding region could be eliminated. A clone pHS 1.7-pGEM is obtained through the insertion of the non-coding fragment *Sph* I/*Pvu* I into the vector pGEM-3Zf(+). From this vector it has been cut out a fragment *Sph* I/*Bam* HI which has been inserted into the expression vector p BI 101.2 utilized for the transient assay with the tobacco and cereal protoplasts. The resulting vector pHSGUS is characterized by the presence of the heat shock promoter of the Hvhsp17 gene. The activity of this promoter has been visualized by a spectrophotometric assay of the GUS activity.

phenotypically normal. PCR analysis was utilized for the identification of the transformation events, by using primers falling into the pHSGUS construct. In the stable events of transformation the HS promoter resulted active in tobacco also, and GUS activity was detected after 2h treatment at 40°C in vascular tissues of regenerated plants. Molecular analyses and histochemical GUS detections are being carried out for studying the structure and integration of the gene into the tobacco genome and its tissue specificity in different events of transformation.

Several genes encode the small heat shock proteins in barley

Genomic Southern analysis showed that several fragments hybridized to Hvhsp18 cDNA probe indicating that various genes encode small HSPs. From six to seven hybridization signals were detected in barley by this probe.

Chromosomal location

Hybridization of Hvhsp18 to *Eco*RI- digested wheat (2n = 6x = 42) genomic DNA from 21 nullisomic-tetrasomic (NT) lines revealed a relatively high number of bands. Four hybridising fragments were clear enough in their behaviour that were located on chromosome arms 3B and 3D as shown by the disappearance of three of those four bands from the mutants N3BT3A and the absence of one band in the mutant N3DT3B (Figure 9). These results are in agreement with data obtained from 2D analysis of *in vivo* translation experiments and with *in*

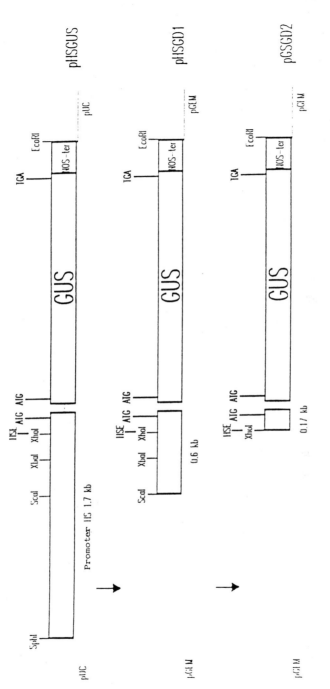

Figure 8

Construction of deletion mutations within the promoter region of the Hvhsp17 gene inserted into the pHSGUS vector. The plasmid pHSGUS was subject to a triple digestion Sca I/*Eco* RI/*Hin* dIII leading to the isolation of a 2700 bp fragment including the promoter with *Sca* I and the ATG start codon 600 bp long. The fragment was inserted into pGEM-3Zf(+) giving rise to pHSGD1. A second plasmid pHSGD2 was obtained by digestion of the insert of pHSGD1 with *Xho* I/*Eco* RI which leads to the isolation of a 2200 bp fragment including the promoter region between *Xho* I and the start ATG and 171 bp long.

vitro translation experiments (Gulli et al. 1993) with the same mutants.

Effect of heat shock on splicing

Preliminary experiments with maize cell protoplasts had shown that the optimal induction of the HSP18 promoter linked to GUS gene in the plasmid pZO1406 is 39°C (2 hours) as compared with the 35S promoter in the plasmid pZO1066. The optimal conditions for GUS induction after transformation with various plasmids carrying or not carrying an intron were determined.

In particular after transformation a treatment at 39°C determined a good induction of the HSP82 promoter both with the plasmid pZO1414, containing the gene GUS only, and with the plasmid pZO1403 containing the gene GUS and an intron of the gene HSP82 of maize. At 41°C the expression of the GUS gene was significantly higher in the plasmid pZO1414 (without intron) than with pZO1403 (with the intron) (Figure 10). Different promoters were analysed for their activity at 39°C and 41°C using the same reporter gene GUS associated or not with intron(s) and it appears quite evident the inhibition at the temperature of 41°C of the expression of those constructs with an intron fused to the GUS gene.

Mutation in the splicing sites (both 3' and 5') showed a strong inhibitory effect upon GUS expression at every temperature.

At the temperature of 45°C the inhibition of the GUS expression observed was very high in the plasmids containing an intron; the inhibition was negligible in the plasmids without the intron or in plasmid in which the GUS gene was under the control of a constitutive promoter (Figure 11).

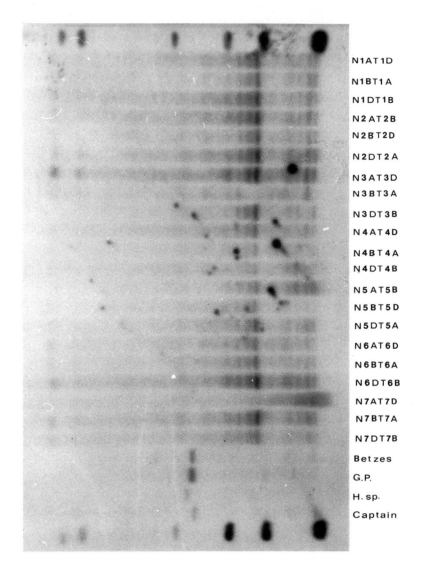

N1AT1D
N1BT1A
N1DT1B
N2AT2B
N2BT2D
N2DT2A
N3AT3D
N3BT3A
N3DT3B
N4AT4D
N4BT4A
N4DT4B
N5AT5B
N5BT5D
N5DT5A
N6AT6D
N6BT6A
N6DT6B
N7AT7D
N7BT7A
N7DT7B
Betzes
G.P.
H. sp.
Captain

Figure 9

Hybridization patterns of the probe Hvhsp17 with the *Eco* RI digested DNAs of nullisomic tetrasomic mutants of wheat and with some cvs of barley (Betzes, G.P., *H.spontaneum* and Captain).

Comparison of GUS Activities 39° C vs 41° C

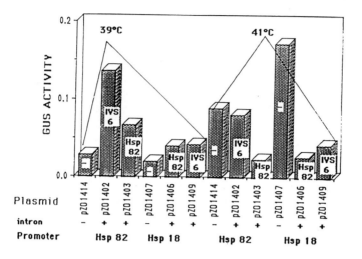

Figure 10

Illustration of the GUS relative activities transiently expressed in maize protoplasts electroporated with various expression cassettes containing or not containing an intron from the ADH1 gene (IVS6) or the intron of the HSP82 gene (HSP82). Moreover the plasmids utilized contained a constitutive (CaMV 35S) or an inducible promoter (HSP82 or HSP18 of maize), the GUS gene and the NOS terminator.

Transformed maize protoplasts were either expressed at 39°C or 41°C for 2 hours before the GUS assay. The GUS activity was determined colorimetrically at 415 nm with p-nitrophenyl-ß glucuronide as substrate for the enzyme ß-glucuronidase.

A pretreatment at 40°C before the exposure at 45°C evidenced in the maize protoplasts expressing a foreign gene, the insurgence of thermotolerance. In fact after the pretreatment at 40°C the expression of the GUS gene at 45°C in plasmids containing an intron and an heat shock promoter increased

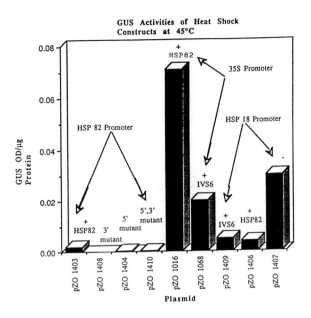

Figure 11

Effect of heat shock treatment at 45°C for 2 hours upon the GUS activities of the different constructs. The plasmid pZO1408, pZO1404 e pZO1410 carried mutations in the 3', 5' and 3', 5' splicing sites of the intron HSP82.

significantly as compared with the direct exposures to 45°C (Figure 12). Further experiments have shown that this heat stress resistant splicing may be due to the utilization of an alternative splicing site.

Isolation and characterization of stress tolerant mutants in *Arabidopsis thaliana*

After the mutagenic treatments several thousands of seedlings were subjected to temperature stress (46°C, 2h) and then allowed to recover at normal temperature for selection of survivors. After the stress the growth medium was changed to avoid accumulation of substances with toxic or inhibitory effects. In this way seven stress resistant mutants have been selected (Figure 13). Of these only two had morphogenetic defects and were also substantially sterile. The remaining five fertile mutants were reproduced and confirmed for their

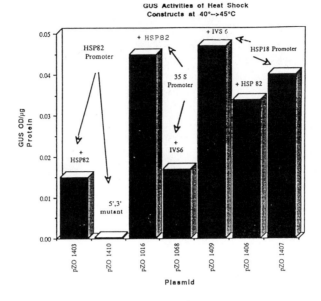

Figure 12

Effect of the pretreatment at 40°C upon the expression of GUS activity in maize protoplasts for 2 hours at 45°C. The plasmids utilized for electroporation of the protoplasts were as those described in Figures 10 and 11.

phenotype. The mutants were reproduced either sexually or vegetatively from explants or through regeneration of calli. The confirmed mutants have been analysed for HSPs synthesis in normal and heat shock conditions. None of the mutants had defects in the synthesis of some HSPs or showed induced synthesis of unusual HSPs (Figure 14).

DISCUSSION AND CONCLUSIONS

The genetic dissection of heat shock system in plants could be undertaken by the analysis of the existing variability in HSPs between cultivars or lines of different species. An example has been presented with barley in which some of the HSPs showed complex patterns of inheritance which were reminiscent of a double recessive epistasis. This could be consistent for instance with the interaction between the promoter of a structural gene for HSP and its Trans

Figure 13

Growth at 25°C of *Arabidopsis thaliana* ecotype Columbia mutants surviving the selection at 46°C for 2 hours. Only few out of these plants survived a second selection round, confirming their mutant phenotype.

Activating Factor or Heat Shock Factor (HSF). Element of this sort has been found in plants and their interaction with the Heat Shock Elements (HSE) of the heat shock genes has been proved (Nover, 1991). On the other hand this approach gave the opportunity of correlating any particular electrophoretic phenotype in terms of HSPs with the physiological behaviour in heat shock conditions determined through physiological parameters or indexes. A more reductionistic approach which started from the isolation of some heat shock genes in cereals has led to the theoretical identification of regulatory signals situated upstream of the hsp gene which in transient condition and in stable transformants reacts quite effectively to the heat shock giving the induction of any gene situated under their control. Another feature of the regulation of the heat shock response is the inhibition of intron splicing which indeed could be overcome by a thermoprotection treatment, i.e. a mild treatment which induce

35° 35° 22° 22°C

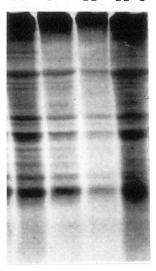

M2 M1 M2 M1

Figure 14

Silver stained electrophoretogram of two *Arabidopsis thaliana* ecotype Columbia stress tolerant mutants (M1 and M2). The two mutants were either growth at 25°C or heat shocked for two hours at 38°C before extraction of proteins and electrophoretic separation (SDS-PAGE).

the hsp genes without damaging the cells. This protection is probably due to the occurring of an alternative splicing site in condition of induced thermotolerance. The implication of these molecular evidences of thermal protection with the global phenomenon of heat stress tolerance has been addressed also by the isolation and analysis of stress tolerant mutants of *Arabidopsis thaliana*.

ACKNOWLEDGEMENTS

The authors want to thank Dr. Patrizia Lauri for preparation of the manuscript. This paper was supported by National Research Council of Italy, Special Project RAISA, Sub-project n°2.

REFERENCES

BILANG, R., SCHROTT, M. (1991) Mesophyll protoplasts of tobacco (*Nicotiana tabacum*): preparation and transformation by direct gene transfer, transient gene expression and plant regeneration. EMBO Advanced Laboratory Course. Gene transfer to plants. ETH (Zürich, Switzerland, September 9-21) pp. 18-27.

BRADFORD, M.M. (1976) A rapid and sensitive method for the qualification of microgram quantities of protein utilizing the principle of protein-dye binding. *Anal. Biochem.* **72**: 248-254.

DARWISH, K., WANG, L., HWANG, C.H., APUYA, N., ZIMMERMAN, J.L. (1991) Cloning and characterization of genes encoding low molecular weight heat shock proteins from carrot. *Plant Mol. Biol.* **16**: 729-731.

FROMM, M., TAYLOR, L., WALBOT, V. (1985) Expression of gene transferred into monocot and dicot plant cells by electroporation. *Proc. Natl. Acad. Sci. U.S.A.* **82**: 5824-5828.

GOPING, I.S., FRAPPIER, J.R.H., WALDEN, D.B., ATKINSON, B.G. (1991) Sequence, identification and characterization of cDNAs encoding two different members of the 18 kDa heat shock family of *Zea mays* L. *Plant Mol. Biol.* **16**: 699-711.

GULLI, M., HARTINGS, H., RAHO, G., PERROTTA, C., DEVOS, K.M., MARMIROLI, N. (1993) Induction of Rab genes in barley seedlings exposed to exogenous abscisic acid (ABA). Submitted for publication to Plant Science.

HELM, K.W., VIERLING, E. (1989) An *Arabidopsis thaliana* cDNA clone encoding a low molecular weight heat shock protein. *Nucl. Acid. Res.* **17**: 7995.

HURKMAN, W.J., TANAKA, C.K. (1986) Solubilization of plant membrane proteins for analysis by two-dimensional gel electrophoresis. *Plant Physiol.* 81:802-806.

JEFFERSON, R.A. (1987) Assaying chimeric genes in plants: the GUS gene fusion system. *Plant Mol. Biol. Report* **5**: 387-405

JOSHI, C.P. (1987) Putative polyadenylation signals in nuclear genes of higher plants: a compilation and analysis. *Nucl. Acid. Res.* **15**: 9627-9640.

KARIN, M., HASLINGER, A., HOLTGREVE, H., RICHARDS, R.I., KRAUTER, P., WESTPHAL, H.M., BEATO, M. (1984) Characterization of DNA sequences through which cadmium and glucocorticoid hormones induce human metallothionein-II$_A$ gene. *Nature* **308**: 513-518.

KEY, J.L., KIMPEL, J.A., NAGAO, R.T. (1987) Hreat shock gene families of soybean and the regulation of their expression. In: Key, J.L. and McIntosh, L. (eds) Plant Gene Systems and their Biology. Alan R. Liss, New York, NY, pp. 87-97.

KYTE, J., DOOLITTLE, R.F. (1982) A simple method for displaying the hydropathic character of a protein. *J. Mol. Biol.* **157**: 105-132.

KVET, J., ONDONDOK, J.P., NECAS, T., JARVIS, P.G. (1971) In: Sestak, Z., Kats, K.J. and Jarvis, P.G. (eds) Plant Photosynthetic Production. Manual of Methods. Junk N.V. Publishers, The Hague, Netherlands, pp. 343-391.

LINDQUIST, S. (1986) The heat-shock response. *Annu. Rev. Biochem.* **55**:1151-1191.

LIPMAN, D.J., PEARSON, W.R. (1985) Rapid and sensitive protein similarities searches. *Science* **227**: 1435-1441.

MARMIROLI, N., PAVESI, A., DI COLA, G., HARTINGS, H., RAHO, G., CONTE, M.R., PERROTTA, C. (1993) Identification, characterization and analysis of cDNA and genomic sequences encoding two different small heat shock proteins in *Hordeum vulgare* L. *Genome*, in press.

MCELWAIN, E.F., SPIKER, S. (1989) A wheat cDNA clone which is homologous to the 17 kD heat shock protein gene family of soybean. *Nucl. Acid. Res.* **17**: 1764.

MURASHIGE, T., SKOOG, F. (1962) A revised medium for rapid growth and bioassays with tobacco tissue cultures. *Physiol. Plant* **15**: 473-497.

NAGY, F., KAY, S.A., CHUA, N.H. (1988) Analysis of gene expression in transgenic plants. In: Gelvin, S.B., Schilperoort, R.A. and Verma, P.S. (eds) Plant Molecular Biology Manual, B4. Kluwer Acad. Publishers, Dordrecht,

The Netherlands, pp. 1-29.

NOVER, L. (1991) Heat Shock Response. CRC Press, Boca Raton, FL.

O'FARRELL, P.H. (1975) High resolution two-dimensional electrophoresis of proteins. *J. Biol. Chem.* **250**: 4007-4021.

RASCHKE, E., BAUMANN, G., SCHOEFFL, F. (1988) Nucleotide sequence analysis of soybean small heat shock protein genes belonging to two different multigene families. *J. Mol. Biol.* **199**: 549-557.

RICHARDS, L.A., OGATA, G. (1958) Thermocouple for vapor pressure measurements in biological and soil systems at high humidity. *Science* **128**: 1089-1090.

ROGERS, S.O., BENDICH, A.J. (1988) Extraction of DNA from plant tissues. In: Gelvin, S.B., Schilperoort, R.A. and Verma, P.S. (eds) Plant Molecular Biology Manual, A6. Kluwer Acad Publishers, Dordrecht, The Netherlands, pp. 1-10.

SAUL, M.W., SHILLITO, R.D., NEGRUTIU, J. (1988) Direct DNA transfer to protoplasts with and without electroporation. In: Gelvin, S.B., Schilperoort, R.A. and Verma, P.S. (eds) Plant Molecular Biology Manual, A1. Kluwer Acad Publishers, Dordrecht, The Netherlands, pp. 1-16.

SAMBROOK, J., FRITSCH, E.F., MANIATIS, F. (1989) Molecular Cloning: A laboratory manual. Cold Spring Harbor Laboratories, Cold Spring Harbor, NY

SEARS, E.R. (1954) The aneuploids of common wheat. *Agric. Exp. Stn. Res. Bull.*, University of Missouri, **572** 1-58.

STADEN, R. (1984) Computer methods to locate signals in nucleic acid sequences. *Nucl. Acid. Res.* **12**: 505-519.

TEERI, T.H., PATEL, G.K., ASPEGREN, K., KAUPPINEN, V. (1988) Chloroplast targeting of neomycin phosphotransferase II with a transit peptide in electroporated barley mesophyll protoplasts plant. *Plant Cell Reports.* **8**: 187-190.

TERZI, V., CATTIVELLI, L., GATTI, A., STANCA, A.M., LORENZONI, C., MARMIROLI, N. (1990) Physiological and molecular studies of heat and drought tolerance in barley. *Bull. Soc. Bot. Fr.* **137**: 73-80.

VIERLING, E. (1991) The roles of heat shock proteins in plants. *Annu. Rev. Plant Physiol. Plant Mol. Biol.* **42**:579-620.

WENG, J., WANG , Z.F., NGUYEN, H.T. (1991) A *Triticum aestivum* cDNA clone encoding a low-molecular-weight heat shock protein. *Plant Mol. Biol.* **17**: 273-275.

INTEGRATION OF ACQUIRED THERMOTOLERANCE WITHIN THE DEVELOPMENTAL PROGRAM OF SEED RESERVE MOBILIZATION.

John J. Burke
USDA-ARS
Plant Stress and Water Conservation Research
Route 3, Box 215
Lubbock, Texas 79401 USA

Abiotic stresses alter seedling metabolism from imbibition to seedling emergence resulting in stand establishment delays, poor stand quality, and often regional catastrophic crop losses. Temperature stress is the most universal of all the stresses associated with germination and stand establishment. The present study evaluated changes in acquired thermotolerance during the developmentally programmed mobilization of seed reserves for early seedling growth in cucumber (*Cucumis sativus* L.). To accomplish this, the development of a new protocol for evaluating thermotolerance was required because of limitations inherent in existing viability assays. The procedure developed for this study took advantage of the temperature sensitivity of chlorophyll synthesis and enabled the detection of cellular injury at temperatures 8 to 10°C below those identified for high temperature-induced injury by the standard triphenyltetrazolium chloride reduction test. Cellular injury at temperatures similar to those reported previously for cellular regrowth studies could be seen with the Chl accumulation test. Developmental changes in inherent and acquired thermotolerance following germination were identified using this protocol. An inherent resistance observed during the first days following planting was lost with increasing seedling age, and the range of temperatures responsible for the induction of acquired thermotolerance narrowed with age. Seedlings grown at different temperatures exhibited differences in the time required for the induction of acquired thermotolerance when exposed to 40°C. In summary, this study has identified developmental changes in the inducibility and magnitude of acquired thermotolerance in cucumber cotyledons through the use of a novel viability assay.

NATO ASI Series, Vol. H 86
Biochemical and Cellular Mechanisms
of Stress Tolerance in Plants
Edited by J. H. Cherry
© Springer-Verlag Berlin Heidelberg 1994

Environmental stresses responsible for crop losses from seed imbibition to seedling emergence are a drain on America's farm economy. Researchers have developed seed pre-planting treatments that lessen the inhibitory effects of some environmental stresses. One such treatment is seed priming in osmotic solutions of PEG 8000[1], mannitol, or K_3PO_4 and KNO_3. These seed treatments have been shown to improve germination in unfavorable thermal environments (Simmonds, 1980; Cantliffe, 1981; Carpenter, 1989; Rumpel and Szudyga, 1978; Heydecker, 1977; Carpenter and Boucher, 1991; Parera and Cantliffe, 1992). Carpenter and Boucher (1991) reported that the benefits from seed priming were found to vary with germination temperature. At the recommended temperature range for germination of pansy seeds, germination percentages were essentially equal between primed and nonprimed seeds. Nonprimed seed germination percentages were shown to decrease with temperatures above or below the 15 to 25°C optimal range. The germination percentages, however, decreased less in the primed seeds. Significantly higher germination percentages were obtained at 10, 30, and 35°C in primed seeds, with better than 5 times the germination percentages at 35°C. The improved performance of primed seeds at temperatures known to induce the heat shock response suggests that alterations in metabolism early in germination may affect plant responses to thermal stress. These results warrant further study into mechanisms responsible for the observed growth improvements.

The scarcity of techniques for monitoring acquired thermotolerance clouds our understanding of the relationship between inherent heat tolerance, acquired thermotolerance, the heat shock response, and seedling development. Protocols commonly used in evaluating acquired thermotolerance include electrolyte leakage analysis, 2,3,5-triphenyltetrazolium chloride reduction, and plant or cellular regrowth following high temperature exposure (Wu and Wallner, 1983; Krishnan et al., 1989; Caldwell, 1993). Wu and Wallner (1983) compared these protocols and concluded that electrolyte leakage and triphenyltetrazolium chloride reduction were not adequate substitutes for regrowth tests in assessing heat injury to cultured plant cells. Electrolyte leakage and triphenyltetrazolium chloride reduction, however, are routinely used to this day because of the absence of substitute procedures.

The present study describes the development of a new protocol for evaluating thermotolerance in germinated cucumber seedlings. This procedure detects cellular injury at

---

[1]Abbreviations: TTC, 2,3,5-triphenyltetrazolium chloride; Chl, chlorophyll; DAP, days after planting; PEG, polyethylene glycol.

temperatures similar to those reported previously from cellular regrowth studies. Injury was detected at temperatures 8 to 10°C below those identified by the triphenyltetrazolium chloride reduction test. Developmental changes in thermal sensitivity following germination were identified using this protocol, as were the effects of different germination temperatures.

MATERIALS AND METHODS

Plant material. Cucumber (*Cucumis sativus* L. cv. Ashley) seeds were planted in vermiculite and grown in the dark at 20°C or 32°C under humid conditions. Seedlings were watered daily and the thermal sensitivity of metabolism responsible for chlorophyll accumulation was evaluated during the first two weeks after planting. Individual cotyledons excised from etiolated seedlings were placed onto Whatman 3 MM chromatography paper moistened with 4.5 ml of water and positioned on 2 x 2.5 inch temperature blocks. The cotyledons on the temperature blocks were then covered with Glad Cling Wrap[2] (a CO_2 permeable polyethylene film). Cotyledon temperatures were measured with a thin-wire thermocouple placed on the cotyledon surface to insure that they were within ±0.5°C of the treatment set temperatures. Initial studies evaluated chlorophyll accumulation in control and heat shocked (40°C for 4 h) cucumber cotyledons. Cotyledons were exposed to light levels of 300 μmol m^{-2} s^{-1} for 24 h at 32°C immediately following dark incubation at a challenge temperature of 48°C for 0, 5, 10, 15, 20, 25, 30, and 35 min. Because growth temperature affects the temperature range for Chl accumulation and the number of days after planting in which chlorophyll accumulation occurs in cucumber cotyledons (Burke and Oliver, 1993), cotyledons from 32°C grown cucumber seedlings at 5 DAP and from 20°C grown cucumber seedlings at 13 DAP were evaluated. Chl accumulation and TTC reduction experiments were replicated 2 to 4 times, and a minimum of 3 cotyledons were evaluated for each treatment within an experiment. The Chl accumulation assay also was evaluated in wheat leaves and cotton cotyledons from seedlings grown for 2 weeks at 20°C. The tissue was incubated at 40°C for 4 h, and challenged for 30 min at 47°C, and incubated at 32°C in the light for 24 h.

---

[2]Mention of a trademark or proprietary product does not constitute a guarantee or warranty of the product by the United States Department of Agriculture, and does not imply its approval to the exclusion of other products that may also be suitable.

The effectiveness of a 30 min incubation at challenge temperatures of 32, 44, 46, 48, 50, 52, 54, or 56°C in preventing Chl accumulation or TTC reduction was evaluated. Cotyledons were excised at 14 DAP from seedlings that had been grown at 20°C. The etiolated cotyledons were preincubated for 4 h at either 32°C or 40°C before exposure to the challenge temperature. The temperature settings of the blocks were adjusted to the assigned challenge temperatures, incubated for 30 min, and the block temperatures adjusted to 32°C. Block temperatures reached set points within 1 min of adjustment. Cotyledons used for TTC reduction were removed from the temperature blocks, placed in 3 ml of 0.8% TTC in 50 mM potassium phosphate buffer at pH 7.0, vacuum infiltrated, and placed in a 32°C water bath for 24 h. The level of TTC reduction was determined as described by Krishnan et al., 1989, following extraction into 95% ethanol. Cotyledons used for the Chl accumulation studies were left on the temperature blocks at 32°C and exposed to light as described above. Chl concentrations were determined according to the procedure of Arnon (1949). The percent injury was determined by using the 32°C challenge temperature as the control.

To evaluate cellular stress perception within the developmentally programmed change in cucumber cotyledons from a protein and lipid-storing heterotrophic organ into a photosynthetic one, the inherent metabolic sensitivity to a 30 min 47°C challenge temperature was evaluated in cotyledons of 20°C seedlings at 7, 9, 11, 13, and 15 DAP. Besides determining the inherent sensitivity, the temperatures that induce acquired thermotolerance to a 30 min 47°C challenge temperature of the metabolic constituents responsible for Chl accumulation were evaluated by preincubating cotyledons for 4 h at 32, 34, 36, 38, 40, 42, 44, and 46°C before the temperature challenge.

The time course for the induction of the acquired thermotolerance was determined for cotyledons from seedlings grown for 14 DAP at 20°C and for 6 DAP at 32°C. Etiolated cotyledons were placed on temperature blocks set for 32°C. The temperatures were adjusted to 40°C for 240, 180, 120, 60, 30, and 15 min. These temperature blocks and one that had been kept at 32°C were then adjusted to 47°C for 30 min. Following the 30 min challenge the temperature blocks were reset to 32°C. The cotyledons on these blocks and on another that had only experienced 32°C (control) were exposed to light and Chl accumulation monitored.

RESULTS

Initial studies evaluated the effect of the duration of a 48°C challenge temperature on the subsequent chlorophyll accumulation at 32°C in control and heat shocked (40°C for 4 h) cucumber cotyledons. Chlorophyll accumulation in control cotyledons of the 13 DAP 20°C seedlings and the 5 DAP 32°C seedlings declined following as little as 5 min at 48°C (Fig. 1). Chl accumulation was blocked completely by a 30 min exposure to 48°C even after 24 h at 32°C. Cotyledons that were heat shocked for 4 h immediately before the 48°C challenge did not loose the ability to accumulate Chl. Similar results were observed in wheat leaves and cotton cotyledons (Fig. 2).

The temperature response curve for preventing Chl accumulation or TTC reduction following a 30 min incubation at challenge temperatures of 32, 44, 46, 48, 50, 52, 54, or 56°C was evaluated. Chl accumulated in cotyledons preincubated at 32°C for 4 h that were challenged at 32 and 44°C (Fig. 3). No Chl accumulation occurred in these control cotyledons when challenged at 46, 48, 50, 52, 54, or 56°C. Protection against the 46 and 48°C challenge temperatures was observed following a 4 h 40°C heat shock pretreatment. TTC reduction in the cotyledons of the control treatment remained high from 32 to 52°C (Fig. 3). The level of TTC reduction declined at 54 and 56°C. Protection against the 54 and 56°C challenge temperatures were observed following a 4 h 40°C heat shock pretreatment.

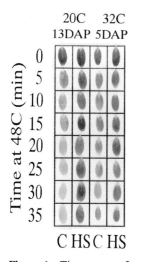

Figure 1. Time course for 48°C inhibition of Chl accumulation at 32°C in control (C) and heat shocked (HS) cucumber cotyledons. Cotyledons were excised from 20°C seedlings 13 DAP and 30°C seedlings 5 DAP.

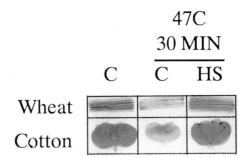

Figure 2. Acquired thermotolerance of Chl accumulation in wheat leaves and cotton cotyledons.

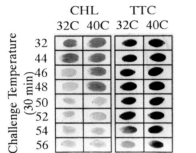

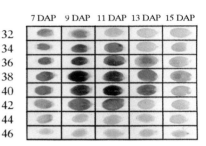

Challenge Temperature 47C for 30 min

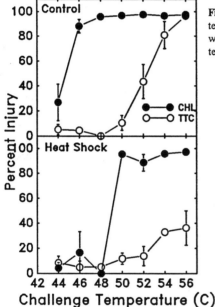

Figure 4. Photograph showing changes in the temperature sensitivity of Chl accumulation associated with developmental age and preincubation temperature.

Figure 3. Temperature sensitivity of Chl accumulation and TTC reduction in control (32°C) and heat shocked (40°C) cotyledons. The percent injury of control and heat shocked cotyledons is presented graphically.

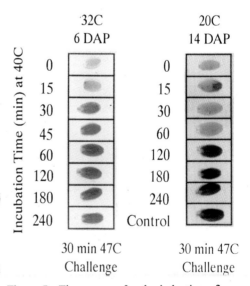

Figure 5. Time course for the induction of thermotolerance in cotyledons from seedlings grown at 32°C for 6 days and 20°C for 14 days.

The levels of inherent and acquired thermotolerance were evaluated in cotyledons from 20°C grown seedlings at 7, 9, 11, 13, and 15 DAP (Fig. 4). Developmental age and preincubation temperature affected the levels of inherent and acquired thermotolerance. A low, yet significant level of inherent thermotolerance was evident in the 7 and 9 DAP cotyledons preincubated at 32°C control temperature. This inherent protection, however, was not apparent at 11, 13 or 15 DAP. An inducible thermotolerance could be seen to peak at 9 and 11 DAP, and declined with age at 13 and 15 DAP. The range of temperatures associated with the induction of thermotolerance was broad at 7 and 9 DAP and narrowed with age. Maximum Chl accumulation occurred following 38 and 40°C preincubations across all dates tested.

The time required for the induction of acquired thermotolerance of Chl accumulation was determined for cotyledons of seedlings grown at 32°C for 6 days and seedlings grown at 20°C for 14 days (Fig. 5). Chl accumulation occurred in cotyledons preincubated at 40°C for 30, 45, 60, 120, 180, and 240 min in cotyledons of the seedlings grown at 30°C and for 120, 180 and 240 min in the seedlings grown at 20°C. The "control" cotyledon shown in Fig. 4. was kept at 32°C for 270 min and was not exposed to the 40°C incubation or the 47°C challenge temperature. The level of Chl accumulation in the "control" cucumber cotyledon was the same as the level in the cotyledons incubated at 40°C for 120, 180, and 240 min, challenged at 47°C for 30 min, and placed in the light at 32°C for 24 h.

DISCUSSION

This paper describes a new and rapid test for monitoring heat injury and acquired thermotolerance in plant tissues that accumulate Chl. This procedure is more sensitive than the TTC reduction assay (Fig. 3; and Caldwell, 1993) or electrolyte leakage (Wu and Wallner, 1983). The inhibition of Chl accumulation following the 46°C treatment is similar in magnitude to the injury reported for pear suspension cultures following a 20 min 46°C treatment (Wu and Wallner, 1983). Cellular regrowth capacity, triphenyltetrazolium chloride reduction, and electrolyte leakage have been used effectively in identifying heat injury in plant tissues (Wu and Wallner, 1983; Krishnan et al., 1989; Caldwell, 1993). Each of these techniques have advantages and disadvantages. Cellular regrowth capacity described by Wu and Wallner (1983) provides a sensitive measure of heat injury but the procedure is time consuming compared to TTC and electrolyte leakage measurements. TTC and electrolyte leakage techniques are rapid and easily scaled up for analysis of large sample numbers, however these techniques have high temperature coefficients and are not adequate substitutes for the regrowth test in assessing heat injury (Wu and Wallner, 1983; Caldwell, 1993). The chlorophyll accumulation test described in this paper provides a rapid and sensitive alternative to previous techniques used to monitor viability. This technique uses an in vivo indicator and eliminates the problems with chemical infiltration associated with the TTC reduction system. The only limitation of the chlorophyll accumulation test is that it requires plant tissues that have the capacity to accumulate Chl.

One aspect of the characterization of acquired thermotolerance in cucumber cotyledons is that it avoids the temperature sensitivity of cell division present in regrowth studies of cultured plant cells. Comparative studies of cotyledons and meristematic tissues may provide insight into the relative proportion of heat injury that is attributable to cell division.

Developmental changes in acquired thermotolerance monitored by the Chl accumulation test have been shown in this study (Fig. 4). A recent review of the literature on heat shock protein (HSP) expression during plant development showed that the induction of HSP synthesis in response to high-temperature stress occurs in most transcriptionally active tissues (Vierling, 1991). The two developmental stages in which full activation did not occur were very early embryo development and pollen germination. The expression of heat shock proteins in the absence of elevated temperatures has been reported for several eukaryotes (Bond and Schlesinger, 1987). In plants, HS mRNA and corresponding protein accumulation during seed development has been characterized (Helm and Abernethy, 1990). The heat shock response has been studied in germinating wheat embryos and has been shown to mount a complete heat shock response from the earliest time of imbibition (Helm et al., 1989). Helm reported that the embryos could synthesize a complete set of heat shock proteins simultaneously with the initiation of imbibition. Of particular interest was their report of several HSPs of 60, 58, 46, 40 and 14 kD that were expressed very early in germination and detectable only at early imbibition times. Abernethy et al. (1989) reported that the embryos were unusually tolerant to high temperatures during early germination; and Helm et al. (1989) reported that this tolerance diminished after 6 to 8 hours of imbibition, coincident with the loss of ability to make these proteins. The mechanism behind the observed inherent and acquired thermotolerance present in cucumber cotyledons determined by the chlorophyll accumulation test (Fig. 4) remains to be determined. Whether the inherent tolerance seen in the 7 and 9 DAP cotyledons preincubated at 32°C is related to the presence of heat shock proteins in the cotyledons is not known. This observation may be similar to the hsc70 species present in mung bean seeds that diminishes after germination (Wang and Lin, 1993). Clearly, the cucumber cotyledon experimental system provides an ideal foundation for a detailed biochemical and molecular analysis of the mechanisms responsible for both inherent and acquired heat tolerance.

Finally, the observation that different induction times for protection of Chl accumulation were required in the cucumber cotyledons grown at 32°C for 6 days and the 20°C grown seedlings for 14 days (Fig. 5) suggests that these cotyledons do not perceive either the 40°C heat shock or the 47°C challenge temperature to the same extent. Using this experimental system, basic questions concerning the nature of temperature stress perception and acquired thermotolerance can be addressed.

In summary, this paper describes developmental changes in the time course for induction of acquired thermotolerance, shows developmental changes in the characteristics of inherent and acquired thermotolerance, and describes a new procedure for monitoring heat injury in tissues able to produce Chl.

Acknowledgements: The author thanks Thomas Mahan and Jacob Sanchez for their technical assistance throughout the course of this research.

REFERENCES

Abernethy RH, Thiel DS , Petersen NS, Helm KW (1989) Thermotolerance is developmentally dependent in germinating wheat seed. Plant Physiol 89:569-576

Arnon DI (1949) Copper enzymes in isolated chloroplasts. Polyphenoloxidase in *Beta vulgaris*. Plant Physiol 24:1-15

Bond U, Schlesinger MJ (1987) Heat shock proteins and development. Adv Genet 24:1-28

Burke JJ, Oliver MJ (1993) Optimal Thermal Environments for Plant Metabolic Processes (*Cucumis sativus* L.). Light harvesting chlorophyll a/b pigment-protein complex of photosystem II and Seedling Establishment in Cucumber. Plant Physiol 102:295-302

Caldwell CR (1993) Estimation and analysis of cucumber (Cucumis sativus L.) leaf cellular heat sensitivity. Plant Physiol **101**: 939-945

Cantliffe DJ (1981) Priming of lettuce seed for early and uniform emergence under conditions of environmental stress. Acta Hort **22**:29-38

Carpenter WJ (1989) *Salvia splendens* seed pregermination and priming for rapid and uniform plant emergence. J Amer Soc Hort Sci **114**:247-250

Carpenter WJ, Boucher JF (1991) Priming improves high-temperature germination in pansey seed. HortSci **26**:541-544

Helm KW, Abernethy RH (1990) Heat shock protein and their mRNAs in dry and early imbibing embryos of wheat. Plant Physiol **93**:1626-1633

Helm KW, Petersen NS, Abernethy RH (1989) Heat shock response in germinating embryos of wheat. Plant Physiol **90**:598-605

Heydecker W (1977) Stress and seed germination: An agronomic view, p.237-276. In: AA Khan (ed.). The physiology and biochemistry of seed dormancy and germination. North Holland Publishing Co., Amsterdam.

Krishnan M, Nguyen HT, Burke JJ (1989) Heat shock protein synthesis and thermal tolerance in wheat. Plant Physiol **90**:140-145

Parera CA, Cantliffe DJ (1992) Priming leek seed for improved germination and emergence at high temperature. HortSci **27**:1077-1079

Rumpel J, Szudyga I (1978) The influence of pre-sowing seed treatments on germinations and emergence of tomato 'New Yorker' at low temperatures. Sci Hort **9**:119-125

Simmonds J (1980) Increasing seedling establishment of *Impatiens wallerana* in response to low temperature or polyethylene glycol seed treatments. Can J Plant Sci **60**:561-569

Vierling E (1991) The roles of heat shock proteins in plants. Ann Rev Plant Physiology Plant Mol Biol **42**:579-620

Wang C, Lin B-L (1993) The disappearance of an hsc70 species in mung bean seed during germination: purification and characterization of the protein. Plant Mol Biol **21**:317-329

Wu M-T, SJ Wallner (1983) Heat stress responses in cultured plant cells. Development and comparison of viability tests. Plant Physiol **72**:817-820

CROSS PROTECTION OF ONE STRESS BY ANOTHER: STRATEGIES IN
POSTHARVEST FRUIT STORAGE

Susan Lurie, Joshua D. Klein, Elazar Fallik
Department of Postharvest Science
Volcani Center, ARO
Bet Dagan 50250 Israel

INTRODUCTION

An ability to modify metabolism and development in response to
changes in the environment is essential to nonmobile organisms such
as plants. These environmental changes can include extremes in
temperature, excess of water causing anaerobic conditions, lack of
water, the presence of salinity or heavy metals. All of these
conditions have been shown to induce altered gene expression as one
response to stress, and in some cases this results in a dramatic
change in protein synthesis (Matters and Scandalios, 1986; Sachs
and Ho, 1986). Stress such as anaerobiosis (Lin and Key, 1967),
water deficits (Hsaio, 1970; Rhodes and Matsuda, 1976), heat shock
(Key et al., 1981), and heavy metals (Orzech and Burke, 1988) all
lead to a rapid shift of polysomes to monosomes and subunits, and
a low rate of total protein synthesis. Among the stresses, heat
shock and anaerobiosis reduce the rate of total protein synthesis
by inhibiting the synthesis of pre-stress proteins, while heat
stress and anaerobic proteins are actively synthesized (Key et al.,
1981; Sachs et al., 1980). The alteration of the protein pattern
is a reflection of both transcriptional (of stress related
proteins) and translational (selective for stress induced mRNAs)
regulation (Key et al, 1981; Matters and Scandalios, 1986; Sachs
and Ho, 1986; Sachs et al., 1980).

There are a number of studies showing that the synthesis of
heat shock proteins will protect cells from deleterious effects of
extreme temperatures, i.e. thermotolerance (see Vierling, 1992).
Fewer studies have documented protection of one stress by exposing
a plant to another stress. This cross protection has been shown
for freezing tolerance and water stress in grain seedlings
(Cloutier and Andrews, 1984; Siminovitch and Cloutier, 1982), heat
stress and salt stress (Harrington and Alm, 1988; Kuznetsov et al.,

NATO ASI Series, Vol. H 86
Biochemical and Cellular Mechanisms
of Stress Tolerance in Plants
Edited by J. H. Cherry
© Springer-Verlag Berlin Heidelberg 1994

1993), heat stress and heavy metal stress (Bonham-Smith et al., 1987; Lin et al., 1984; Orzech and Burke, 1988), and water stress and heat stress (Bonham Smith et al., 1987). The explanations for these cross protections have been many and varied, and did not always involve the synthesis of specific stress proteins. The interaction between water stress and freezing tolerance was suggested to be due to a diversion of nutrients from growth to making membranes and augmented protoplasm (Siminovitch and Cloutier, 1982). The interaction between heat stress and salt or heavy metal stress may involve the accumulation of small molecular weight compounds such as putrescine, osmotin-like proteins, proline or linear polymers of glutathione (Bohnam-Smith et al., 1989; Kuznetsov et al., 1993; Orzech and Burke, 1988). In some studies protection was achieved without the synthesis of stress-specific proteins (Bonham-Smith et al., 1987).

Therefore, although cross protection of one stress by another has been demonstrated in a number of plants for various stresses, a commonality of action has not been found. It may be that a plant can have a number of ways of dealing with a stress situation and particular studies focus on one of the methods to the exclusion of others. Alternatively, different plants may have different mechanisms of defense for the same stress. What is still missing is an overall picture tying the various studies, results and explanations together into a coherent whole.

Harvested fruits differ from most tissue used to investigate stress situations in that they are no longer growing, and as such may have more limited means at their disposal to respond to stress situations. Nonetheless, fruits are often exposed to stress in the form of low temperature storage, sometimes combined with controlled atmosphere. These conditions can lead to tissue damage in the form of chilling injury and storage disorders. If the fruit is climacteric and not stored at a low temperature, ripening will continue during storage. Therefore, a compromise is often to hold fruit at a temperature which slows ripening without causing storage disorders, thus limiting the possible period of storage. In an attempt to increase the effective storage life of climacteric fruits, such as apples and tomatoes, we have been trying prestorage heat stress treatments (Klein and Lurie, 1990; Klein and Lurie, 1992; Lurie and Klein, 1991). The results show that this stress can: a) prevent or delay the development of physiological disorders, b) reversibly inhibit fruit ripening, c) reduce pathogen infection. The underlying connection among these three major effects appears to be that they are controlled by changes in gene expression during and after the stress treatment, whether by inhibition of normal protein synthesis or continued accumulation of stress proteins.

PHYSIOLOGICAL DISORDERS

Apples, which can be stored at low temperatures for months if harvested at the proper stage of ripeness, will develop disorders of both peel and flesh under certain conditions. One of the most dramatic of these disorders is superficial scald, causing browning of the fruit peel. The development of this disorder has been correlated with the oxidation of α-farnesene to conjugated tirenes, compounds which accumulate in apple peel and cuticle during storage (Anet and Coggiola, 1974). The oxidized products are thought to cause membrane damage leading to leakage of phenols from the vacuole and their polymerization by polyphenoloxidase (PPO) causing the brown discoloration (Bramlage and Meir, 1990). The commercial practice to prevent this disorder is a prestorage dip in an antioxidant such as diphenylamine (DPA) or ethoxyquin (Smock, 1961). However, we have found that giving apples a period of heat stress before storage will protect them from developing superficial scald (Table 1, Lurie et al., 1990).

It is clear that heat stress at 38°C requires a longer time of treatment to be as effective as a higher temperature. One day of temperature stress at 38°C has no effect on scald development, while one day at 42°C partially ameliorates the disorder. In addition, DPA prevents scald development by inhibiting the oxidation of α-farnesene to conjugated trienes, while the heat treatments decrease the accumulation of α-farnesene.

Table 1. Development of superficial scald on 'Granny Smith' apples after 4 months of 0°C storage plus 1 week at 20°C, and peel concentration of α-farnesene and conjugated trienes

Treatment	Scald (%)	Conjugated Trienes (μmoles/cm^2 peel)	α-Farnesene (μmoles/cm^2 peel)
Untreated	82	18.7	100
12 h 46°C	22	2.6	53
24 h 42°C	17	3.5	55
24 h 38°C	75	17.7	101
72 h 38°C	2	2.5	52
DPA	3	2.4	99

The activity of polyphenol oxidase is also affected by heat treatment to a similar extent as DPA dip (Table 2).

Table 2. Activity of polyphenoloxidase (OD 420 min^{-1} mg protein^{-1}) in the skin of 'Granny Smith' apples after 4 months of 0°C storage.

Treatment	Polyphenoloxidase
Control	1.38
12 h 46°C	0.56
24 h 38°C	1.60
72 h 38°C	0.47
DPA	0.61

Our interpretation of these results is that the heat stress treatment decreases superficial scald by inhibiting the synthesis of α-farnesene and polyphenoloxidase, in contrast to DPA which exerts an antioxidative effect. We have not, however, demonstrated inhibition of synthesis directly.

Tomatoes, in contrast to apples, can be stored for only a couple of weeks at low temperature before they develop the physiological disorder of chilling injury. A prestorage heat treatment can also prevent this disorder (Table 3).

Table 3. The development of chilling injury as evidenced by necrotic areas, the absence of color development, and susceptibility to rots on mature green tomatoes stored for 3 weeks at 2°C plus 5 days at 20°C.

Treatment	Necrotic Areas (%)	Red Fruit (%)	Rots (%)
Unheated	60	30	30
72 h 38°C	0	90	5

Chilling injury is thought to be a membrane related disorder (Lyons, 1973) and there is a large body of evidence that alteration in the amount and composition of membrane phospholipids and sterols can protect tissues from chilling injury (Guy, 1990). Recently, Murata and coworkers (1992) have caused overexpression of a lipid acyl transferase in tobacco and demonstrated that because of the position of the unsaturated fatty acids in the glycolipids these plants are more resistant to chilling injury. What we have observed in tomato fruit with chilling injury is that there is a decrease in total phospholipid from the microsomal membrane fraction with no change in sterol content (Table 4). This decrease in phospholipid is accompanied by an increase in leakage in fruit discs. A prestorage heat treatment prevents both these phenomena. Again, the hypothesis is that as a consequense of the heat stress synthesis of enzyme(s) responsible for phospholipid degradation is inhibited.

Table 4. Microsomal membrane phospholipid and sterol content and leakage of K^+ from fruit discs in tomatoes stored for 3 weeks at 2°C.

Treatment	Phospholipid (μg.100 mg DWt)	Sterol	K^+ Leakage (% of total)
Unheated	804	190	49
72 h 38°C	1350	178	34

FRUIT RIPENING

In contrast to the effect of heat stress on physiological disorders where the mode of action is still only a hypothesis, the effect on fruit ripening has been shown to be correlated to gene expression. Picton and Grierson (1988) found that when tomatoes were held at 40°C the expression of a number of ripening related mRNAs were inhibited. We have found that these mRNAs continued to be supressed during storage following a heat treatment and expression recovered when the fruits were removed to 20°C (Figure 1).

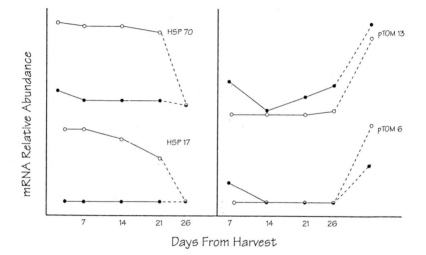

Figure 1. Relative abundance of pTOM6, pTOM13, HSP17 and HSP70 mRNA in tomatoes heated (filled circles) or unheated (open circles) before 2°C storage. Dotted line represents removal of the tomatoes to 20°C after 21 days of storage.

Both pTOM6 and pTOM13, which are cDNA clones for polygalacturonase (PG) and ACC oxidase, respectively, are inhibited in tomatoes during heat stress and remain low during subsequent storage at 2°C. In unheated tomatoes PG mRNA expression is also inhibited by low temperatures, but ACC oxidase message is present during storage.

Two messages for heat shock proteins, HSP17 and HSP70, were expressed to a high degree during heat stress and their mRNA coninued to be present during 2°C storage.

One consequense of the inhibition of ripening related gene expression during storage is that ripening processes, which continue, albiet slowly, at low temperatures, do not occur in the heat stressed fruit, so that at the end of storage unheated fruits are more advanced in ripeness than the heated fruit. This is dramatic in the case of apple firmness (Figure 2).

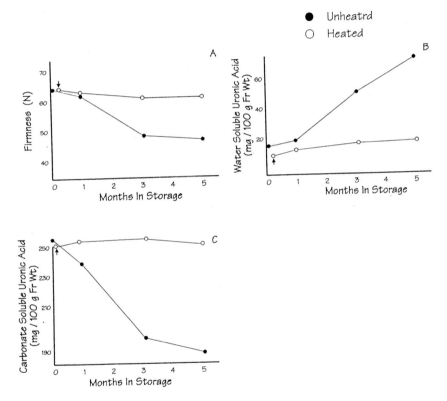

Figure 2. Changes in firmness (A), water soluble pectin (B) and carbonate soluble pectin (C) in Golden Delicious apples during 5 months of 0°C storage. Open circles, heated fruit; closed circles, unheated fruit.

Firmness decreased in unheated 'Golden Delicious' apples stored at 0°C and correlated well with cell wall changes - increase in water soluble- and decrease in carbonate soluble-pectin. The heated fruit did not soften in storage, but ripened normally after removal from storage.

Therefore, the effect of heat stress before storage on lengthening storage and shelf life appears to be due to its inhibition of ripening processes which require gene expression and new protein synthesis. These processes, which include ethylene production and cell wall softening, are repressed by heat stress and the repression is maintained during low temperature storage, being relieved when the fruits are removed from storage.

PATHOGEN INFECTION

As fruits ripen, they become more prone to pathogen attack. This is due to a number of factors, among them disappearance of antifungal compounds from the tissues and the weakening of the physical barrier of the cell wall as the fruit softens. By inhibiting the ripening processes the heat stress treatment will help lower the possibility of fungal invasion. In certain cases a stress treatment will induce the formation of antifungal compounds, phytoalexins, similarly to the effect of pathogen invasion itself. Heat stress of grapefruit fruits will enhance the production of the antifungal compound scoparone in the peel (Kim, et al., 1991). Stress will also induce phenylalanine ammonia lyase (PAL) synthesis in citrus, tomatoes and other fruits (Golumb et al., 1984; Lurie, unpublished). This is one of the key enzymes in the production of many compounds with antifungal activity. In addition, a stress can raise the level of preformed antifungal compounds. Avocados contain an antifungal compound which decreases during ripening (Prusky and Keen, 1993). Giving the fruit a high CO_2 treatment for 24 h before storage led to enhanced levels of this compound and a delay in development of *Colletotrichum gloeosporiodes* rots (Prusky et al., 1991).

Another mode of action of heat is to inhibit either germination or growth of fungal spores. The heat treatment has been found to be effective in abolishing infection of *Botrytis cinerea* on tomatoes (Fallik et al., 1993). The germination of spores is inhibited within the first 24 h at 38°C, and after longer times of heat stress the development of germinated spores is inhibited (Figure 3). Thus, a heat stress can be beneficial in reducing pathogen infection either by enhancing tissue resistance, inhibiting fungal development, or both, depending on the fruit and pathogen investigated.

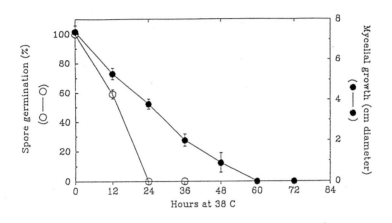

Figure 3. Effect of increasing times at 38°C on spore germination and mycelial growth of *Botrytis cinerea* grown at 25°C.

CONCLUSIONS

The means by which a prestorage heat stress will protect fruit tissue against low temperature, pathogen infection and delay ripening appears to be due to its effect on gene expression. Many genes, including those involved in ripening are inhibited, while others including heat shock proteins are activated. Upon closer examination this may be the commonality of action of cross protection in all systems. The difference may be as to the relative importance of the subset of stress proteins synthesized in protection, or inhibition of syntheses of particular proteins. For example, the effect of heat stress on ripening clearly is due to the inhibition of proteins involved in ripening. However, it is unclear to what extent the protection given by heat stress against chilling injury is due either to inhibition of enzymes involved in lipid breakdown, or the persistance of heat shock proteins during the low temperature. Heat shock proteins have a protective effect in thermotolerance, and may have a similar function in low temperature.

Other stress prestorage treatments can affect fruit storage similarly to heat stress. An anaerobic treatment has been found to prevent chilling injury in avocados (Pesis, et al., 1993) and to delay ripening in nectarines (Lurie and Pesis, 1992). Ultraviolet irradiation causes the accumulation of scoparone in grapefruit similarly to heat stress (Kim et al., 1991). Therefore, this type of prestorage manipulation shows promise in a number of ways, from reducing the use of fungicides to extending storage without quality loss.

REFERENCES

Anet E, Coggiola I (1974) Superficial scald a functional disorder of stored apples. X. Control of a-farnesene oxidation. J Sci Fd Agric 25:793-798.

Bonham-Smith PC, Kapoor M, Bewley JD (1987) Establishment of thermotolerance in maize by exposure to stresses other than a heat shock does not require heat shock synthesis. Plant Physiol. 85:575-580.

Bramlage WJ, Meir S (1990) Chilling injury of crops of temperate origin. In: Chilling Injury of Horticultural Crops (CY Wang, ed) CRC Press, Boca Raton, FL pp 37-50.

Cloutier Y, Andrews CJ (1984) Efficiency of cold hardiness induction by desiccation stress in four winter cereals. Plant Physiol 76:595-598.

Fallik E, Klein J, Grinberg S, Leomniec E, Lurie S, Lalazar E (1993) Postharvest heat treatment of tomatoes: Effect on Botrytis cinerea and fruit ripening. Plant Disease, in press.

Golumb A, Ben Yehoshua S, Sarig Y (1984) High density polyethylene wrap improves wound healing and lengthens shelf life of mechanically harvested grapefruit. J Amer Soc Hort Sci 109:155-159.

Guy CL (1990 Cold acclimation and freezing stress tolerance: role of protein metabolism. Annu Rev Plant Physiol Plant Mol Biol 41:187-223.

Harrington HM, Alm DM (1988) Interaction of heat and solt shock in cultured tobacco cells. Plant Physiol 88:618-625.

Hsaio TC (1970) Rapid changes in levels of polyribosomes in Zea mays in response to water stress. Plant Physiol. 46:281-285.

Key JL, Lin CY, Chen YM (1981) Heat shock proteins of higher plants. Proc Natl Acad Sci USA 78:3526-3530.

Kim JJ, Ben Yehoshua S, Shapiro B, Henis Y, Carmeli S (1991) Accumulation of scoparone in heat treated lemon fruit inoculated with *Penicillium digitatum* Sacc. Plant Physiol. 97:880-885.

Klein JD, S Lurie (1990) Prestorage heat treatment as a means of improving postharvest quality of apples. J Amer Soc Hort Sci 115:265-269.

Klein JD, S Lurie (1992) Heat treatment for improved postharvest quality of horticultural crops. HortTech 2:316-320.

Kuznetsov V, Rakutin V, Boisova N, Rotschupkin B (1993) Why does heat shock increase salt resistance in cotton plants. Plant Physiol Biochem 31:180-188.

Lin CY, Key JL (1967) Dissociation and reassembly of polyribosomes in relation to protein synthesis in the soybean root. J Mol Biol 26:237-247.

Lin CY, Roberts JK, Key JL (1984) Acquisition of thermal tolerance in soybean seedlings. Plant Physiol 74:152-160.

Lurie S, JD Klein (1991) Acquisition of low temperature tolerance in tomatoes by exposure to high temperature stress. J Amer Soc Hort Sci 116:1007-1012.

Lurie S, JD Klein, R Ben Arie (1990) Prestorage heat treatment of apples as a means to reduce superficial scald. J Hort Sci 65:317-321.

Lurie S, Pesis E (1992) The effect on peach and nectarine quality and softening of a postharvest acetaldehyde vapor treatment or anaerobic conditions. Postharvest Biol and Technol 1:317-326.

Lyons JM (1973) Chilling injury in plants. Annu Rev Plant Physiol 24:445-466.

Matters GL, JG Scandalios JG (1986) Changes in plant gene expression during stress. Dev Genetics 7:167-175.

Murata N, Ishizaki-Nishizawa O, Higashi S, Hayashi H, Tasaka Y, Nishida I (1992) Genetically engineered alteration in the chilling sensitivity of plants. Nature 356:710-713.

Orzech KA, Burke JJ(1988) Heat shock and the protection against metal toxicity in wheat leaves. Plant Cell Envir 11:711-714.

Pesis E, Marinansky R, Zauberman G, Fuchs Y (1993) Reduction of chilling injury symptoms of stored avocado fruit by prestorage treatment with high nitrogen atmosphere. In: Postharvest (M Saltveit, ed.) Acta Hortic., in press.

Picton S, Grierson D (1988) Inhibition of expression of tomato-ripening genes at high temperature. Plant Cell Envir 11:265-272.

Prusky D, NT Keen (1993) Involvement of preformed antifungal compounds in the resistance of subtropical fruits to fungal decay. Plant Dis 77:114-119.

Prusky D, RA Plumbley, I Kobiler (1991) Modulation of natural resistance of avocado fruits to *Colletotrichum gloeosporioides* by CO_2 treatment. Physiol Mol Plant Path 39:325-334.

Rhodes PR, Matsuda K (1976) Water stress, rapid polysome reductions and growth. Plant Physiol 58:631-635.

Sachs MM, Freeling M, Kimoto RO (1980) The anaerobic proteins of maize. Cell 20:761-767.

Sachs MM, Ho T-HD (1986) alteration of gene expression during environmental stress in plants Annu Rev Plant Physiol 37:363-376.

Siminovitch D, Cloutier Y (1982) Twenty four hour induction of freezing and drought tolerance in plumules of winter rye seedlings by dessication stress at room temperature in the dark. Plant Physiol 69:250-255.

Smock RM (1961) Methods of scald control on apple. Bull Cornell Univ Agric Exp Sta 970:1-55.

Vierling E (1991) The role of heat shock proteins in plants. Annu Rev Plant Physiol Plant Mol Biol. 42:579-620.

Photosynthesis, Epicuticular Wax and Lipid Changes In Cowpea Cultivars Grown Under Hyperthermic Conditions

Henry Daniell, Guda Chittibabu, Narendra K. Singh, John D. Weete and Joe H. Cherry
Department of Botany and Microbiology
Auburn University, Auburn, AL 36849-5407

INTRODUCTION

The cellular mechanisms of long-term thermoadaptation and acquisition of thermotolerance in plants have not yet been explored. Characteristic changes in membrane lipid are associated with exposure to elevated temperatures, such as reduced $C_{18:3}$, and increased $C_{18:2}$ and $C_{16:0}$, i.e. an overall reduction in unsaturation (Pearcy, 1979; Raison et al., 1982). Using mutants of *Arabidopsis thaliana* (a "$C_{16:3}$" plant) deficient in various aspects of lipid metabolism, Somerville and his associates have associated specific shifts in lipid composition with thermotolerance (Somerville and Browse, 1991). With a mutant deficient in n-6 desaturase activity, which accumulated $C_{18:1}$ and $C_{16:1}$ in acyl lipids, Hughly et al. (1989) showed that lipid unsaturation directly affects the thermal stability of photosynthetic membranes. Enhanced thermotolerance was also exhibited in a mutant deficient in $C_{16:0}$ unsaturation (Kunst et al., 1989b). Results of studies where the membrane lipids were catalytically hydrogenated led Thomas et al. (1986) to suggest that decreased unsaturation raises the temperature at which non-bilayer forming lipids (MGDG) phase separate to non-bilayer structures which disrupt the thylakoid structure. Mutants deficient in plastid glycerol-3-phosphate acyl-transferase activity show a 15-20% reduction in phosphatidyl glycerol (PG) and minor alterations in other complex lipids (Kunst et al., 1988), and exhibit "$C_{18:3}$" lipid metabolism, grew slightly more rapidly at elevated temperatures (Kunst et al., 1989c).

Orr and Raison (1987) studied the relation between the phase transition of polar lipids from thylakoids of oleander grown at 20°C and 45°C and changes in the fatty acid composition of molecular species of PG. Although the composition of lipid classes was essentially the same in thylakoid membranes of plants grown at both temperatures, there were increases in the transition temperatures of both PG and sulfolipid. They concluded that the polar lipid phase transition appeared to be a product of mixing of both high and low melting point lipids.

The impact of a shift in favor of less unsaturation with elevated temperature is believed to improve the thermal stability of membranes. Membranes of plants grown at 45°C are less fluid than those of plants grown at 20°C (Raison et al., 1982). Differences in the thermal stability of photosystem II electron transport located in the chloroplast membrane can be demonstrated by chlorophyll fluorescence emission as a function of temperature. Thermal stability, as reflected by the temperature at which fluorescence increased with increasing temperature, increased 10°C for a 25°C difference in growth temperature (20°C → 45°C) in *Nerium oleander* leaves (Raison, 1985). From physical data, Raison (1985) suggested that the stability to high temperature stress is related to a balance between hydrophobic interactions holding lipids and proteins in the bilayer and the hydrophilic interactions between these components and water at the bilayer interface.

NATO ASI Series, Vol. H 86
Biochemical and Cellular Mechanisms
of Stress Tolerance in Plants
Edited by J. H. Cherry
© Springer-Verlag Berlin Heidelberg 1994

In order to gain insight into the mechanism of thermotolerance, we have studied two cultivars of cowpea differing in their ability to survive hyperthermic growth conditions. Seedlings of CB-5 (California Blackeye Pea #5) and NJC-1 (a cultivar identified in a field in Niger) germinated at 26°C were grown under hyperthermic conditions at 42°C with abundant supply of water. Changes taking place in variable fluorescence, fluorescence quenching, stomatal conductance, epicuticular wax formation and lipid composition are reported here.

MATERIALS AND METHODS

Plant Material and Growth Conditions:

Two cultivars of cowpea, *Vigna unguiculata*, differing in their ability to survive hyperthermic conditions were used. California Blackeye pea #5 (CB-5) is sensitive to heat stress. A cultivar identified from a field in Niger (NJC-1) is capable of growing at higher temperatures. Seedlings of these two cultivars were germinated at 26°C in a growth chamber for 10 days. For exposure to hyperthermic stress, germinated seedlings were transferred to another growth chamber maintained at 42°C. Plants were provided with plenty of water and ideal growth conditions. Elevated growth temperature was the only known environmental variation between plants grown at 26°C and 42°C. When measurements were made using the Chlorophyll Fluorescence Measurement System or the Porometer in intact leaves, every effort was made not to alter existing growth conditions or damage the plants.

Stomatal conductance:

Stomatal conductance (mmol m^{-2} sec^{-1}), the rate of transpiration (mmol m^{-2} sec^{-1}), leaf temperature (measured by a thermocouple in contact with the leaf in the cuvette), relative humidity inside the cuvette (%) and photosynthetically active radiation (PAR, μE m^{-2} sec^{-1}) were measured in a LI-1600 steady state porometer (LI-COR, Inc.) according to manufacturer's instructions. Three readings were recorded for each measurement. The primary leaf was used as the sample for all measurements. Measurements were made from both leaf surfaces.

Chlorophyll fluorescence:

Chlorophyll fluorescence measurements were made using a Morgan CF-1000 portable chlorophyll fluorescence measurement system. The instrument is centered around a unique optical block which utilizes a red reflecting, dichroic mirror set at an angle of 45°, to separate fluorescence from actinic light. The instrument measures both fast and slow fluorescence kinetics, but is particularly appropriate for measuring fast kinetics, capable of storing 2,000 sets of data. The fast sample time (20 μs for the first 40 ms, followed by a 1.0 ms rate for 4 seconds, followed by a 50 ms rate for up to 5 minutes) and strong variable actinic light (0-1,000 μmol m^{-2} s^{-1}), result in a high resolution capability and an excellent definition of F_0 (non-variable fluorescence) and F_m (maximal fluorescence). The built-in microprocessor calculates the following fluorescence parameters: F_0 - Non-variable fluorescence (efficiency of light absorption); F_m - Maximal fluorescence (absorbed light plus electron flow from photosystem II to Q); F_v - Variable fluorescence (electron flow from photosystem II to Q); F_v/F_m - Photochemical efficiency of PSII; $t_{1/2}$ - Half rise time from F_0 - F_m (time required to reduce the primary electron acceptor Q); F_t - Terminal fluorescence value (draining of electrons from Q); F_q - Fluorescence quenching capacity (photosystem I efficiency). Dark acclimation cuvettes with a shutter gate operation allows the fiber optic to be inserted without any light hitting the sample. The primary leaf was used in all measurements. Three readings were recorded for upper or lower surface of each leaf.

Scanning electron microscopy

Primary leaves from NJC-1 grown at 26⁰C and 42⁰C for 30 days and CB-5 grown at 26⁰C for 30 days were fixed at 4⁰C in the dark for 24 hrs. The fixation buffer contained 0.05M sodium cacodylate (pH 7.0) and 4% glutaraldehyde. Fixed samples were then dehydrated in a graded series of ethanol (10-100%) at 4⁰C for 10 min at each grade. The last dehydration with 100% ethanol was repeated twice at 4⁰C and once at room temperature. Ethanol was removed by critical point drying. Dried samples were attached to aluminum specimen mounts using colloidal silver, after identifying the upper or lower surface using a light microscope. The samples were sputter coated twice with gold-palladium in a SeeVac (Auto Conduct Vac IV) unit and were stored in a desiccator until viewed. Specimens were examined in a digital scanning electron microscope (Zeiss DSM 940) at 5 kV and a working distance of 12-13 mm.

Lipid extraction and analysis

Wax was removed from the surfaces of primary leaves by submerging them in chloroform for 45 s each. The crude wax residue was recovered by evaporating the solvent under nitrogen. Total lipids were extracted from the de-waxed leaves using the procedure of Bligh and Dyer (1959). The wax was separated into individual classes by thin-layer chromatography (TLC) on 20 cm x 20 cm glass plates coated with silica gel 60 (250 μm) with chloroform as the developing solvent. The waxes were visualized with iodine vapor. The wax classes were tentatively identified by comparing their TLC migration with those authentic standards. For further analysis by gas chromatography, wax classes were removed from the silica gel with chloroform: methanol (1:1 by vol) after TLC.

Individual components of the wax classes were analyzed using a Hewlett Packard gas chromatograph equipped with a 30 m x 0.25 mm fused silica capillary column coated with SE-54 (J&W, Folsom CA.). The column temperature was programmed from 150°C to 290°C at 4°C/min. Individual components were tentatively identified by comparison of their retention times with those of authentic standards. Free fatty acids of the wax and total fatty acids of the lipid were analyzed as their methyl ester derivatives using a Varian 3300 gas chromatograph equipped with a DB225 fused silica column (30 m x 0.25 mm) programmed from 140°C at 2°C/min (140°C-160°C), 10°C/min (160°C-180°C) and 1°C/min (180°C-220°C). Fatty acids were identified as before.

RESULTS AND DISCUSSION

Both NJC-1 and CB-5 showed similar growth rates at 26°C. Leaves of CB-5 when grown at 42°C lost their pigments and continued to wilt until they became completely dry on the eighth or tenth day. On the other hand, NJC-1 continued to grow at 42°C putting out several new

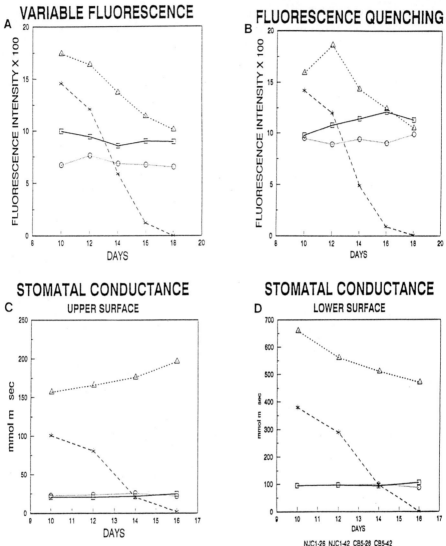

Figure 1. Measurement of various indicators of photosynthesis. Figure 1A indicates variable fluorescence and figure 1B indicates fluorescence quenching observed in CB-5 and NJC-1 grown at 26°C and 42°C. Fluorescence intensity is shown in relative units (x 100). Stomatal conductance of the upper (1C) and lower (1D) surface of CB-5 and NJC-1 grown at 26°C and 42°C.

leaves, most of which remained white for a few days before turning green. NJC-1 continued to grow for several weeks at 42°C although the growth rate was significantly lower than NJC-1 grown at 26°C.

Chlorophyll Fluorescence:

During the process of light absorption and energy transduction, a small portion of light is re-emitted from the chlorophylls associated with the reaction centers. This re-emitted light is termed chlorophyll fluorescence and is a strong, narrow optical signal with a peak at 685 nm. At physiological temperatures, the chlorophylls associated with Photosystem II are mainly responsible for a majority of the fluorescence emission. Upon a dark-light transition, rapid changes in chlorophyll fluorescence intensity occur within the first few moments of illumination. During the first few seconds of illumination, there is a fast rise phase (O,l,D,P) from an initial level (O) to the maximum fluorescence level (F_m). After reaching maximal level, fluorescence decreases to a quasi steady-state (S), followed by a second rise (M), from where it decays slowly to a terminal steady-state level (T). These complex kinetics are related to changes in photosynthetic function and the structural integrity of the thylakoid membrane system. For example, the ratio of variable fluorescence (F_v) to maximal fluorescence (F_m) is highly correlated to the quantum yield of Photosystem II, and hence is an estimation of quantum efficiency. The characteristics of chlorophyll fluorescence induction curves are modified by any factor which directly or indirectly affects photosynthetic metabolism. It follows, therefore, that measurements of chlorophyll fluorescence give unique possibilities to study environmental effects on the chloroplast thylakoid membrane system and can provide information on the course and mechanism of stress damage to the photosynthetic apparatus. For a detailed discussion on chlorophyll fluorescence see the following references: Duysens and Sweers, 1963; Daniell et al., 1981; Kulandaivelu and Daniell, 1980.

Variable Fluorescence:

Both CB-5 and NJC-1 when grown at 26°C, showed similar fluorescence kinetics. In CB-5, when grown at 26°C, F_o varied from 618 to 707 in the lower surface and 717 to 762 in the upper surface, indicating that there is a small variation in the non-variable fluorescence between the two leaf surfaces; F_m varied from 1395 to 1525 in the upper surface and between 1349 and 1380 in the lower surface; the quantum yield (F_v/F_m) varied from 0.45 to 0.52 in the upper surface to 0.48 to 0.54 in the lower surface. These values did not significantly change during growth of CB-5 at 26°C over the period of this study (Fig. 1 A). In NJC-1 grown at 26°C, F_o varied from 712 to 801 in the upper surface and 796 to 840 in the lower surface, again indicating that there is a small variation in non-variable fluorescence between the two leaf surfaces; F_m varied between 1276 to 1796 in the upper surface and between 1708 and 1933 in the lower surface; the quantum yield varied (F_v/F_m) from 0.55 to 0.67 in the upper surface and 0.46 to 0.53 in the lower surface. These values did not change significantly during growth of NJC-1 at 26°C.

However, when CB-5 germinated at 26°C was grown at 42°C, variable fluorescence, a direct indicator of quantum yield of photosystem II (F_m minus F_o) declined steadily; no variable fluorescence was detected after eight days of growth at 42°C. On the other hand, although NJC-

1 showed initial decline of variable fluorescence in the first few days, this decline stabilized upon continued growth at 42^{0}C (Fig. 1 A). The intensity of variable fluorescence observed in NJC-1 after eight days of growth at 42^{0}C was still higher than the variable fluorescence observed in control plants grown at 26^{0}C, indicating that photosystem II was still intact in the chloroplasts and that it was fully functional.

Fluorescence Quenching:

Fluorescence quenching was measured for two minutes; this is generally interpreted as the ability of photosystem I to drain electrons from the primary electron acceptor and therefore has been used as an index of photosystem I efficiency (Duysens and Sweers, 1963; Daniell et al., 1981). When CB-5 was grown at 26^{0}C, terminal fluorescence values (F_t) varied from 392 in the upper surface to 397 in the lower surface; in NJC-1 grown at 26^{0}C, terminal fluorescence varied from 295 in the upper surface to 271 in the lower surface. Fluorescence quenching capacity showed no significant variation when CB-5 and NJC-1 were grown at 26^{0}C (Fig.1 B). When CB-5 was grown at 42^{0}C, fluorescence quenching capacity decreased steadily and no fluorescence quenching was detected after eight days of growth at 42^{0}C. On the other hand when NJC-1 was grown at 42^{0}C, fluorescence quenching capacity showed a slight increase in the first two days followed by a slight decline; even after ten days of growth at 42^{0}C, NJC-1 showed as much fluorescence quenching capacity as the control grown at 26^{0}C. Thus, the photosystem I is intact and functional in NJC-1 after continued growth at 42^{0}C whereas CB-5 showed deterioration of photosystem I activity.

DCMU induced increase in chlorophyll fluorescence has been used in the past as an index to study photosynthetic oxygen evolution in leaves, chloroplasts and algae (Kulandaivelu and Daniell, 1980). Upon the addition of DCMU (by infiltration), terminal fluorescence increased from 392 to 1605 (upper surface) and from 397 to 1709 (lower surface) in CB-5 grown at 26^{0}C. Similarly, in NJC-1 grown at 26^{0}C, the terminal fluorescence increased from 295 to 1865 (upper surface) and from 271 to 1989 (lower surface). Thus, DCMU induced increase in chlorophyll fluorescence was normal in both NJC-1 and CB-5 plants grown at 26°C; however, after eight days of growth at 42°C, no DCMU induced increase in chlorophyll fluorescence was detected in CB-5. On the other hand, DCMU induced increase in chlorophyll fluorescence observed in NJC-1 plants grown at 42°C was higher than control plants grown at 26°C, indicating that the photosynthetic oxygen evolving apparatus is intact and functional.

Stomatal Conductance:

Stomatal conductance and the rate of transpiration were measured in both leaf surfaces. Leaf temperature was the same as cuvette temperature or the growth chamber temperature (26^{0}C or 42^{0}C). Humidity varied from 16 to 28% in the growth chambers during measurement. Photosynthetic Active Radiation varied from 12 to 14 μE m^{-2} sec^{-1} during measurements and the pressure was steady at 97.09 kPa. Stomatal conductance varied significantly between the upper and lower surface in both CB-5 and NJC-1. When grown at 26^{0}C, both CB-5 and NJC-1 showed a four fold higher rate of transpiration in the lower surface than in the upper surface (Fig. 1 C, D). This correlates well with the presence of about 33 stomates per mm^2 in the upper surface and about 145 stomates per mm^2 in the lower surface in both CB-5 and NJC-1 (see Fig. 2 for

Figure 2: Scanning electron micrographs of upper or lower surface of leaves of CB-5 or NJC-1 grown at 26°C or 42°C. A) CB-5, upper surface, 26°C, 2000X; B) NJC-1, upper surface, 26°C, 2000X; C) NJC-1, upper surface, 42°C, 2000X; D) NJC-1, upper surface, 42°C, 3000X; E) CB-5, upper surface, 26°C 3000X; F) NJC-1, upper surface, 26°C, 3000X; G) NJC-1, upper surface, 42°C, 3000X; H) NJC-1, lower surface, 26°C, 300X; I) NJC-1, lower surface, 42°C, 300X; J,K,L) NJC-1, lower surface, 42°C, 2000X.

details). However, this 4:1 ratio of stomatal conductance and the rate of transpiration were significantly altered when the plants were grown at 42^0C. When CB-5 was grown at 42^0C, stomatal conductance increased four fold initially; but upon continued growth at 42^0C, the rate of transpiration declined steadily; no transpiration could be measured from either leaf surface after eight days of growth at 42^0C because leaves were dry and crumbled when placed inside the cuvette (Fig. 1 C, D). Interestingly, the rate of transpiration increased in the upper surface of NJC-1 when grown at 42^0C and declined slightly in the lower surface, eventually altering the ratio from 4.2:1 at 26°C to 2.3:1 at 42^0C (Fig. 1 C, D). The net rate of transpiration did not significantly change during continued growth of NJC-1 at 42^0C (Fig. 1 C, D). This may be one of the key differences between NJC-1 and CB-5 conferring differential tolerance to heat stress.

<u>Scanning electron microscopy</u>:

When fixed leaf samples were observed by scanning electron microscope, there were essentially no morphological changes in the structure of stomates or in their distribution between CB-5 and NJC-1. The upper leaf surface of CB-5 and NJC-1 grown at 26°C were also very similar (Fig. 2 A, B), but the distribution of the epicuticular wax in NJC-1 grown at 42^0C was quite different, i.e. the epicuticular wax structures were localized in dense aggregates (Fig. 2 C). Examination of the upper leaf surface of NJC-1 grown at 42^0C at higher magnification showed the absence of epicuticular wax and the occurrence of aggregated clusters of scales or linear chains of scales (Fig. 2 D, E, F, G). Clearing of epicuticular wax in the upper leaf surface of NJC-1 grown at 42^0C may correlate with the observed increase in the rate of transpiration (Fig. 1 C, D); however, no attempt was made in this study to distinguish cuticular transpiration from stomatal transpiration.

There was also essentially no difference in stomatal structure or distribution on the lower surfaces of leaves from CB-5 and NJC-1 grown at 26^0C. In contrast to the upper leaf surface, there was no epicuticular wax on the lower surface (Fig. 2 H). However, epicuticular wax was evident even at a low magnification on the lower leaf surface from NJC-1 grown at 42^0C (Fig. 2, I). At higher magnification, a variety of epicuticular wax structures were observed on the lower leaf surface that varied in shapes from scales to discs (Fig. 2 J, K, L). Epicuticular wax formation in the lower leaf surface in NJC-1 grown at 42°C may correlate with a gradual decrease in the rate of transpiration (Fig. 1 C, D). All these observations clearly demonstrate substantial changes in the epicuticular wax appearance and distribution in NJC-1 when grown at 42^0C. These observations could not be made in CB-5 grown at 42^0C because the leaf tissues crumbled during fixation or processing.

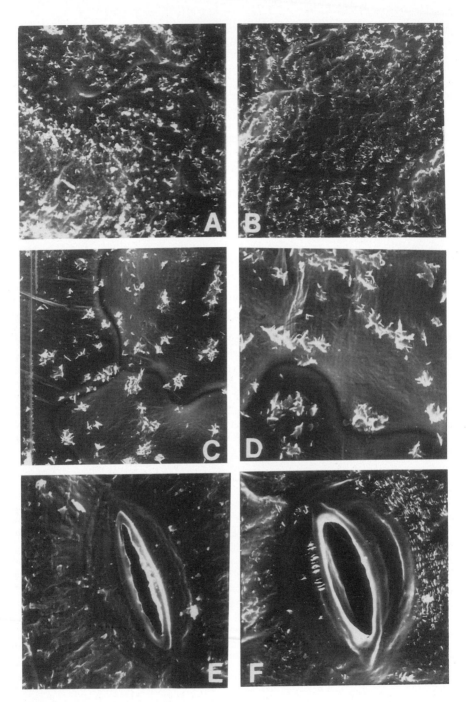220

FIGURE 2

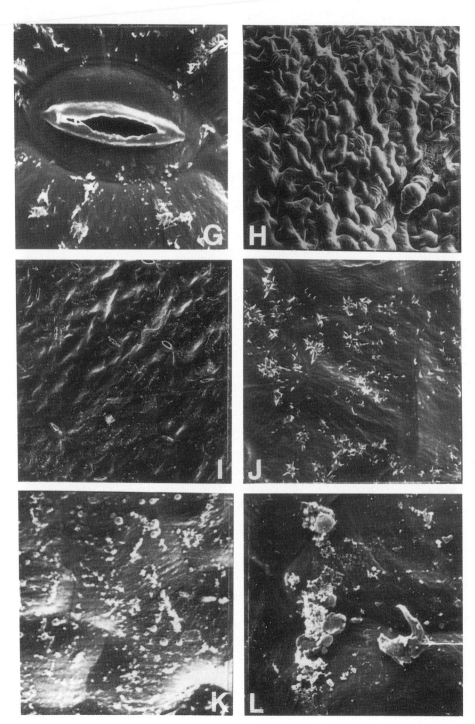

FIGURE 2

222

Table 1. Wax and total lipid content of cowpea cultivars grown at 26°C and 42°C.

Cultivar	Cultivation Temperature	Total lipid (mg/g fresh wt.)	Total wax (mg/g fresh wt.)
CB-5	26°C	6.0	0.4
NJC-1	26°C	6.5	0.4
NJC-1	42°C	8.1	0.5

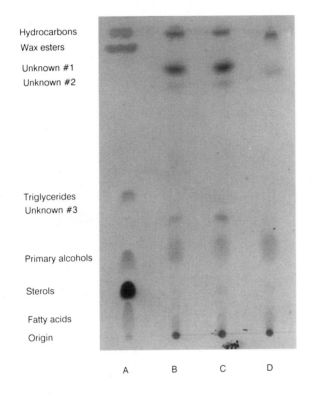

Hydrocarbons
Wax esters
Unknown #1
Unknown #2

Triglycerides
Unknown #3

Primary alcohols

Sterols

Fatty acids

Origin

A   B   C   D

Figure 3. TLC separation of reference compounds (Lane A) and epicuticular wax components of CB-5 grown at 26°C (B), NJC-1 grown at 26°C (C), and at 42°C (D). The reference compounds were: hydrocarbons=octacosane (C_{28}), waxester=(C_{32}), triglyceride=triolein, primary alcohol=octadecanol (C_{18}), sterol=cholesterol, and fatty acid=stearic acid ($C_{18:0}$).

Leaf surface wax and lipid composition:

In view of results from observations of the leaf surfaces by SEM, preliminary analyses of wax components were conducted. There were essentially no differences in the amount of wax present on CB-5 and NJC-1 primary leaves from plants grown at 26°C. However, the total wax content of NJC-1 grown at 42°C was about 10% greater than the leaves from corresponding plants grown at the lower temperature (Table 1).

Based on the TLC separation patterns, there were also no apparent qualitative differences in the wax composition of primary leaves from the CB-5 and NJC-1 cultivars grown at 26°C (Fig. 3). The wax classes tentatively identified included aliphatic hydrocarbons, primary aliphatic alcohols, sterols and free fatty acids. In addition, a major unidentified component (unknown #1) that migrated just behind the wax ester standard, a minor component (unknown #2) that migrated slightly below unknown #1, and component that migrated just above the primary alcohol fraction   (unknown #3) were detected. Unknowns #1 & #2 migrated in the region where aliphatic ketones would be expected and unknown #3 was in the region where secondary alcohols might be expected (Kolattukudy, 1970; Tulloch, 1976). There were substantial changes in wax composition in NJC-1 grown at 42°C, i.e. the three unknown components of the wax were greatly reduced in amount or absent (Fig. 3).

Some of the wax components were analyzed further by gas chromatography. The aliphatic hydrocarbon distribution in both cultivars was similar with C27, C29, C31 and C33 as the predominant chain lengths detected (Table 2). This distribution is very typical of surface wax from most higher plants (Tulloch, 1976), in this case with C31 at over 40% of the total hydrocarbons; but unlike another pea (*Pisum sativum*) that has C31 as 99% of the hydrocarbon fraction (Kolattukudy, 1970). Hydrocarbons from NJC-1 grown at 42°C were similar in chain length distribution but the relative proportions of the relatively shorter chain components such as C23 through C29 and C33 increased at the expense mainly of C31 (Table 2). The primary alcohol fraction of the surface wax from both cultivars ranged in carbon chain length from C24 to C32 with C26 and C28 being the predominant homologues (Table 2). The two cultivars differed in that C28 and C30 were found in higher relative proportions in NJC-1 over CB-5, however the latter cultivar contained about twice the relative amount of C26 as the former. Growth of NJC-1 at 42°C showed higher relative proportions of C29, C30 and C32 at the expense of C28 (Table 2). Predominance of C26 and C28 is also typical of leaf surface wax (Tulloch, 1976).

Saturated fatty acids were most abundant in the free fatty acid fraction isolated from the region of the TLC plate from and including the origin to just below the sterol fraction (Fig. 3). In both cultivars, C16:0 plus C18:0 represented about 55% of the free fatty acids detected (Table 2). Growth of NJC-1 at 42°C resulted in an accumulation of C18:0 over that in corresponding plants grown at 26°C and a reduction in the relative proportions of C18:1 and C18:3.

Clearly, the substantial changes in wax composition described above might be expected to account for the structural differences in the epicuticular wax observed by scanning electron microscopy. However, whether these changes are only a consequence of exposure to high temperature, or if they are involved in the acclimation to the elevated temperature is not known. The relatively small increase in wax content of the primary leaves from NJC-1 grown at 42°C would not be expected to contribute substantially to resistance to cuticular transpiration which might be required for growth at higher temperatures. However, in this study the relative humidity of the growth chamber in which the plants were grown at 42°C was maintained at relatively high levels and therefore the leaves were not subjected to undue water stress that might be expected at elevated temperatures. Thus, a substantial increase in surface wax might not be expected under the growth conditions used in this study. However, according to Juniper and Jeffree (1983) there is no clear association between surface wax and xerophytic success, but certainly if wax deposition is prevented cuticular transpiration increases substantially.

Table 2. Hydrocarbon, alcohol, and free fatty acid composition of the primary leaf surface wax of cowpea cultivars CB-5 and NJC-1 grown at 26°C and NJC-1 grown at 42°C.

Chain Length	CB-5 (26°C)			NJC-1 (26°C)			NJC-1 (42°C)		
	HC	AL	FFA	HC	AL	FFA	HC	AL	FFA
$C_{14:0}$	-	-	6.9	-	-	5.3	-	-	4.3
$C_{15:0}$	-	-	-	-	-	-	-	-	-
$C_{16:0}$	-	-	40.8	-	-	35.1	-	-	34.6
$C_{17:0}$	-	-	-	-	-	-	-	-	-
$C_{18:0}$	-	-	14.9	-	-	19.6	-	-	31.6
$C_{18:1}$	-	-	22.4	-	-	19.6	-	-	11.1
$C_{18:2}$	-	-	7.6	-	-	14.4	-	-	13.7
$C_{18:3}$	-	-	7.4	-	-	6.0	-	-	4.4
$C_{19:0}$	-	-	-	-	-	-	-	-	-
$C_{20:0}$	-	-	-	-	-	-	-	-	-
$C_{21:0}$	-	-	-	-	-	-	-	-	-
$C_{22:0}$	-	-	-	-	-	-	-	-	-
$C_{23:0}$	1.5	-	-	2.1	-	-	4.1	-	-
$C_{24:0}$	1.0	2.5	-	1.1	1.2	-	3.0	1.1	-
$C_{25:0}$	1.9	0.4	-	1.8	-	-	3.1	0.2	-
$C_{26:0}$	1.9	30.2	-	1.3	15.2	-	-	14.3	-
$C_{27:0}$	7.0	1.9	-	10.6	1.3	-	12.8	2.2	-
$C_{28:0}$	1.1	59.6	-	2.3	72.1	-	1.8	66.5	-
$C_{29:0}$	14.5	0.9	-	17.0	1.2	-	22.9	2.1	-
$C_{30:0}$	1.3	4.6	-	2.0	7.1	-	-	9.7	-
$C_{31:0}$	48.0	-	-	41.1	-	-	30.3	-	-
$C_{32:0}$	2.5	-	-	3.3	1.7	-	-	3.9	-
$C_{33:0}$	19.3	-	-	17.3	-	-	22.1	-	-

HC=hydrocarbons, AL=primary alcohols, FFA=free fatty acids.

Lipids were extracted from the de-waxed leaves of CB-5 and NJC-1 cultivars grown at 26°C and NJC-1 grown at 42°C. The total lipid content of the two cultivars was essentially the same when grown at the lower temperature, but it was about 25% higher in NJC-1 grown at 42°C compared to that of NJC-1 grown at 26°C (Table 1). The fatty acid composition and degree of unsaturation was as might be expected for photosynthetic tissues with C18:3 being the predominant fatty acid (Ohlrogge et al. 1979), and were similar in both cultivars (Table 3).

Growth of NJC-1 at 42°C showed considerable shifts in the relative proportions of the fatty acids with a slight decrease in the degree of unsaturation manifested mainly as a 19% decrease in C18:3, a 96% increase in C18:2 and 48% increase in C16:0. An increase in unsaturation is an expected response to growth at a temperature higher than that considered optimum for growth (Kunst et al.,1989b).

Table 3. Fatty acid composition of total lipids from CB-5 and NJC-1.

Fatty acid	CB-5(26°C)	NJC-1(26°C)	NJC-1(42°C)
$C_{16:0}$	19.00	17.80	23.50
$C_{16:1\Delta}^{3t}$	2.74	2.44	1.16
$C_{16:1}$	2.13	2.97	0.60
$C_{16:2}$	4.40	3.99	1.82
$C_{16:3}$	trace	trace	trace
$C_{18:0}$	1.34	1.63	2.53
$C_{18:1}$	1.57	1.33	3.54
$C_{18:2}$	6.46	4.94	14.29
$C_{18:3}$	62.36	64.90	52.58
Total lipid (mg/gm)	6.00	6.50	8.04
Δ/mole	2.04	2.09	1.91

In addition, the relative proportion of C16:1 3t was 50% lower in the lipid of NJC-1 grown at 42°C than that from corresponding plants grown at 26°C, implicating phosphatidyl-glycerol (PG) in the response to high temperature because this fatty acid occurs exclusively in this phospholipid (Table 3). The relevance of this is that PG seems to be involved as a key molecule in the cold temperature acclimation of plants (Murata, 1983; Murata and Yamaya, 1984; Norman et al., 1984; Roughan, 1985), and therefore might be expected to be involved in adaptation to changes in temperature in general.

ACKNOWLEDGEMENT

The authors are grateful to Drs. S.C. Weller (Purdue University), R. Mitchell (Auburn University) for providing Morgan CF-1000 and LI-COR 1600 and C. C. Chang for providing assistance in photosynthesis measurements.

BIBLIOGRAPHY

Bligh EG, Dyer WJ (1959) A rapid method for total lipid extraction and purification. Can J Biochem Physiol **37**: 911-917

Daniell H, Sarojini G, Kumarachinnayan P, Kulandaivelu, G (1981) Action of propanil on *in vivo* chlorophyll fluorescence in *Echinochloa crus-galli* and rice. Weed Res **21**: 171-177

Duysens LNM, Sweers HE (1963) Mechanisms of two photochemical reactions in algae as studied by means of fluorescence *In* J Ashida, eds, Microalgae and photosynthetic bacteria. Univ. of Tokyo Press, pp 353-372

Hugly S, Kunst L, Browse J. Somerville C (1989) Enhanced thermal tolerance of photosynthesis and altered chloroplast ultrastructure in a mutant of *Arabidopsis* deficient in lipid desaturation. Plant Physiol **90**: 1134-1142

Juniper BE, Jeffree CE (1983) Plant Surfaces, Edward Arnold Publishers, London, pp 27

Kolattukudy PE (1970) Composition of the surface lipids of pea leaves. Lipids **5**: 398-402

Kulandaivelu G, Daniell H (1980) DCMU induced increase in chlorophyll $\underline{a}$ fluorescence intensity - an index of photosynthetic oxygen evolution in leaves, chloroplasts and algae. Physiol Plant **48**: 385-388

Kunst L, Browse J, Somerville C (1988) Altered regulation of lipid biosynthesis in a mutant of *Arabidopsis* deficient in chloroplast glycerol-3-phosphate acyltransferase activity. Proc Natl Acad Sci USA **85**: 4143-4147

Kunst L, Browse J, Somerville C (1989a) Altered chloroplast structure and function in a mutant of *Arabidopsis* deficient in plastid glycerol-3-phosphate acyltransferase activity. Plant Physiol **90**: 846-853

Kunst L, Browse J, Somerville C (1989b) Enhanced thermal tolerance in a mutant of *Arabidopsis* deficient in palmitic acid unsaturation. Plant Physiol **91**: 401-408

Murata N (1983) Molecular species composition of phosphatidylglycerols from chilling-sensitive and chilling-resistant plants. Plant Cell Physiol **24**: 81-86

Murata N, Yamaya J (1984) Temperature dependent phase behaviour of phosphatidylglycerols from chilling-sensitive and chilling-resistant plants. Plant Physiol **74**: 1016-1024

Norman HA, McMillan C, Thompson GA (1984) Phosphatidylglycerol molecular species in chilling-sensitive and chilling-resistant populations of *Avicennia germinans*. Plant Cell Physiol **25**: 1437-1444

Ohlrogge JB, Kuhn KN, Stumpf PK (1979) Subcellular localization of acyl carrier protein in leaf protoplasts of *Spinacia oleracea*. Proc Natl Acad Sci USA **76**: 1194-1198

Orr GR, Raison JK (1987) Compositional and thermal properties of thylakoid polar lipids of *Nerium oleander* in relation to chilling sensitivity. Plant Physiol **84**: 88-92

Pearcy RW (1978) Effect of growth temperature on the fatty acid composition of the leaf lipids in *Atriplex lentiformis*. Plant Physiol **61**: 484-486

Raison JK (1985) Alterations in physical properties of thermal response of membrane lipids: Correlations with acclimation to chilling and high temperature. *In* JB St. John, E Berlin, PC Jackson, eds, Frontiers of membrane research in agriculture, Rowman and Allandale, Totowa, NJ, pp 383-401

Raison JK, Roberts JKM, Berry JA (1982) Correlations between the thermal stability of chloroplast thylakoid membranes and the composition and fluidity of their polar lipids upon acclimation of *Nerium oleander* to growth temperature. Biochim Biophys Acta **688**: 218-228

Roughan PG (1985) Phosphatidylglycerol and chilling sensitivity in plants. Plant Physiol **77**: 740-746

Sommerville CR, Browse J (1991) Plant lipids: metabolism, mutants and membrane. Science **252**: 80-87

Thomas PG, Dominy PJ, Vigh L, Mansourian AR, Quinn PJH, Williams WP (1986) Increased thermal stability of pigment protein complexes of pea thylakoids following catalytic hydrogenation of membrane lipids. Biochim Biophys Acta **849**: 131-140

Tulloch AP (1976) Chemistry of waxes of higher plants. In PE Kolattukudy, eds, Chemistry and Biochemistry of Natural Waxes, Elsevier, Amsterdam, pp. 235-287.

ALTERED GENE EXPRESSION IN THERMOADAPTED CULTURED CELLS OF COWPEA

Joe H. Cherry[+], Randall R. Mayer[*], Kathleen Heuss-LaRosa[#], P. Maheshwara Reddy[+], and Narendra K. Singh[+]

Abstract

Suspension cell cultures of cowpea (*Vigna unguiculata* L.) were adapted to grow gradually at a normally non-permissive temperature of 38°C. Proteins expressed were compared to those of thermoadapted cells maintained at 38°C for over 20 growth cycles to those of unadapted cells maintained at 26°C, and unadapted cells heat-shocked at 42°C. The expression of mRNAs, and differentially expressed cDNAs were also compared for each of these cultures. Unadapted cells transferred to an elevated temperature produced the normal array of heat shock proteins, while thermoadapted cells produced large numbers of proteins present in unadapted cells and several novel polypeptides not present in unadapted cells or heat shocked cells. Twenty-six differentially expressed cDNA clones were isolated using subtracted probes from a cDNA library produced from mRNA isolated from thermoadapted cells. These clones represent eight hybridization groups and six groups of cDNA clones exhibit increased level of gene expression in thermoadapted cells. We suggest that these non-hsp genes are involved in the long-term tolerance to hyperthermic stress.

INTRODUCTION

Changes in the environment often produce an alteration in the expression of genes in plants. Although several stress factors have been investigated, one of the best studied responses of plants to environmental stress is the heat shock response. The heat shock response is associated with the rapid synthesis of new polypeptides concomitant with the inhibition of normal protein synthesis. An extensive literature demonstrates the universality of this response from bacteria to man (Schlesinger et al., 1982). The protective function of heat shock proteins (HSP) in stabilizing cellular components during heat stress, and the role of HSP in the recovery and thermotolerance of cells have been experimentally demonstrated.

[+]Department of Botany and Microbiology, Auburn University, Auburn, AL 36849-5407
[*]Horticulture Department, University of Maryland
[#]USDA/ARS, Bldg. 006, Beltsville, Maryland

NATO ASI Series, Vol. H 86
Biochemical and Cellular Mechanisms
of Stress Tolerance in Plants
Edited by J. H. Cherry
© Springer-Verlag Berlin Heidelberg 1994

Several reviews on heat-shock response in plants (Kimpel and Key 1985; Nagao et al., 1986 and 1990; Nagao and Key 1989; Ellis 1990, Brodl 1990, Schoffl et al., 1986; Vierling 1991); in other systems (Craig et al., 1985; Lindquist 1986; Lindquist and Craig 1988; Pelham 1989 a,b; Schlesinger 1990), reviews and published monographs (Newmann et al., 1989; Nover 1991) and authors in this volume have summarized and eloquently discussed regulation and functional aspects of heat shock protein and acquisition of thermotolerance by organisms.

Although HSP remain a focal point for experimentation on the mechanisms of short-term thermotolerance, we have been interested in the cellular mechanisms of sustained thermotolerance in plants. This study was initiated with cultured cells of tobacco (Kanabus et al., 1984) and cowpea (Heuss-LaRosa et al., 1987, Cherry et al., 1989, Mayer et al., 1990, Dylewski et al., 1991) and has included thermotolerant cultivars of cowpea selected from a field in Niger (Daniell et al., this volume). This report indicates that thermoadapted cells of cowpea synthesize a limited spectrum of HSP and several novel non-HSP polypeptides in addition to polypeptides present in unadapted cells. Differentially expressed cDNA clones, apparently representing non-hsp genes have been isolated. This communication describes changes in gene expression in cultured cowpea cells in response to sustained growth at 38°C, a normally nonpermissive growth temperature.

MATERIALS AND METHODS

Cell Culture:

Cell suspensions of *Vigna unguiculata* cv California Blackeye Pea #5 (CB-5) were initiated from seedling hypocotyl segments and cell suspension stocks were routinely maintained in Marashige and Skoog based medium without casein hydrolysate as described by Heuss-LaRosa et al., (1987). The growth temperature for control cell cultures was 26°C. For thermoadapted cultures, cells were adapted to 38°C by increasing incubation temperatures in increments of 2°C over several generations of culture growth (Heuss-LaRosa et al., 1987).

Labeling Conditions:

Experiments were performed using cell lines maintained at 26°C and at 38°C for more than 20 cycles. All cell lines were in mid-log phase of growth (approximately 7 days after inoculation) at the time of labeling. Three millimeter samples of the cell suspensions in 50 ml flasks were labelled with L-4,5-[³H] leucine at the appropriate temperature after 30 min acclimation in a reciprocating shaker. At the end of the 30 min. incubation period, 20 ml cold (-20°C) acetone was added per flask and the samples were stored at -20°C. Acetone-insoluble material was collected on Whatman No. 1 filter paper, washed once with 20 ml of cold (-20°C) acetone, and lyophilized. Control cells maintained at 26°C were labeled at 26°C (26/26) and heat shocked at 42°C (26/42). Thermoadapted cells were labeled at 38°C (38/38) and heat shocked at 45°C (38/45).

In vitro translation of poly (A)⁺ RNA:

Poly A⁺ RNA was isolated from unadapted, 38°C adapted, and heat shocked cowpea cell cultures according to the phenol-SDS extraction method (Sambrook et al., 1989) followed by oligo (dT)-cellulose chromatography . Poly A⁺ RNA from the different cell lines was subjected to *in vitro* translation using a rabbit reticulocyte lysate system (Bethesda Research Laboratories) for 1 hr at 30°C, labeling with L-[³⁵S] methionine. The translation products were resolved by SDS-PAGE on a 10 to 16% gradient polyacrylamide gel, and exposed to Kodak XAR 5 film after treatment with Fluorenhance.

Two-dimensional electrophoresis:

Lyophilized cells were sonicated briefly in buffered phenol. An equal volume of extraction buffer (0.7 M sucrose, 50 mM Na_2EDTA, 50 mM Tris-HCl (pH 8.0), 2% β-mercaptoethanol, 0.1 M KCl) was added and mixed. The phenol phase was removed and eight volumes of chilled methanol containing 0.1 M ammonium acetate was added. After overnight precipitation of proteins at -20°C, the proteins were pelleted at 10,000g for 15 min, washed with cold (-20°C) acetone, air dried and resuspended in sample buffer (1% NP-40, 9 M urea, 2% Pharmalyte ampholytes (3-10), 2% β-mercaptoethanol). The clear supernatant was used for 2 dimensional polyacrylamide gel electrophoresis using the method of O'Farrell (1975). Equal amounts of radioactivity were loaded onto the basic end of the isoelectric focusing gels containing 2% Pharmalyte ampholytes, mixture of H 3-10 and pH 5-8. Electrophoresis was performed at 400 V for eight hours and then 800 V for one hour. Second dimension SDS-PAGE was then performed on gels containing a 10 to 16% acrylamide gradient. Gels were treated for fluorography before drying.

Isolation of differentially expressed cDNA clones:

Poly A$^+$ RNA from thermoadapted cells was used to construct a λ Zap cDNA library according to manufacturers protocol (Stratagene Manual, 1990). Recombinant cDNA clones (5 x 10^5) were screened with ^{32}P labeled subtracted probes. The probes were produced by subtracting the common pool of mRNA between unadapted cells and heat shocked cells from A$^+$ RNA of the thermoadapted cells using Invitrogen subtractor II. Single stranded cDNA from thermoadapted cells (38/38) labeled with ^{32}P was hybridized to a 10-fold excess of biotenylated poly A$^+$RNA from unadapted (26/26) and heat-shocked (26/42) cells. The selected cDNA clones were converted into phagemids by *in vivo* excision (Short et al., 1988). The inserts in the phagemids were labeled with random primer and cross-hybridized to place cDNA clones in groups. The ^{32}P labeled DNA insert representing each group was used in determination of expression of the genes by RNA slot blot analysis (Sambrook et al., 1989).

RESULTS AND DISCUSSION

Under heat shock conditions, cowpea cells produce at least 32 HSP, most of which are low molecular mass HSP ranging in size from 15 to 30 kDa (Fig. 1B). Thermoadapted cells 38/38) in mid-log phase of growth produce a pattern of protein synthesis somewhat intermediate between unadapted and heat-shocked cells (Fig. 1C). As many as sixteen of the HSP as well as several novel proteins not seen in cells grown at 26/26 are synthesized in thermoadapted cells. Several proteins are also down-regulated upon adaptation to heat (Fig. 1C).

Cells which have been adapted to 38^{0}C produce as many as 20 proteins which are not readily detected in two-dimensional electrophoretic gels of unadapted cells labeled at 26^{0}C (Fig. 1A). It appears that these polypeptides are not all HSP, but may be important for sustained growth and development at elevated temperatures, in contrast to the transient HSP responses which may be more important in protection or repair of heat-induced damage caused by sudden changes in growth temperature.

Based on SDS-PAGE, we have previously shown that a 70 kDa and an 80 kDa HSP, are continuously produced in cells which have adapted to high temperature (Heuss-LaRosa et al, 1987). It now appears that several other HSP-like polypeptides may have roles in heat tolerance along with several non-HSP. Most notably, a 30 kDa non-HSP with an apparent pI of 6.2 and a pair of 22 kDA HSP with apparent pI's of 5.6 and 5.7 label heavily in

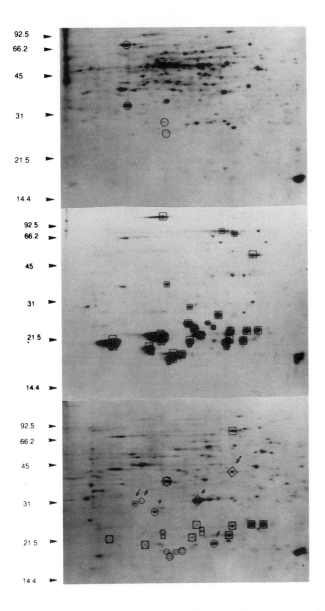

Figure 1. 2-D gel electrophoresis of *in vivo* labeled proteins from (A) unadapted cells cultured and labelled at 26°C (26/26), (B) heat-shocked cells cultured at 26°C and labelled at 42°C (26/42), and (C) thermoadapted cells cultured and labelled at 38°C (38/38). Polypeptides with ○ are associated with thermoadapted cells, □ heat shock proteins, and ◇ increased over unadapted cells.   The pH gradient ranged from 8.0 → 5.0.

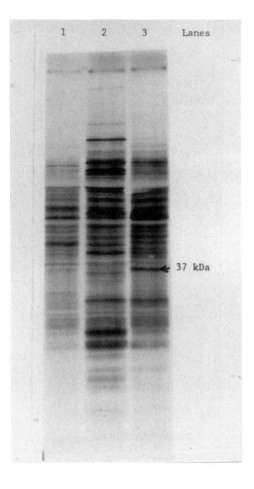

Figure 2. SDS-PAGE of ³⁵S-methionine labeled *in vitro* translation products
of mRNA from unadapted cells (lane 1), heat-shocked cells (lane 2) and thermoadapted cells
(lane 3). Arrows represent a major 37 kDa polypeptide in thermoadapted cells.

adapted cells and comprise a major part of the Coomassie blue-staining protein in gels (not

shown). The 30 kDa protein may be the same as the non-modified, 37 kDa *in vitro*

translation product (Fig 2). Accumulation of these proteins, as evidenced from the

Coomassie staining, by cells in mid-log growth phase may indicate specific roles for these

proteins in acquired thermotolerance. The 70 and 80 kDa HSP, while appearing as major

235

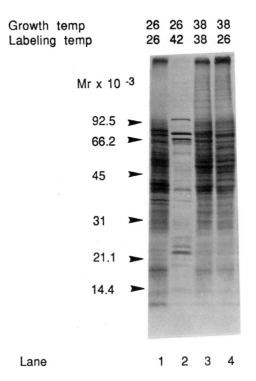

Growth temp   26 26 38 38
Labeling temp   26 42 38 26

$Mr \times 10^{-3}$

92.5 ➤
66.2 ➤
45 ➤
31 ➤
21.1 ➤
14.4 ➤

Lane   1   2   3   4

Figure 3. SDS-PAGE of *in vivo* labeled polypeptide from unadapted cells (26°C), (lane 1), heat shocked cells (42°C) (lane 2), thermoadapted cells (38/38), (lane 3) and thermoadapted cells grown for two genrations at 26°C and labeled at 26°C (38/26), (lane 4).

staining bands in the 38/38 one-dimensional gels, are more difficult to detect in the two-dimensional gels. The 70 kDa HSP is known to be a multigene family in *Drosophila* (Artavanis-Tsakonas et al., 1979; Craig et al., 1983) and *Saccharomyces cerevisiae* (Craig et al., 1985; Ingolia et al., 1982). *Saccharomyces cerevisiae* produces eight 70 kDa proteins ranging from 45 to 94% identical sequences, three of these were induced by heat, three of these were unaffected by heat and the remaining two were repressed by heat. It follows that several different 70 kDa proteins could comigrate to give a falsely high labeling pattern. When separatation is based on their pI's, however, several smaller spots are seen assuming that amino acid substitutions would alter the pI of the protein. A similar situation may exist for the cowpea 80 kDa protein. *Saccharomyces cerevisiae* has two genes for its 83 kDa HSP, one is constitutive and one is inducible (Lindquist, 1986). In cowpea, several closely migrating bands are present in one-dimensional gels. It is also possible that some of the 70 or 80 kDa protein has a pI outside the effective pH range of the isoelectric focusing gels. However, based on published pI's of other similar mass HSP and based on two-dimensional gels using a wider range of ampholytes, this seems less plausible.

HSP appears to be an important factor in the ability of cowpea cells to survive sustained hyperthermic stress. Thermoadapted cells labeled at 45°C (38/45) produced only normal HSP (Heuss-LaRosa et al., 1987). This is consistent with the idea that the heat shock response results from the temperature differential between a preexisting temperature and a subsequent elevated temperature. However, *in vivo* labeling of (38/38) cells (Fig 1C), Coomassie blue staining of cellular proteins (not shown) and *in vitro* translations products of poly A$^+$ RNA from 38/38 cells (Fig 2) clearly show a reduction in the number of HSP and corresponding mRNA in 38/38 cells and the simultaneous appearance of new polypeptides.

Among the *in vitro* translation products, a major 37 kDa polypeptide was detected in thermoadapted cells (Fig 2). In a somewhat analogous situation, the *htr* B gene in *E. coli* has been shown to encode a 35 kDa membrane-associated protein that is essential for bacterial viability at 33°c (Karow et al., 1991). At a lower temperature, the htr B gene product is dispensable. At present, we do not know if the 37 kDa *in vitro* translation product from thermoadapted cowpea cells is related to the htr B gene.

Furthermore, when thermoadapted cells (38/38) are transferred to a 26°C growth temperature (38/26) culture, growth resumed after a few day longer lag period than when 26°C unadapted cells (26/26) were grown at 26°C. *In vivo* labeling of proteins in (38/26) cells showed some differences in the polypeptide profile (Fig 3). These changes are mostly reductions in the labeling of proteins in the molecular weight range of 31-38 kDa, 50-60 kDa and 70-80 kDa. It is interesting to note that thermoadapted cells (38/38) that were returned and maintained for 10 cycles at 26°C retained the capacity to grow at 38°C. From these results we suggest a stable alteration in the ability of thermoadapted cells to tolerate hyperthermic stress.

Table 1.

COMPARISON OF IN VIVO RATES OF AMINO ACID BIOSYNTHESIS OBSERVED IN
CB5-26/26, CB5-26/42 AND CB5-38/38 CULTURES (From Mayer 1989)

Amino Acid	CB5-26/26	CB5-26/42	CB5-38/38
	in vivo synthesis rate (nmol/h gfw)		
Alanine	1250 [1] [a]	2000 [1.6]	3000 [2.4]
Asparagine	17 [1]	17 [1.0]	150 [8.8]
Aspartate	500 [1]	700 [1.4]	1000 [2.0]
B-alanine	1 [1]	3.5 [3.5]	8 [8.0]
GABA	12.5 [1]	800 [64.0] [b]	100 [8.0]
	--	100 [8.0] [c]	--
Glutamate (GDH)	500 [1]	1800 [1.2]	1500 [1.0]
Glutamate (GOGAT)	700 x 2 [1]	100 x 2 [1.6]	1700 x 2 [2.4]
Glutamine	750 [1]	1200 [1.6]	1900 [2.5]
Isoleucine	10 [1]	60 [6.0]	80 [8.0]
Leucine	30 [1]	150 [5.0]	200 [6.7]
Proline	12.5 [1]	33.5 [2.7]	80 [6.4]
Threonine	7.5 [1]	6 [0.8]	150 [20.0]
Valine	35 [1]	70 [2.0]	180 [5.1]

[a] values in brackets refer to fold increases above CB5-26/26 rates.
[b] GABA synthesis rate at t= 0-2 hours. [c] GABA synthesis rate at
t= 2 - 24 hours.

It is evident from these experiments that non-HSP proteins in thermoadapted cells
may function in the survival of cells beyond the known protective function of HSP.
Although we do not know their nature or biological role in long-term thermoadaption, it can
be assumed that some of these proteins are likely to be enzymes acting at specific steps in
biochemical pathways. Using a ^{15}N labeled nitrogen source, the turnover of ^{15}N amino acids
was followed by GC-mass spectrometry in unadapted, heat shocked and thermoadapted cells,
and in their medium the turnover of ^{15}N amino acids in proteins was also determined (Mayer,
1990). A dramatic change in the rate of metabolism for amino acids metabolism was evident
in cowpea cells under heat shock (Mayer et al., 1990) and in thermoadapted cells (Table I;
Mayer, 1989). It is possible that some of the polypeptides present in 38/38 cells may be
involved in the increased biosynthesis of amino acids such as alanine, threonine, isoleucine,
valine, proline, asparagine, and gamma-aminobutyric acid.

Also, we have shown that thermoadaptation of CB-5 cells alters their ultrastructural morphology (Dylewksi et al., 1991). During heat shock, there is considerable loss of ER and ribosomes. Other ultrastructural changes which occur in thermoadapted cells involve the loss of starch grains from plastids, swollen nucleoli and mitochondria, cytoplasmic invagination into nuclei and formation of an electron dense layer appressed to the tonoplast. It is likely that changes in the level of fatty acid unsaturation in the membranes of thermoadapted cells causes shifts in the phase transition temperature of these membranes resulting in the ability of these cells to survive at nonpermissive temperatures. If such is the case, some of the polypeptides in thermoadapted cells may be involved in changing the degree of fatty acid unsaturation.

In support of this hypothesis, we have demonstrated alterations in the levels of fatty acids in thermotolerant cultivars of cowpea, NJC-1 during growth at high temperature (Daniell et al., this volume). Changes in the lipid unsaturation and protein:lipid ratio in the membrane resulting in alternations in membrane fluidity have been implicated in long-term thermotolerance (Vigh et al., this volume; Suss and Yordanov 1986). Also, reduction of membrane lipids in the signal transduction pathway of temperature perception and thermoterance have been proposed (Vigh et al., 1993). Increased levels of fatty acid and membrane phospholilpid saturation contribute to thermotolerance in *Arabidopsis* (Hugly et al., 1989; Kunst et al., 1989).

We initiated a study to understand the biological roles of the proteins associated with thermoadapted cells and chose the route of obtaining differentially expressed cDNA clones. Twenty-six putative cDNA clones were selected. These clones were arbitrarily designated as TAP, representing genes involved in thermoadaptation. Using labeled cDNA inserts as probes, selected cDNAs were cross-hybridized to each other and were placed in 8 hybridization groups. One cDNA representing each group was used to determine unique expression of corresponding genes by RNA slot blot analysis where equal amounts of poly A^+ RNA from 26/26, 38/38 and 26/42 cells were used in slots. After RNA blot analysis using different cDNA clones, six TAP cDNAs demonstrate up regulation in the thermoadapted cells. (Fig 4). The remaining two groups of TAP cDNAs appear to be upregulated in both heat-shocked cells (26/42) and thermoadapted cells (38/38). The size of the cDNA inserts range from 0.6 kb to 2.0kb in these clones. We are in the process of

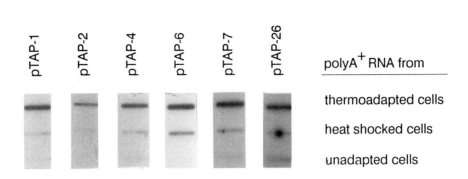

Figure 4. Slot blot analysis of poly A⁺ RNA from unadapted, heat shocked and thermoadapted cells using one representative cDNA clones. 1μg poly A⁺ RNA was applied in each lane and hybridized to ^{32}P labeled cDNA probes.

obtaining the nucelotide sequence and characterizing in more detail the expression of the corresponding genes.

From these results we suggest that during thermoadaption of cowpea cells a novel set of genes are expressed which are not regulated in the same way as HSP genes and the products of these genes may have important roles in sustained thermotolerance in plant cells.

LITERATURE CITED

Artavanis-Tsakonas, S., P. Schedl, L. Moran, J. Lis. 1979. Genes for the 70,000 Dalton heat shock protein in two cloned D. melanogaster DNA segments. Cell 17:9-18.

Brodl, M. R. 1990. Biochemnistry of heat shock response in Plants. In: Environmental Injury to Plants., Ed. F. Katterman, Academic Press, pp. 113-135.

Cherry, J.H., Heuss-LaRosa, K. and Mayer, R.R. 1989. Adaptation of thermotolerance in cowpea suspension cultures. In: Environmental Stress in Plants: Biochemical and Physiological Mechanism, ed. J.H. Cherry NATO-ASI Series. Vol. 19. Springer-Verlag. pp.335-369.

Craig, E.A., Ingolia, T.D., and Manseau, L.J. 1983. Expression of Drosophila heat-shock cognate genes during heat shock and development. Dev. Biol. 99:418-426.

Craig, E.A., Salter, M.R., Boorstein, W.R. and Palter, K. 1985. Expression the S. cerevisiae hsp 70 multigene family. UCLA Symp. Mol. Cell Biol. (NS) 30:659-668. New York: Liss.

Dylewski, D.P., N.K. Singh and J.H. Cherry. 1991. Effects of heat shock and thermoadaptation on the ultrastructure of cowpea (Vigna unguiculata) cells. Protoplasma 163:125-135.

Ellis, J.R. 1990. Molecular chaperones: the plant connection. Science 250:954-959.

Heuss-LaRosa, K. , Mayer, R.R. and Cherry, J.H. 1987. Synthesis of only two heat shock proteins is required for thermoadaptation in cultured cowpea cells. Plant Physiol. 85:4-7.

Hugly, S., L. Kunst, J. Browse, C. Somerville. 1989. Enhanced thermotolerance and altered chloroplast ultrastructure in mutants of Arabidopsis deficient in lipid unsaturation. Plant Physiol. 90:1134-1142.

Ingolia, T.D., Salter, M.J., and Craig, E.A. 1982. Saccharomyces cerevisiae contains a complex multigene family related to the major heat shock-inducible gene of Drosophila. Mol. Cell Biol. 2:1388-1389.

Kanabus, J., Pikaard, C.S., and Cherry, J.H. 1984. Heat shock proteins in tobacco cell suspensions during growth cycle. Plant Physiol. 75:639-644.

Karow, M., Fayet, O., Cigielska, A., Ziegalhoffer, T. and Gregopoulus, C. 1991. Isolation and characterization of Escherichia coli htr B gene whose product is essential for bacterial viability above 33^0 in rich media. J. Bacteriol. 173:741-750.

Kimpel, J.A. and Key, J.L. 1985. Heat shock in plants. Trends Biochem. Sci. 10:353-57.

Kunst, L., J. Browse, C. Somerville. 1989. Enhanced thermotolerance in a mutant of Arabidopris deficient in pelmitic acid unsaturation. Plant Physiol. 91:401-408.

Lindquist, S. 1986. The heat-shock response. Annu. Rev. Biochem. 55:1151-1191.

Lindquist, S. and Craig, E.A. 1988. The heat shock proteins. Annu. Rev. Genet. 22:631-677.

Mayer, R.R. 1989. Metabolic responses of cowpea cells to high temperature stress. M.S. thesis. Purdue University, Indiana.

Mayer, R.R., Cherry, J.H. and Rhodes, D. 1990. Effects of heat shock on amino acid metabolism of cowpea cells. Plant Physiol. 94:796-810.

241

Nagao, R.T. and Key, J.L. 1989. Heat shock protein genes of plants. In "Cell Culture and Somatic Cell Genetics of Plants" (I.K. Vasil and J. Schell, eds.), pp.297-328. Academic Press, San Diego, California.

Nagao, R.T., Kimpel, J.A., Vierling, E. and Key, J.L. 1986. The heat shock response: a comparative analysis. In Oxford Surveys of Plant Molecular & Cell Biology, ed. B.J. Miflin, 3:384-438. Oxford Univ. Press.

Nagao, R.T., Nimpel, J.C. and Key, J.L. 1990. Molecular and cellular biology of heat shock response. Advances in Genetics. 28:235-274.

Newmann, D. , Nover, L. , Parthier, B. , Rieger, R., Scharf, K.D., 1989. Heat shock and other stress response systems of plants. Biol. Zentralbl. 108:1- 156.

Nover, L. (ed.) 1991. Heat Shock Response. CRC Press.

O'Farrell, P.H. 1975. High resolution two dimensional electrophoresis of proteins. J. Biol. Chem. 250:4007-4021.

Pelham, H.R.B. 1989(a). Control of protein from E.R. Ann. Rev. Cell Biol. 5:1-23.

Pelham, H.R.B. 1989(b). Heat shock and the sorting of luminal ER proteins. EMBO J. 8:3171-3176.

Sambrook, J., E.F. Fritsch and T. Maniatis. 1989. Molecular cloning: A laboratory manual. Cold Sring Harbor Laboratory Press. CSH, N.Y.

Schlesinger, M., Ashburner, M. and Tiessieres, A. (eds). 1982. Heat shock from Bacteria to Man. Cold Spring Harbor Laboratory Press. CSH., N.Y.

Schlesinger, M.J. 1990. Heat shock proteins. J. Biol. Chem. 265:12111-12114.

Schoffl, F., G. Bawmann, E. Raschke and M. Bevan. 1986. The expression of heat-shock genes in higher plants. Phil. Trans. R. Soc. Lond. B. 314:458-468.

Short, J.M. , Fernandez, J.M., Sorge, J.A. and Huse, W.D. 1988. Zap: A bacteriophage expression vector in vivo excision properties. Nucl. Acids Res. 16:7583-7599.

Suss, K.H. and I. Yordanov. 1986. Biosynthetic cause of in-vivo acquired thermotolerance of photosynthetic light reactions and metabolic responses of chloroplasts to heat stress. 81:192-199.

Vierling, E. 1991. The roles of heat shock proteins in plants. Annu. Rev. Plant Physiol. Plant Mol. Biol. 579-620.

Vigh, L., D.A. Los, I. Horvath and N. Murata. 1993. The primary signal in the biological perception of temperature: Pd-catalyzed hydrogenation of memebrane lipids stimulated the expression of <u>des</u> A gene of <u>Synechocystis</u> PCC 6803. Proc. Natl. Acad. Sci. USA 90:9090-9094.

Chapter 2

DROUGHT STRESS

CLONING OF A DNA FRAGMENT ENCODING γ-GLUTAMYL KINASE AND γ-GLUTAMYL PHOSPHATE REDUCTASE FROM A TOMATO cDNA LIBRARY

Mario G. García-Ríos[1], Laszlo N. Csonka[1], Ray A. Bressan[2], P. Christopher LaRosa[2,3], and José Hanquier[1,4]

[1]Department of Biological Sciences, Purdue University, West Lafayette, IN 47907-1392 U.S.A.
[2]Department of Horticulture, Purdue University, West Lafayette, IN 47907 U.S.A.
[3]Present address: Plant Molecular Biology Laboratory, USDA, Bld. 006, Baltimore Ave., Beltsville, MD 20705-2350 U.S.A.
[4]Present address: Eli Lilly & Co., Indianapolis, IN 46285 U.S.A.

ABSTRACT

We cloned the gene encoding the first enzyme of proline synthesis from tomato (*L. esculentum*) by complementation of a *proB* mutation in *E. coli* with a λgt-11 cDNA library of tomato fruit. We obtained seven phages which were able to restore proline prototrophy to the *proB* mutant. The insert from one of the complementing phages, *PRO1*, was subcloned into plasmid pBluescript IIKS+. When transferred into a mutant Δ*proBA* (which is deficient in both the first and second enzymes of proline biosynthesis, γ-glutamyl kinase and γ-glutamyl phosphate reductase), the *PRO1* gene was able to complement this deletion mutation. Assays of the coupled γ-glutamyl kinase/γ-glutamyl phosphate reductase activity of *E. coli proB* or Δ*proBA* mutant strains revealed that these strains have elevated levels of both γ-glutamyl kinase and γ-glutamyl phosphate reductase. These results indicate that the *PRO1* gene is a hybrid locus which specifies both enzymatic activities.

Nucleotide sequence analysis revealed that the 5' portion of the *PRO1* locus contains an 849 bp open reading frame which has the coding capacity for a protein with extensive amino acid sequence similarities to γ-glutamyl kinase from *E. coli*, yeast and Δ[1]-pyrroline-5-caroboxylate synthetase from moth bean, and the 3' region specifies a 1287 bp open reading frame with the coding capacity for a protein with extensive similarity to γ-glutamyl phosphate reductase from *E. coli* and Δ[1]-pyrroline-5-carboxylate synthetase from moth bean. Surprisingly, there is a TAA translation termination codon within the *PRO1* region between the open reading frames that encode γ-glutamyl kinase and γ-glutamyl phosphate reductase. The structure of the *PRO1* mRNA, which resembles polycistronic messages in prokaryotes, is unusual in eukaryotes because of difficulties in continuing translation across translation termination codons. As expected, in *E. coli*, the TAA codon is recognized as a translation

NATO ASI Series, Vol. H 86
Biochemical and Cellular Mechanisms
of Stress Tolerance in Plants
Edited by J. H. Cherry
© Springer-Verlag Berlin Heidelberg 1994

terminator, so that in this host, the *PRO1* mRNA is translated into two proteins with apparent molecular masses of 30 and 45 kd. We do not know the nature of the translation product of the *PRO1* gene in plants: whether only the 30 kd γ-glutamyl kinase encoded at the 5' end of the mRNA is made, whether the downstream γ-glutamyl phosphate reductase is translated as a separate polypeptide as a result of internal reinitiation, or whether the translation product is a long hybrid protein composed of γ-glutamyl kinase/γ-glutamyl phosphate reductase, which arose as a result of editing the TAA codon from the mRNA, by splicing or skipping over it by translational frameshifting.

INTRODUCTION

A large variety of plants accumulate the compatible solute proline upon exposure to conditions of high salinity, osmolality, or drought (Delauney and Verma 1993). The function of proline and other compatible solutes is thought to be the maintenance of turgor in environments of reduced water potential. Despite the importance of proline in osmotic adjustment, its biosynthetic pathway in plants has been elucidated only recently. In bacteria and fungi, proline is synthesized from glutamate via four reactions (Fig. 1): the first, second and fourth reactions are catalyzed by γ-glutamyl kinase, γ-glutamyl phosphate reductase, and Δ^1-pyrroline-5-carboxylate reductase, respectively, and the third step proceeds spontaneously (Csonka and Baich 1983; Li and Brandriss 1992). In *Escherichia coli*, the genes encoding the three enzymes, in sequential order, are called *proB*, *proA*, and *proC*.

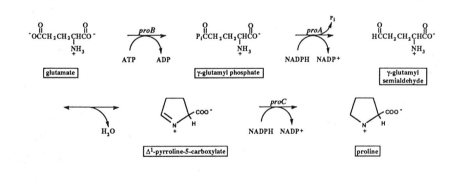

Figure 1. The proline biosynthetic pathway in bacteria and fungi.

Δ^1-Pyrroline-5-carboxylate reductase has been demonstrated in a number of plants (Huber 1974; Treichel 1986; Delauney and Verma 1990; LaRosa et al. 1991; Szoke et al. 1992; Williamson and Slocum 1992), and it has been generally accepted that the last step of the proline biosynthetic pathway is the same as in bacteria. However, until recently, it was not clear whether Δ^1-pyrroline-5-carboxylate is synthesized by the reactions found in bacteria and fungi, or whether it is made from ornithine via ornithine transaminase (Csonka and Baich 1983; Delauney and Verma 1993).

In order to identify the proline biosynthetic pathway in plants and to study its regulation, we cloned in *E. coli* the gene for the first enzyme of proline synthesis from tomato (*Lycopersicon esculentum*). The analysis of the gene revealed that the cDNA clone that we obtained contains information for two proteins that show sequence similarity to the *E. coli* *proB* and *proA* gene products. Thus, the tomato genetic locus for the first enzyme of proline biosynthesis is a hybrid site which can direct in *E. coli* the synthesis of two enzymatic activities corresponding to the first and second enzymes of proline synthesis. Our results have been published in preliminary form in the Abstracts of the Third International Congress of Plant Molecular Biology (García-Ríos et al. 1991). Hu et al. (1992) reported the cloning of the first gene for proline synthesis from moth bean (*Vigna aconitifolia*), which also proved to specify a hybrid γ-glutamyl kinase/γ-glutamyl phosphate reductase (Δ^1-pyrroline-5-carboxylate synthetase).

RESULTS

Cloning of γ-glutamyl kinase gene from tomato. We isolated the gene for the first enzyme of proline synthesis from tomato by *in vivo* complementation of a *proB* mutation in *E. coli*. A 500-fold amplified phage λgt-11 cDNA library of ripening tomato (*L. esculentum*) fruits, obtained as the generous gift of Dr. Robert Fisher (University of California, Berkeley), was infected at a multiplicity of 1 phage/cell into *E. coli* strain G13 (*proB thr leu thi lac rpsL* λ⁻ F⁻) and G9 (*proA thr leu thi lac rpsL* λ⁻ F⁻). The infected cells were cultured on minimal maltose medium containing the required supplements threonine, leucine, and thiamine-HCl. Because the foreign genes in the expression vector λgt-11 are under the control of the *lac* promoter (Ausubel et al. 1989) we included in the medium 1 mM IPTG, to induce the *lac* promoter. Uninfected control cultures of the recipient strains G13 and G9 did not give any proline prototrophic revertants under the same conditions. We isolated seven, not necessarily independent, derivatives of strain G13 which had acquired proline prototrophy (Pro⁺) on proline-deficient medium, each of which carried a λ prophage. We obtained four, not necessarily independent transductants of strain G9 which became Pro⁺; these have not been characterized extensively, but are of interest because they could contain λ phage carrying the tomato homologue of γ-glutamyl phosphate reductase. Phage DNA was isolated from two of the Pro⁺ transductants of G13; restriction mapping revealed that one carried a 3.0 kbp insert, subsequently, named *PRO1*, and the second a 3.7 kbp insert, named *PRO7*. The *PRO1* insert was transferred into plasmid pBluescript IIKS⁺ (Stratagene), resulting in plasmids pPRO1-1 and pPRO1-6 which carry the insert in two opposite orientations. In Southern analysis of total DNA from tomato and tobacco, the *PRO1* insert hybridized as a single copy sequence to DNA from both species (data not shown).

The locus *PRO1* encodes both γ-glutamyl kinase and γ-glutamyl phosphate reductase. Unexpectedly, when we transformed plasmids pPRO1-1 and pPRO1-6 into *E. coli* strains that carry a deletion of both the *proB* and the *proA* genes, such as HB101 and CSH26 (Mahan and Csonka 1983), we found that the plasmids could complement this double mutation. This result could have arisen either if the insert on the plasmids specifies both γ-glutamyl kinase and γ-glutamyl phosphate reductase, or if it carried a suppressor gene encoding some enzyme that supplies Δ¹-pyrroline-5-carboxylate by some other biochemical route (for example, ornithine transaminase [Delauney and Verman 1993; Delauney et al. 1993] or ornithine acetyl transferase [Csonka and Baich 1983]). In order to distinguish between these possibilities, we performed assays for the coupled γ-glutamyl kinase/γ-glutamyl phosphate reductase activity in crude extracts of *E. coli proB* or Δ*proBA* mutants carrying the cloned insert. We found that the strains carrying the cloned gene had an elevated Δ¹-pyrroline-5-carboxylate synthetase activity (Table I). Furthermore, this Δ¹-pyrroline-5-carboxylate synthetase activity was sensitive to feedback inhibition by proline (Table 1), which supports the idea that *PRO1* encodes a physiologically relevant proline biosynthetic enzyme.

There is a translation termination codon within the *PRO1* locus. We determined the nucleotide sequence of both strands of the *PRO1* gene. The sequence analysis revealed two protein domains: an γ-glutamyl kinase domain consisting of 283 amino acids at the 5' end and a γ-glutamyl phosphate reductase domain consisting of 429 amino acids at the 3' end. The γ-glutamyl kinase domain has a 66-69% amino acid sequence similarity to the γ glutamyl kinases of *E. coli* and yeast, and a 69% similarity to the γ-glutamyl kinase domain of the moth bean Δ¹-pyrroline-5-carboxylate reductase. The γ-glutamyl phosphate reductase portion exhibited a 69% and 66% similarity to the corresponding enzyme from *E. coli* and moth bean, respectively.

Table 1. Coupled γ-glutamyl kinase, γ-glutamyl phosphate reductase assays in *E. coli* strains carrying the tomato *PRO1* clone

	Activity in crude *E. coli* extracts (nmol NADPH consumed/min/mg protein)			
Strain (genotype)	G13-1 (*proB*/λgt11*PRO1*) [*PRO1* in single copy])	G13 (*proB*)	KC562 (Δ*proBA*/pKS-*PRO1*) [*PRO1* in multi-copy])	HB101 (Δ*proBA*)
Assay Conditions:				
Complete[1]	43.8±1.6	5.2±0.2	340.0	1.2
-ATP	3.8±1.2	3.9±1.3	2.0	--
+100 mM proline	8.9±0.2	5.6±0.4	1.5	--

[1]The assay involved the ATP-dependent conversion of glutamate to γ-glutamyl phosphate, followed by reduction to γ-glutamyl semialdehyde by NADPH. The reaction rate was monitored as the rate of disappearance of NADPH at A_{340}. Complete assay mixtures contained 100 mM Tris-Cl, 25 mM $MgCl_2$, 100 mM Na-glutamate, 5 mM ATP, 0.4 mM NADPH, pH 7.2, 10-50 µg cell protein; assays carried out at 30°.

To our surprise, we found an in-frame TAA codon in the tomato *PRO1* cDNA clone between the two segments that specify γ-glutamyl kinase and γ-glutamyl phosphate reductase. Normally, TAA is recognized as a translation termination codon. This TAA codon is followed after five bases (that is at a -1 or +2 frameshift with respect to the γ-glutamyl kinase) by a potential translation start ATG codon, which begins the second long open reading frame specifying γ-glutamyl phosphate reductase. The structure of the *PRO1* cDNA clone is shown in Fig. 2. Unlike the *PRO1* gene we obtained from the tomato cDNA library, the moth bean gene for Δ1-pyrroline-5-carboxylate synthetase isolated by Hu et al. (1992) does not contain an internal TAA, but rather it specifies a single hybrid γ-glutamyl kinase, γ-glutamyl phosphate reductase polypeptide.

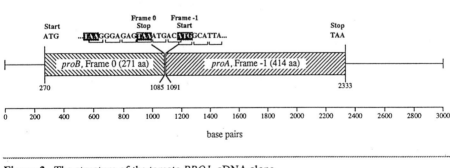

Figure 2. The structure of the tomato *PRO1* cDNA clone.

We were very concerned that the TAA codon within the *PRO1* locus might be a cloning or sequencing artifact. For the following reasons, however, we believe that this is not the case. First, we analyzed the expression of *PRO1* in *E. coli* maxi cells, and found that the recombinant plasmid directed the synthesis of two distinct proteins, of 30 and 45 kd (Fig. 3).

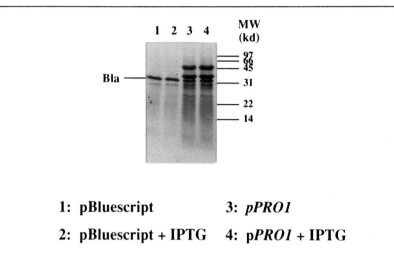

1: pBluescript   3: *pPRO1*

2: pBluescript + IPTG   4: pPRO1 + IPTG

Figure 3. The expression of the tomato *PRO1* gene in *E. coli* maxi cells. Plasmid p*PRO1* was transformed into *E. coli* strain IT2761 [*Δ(srl-recA)-306 sulA11 Δlac-169 thr-1 leu-6 his-4 thi-1 argE3 ilv(*ts) *galK2 rpsL31*], cells were irradiated with UV light, labeled with ^{35}S-methionine, and the proteins synthesized were analyzed on SDS-PAGE gels. See ref. 18 for details of the "maxi cell" procedure.

The results in this figure demonstrate that, as expected in *E. coli,* the TAA codon is recognized as a translation terminator, so that the *PRO1* locus is translated as two separate proteins of the approximate masses predicted from the sequence analysis. Second, we obtained a large collection of 5' and 3' deletions of the *PRO1* gene for the sequence determination. We introduced these deletion derivatives on plasmid pBluescript IIKS⁺ into *proB* and *proA* point mutant *E. coli* strains and tested their ability to complement the mutations. The results are in Fig. 4. According to this analysis, the *PRO1* locus clearly consists of two independent complementation units (or cistrons): a 5' portion which can complement *proB* mutations and a 3' portion which can complement *proA* mutations. Importantly, the boundary between the two complementation units is in close agreement with the position of the TAA codon in the nucleotide sequence. Third, we used sequencing primers near the TAA codon of *PRO1* to determine the sequence of the second, independent λgt-11 clone, *PRO7*, which was able to complement the *E. coli proB* mutation. The sequence of this clone matched completely that of *PRO1* in the region of interest, including the TAA codon at the identical position. Fourth, we used the PCR procedure to amplify a 780 bp region from chromosomal DNA from tissue culture-tomato cells (cultivar *Flacca*). The sequence of the amplified region agreed with that of the same portion of the *PRO1* cDNA clone. Thus, these results demonstrate the TAA codon is present in not only two independent cDNA clones, but also in the tomato genome.

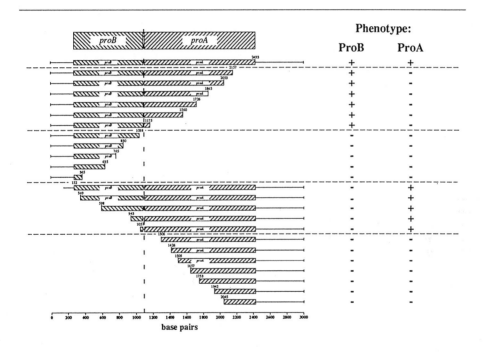

Figure 4. Complementation analysis of the tomato *PRO1* locus. The indicated deletion derivatives were introduced on pBluescript IIKS+ into E. coli *proB* mutant strain G13 (16) and *proA* mutant strain G9 (16). The ability of the plasmids to complement the mutations was assessed by testing whether the strains carrying the indicated deletion derivatives could grow in the absence of proline. The ProB and ProA phenotypes refer to the ability (+) or inability (-) of the indicated deletion derivatives to complement the *proB* and *proA* mutations.

DISCUSSION

We had isolated a DNA fragment that specifies γ-glutamyl kinase and γ-glutamyl phosphate reductase from a cDNA library made from tomato mRNA's. Because the insert thus obtained hybridizes as a single copy gene to tomato and tobacco genomic DNA, and because we were able to amplify an internal position of it with PCR from tomato genomic DNA, we conclude that the *PRO1* locus with the internal TAA codon is present in the *L. esculentum* chromosome.

The TAA codon within the *PRO1* gene is unexpected. To our knowledge, a translation termination codon within a nuclear gene that encodes a functional biosynthetic enzyme is unprecedented. Although there are examples of translation termination codons within eukaryotic mRNA's, these are either near the 5' end or found within viral or retrotransposon encoded messages (Kozak 1992). In most of the examples of translation termination codons in viral mRNA's, these are bypassed by a -1 translation frameshift, which results in a hybrid polypeptide made up of fragments upstream and downstream of the translation terminated codon (Jacks and Varmus 1985; Bredenbeck et al. 1990; Prüfer et al. 1992). There are, however, limited examples of polycistronic mRNA's which encode two proteins, in which the 3' protein is initiated at an AUG codon located within or downstream of the coding region of the 5' protein (Chang et al. 1989; Kozak 1992).

Because we isolated the *PRO1* gene from a cDNA library, we can conclude that this gene is transcribed in tomato fruit. Moreover, because the cloned gene can complement both *proB* and *proA* mutations in *E. coli*, the product(s) of this gene are functional enzymes, at least in the bacterial host. It is intriguing how the TAA codon might be acted on in the plants. If it is recognized as a translation terminator, *PRO1* could direct the synthesis of at least a functional γ-glutamyl kinase encoded at the 5' portion, whether or not the downstream region potentially encoding γ-glutamyl phosphate reductase is translated. It is possible that the sequences encoding γ-glutamyl phosphate reductase are translated as a result of an internal reinitiation event at the ATG codon after the TAA. However, there are other possibilities, which we will investigate. For example, it is conceivable that the TAA codon is removed by mRNA splicing. Alternatively, it could be that the translation termination codon is avoided by a -1 translational frame shift which would generate a long polyprotein consisting of γ-glutamyl kinase and γ-glutamyl phosphate reductase, along the lines seen in some viruses (Jacks and Varmus 1985; Bredenbeck et al. 1990; Prüfer et al. 1992). Translation frameshifts have been observed to occur at regions containing polyU or UA tracts (Jacks and Varmus 1985; Bredenbeck et al. 1990; Prüfer et al. 1992), which are not present in the tomato *PRO1* sequence. However, there is a stretch of 9 nucleotides containing 5 G's and 4

A's immediately preceding the TAA codon at the end of the γ-glutamyl kinase portion of the *PRO1* gene (Fig. 2), which possibly might be involved in a frameshift. If such a frameshift is involved in the translation of the γ-glutamyl phosphate reductase portion of the tomato *PRO1* locus, then it must occur within 7 bases upstream of the TAA codon ending the γ-glutamyl kinase region, because there is another TAA codon in the same reading frame as the γ-glutamyl kinase 7 bases upstream of the translation termination site of γ-glutamyl kinase (Fig. 2). Lastly, it could be that despite the fact that the *PRO1* locus we had cloned from the cDNA library is transcribed, it is not the actual gene for a functional enzyme, but perhaps there is a different locus which specifies the authentic γ-glutamyl kinase/γ-glutamyl phosphate reductase.

ACKNOWLEDGMENTS

This work was supported by the McKnight Foundation. We thank S. Weiss for critical comments on the manuscript.

LITERATURE CITED

Ausubel FM, Brant R, Kingston RE, Moore DD, Seidman JG, Smith JA, Struhl K (1989) Current Protocols in Molecular Biology. John Wiley and Sons, New York

Bredenbeck PJ, Pachule CJ, Noten AFH, Charité J, Luytes W, Weiss SR, Spaan W (1990) The primary structure and expression of the second open reading frame of the polymerase gene of the coronavirus MHV-A5g; a highly conserved polymerase is expressed by an efficient ribosomal frameshifting mechanism. Nucl Acids Res 18:1825-1831

Chang L-J, Pryciak P, Ganem D, Varmus HE (1989) Biosynthesis of the reverse transcriptase of hepatitis B viruses involves *de novo* translational initiation not ribosomal frameshifting. Nature 337:364-368

Csonka LN, Baich A (1983) Proline biosynthesis. *In* KM Herrmann, RL Somerville, eds, Amino Acids, Biosynthesis and Genetic Regulation. Addison-Wesley, Reading, MA, pp 35-51

Delauney AJ, Verma DPS (1990) A soybean Δ^1-pyrroline-5-carboxylate reductase gene was isolated by functional complementation in *Escherichia coli* and is found to be osmoregulated. Mol Gen Genet 221:299-305

Delauney AJ, Verma DPS (1993) Proline biosynthesis and osmoregulation in plants. The Plant Journal 4:215-223

Delauney AJ, Hu CAA, Kisher PBK, Verma DPS (1993) Cloning of ornithine δ-aminotransferase cDNA from *Vigna* aconitifolia by trans-complementation in *Escherichia coli* and reulgation of proline synthesis. J Biol Chem 268:18673-18678

García-Ríos MG, LaRosa PC, Bressan RA, Csonka LN, Hanquier J (1991) Cloning by complementation of the γ-glutamyl kinase gene from a tomato expression library.

Abstracts, Third International Congress, International Society for Plant Molecular Biology. Tucson, Arizona, #1507

Hu CA, Delauney AJ, Verma DP (1992) A bifunctional enzyme (Δ^1-pyrroline-5-carboxylate synthetase) catalyzes the first two steps of proline biosynthesis in plants. Proc Natl Acad Sci USA 89:9354-9358

Huber W (1974) Influence of NaCl and abscisic acid treatment on proline metabolism and some further enzymes of amino acid metabolism in seedlings of *Pennisetum typhoides*. Planta 121:225-235

Jacks T, Varmus H (1985) Expression of the Rous sarcoma virus *pol* gene by ribosomal frameshifting. Science 230:1237-1242

Kozak M (1992) A consideration of alternative models for the initiation of translation in eukaryotes. Crit Rev in Biochem and Mol Biol 27:385-402

Krueger R, Jager H, Hintz J, Palich E (1986) Purification to homogeneity of pyrroline-5-carboxylate reductase of barley. Plant Physiol 80:142-144

LaRosa PC, Rhodes D, Rhodes JC, Bressan RA, Csonka LN (1991) Elevated accumulation of proline in NaCl-adapted tobacco cells is not due to altered Δ^1-pyrroline-5-carboxylate reductase. Plant Physiol 96:245-250

Li W, Brandriss MC (1992) Proline biosynthesis in *Saccharomyces cerevisiae*: molecular analysis of the *PRO1* gene, which encodes γ-glutamyl kinase. J Bacteriol 174:4148-4156

Mahan MJ, Csonka LN (1983) Genetic analysis of the *proBA* genes of *Salmonella typhimurium*; physical and genetic analyses of the cloned *ProB$^+$A$^+$* genes of *Escherichia coli* and of a mutant allele that confers proline overproduction and enhanced osmotolerance. J Bacteriol 156:1249-1262

Prüfer D, Tacke E, Schmitz J, Kull B, Kaufmann A, Rohde W (1992) Ribosomal frameshifting in plants: a novel signal directs the -1 frameshift in the synthesis of the putative viral replicase of potato leafroll luteovirus. EMBO J 11:1111-1117

Ragapati PJ, Stewart CR, Hack E (1989) Pyrroline-5-carboxylate reductase is in pea (*Pisum sativum* L.) leaf chloroplasts. Plant Physiol 91:581-586

Rhodes D, Myers AC, Jamieson G (1981) Gas chromatography-mass spectrometry of N-heptafluorobutyryl isobutyl esters of amino acids in the analysis of the kinetics of [^{15}N]H$_4^+$ assimilation in *Lemna minor* L. Plant Physiol 68:1197-1205

Rodriguez RL, Tait RC (1983) Recombinant DNA Techniques: An Introduction, pp 192-194. The Benjamin/Cummings Publishing Co., Inc., Menlo Park, CA

Szoke A, Miao G-H, Hong Z, Verma DPS (1992) Subcellular localization of Δ^1-pyrroline-5-carboxylate reductase in root/nodule and leaf of soybean. Plant Physiol. 99: 1642-1649

Treichel S (1986) The influence of NaCl on Δ^1-pyrroline-5-carboxylate reductase in proline-accumulating cell suspension cultures of *Mesembryanthemum nodiflorum* and other halophytes. Plant Physiol 67:173-181

Williamson CL, Slocum RD (1992) Molecular cloning and evidence for osmoregulation of the Δ^1-pyrroline-5-carboxylate (*proC*) gene in pea (*Pisum sativum* L.) Plant Physiol 100:1464-1470

REGULATION OF GENE EXPRESSION IN RESPONSE TO DROUGHT AND OSMOTIC SHOCK *

Stefania Grillo, Antonello Costa[1], Marina Tucci[1], Antonella Leone
Research Center for
Vegetable Breeding,
CNR, Portici, Italy

Abstract

We have studied the response of potato cell suspension culture to PEG-mediated low water potential. Gradual adaptation to high concentrations of PEG 8000 (20%) did not affect growth and endogenous ABA content of potato cells. On the contrary, direct exposure to 20% PEG (osmotic shock) drastically reduced growth in unadapted cells and also induced a six-fold increase in ABA level within 5 days from the imposition of the stress. 2D-electrophoretic pattern of *in vivo* labeled proteins of adapted cells was very similar to 'that of control cells. Only few polypeptides were down-regulated and the synthesis of at least 31 individual polypeptides was increased upon adaptation. Osmotic shock as well as ABA treatment of unadapted cells determined an overall reduction of protein synthesis. Most of the induced polypeptides in PEG-shocked cells were also found

---

[1]Department of Agronomy and Plant Genetics, University of Naples, Portici, Italy

*Researches supported by the Italian Ministry of Agriculture in the framework of the project "Resistenze genetiche delle piante agrarie agli stress biotici e abiotici"

Contribution no. 109 of Research Center for Vegetable Breeding, CNR, Portici, Italy

NATO ASI Series, Vol. H 86
Biochemical and Cellular Mechanisms
of Stress Tolerance in Plants
Edited by J. H. Cherry
© Springer-Verlag Berlin Heidelberg 1994

to be induced in unadapted cells upon treatment with exogenous ABA. However, it was possible to identify some polypeptides specifically induced in osmotically shocked cells, a group of which was also detected in adapted cells. The comparison of protein electrophoretic patterns demonstrated that most changes observed in PEG adapted cells were not mediated by ABA. Long-term adaptation of potato cells to PEG enhanced both the transcript and protein level of osmotin, a 26 kD protein reported to accumulate in salt-adapted tobacco cells. In osmotic shocked cells a dramatic increase of osmotin transcripts was also detected, which however did not correlate with the level of the protein.

Introduction

Plants have developed specific mechanisms which allow them to tolerate seasonal and abrupt temporary environmental changes. Two components of plant cell response to decreased water potential may be identified: a short-term response when cells are abruptly transferred to lower water potential environments and a long-term response when cells are allowed to adapt to continuos growth at low water potential. The short term response implies a ready recognition of the newly imposed environment by the cell and the activation of a signal pathway which triggers a cascade of molecular and biochemical events that ultimately causes changes in the expression of the genes (Ho and Sachs, 1989). Despite their importance in the perception of the stress conditions, early responses may be only transient and not relevant in the long-term adjustment of the plants to a new environment. Long-term exposure to an altered environment which varies gradually in intensity elicits changes in gene expression which ensure the cells not only survive but also recover normal cellular functions (Matters and Scandalios, 1986). A gradual adaptation avoids significant cellular damage whereas shock conditions often lead to irreversible damage of the cellular components, such as cellular membranes (Stewart, 1989).

We have compared the response of potato cells to shock conditions *versus* long-term and gradual adaptation to osmotic stress mediated by PEG 8000, which reduces free water concentration extracellularly. The specific aims of this research were to identify modifications of gene expression in potato cells during adaptation to water stress.

Materials and methods

Cell culture and stress treatments

Cell suspension cultures were obtained from leaf callus of the potato (*Solanum tuberosum*) dihaploid clone SVP11 (kindly provided by Dr. T. Cardi, Italy) and subcultured weekly. Cells were maintained in modified MS medium (Tavazza *et al.*, 1988) in a rotary shaker at 28 °C in the dark. Cells were allowed to adapt to low water potential by subcultures into media with increasing concentrations of PEG 8000 from 0 to 5, 10, 15, 20% (w/v). Cells were maintained at each PEG concentration for at least 2 subculture passages. Shock conditions were imposed on unadapted cells by transferring the cells to a medium containing 20% PEG for 72 h. In other experiments, unadapted cells were treated with 100μM ABA for 24 h. Cell growth (fresh and dry weight) and vitality (fluoresceine diacetate staining) were measured during the cell subculture cycle.

In vivo protein labeling and two-dimensional gel electrophoresis (2D-PAGE)

Unadapted (control and shocked) and PEG adapted potato cells in the exponential phase of growth were labeled with ^{35}S-methionine (50-100 μCi/ml) for 3 h. Cells were harvested by centrifugation and immediately homogenized in liquid nitrogen. Proteins were extracted with TCA-acetone according to Granier (1988). The acetone precipitated proteins were suspended in lysis buffer containing 9.5M Urea, 3% (w/v) CHAPS (SIGMA), 2% ampholites (4 parts pH 5-8 and 1 part pH 3.5-10, Pharmacia). Equal amounts of TCA precipitable radioactive proteins (250-500,000 cpm) were separated by 2D-PAGE according to O'Farrell (1975) with modifications by Hochstrasser *et al.* (1988). After the run, gels were fixed, processed for fluorography (Bonner and Laskey, 1974), dried and exposed to X-ray film for 72 h. All experiments were replicated at least once and only reproducible changes in labeled proteins were reported.

ABA quantification

ABA was quantified using a competitive RadioImmunoAssay, as described by Quarrie *et al.* (1988). The monoclonal Mac62 antibody was kindly provided by Dr. S. Quarrie (U.K.).

RNA analysis

Total RNA was prepared by the acid guanidinium thiocyanate phenol chloroform method (Chomeczynski and Sacchi, 1987) and separated overnight in agarose denaturing gel. Gels were blotted in SSC 20X onto NC filters and RNA fixed by UV crosslinking. Filters were hybridized with [α–^{32}P]dCTP labeled tobacco osmotin cDNA clone pOC (Singh *et al.*, 1989). Hybridization and washes were performed by standard protocols (Ausubel *et al.*, 1989). Fold-increase in osmotin transcripts was estimated by scanning films with a laser densitometer (LKB). Transcript size was determined by comparison with Gibco BRL molecular standards (0.24 to 9.5 Kb RNA ladder).

Protein immunoblotting

Proteins were extracted in sample buffer (62.5 mM Tris pH 6.8, 5% glycerol, 2% SDS, 40 mM DTT), quantified, separated electrophoretically on a 12.5% polyacrylamide gel according to Laemmli (1970) and electroblotted onto nitrocellulose membrane. Osmotin protein was detected on western blots reacted with anti-tobacco osmotin antibodies, obtained from Dr. R. Bressan (USA).

Results

Cell growth and adaptation to low water potential

The growth (dry weight) of unadapted control cells in the standard medium increased exponentially reaching a maximum after 7 days. Except for a slight prolonged log phase, growth of cells gradually adapted

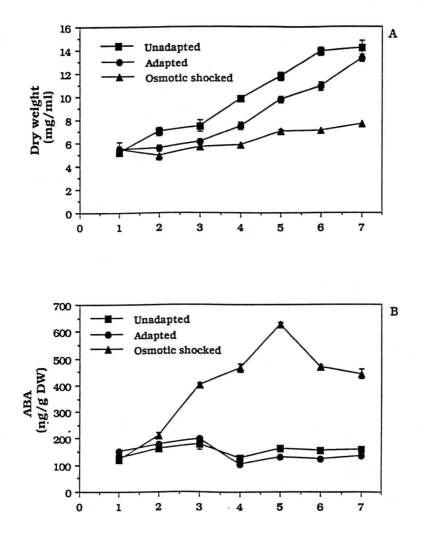

Fig. 1. Growth (A) and endogenous ABA content (B) of unadapted, adapted and osmotic shocked potato cells over the culture period. Values are means of three replicates.

to low water potential was not significantly different from that of control cells (Fig. 1 A). In contrast, growth was drastically inhibited in unadapted cells osmotically shocked by abrupt transfer to a medium containing 20% PEG. Compared to unadapted cells, PEG adapted cells also exhibited enhanced resistance to high NaCl concentrations (data not shown).

ABA quantitation

No significant difference was found in the endogenous ABA level of unadapted and adapted cells (Fig. 1 B). The endogenous level peaked at day 3 of the growth cycle both in unadapted and adapted cells. In unadapted cells directly exposed to 20% PEG, ABA level continuously increased during the growth cycle. Maximum ABA accumulation (a six-fold increase over the basal level) was observed within 5 days from the imposition of the stress. After this peak the cellular ABA content declined.

Stress-induced changes in gene expression

Polypeptides synthesized in unadapted, PEG adapted, osmotically shocked and ABA-treated potato cells were radiolabeled and separated by 2D-PAGE. All treatments had a marked effect on 2D-electrophoretic patterns of extractable proteins (Fig 2. A and B). A summary of the major polypeptides changes and their relative abundance is reported in Table 1.

Adaptive versus shock response. Gradual adaptation to low water potential did not affect the synthesis of the majority of the polypeptides, which was maintained at a constitutive level. Only few polypeptides were identified whose synthesis was down-regulated upon gradual adaptation to PEG 20% (polypeptides # 60-62 and # 73). Furthermore, the synthesis of at least 31 individual polypetptides was increased upon adaptation. Some of them were newly synthesized, such as group # 27-30 and the polypeptide of molecular mass 13.6 kD (# 24). Among the up-regulated polypeptides it was possible to distinguish different sets on the basis of the kinetics of induction during gradual adaptation to increasing concentrations of PEG. Protein synthesis was inhibited as soon as cells were transferred to medium containing 5% PEG and gradually restored during the adaptive period. The majority of the polypeptides up-regulated in cells adapted to 20% PEG were already present in an earlier stage of adaptation (10%

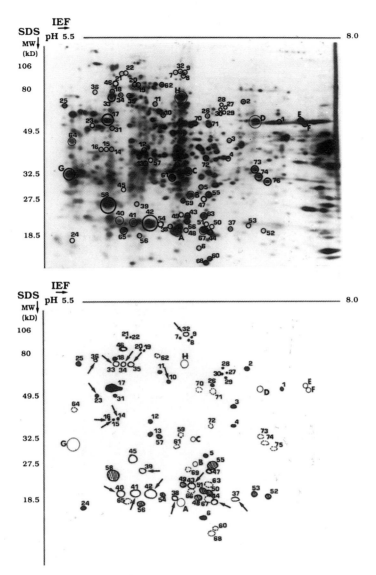

Fig. 2 A. *In vivo* synthesised proteins from unadapted potato cells. [35S] -
      methionine labeled proteins were separated first by IEF and then
      in a 8-13% gradient SDS gel.
   B. Schematic representation of the protein profile reported in A.
      Major proteins are numbered and indicated as follows:
          (●) proteins up-regulated upon adaptation
          (○) proteins up-regulated after osmotic shock
          (⊘) proteins up-regulated upon ABA treatment
          (⊙) proteins down-regulated upon adaptation, ABA or osmotic
              shock treatments
   Arrows indicate ABA mediated up-regulation of proteins induced by
   adaptation and/or osmotic shock.
   Major housekeeping proteins are indicated by capital letters (A-H)

Table 1 - Major *in vivo* labeled polypeptides which change upon gradual adaptation
to PEG-mediated low water potential, osmotic shock (PEG 0→20%) or
ABA (100 μM) treatment of potato cells (+ indicates relative amount;
- indicates absence of the protein).

Polypeptide		Unadapted			Adapted		
No. (Fig. 2)	MW (kD)	Control	Osmotic shock	ABA	PEG 5%	PEG 10%	PEG 20%
1	49.2	+	-	+	+	+	++
2	61.7	+	+	+	+	+	+++
3	40.6	+	+	+	+	+	+++
4	33.6	+	++	+	+	+	+++
5	24.8	+	+++	+	+	+++	+++
6	13.3	+	+	-	+	+	+++
7	82.2	+	+++	-	+	+	+++
8	79.3	+	+++	+	+	+	++
9	82.2	+	++	+	+	++	+++
10	51.7	+	+	+++	+	+	+++
11	58.4	+	++	+	+	++	+++
12	35.0	+	++	-	+	++	+++
13	30.4	+	+	-	+	++	+++
14	35.3	+	+	++	+	++	++
15	35.3	+	+	++	+	++	++
16	35.3	+	+	+	+	++	++
17	49.0	++	++++	-	++	+++	++++
18	65.9	++	-	-	++	++	+++
19	70.5	+	+	+	+	+	++
20	70.5	+	+	+	+	+	++
21	79.5	+	+	+	+	+	++
22	80.4	+	+	+	+	+	++
23	44.8	-	-	+++	-	++	++
24	13.6	-	-	-	-	++	+++
25	59.4	+	++	+	+	+	+++
26	52.6	+	+	+	+	++	++
27	59.7	-	-	-	-	-	++
28	61.7	-	-	-	-	-	++
29	57.4	-	-	-	-	-	++
30	59.0	-	-	-	-	-	++
31	45.3	+	+	+	+++	+++	+++
32	82.7	-	+++	+++	-	-	-
33	63.1	+	++++	++++	+	+	+
34	63.8	+	++++	++++	+	+	+
35	63.4	+	++++	++++	+	+	+
36	65.2	+	+++	+++	+	+	+
37	16.0	++	+++	+++	++	++	++
38	13.6	-	+++	+++	-	-	+
39	21.1	-	+++	++++	-	-	++
40	16.9	+	++++	++++	+	+	+
41	15.9	+	++++	++++	+	+	+
42	15.9	++	++++	++++	++	++	++

263

(cont)

Polypeptide		Unadapted			Adapted		
No. (Fig. 2)	MW (kD)	Control	Osmotic shock	ABA	PEG 5%	PEG 10%	PEG 20%
43	18.2	++	+++	++++	++	++	++
44	15.2	+	+++	+++	+	+	+
45	24.0	-	+++	-	-	-	-
46	71.7	+	+++	+	+	+	+
47	20.9	-	++	-	-	-	-
48	15.8	+	+	+++	-	-	-
49	18.7	-	+	+++	-	-	-
50	16.8	-	+	+++	-	-	-
51	17.1	-	+	+++	-	-	-
52	16.2	+	+	+++	+	+	+
53	16.4	+	+	+++	+	+	+
54	16.8	+	+	+++	+	+	+
55	23.0	++	++	++++	++	++	++
56	14.5	+	+	+++	+	+	+
57	30.3	+	+	+++	+	+	+
58	20.3	+++	+++	++++	+++	+++	+++
59	30.5	+	+	+++	+	+	+
60	11.9	++	+	+	+	+	+
61	26.7	+++	-	-	+	+	+
62	67.8	+++	+	+	+	+	+
63	18.5	++	-	-	+	+	++
64	38.6	+++	+	+	++	++	+++
65	15.2	+++	-	-	+++	+++	+++
66	17.3	++	-	+	++	++	++
67	16.0	+++	+	+	+++	+++	+++
68	11.4	++	+	+	++	++	++
69	21.3	++	+	+	++	++	++
70	48.1	+++	+	+	+++	+++	+++
71	40.6	+++	+	+	++	++	++
72	33.3	++	+	+	++	++	++
73	29.5	+++	+++	+++	+++	+++	+
74	27.7	+++	-	-	+++	+++	+++
75	26.5	+++	-	-	+++	+++	+++
A	15.4	+++	+++	+++	+++	+++	+++
B	22.7	++	++	++	+++	+++	++
C	28.8	++	++	++	++	++	++
D	49.8	++	++	+	++	++	++
E	49.8	+++	++	+	+++	+++	+++
F	47.9	+++	++	+	+++	+++	+++
G	26.7	+++	++	++	+++	+++	+++
H	64.1	++	+	++	++	++	++

PEG, corresponding to approx. -1.5 MPa), but their relative amount increased with the further decrease of water potential (# 5, 9, 11-16, 18, 23-24, 26). Some other proteins were induced only in a later stage of the adaptive process when cells were exposed to 20% PEG (# 1-4, 6-8, 10, 17, 19-22, 25, 27, 30). One polypeptide of molecular mass 45.3 kD (# 31) was early induced in cells exposed to 5% PEG.

While gradual adaptation allowed the maintenance of a normal cellular protein synthesis, osmotic shock conditions imposed on unadapted cells transferred directly to PEG 20% drastically affected protein synthesis Although the stress reduced the protein synthesis rate, the accumulation of at least 15 polypeptides (# 32-47) was consistently enhanced. Despite the overall difference in net protein synthesis of adapted cells *versus* shocked cells, it was possible to identify common polypeptides which were induced both by low water potential both in gradually adapted and shocked cells (# 4, 5, 7, 8, 9, 11, 12, 17, 25).

Role of ABA. We have not found any significant difference in the endogenous ABA accumulation of adapted cells compared to unadapted control cells, while a six-fold increase was observed in PEG-shocked cells (Fig. 1B). Therefore, we addressed the question whether or not this hormone was involved in the modification of the pattern of polypeptides we observed in gradually adapted or shocked-cells. Unadapted cells were treated for 24 h with ABA 100 µM. The quantity of ABA detected intracellularly upon this treatment was 637 ng g^{-1} DW, comparable to the maximum value found in PEG-shocked cells at day 5 of the cell cycle. ABA treatment caused a shut off of protein synthesis, resembling what we observed in PEG-shocked cells. ABA typically induced the synthesis of low mol wt polypeptides of molecular mass 14-20 kD, most of which were also present in PEG shocked cells (# 32-44). Interestingly, the majority of the down-regulated polypeptides in PEG-shocked cells were also repressed in ABA-treated cells (# 60-72 and 74-75). Unique polypeptides were detected in ABA-treated cells (# 48-51) which were absent both in adapted and in PEG-shocked cells. In addition, we identified proteins (# 45, 46, 47) which specifically accumulated upon shock condition and whose induction was not ABA-mediated. Comparative analysis of 2D-gel profiles of ABA-treated and PEG-adapted cells revealed that most of the changes in protein synthesis found in adapted cells were not ABA-regulated.

Osmotin transcript and protein accumulation

We studied the expression of osmotin, a 26 kD protein which accumulates in tobacco cells adapted to high NaCl or low water potential (Bressan *et al.*, 1987) upon short-term osmotic shock and long-term adaptation of potato cells (Fig. 3 A, B). A constitutive level of osmotin transcripts were detected in control cells. Inhibition of transcription and/or decrease in message stability was observed as soon as cells were transferred to medium containing 5% PEG, whereas a 3-fold increase in the transcript level was observed in cells adapted to 10% or 20% PEG. Osmotic shock stress elicited a dramatic increase in osmotin transcription, associated with the observed accumulation of cellular endogenous ABA.

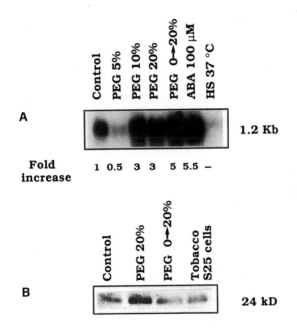

Fig. 3. Osmotin transcript (A) and protein (B) accumulation in unadapted, PEG adapted, ABA treated or osmotic shocked (PEG 0 → 20%) potato cells. Tobacco osmotin cDNA clone and anti-tobacco osmotin antibodies were used for detection.

This result was confirmed by the induction of osmotin transcription by exogenous application of ABA to unadapted cells (Fig. 3 A). Tobacco anti-osmotin antibodies immunodetected a single polypeptide of 24 kD in potato cells. Osmotin protein accumulated at a very high level in adapted cells, whereas in shocked cells the increase in osmotin transcript did not correlate with the accumulation of the protein (Fig. 3 B).

Discussion

Drought negatively affects world crop yield and the need for selecting tolerant varieties has been repeatedly emphasized (for a general review see Blum, 1988). Drought tolerant genotypes often exhibit poor yield performance, due to negative pleiotropic effects of the characters conferring resistance (e.g. reduced leaf area). However, from a breeding point of view a genotype has to be considered tolerant if it is able to withstand stress while ensuring a profitable yield. Much evidence indicates that genes conferring resistance can be found in the genome of plants which are not typically tolerant. The expression of these genes can be modulated upon stress and their activation may help the plant to cope with stress (Ho and Sachs, 1989). The plant ability to overcome the stress is dependent on its intensity and duration. A too severe stress can lead to a complete disgregation of the cellular constituents and ultimately to plant death. On the other hand, if the stress is imposed gradually then the plants are able to undergo metabolic and biochemical changes that enable them to tolerate even more intense stress events. This phenomenon is known as adaptation and permits the maintenance of cellular homeostasis under stressful conditions (Levitt, 1980). At a cellular level two components of the plant response to stress may be identified: a) a rapid response which is firstly involved in monitoring and transducing intracellularly the stress signal and secondly in buffering as quickly as possible the probable damage of cellular constituents (stress response) and b) a long-term response which is mainly related to the recovery of normal cellular metabolism under continuous stress conditions.

We used potato cell suspension cultures in order to study short-term stress response *versus* long-term adaptation to low water potential. The use of cell culture allowed to focus on the physiological and biochemical cellular processes that are the basis for the maintenance of cellular homeostasis under stress conditions. Despite the over-exemplification of the complex physiological mechanisms of the response to drought at the whole plant level, the validity of using a cellular system is strongly supported by the evidence that plants regenerated from tolerant cells exhibit tolerance *in vivo* (Bressan *et al.*, 1987; Sumaryati *et al.*, 1992). Gradual adaptation to low water potential as well as water stress shock was created in the culture medium by addition of high molecular weight PEG. This osmoticum does not penetrate into cells and reduces water availability extracellularly, thus mimicking what naturally occurs in an arid environment (Rains, 1989).

Effects of shock and adaptation on cell growth

PEG-adapted cells were able to sustain active growth at very low values of water potential (approx. -2.0 MPa), without any significant loss in cell vitality. When unadapted cells were directly transferred to medium containing a high concentration of PEG no appreciable gain in dry weight was observed. Vitality was barely affected (70%) thus demonstrating that the osmoticum agent itself or the concentration we used was not toxic for the cells. The structure of the principal cellular components appeared to be undamaged in adapted cells as observed by electron microscopy, while a partial disruption of organelle membranes was found in shocked cells (data not shown). Potato PEG-adapted cells exhibited also salt tolerance, as already reported for carrot (Fallon and Phillips, 1989), tomato (Bressan *et al.*, 1981), potato (Sabbah and Tal, 1990) and *N. plumbaginifolia* (Sumaryati *et al.*, 1992). These data indicate that osmotic adjustment may be a common mechanism at the basis of the cell ability to adapt to salt or low water potential stress. Growth of unadapted cells was completely inhibited by NaCl, suggesting that osmotic adjustment can be acquired only upon gradual adaptation to low water potential (Binzel *et al.*, 1989).

<u>Stress-induced changes in gene expression</u>

Adaptive proteins. We found that in potato cells gradually adapted to grow at very low water potential protein pattern was very similar to that of unstressed control cells. The synthesis of very few polypeptides was repressed but five new proteins (*e.g.* a polypeptide of mol wt 13.6 kD) and at least 26 pre-existing proteins were identified whose synthesis was enhanced in potato adapted cells. In recent years attention has been focused on over-expression of specific polypeptides in cells adapted to different stresses (Singh *et al.*, 1987; Guy, 1990; Cherry *et al.*, 1989), in order to identify genes whose expression could be associated with cellular adaptation to unfavorable environments. Even though their function is unknown, it has been hypothesized that the induced proteins may be related to the activation of metabolic pathways which allow the maintenance of the cellular turgor or of the integrity of cellular and subcellular structures . Compared to the extensive data available for short term response to stress (Ho and Sachs, 1989; Skriver and Mundy, 1990), very few genes or proteins induced upon adaptation have been sufficiently characterized. As an example, induction of a hsp70 and ubiquitin mRNAs in rice cells (Borkird *et al.*, 1991 a) and osmotin in tobacco cells (Bressan *et al.*,1987) has been reported upon long term adaptation to PEG-mediated low water potential.

From our data it was apparent that the ability of potato cells to adapt to changing environments is achieved through a gradual process. After a drastic inhibition of most of the housekeeping proteins in the first steps of the adaptive process (cells treated with 5% PEG), overall protein synthesis was almost completely recovered during the subsequent steps of adaptation. The recovery correlates with a stepwise increase of the induced proteins. Besides, additional polypeptides were induced in the cells adapted to grow at more severe water stress (PEG 20%). Taken together these data indicate that cells are able to switch on quantitative and qualitative changes according to stress intensity and duration.

Role of ABA. Substantial evidence indicates that some of the biochemical and molecular changes induced during water stress are mediated by ABA (Skriver and Mundy, 1990). We have found that ABA accumulates in unadapted shocked cells, while no considerable variation was detected in the endogenous ABA level of adapted cells. These data strongly suggests that this hormone is involved in short-term response rather than in the maintenance of long-term acquired adaptation, although they do not exclude an involvement of ABA in the first stages of the adaptive process. A similar mechanism has been already described as being at the basis of the ability of the plants to cold acclimate. After a temporary accumulation of ABA in the first few days of cold acclimation, ABA tends to decrease (Chen and Li, 1987) and, nevertheless, cells acquire the ability to tolerate more severe cold stress. The hormone may therefore have a primary role in transducing intracellularly the stress signal or in the rapid induction of genes whose products may protect cellular structures from irreversible damage, as suggested for *lea* genes (Dure *et al.*, 1989) or dehydrins (Close *et al.*, 1989). Our data on changes in gene expression in adapted, shocked or ABA-treated cells give additional information on the role of ABA in short-term response to stress. Many of the polypeptides which were down-regulated or up-regulated in shocked cells exhibited the same pattern of expression in ABA-treated cells. Low mol wt polypeptides were typically induced both in ABA and in shocked cells, in which accumulation correlated with a six-fold increase in the endogenous ABA level. Few polypeptides induced in shocked cells were not detected in ABA-treated cells, suggesting that factors other than ABA modulate the stress response. Similar results have been reported for genes induced in rice cells adapted to PEG whose expression was found to be non-ABA-regulated and in some cases even ABA-suppressed (Borkird *et al.*, 1991 a; Borkird *et al.*, 1991 b). Another set of polypeptides includes those which are common to shock and adaptation. This class may represent non-ABA-induced proteins which accumulate specifically in response to low water potential and are involved in the maintenance of cellular homeostasis under prolonged stress.

References

Ausubel FM, Brent R, Kingston RE, Moore DD, Seidman JG, Smith JA, Struhl K, eds (1989) Current Protocols in Molecular Biolog, vol **1** and **2**, John Wiley & Sons, New York

Binzel ML, Hess FD, Bressan RA, Hasegawa PM (1989) Mechanisms of adaptation to salinity in cultured glycophyte cells. In: Cherry JH (ed) Environmental Stress in Plants. NATO ASI Series G: Ecological Sciences, vol. **19**. Springer Verlag, Berlin, pp 139-157

Blum A (1988) Drought resistance. In: Blum A (ed) Plant Breeding for Stress Environments. CRC Press, Boca Raton, pp 43-77

Bonner WM, Laskey R (1974) A film detection method for tritium labeled proteins and nucleic acids on polyacrilamide gels. Eur J Biochem **46**: 83-88

Borkird C, Simoens C, Villarroel R, Van Montagu M (1991) **a** Gene expression associated with water-stress adaptation of rice cells and identification of two genes as hsp 70 and ubiquitin, Physiol Plant **82**: 449-457

Borkird C, Claes B, Caplan A, Simoens C, Van Montagu (1991) **b** Differential expression of water-stress associated genes in tissues of rice plants. J Plant Physiol **138**: 591-595

Bressan RA, Hasegawa PM, Handa AK (1981) Resistance of cultured higher plant cells to polyethylene glycol-induced water stress. Plant Sci Let **21**: 23-30

Bressan RA, Singh NK, Handa AK, Mount R, Clithero J, Hasegawa PM (1987) Stability of altered genetic expression in cultured plant cells adapted to salt. In: Monti L, Porceddu E (eds) Drought Resistance in Plants: Physiological and Genetic Aspects. Comm Europ Commun, Luxembourg, pp 41-58

Chen THH, Li PH (1987) *In vitro* Induction of Cold Acclimation in Potato. In: Bajaj YPS (ed) Biotechnology in Agriculture and Forestry **3**. Potato. Springer Verlag, Berlin, pp 256-267

Cherry JH, Heuss-LaRosa K, Mayer RR (1989) Adaptation of thermotolerance in cowpea suspension cultures. In: Cherry JH (ed) Environmental Stress in Plants.NATO ASI Series G: Ecological Sciences, vol **19**. Springer Verlag, Berlin, pp 355-369

Chomeczynski P, Sacchi N (1987) Single-step method of RNA Isolation by Acid Guanidinium Thiocyanate Phenol Chloroform Extraction. Anal Biochem **162**: 156-159

Close TJ, Kortt AA, Chandler PM (1989) A cDNA based comparison of dehydration-induced proteins (dehydrins) in barley and corn. Plant Mol Biol **13**: 95-108

Dure L, Crouch M, Harada S, Ho THD, Mundy J, Quatrano R, Thomas T, Sung ZR (1989) Common aminoacid sequence domains among the LEA proteins of higher plants. Plant Mol Biol **12**: 475-486

Fallon KM, Phillips R (1989) Responses to water stress in adapted and unadapted carrot cell suspension cultures. J Exp Bot **40**: 681-687

Granier F (1988) Extraction of plant proteins for two dimensional gel electrophoresis. Electrophoresis **9**: 712-718

Guy CL (1990) Cold acclimation and freezing stress tolerance: role of protein metabolism. Annu Rev Plant Physiol Plant Mol Biol :: 187-223

Ho THD, Sachs MM (1989) Environmental control of gene expression and stress proteins in plants. In: Jones HG, Flowers TJ, Jones MB (eds) Plants under stress. Cambridge University Press, Cambridge, pp 156-180

Hochstrasser DF, Harrington MG, Hochstrasser AC, Miller MJ, Merril CR (1988) Methods for increasing the resolution of two-dimensional protein electrophoresis. Anal Biochem **173**: 424-425

Laemmli UK (1970) Cleavage of structural proteins during the assembly of the head of bacteriophage T4. Nature **227**: 680-685

Levitt J (1980) Responses of plants to environmental stresses vol. **1**. Chilling, Freezing, and High Temperature Stresses. Academic Press Inc., New York

Matters GL, Scandalios JG (1986) Changes in plant gene expression during stress. Dev Genet **7**: 167-175

O'Farrell PH (1975) High resolution two-dimensional gel electrophoresis of proteins. J Biol Chem **250**: 4007-4021

Quarrie SA, Whitford PN, Appleford NEJ, Wang TL, Cook SK, Henson IE, Loveys BR (1988) A monoclonal antibody to (s)-abscisic acid: its characterization and use in a radioimmunoassay for measuring abscisic acid in crude extract of cereal and lupin leaves. Planta **171**: 330-339

Rains DW (1989) Plant tissue and protoplast culture: applications to stress physiology and biochemistry. In: Jones HG, Flowers TJ, Jones MB (eds) Plants under stress. Cambridge University Press, Cambridge, pp 181-196

Sabbah S, Tal M (1990) Development of callus and suspension cultures of potato resistant to NaCl and mannitol and their response to stress. Plant Cell, Tissue and Organ Culture **21**: 119-128

Singh NK, Bracker CA, Hasegawa PM, Handa AK, Puchel S, Hermodson MA, Pfankoch E, Regnier FE, Bressan RA (1987). Characterization of osmotin - a thaumatin - like protein associated with osmotic adaptation in plant cells. Plant Physiol **85**: 529-536

Singh NK, Nelson DE, Kuhn D, Hasegawa PM, Bressan RA (1989) Molecular cloning of osmotin and regulation of its expression by ABA and adaptation to low water potential. Plant Physiol **90**: 1096-1101

Skriver K, Mundy J (1990) Gene expression in response to abscisic acid and osmotic stress. Plant Cell **2**: 503-512

Steward GR (1989) Desiccation injury, anhydrobiosis and survival. In: Jones HG, Flowers TJ, Jones MB (eds) Plants under stress. Cambridge University Press, Cambridge, pp 115-130

Sumaryati S, Negrutiu I, Jacobs M (1992) Characterization and regeneration of salt- and water-stress mutants from protoplast culture of *Nicotiana plumbaginifolia* (Viviani) Theor Appl Genet **83**: 613-619

Tavazza R, Tavazza M, Ordas RJ, Ancora G, Benvenuto E (1988) Genetic transformation of potato (*Solanum tuberosum*): an efficient method to obtain transgenic plants. Plant Sci **59**: 175-181

GENE EXPRESSION DURING WATER STRESS

Dorothea Bartels, Don Nelson, Pekka Heino, Detlef Michel, Antonella Furini, Giovanni Bernacchia, Riccardo Velasco, Renza Roncarati, Ralph Elster, Gerhard Schwall, Josefa Alamillo, Katharina Schneider, Francesco Salamini
Max-Planck-Institut für Züchtungsforschung
Carl-von-Linné-Weg 10
D-50829 KÖLN, Germany

Introduction

The severity of the water loss, the developmental stage and the physiological condition determine the response of plants to water stress of affected plants. We initiated research with the objective to identify gene products which may contribute to water stress tolerance using a resurrection plant and barley embryos as model systems. The unique ability of resurrection plants to withstand severe water loss greater than 90% makes them a suitable system to study water stress tolerance: upon rewatering these plants recover quickly from the stress (Gaff 1971, Bartels et al. 1990). An attractive feature of the resurrection plant *Craterostigma plantagineum* is that we can analyse desiccation tolerance in undifferentiated callus tissue, too. Callus withstands rapid dehydration if it is treated with the plant hormone abscisic acid (ABA) prior to the drying treatment (Bartels et al. 1990). This makes two experimental systems available from the same plant for the isolation of molecular components relevant to desiccation. Initially we isolated a large number of cDNA clones which are induced upon dehydration and/or ABA treatment (Bartels et al. 1990, Piatkowski et al. 1990, Bartels et al. 1992).

Results and Discussion

<u>Analysis of cDNA clones</u>: Initially we isolated a large number of cDNA clones which were induced during dehydration using a

NATO ASI Series, Vol. H 86
Biochemical and Cellular Mechanisms
of Stress Tolerance in Plants
Edited by J. H. Cherry
© Springer-Verlag Berlin Heidelberg 1994

simple differential screening approach. A characteristic
feature of these clones is their rapid induction during the
first hours of water loss. The majority of the abundant cDNA
clones could be classified into eleven sequence-related
families based on cross hybridization studies. DNA sequence
information is available on all the abundant cDNA clones. As a
result of the sequence analysis the majority of the
transcripts are related to Lea (= late embryogenesis abundant)
type genes (Galau et al. 1986). Lea genes are characterized by
the following features: abundantly expressed in mature
embryos, transcripts can be prematurely induced by ABA in
embryos or they may be responsive to osmotic stress imposed on
vegetative tissues (Skriver and Mundy 1990, Bray 1991). The
deduced proteins are usually hydrophilic and several contain
regions which most likely exist as α-helices. According to
their sequence motifs a number of different groups of Lea
genes have been defined (Baker et al. 1988, Dure et al. 1989,
Dure 1993). Desiccation induced transcripts from *Craterostigma*
can be assigned to at least seven different types of Lea
genes. The degree of homology varies: conservation may be
restricted to particular sequence motifs and sequence
structures (e.g. pcC11-45, Bartels et al., unpublished) or
genes can be highly identical like the recently discovered
homology between pcC27-45 (Piatkowski et al. 1990) and the
cotton Lea 14 gene (Galau et al. 1993). The high expression of
these different Lea type genes in dehydrated vegetative
tissues of *Craterostigma* suggests that similar metabolic
processes are occurring in the maturation phase of seeds to
which desiccation tolerance is restricted during the life
cycle of higher plants.

Functional aspects: The function of these Lea gene products is
up to now unknown, but the tools to investigate functional
aspects are available. Hypotheses for the role of Lea proteins
during dehydration have been proposed by Dure et al. 1989 and
Dure 1993. It is remarkable that the Lea type genes are not

only conserved among higher plants but immunologically related proteins have also been observed in lower eucaryotes like e.g. in cyanobacteria (Close and Lammers 1993). Overexpression of three different *Craterostigma* proteins in tobacco did not lead to altered phenotypes of the transformed tobacco plants (Iturriaga et al. 1992). This suggests that these proteins do not interfere with the normal metabolism and makes it unlikely that they encode proteins with enzymatic properties. Characteristic for the D11 group is that nearly all genes assigned to this group contain a stretch of serines. The serine residues in the D11-maize proteins expressed in embryos are phosphorylated (Goday et al. 1988). By *in vivo* labelling and immunoprecipitation it was shown that also the homologous proteins in *Craterostigma* leaves are phosphorylated. Whether the phosphorylation is essential for the function of the proteins is unclear; it is also not known whether the Lea D11 proteins are always phosphorylated. The study of mutated proteins will answer these questions. All the *Craterostigma* proteins encoded by the cDNAs were analysed by *in vivo* labelling whether they may occur in phosphorylated forms. With the exception of pcC6-19 and the related pcC27-04 none of the other proteins was found to be phosphorylated.

Regulation of gene expression: All the desiccation induced genes from *Craterostigma* can also be induced by ABA. In the callus the ABA treatment is connected with the tolerant phenotype. An important aspect of this research is to understand the signal transduction starting with the dehydration stress and resulting in gene expression most likely involving ABA in the signalling chain. One approach to answer these questions is the isolation of promoters in order to identify important regulatory elements. Promoters from three different genes (CDeT27-45, CDeT6-19, cDeT11-24) were selected for this research. A functional analysis of the three promoters should reveal common regulatory principles. The experimental approach for this analysis is the construction of

276

unpublished), this suggests a partially different activation pathway.

Conclusion

Analysis of gene expression and the promoter analysis available up to now suggests that pathways leading to desiccation tolerance in the resurrection plant *Craterostigma plantagineum* is very similar to the ones found for seeds in higher plants. It further proposes that the genetic information necessary for desiccation tolerance in vegetative tissues is possibly available in many higher plants. Desiccation tolerance is a multifactorial phenomenon and it might as one part require the expression of an array of different Lea genes; however, it is expected that also genes involved in the synthesis of osmolytes and membrane channel proteins are essential.

References

Baker JC, Steele C, Dure III L. (1988) Sequence and characterization of 6 Lea proteins and their genes from cotton. Plant Mol Biol 11:277-291.

Bartels D, Hanke C, Schneider K, Michel D, Salamini F (1992) A desiccation-related Elip like gene from the resurrection plant *Craterostigma plantagineum* is regulated by light and ABA. EMBO J 11:2771-2778.

Bartels D, Schneider K, Terstappen G, Piatkowski D, Salamini F (1990) Molecular cloning of ABA-modulated genes from the resurrection plant *Craterostigma plantagineum* which are induced during desiccation. Planta 181:27-34.

Bartels D, Velasco R, Schneider K, Forlani F, Furini A, Salamini F (1993) Resurrection plants as model systems to study desiccation tolerance in higher plants. In: Mabry T, Nguyen H, Dixon R eds, Applications and prospects of biotechnology for arid and semi-arid lands. In press.

Bray E (1991) Regulation of gene expression by endogenous ABA during drought stress. In: Abscisic acid physiology and biochemistry (Davies WJ, Jones HG, eds), Bios Scientific Publishers, Oxford, UK.

Close TJ, Lammers PJ (1993) An osmotic stress protein of cyanobacteria is immunologically to plant dehydrins. Plant Physiol 101:773-779.

Curry J, Walker-Simmons MK (1993) Unusual sequence of group 3 LEA (II) mRNA inducible by dehydration stress in wheat. Plant Mol Biol 21:907-912.

Dure III L (1993) A repeating 11-mer amino acid motif and plant desiccation. The Plant Journal 3:363-369.

Dure III L, Crouch M, Harada J, Ho T-H, Mundy J, Quatrano R, Thomas T, Sung ZR (1989) Common amino acid sequence domains among the Lea proteins of higher plants. Plant Mol Biol 12:475-486.

Franz G, Hatzopoulos P, Jones TJ, Krauss M, Sung ZR (1989) Molecular and genetic analysis of an embryonic gene, DC 8 from *Caucus carota* L. Mol Gen Genet 218:143-151.

Gaff DF (1971) Desiccation tolerant flowering plants in Southern Africa. Science 174:1033-1034.

Galau GA, Hughes DW, Dure III L (1986) Abscisic acid induction of cloned cotton late embryogenesis abundant (Lea) mRNAs. Plant Mol Biol 7:155-170.

Galau GA, Wang H, Hughes DW (1993) Cotton Lea5 and Lea14 encode atypical late embryogenesis-abundant proteins. Plant Physiol 101:695-696.

Goday A, Sanchez-Martinez D, Gómez H, Puigdomènech P, Pagès M (1988) Gene expression in developing *Zea mays* embryos: Regulation by abscisic acid of a highly phosphorylated 23- to 25-kD group of proteins. Plant Physiol 88:564-569.

Harada JJ, DeLisle AJ, Baden CS, Crouch ML (1989) Unusual sequence of an abscisic acid-inducible mRNA which accumulates late in *Brassica napus* seed development. Plant Mol Biol 12:395-401.

Hong B, Uknes SJ, Ho T-hD (1988) Cloning and characterization of a cDNA encoding a mRNA rapidly induced by ABA in barley aleurone layers. Plant Mol Biol 99:495-506.

Iturriaga G, Schneider K, Salamini F, Bartels D (1992) Expression of desiccation-related proteins from the resurrection plant *Craterostigma plantagineum* in transgenic tobacco. Plant Mol Biol 20 20:555-558.

Michel D, Salamini F, Bartels D, Dale P, Baga M, Szalay A (1993) Analysis of a desiccation and ABA-responsive promoter isolated from the resurrection plant *Craterostigma plantagineum*. The Plant Journal 4:29-40.

Nordin K, Vahala T, Palva ET (1993) Differential expression of two related, low-temperature-induced genes in *Arabidopsis thaliana* (L.) Heynh. Plant Mol Biol 21:641-653.

Piatkowski D, Schneider K, Salamini F, Bartels D (1990) Characterization of five abscisic acid-responsive cDNA clones isolated from the desiccation-tolerant plant *Craterostigma plantagineum* and their relationship to other water-stress genes. Plant Physiol 94:1682-1688.

Skriver K, Mundy J (1990) Gene expression in response to abscisic acid and osmotic stress. The Plant Cell 2:503-512.

Yamaguchi-Shinozaki K, Shinozaki K (1993) Characterization of the expression of a desiccation-responsive *rd29* gene of *arabidopsis thaliana* and analysis of its promoter in transgenic plants. Mol Gen Genet 236:331-340.

MOLECULAR GENETIC APPROACHES TO IMPROVING HEAT AND DROUGHT STRESS TOLERANCE IN CROP PLANTS

Henry T. Nguyen* and Chandrashekhar P. Joshi
Plant Molecular Genetics Laboratory, Mail Stop 2122,
Department of Agronomy, Horticulture and Entomology
Texas Tech University and Texas Agricultural Experiment Station
Lubbock, Texas 79409-2122, USA

Introduction

Recent advances in plant biotechnology may play crucial roles in the development of superior crop plants with better environmental stress tolerance and higher yield potential under stressful conditions. Appropriate research strategies are, however, required to harvest the fruits of this recent explosion of information in the field of molecular genetics. Most crop plants such as wheat, rice, corn, sorghum, soybean and cotton suffer from heat and drought stress conditions during their normal growth cycles and their yield as well as quality are seriously affected by such environments. Designed genetic modifications of existing crop plants may provide a suitable means to develop plants with improved productivity and quality in stressful environments. However, no cultivar in any crop species has yet been developed and released by introducing a specific stress tolerance gene or by using a molecular marker-aided selection. The main reason for this deficiency is the polygenic nature of the stress tolerance traits which makes it difficult to identify a single gene conferring stress tolerance in crop plants. Most of the data collected in stress tolerance research is correlative and cause and effect relationships have rarely been addressed. In this review, we present the current status of knowledge in the field of molecular biology and molecular markers for heat and drought stress tolerance in crop plants.

Complexity of stress response

Plants adopt several ways to respond to environmental stresses. For any crop species grown in a stressful environment, it is important for us to understand the physiological mechanisms of stress resistance in terms of stress avoidance or stress tolerance as discussed by Blum (1988) and Ludlow (1993). At the cellular level, plants attempt to alleviate the damaging effects of stress by altering their metabolism to cope with the stresses. Major alterations in patterns of gene

NATO ASI Series, Vol. H 86
Biochemical and Cellular Mechanisms
of Stress Tolerance in Plants
Edited by J. H. Cherry
© Springer-Verlag Berlin Heidelberg 1994

expression are known to occur at the early stages of stress. Some of these changes are thought to provide a long-term protection against stress damage. If stress persists for a longer time, more dramatic effects are visible in a plant's phenotype.

Some basic knowledge can be obtained by studying model plants like *Arabidopsis*. However, it should be remembered that ultimately we want to improve crop plants. How we translate the information obtained from model plants to crop plants needs to be addressed now prior to investing too much energy in model plants. Stress responses could be different in crop plants as compared to *Arabidopsis* which has not been exposed to intense selection pressure and recurrent breeding for higher yield potential or other agronomic traits.

Plants respond to and protect themselves from these stresses depending upon their genetic makeup. Significant genetic variability exists in a species which can be employed to understand the mechanisms of stress resistance with the ultimate goal of producing stress tolerant crop cultivars. Understanding the physiological, molecular and biochemical effects of these stresses constitutes the first step towards the development of strategies for designing stress resistant genotypes. Several aspects of physiological and molecular responses to heat and water deficit stress have been described in many recent reviews (e.g. Blum, 1988; Ellis, 1991; Howarth, 1991; Ludlow 1993; McCue and Hanson, 1990; Nguyen and Joshi, 1993a; Skriver and Mundy, 1990; Vierling, 1991).

Molecular responses to high temperature stress: heat shock proteins

Preferential synthesis of heat shock proteins (HSPs) is an instant response to heat stress exhibited by plants and other organisms (Vierling, 1991; Nguyen and Joshi, 1993b). Plant HSPs are commonly classified into high molecular weight (in the range of 70-110 kDa) and low molecular weight (in the range of 15-30 kDa) groups. Hundreds of cDNAs and genomic DNAs encoding heat shock proteins have been characterized from a large number of crop plants as well as model plant species. This kind of research activity is indicative of the excitement in this field. However, the roles of these genes and physiological functions of stress proteins need to be urgently investigated. High molecular weight (HMW) HSPs can be further classified into several groups, such as HSP90, HSP70 and HSP60 families (Vierling, 1991). These features are similar in plants and animals. However, plants are unique in producing abundant amounts of low molecular weight (LMW) HSPs ranging from 15-30 kDa which are encoded by multigene families (Nover et al., 1989). Although the physiological function of these proteins are not known, the universality of this response in higher plants suggests that these HSPs may perform some important function. LMW HSP genes have been isolated from a variety of crop plants such as soybean, pea, carrot, wheat, rice and maize. We have recently discussed the structural complexities of these HSP families, their cellular distribution and probable roles and interested readers may refer to these reviews (Nguyen et al., 1992; Nguyen and Joshi, 1993a; Nguyen and Joshi, 1993b).

Acquired thermotolerance is defined as an organism's ability to survive a lethal heat shock after exposure to a milder heat shock (~10°C above normal growth temperature). The critical set of experiments to link HSP production with acquired thermotolerance and plant productivity is still an under explored avenue in scientific literature. Simultaneous production of HSPs and induction of thermotolerance are indicative of importance of HSPs in thermotolerance (Kimple and Key, 1985; Krishnan et al., 1989; Vierling, 1991; Jorgensen et al., 1992). Moreover, their distribution within the cell suggests that they may play specific physiological roles. However, this circumstantial evidence is not enough to convince us that HSPs are causatively related to heat tolerance.

Molecular changes in plants in response to drought stress: water deficit induced proteins

Although drought stress and heat stress occur simultaneously in the field conditions, molecular mechanisms of stress response are distinctly different. Heat stress response includes a rapid induction of transient synthesis of HSPs by selectively directing new protein synthesis towards exclusive HSP production. Drought stress response, on the contrary, continues the normal protein synthesis for significantly longer periods during water deficit stress. The main research challenge is to identify key stress responsive proteins which provide stress resistance among the hundreds of proteins induced by stress. Several environmental stresses such as water deficit stress, desiccation stress, salt or osmotic stress, cold stress and developmental stages such as late embryogenesis show similar molecular changes and all these may be included under 'water deficit stress' involving osmotic stress and dehydration. It is possible that many of these changes are related to ABA (abscisic acid) accumulation in tissues under water deficit stress (Skriver and Mundy, 1990), although there has been a report which found no correlation between translatable mRNA species induced by water deficit stress and externally applied ABA (Creelman et al., 1990). While the functions of water deficit stress-induced genes have not been ascertained, one water deficit stress-induced protein in soybean has been determined to be associated with cell wall development

(Bozarth et al., 1987). This protein has been hypothesized to be involved with cell wall loosening to allow continued growth during water deficit stress.

Water deficit stress-responsive genes from plants may be classified into two major classes, namely, turgor and ABA responsive genes (Dure et al., 1989; Skriver and Mundy, 1990; Guerrero et al., 1990). Three genes from pea have been studied which are induced by reduction in turgor and have been designated as turgor responsive genes (Guerrero et al.,1990). ABA treatment did not modulate the RNA accumulation patterns for these genes and this group of genes is recognized as a separate group of early genes induced by water stress. These three genes showed homology to three diverse group of proteins. The ABA responsive genes have been named in a variety of ways depending upon the developmental stage or the external stimuli applied. Thus, LEA (late embryogenesis abundant), RAB (responsive to ABA) WSP (water stress proteins) and dehydrins (dehydration induced proteins) constitute a group of similar proteins. Several genes encoding WSPs have been isolated and characterized (see Joshi et al., 1992). The diversity of reported proteins indicates the complexity of the water stress response phenomenon in plants and a great deal of effort is required to unravel these processes of water deficit stress response.

Molecular strategies for genetic improvement of stress tolerance traits

There is a great paucity of appropriate genetic experiments involving stress tolerance traits in plants. First, a tight genetic linkage between physiological traits indicative of heat and drought stress tolerance and stress induced proteins has not been established convincingly. Secondly, mutations and complementations that are commonly used for confirming the function of a protein in bacteria or yeasts, are not accessible in crop plants due to the complexities of the genomes and complex organization of most of the stress responsive genes. HSP or WSP genes generally belong to multigene families and creating a mutation that will hamper certain a function and complement it by reintroduction of an intact gene copy is difficult in most cases. Silencing a gene using antisense technology for confirming it's function is more difficult for common stress protein genes than for other single copy or low copy genes due to high level of homology between DNA sequences. We definitely need more insights into the mechanisms of antisense RNA silencing before this technique becomes routine in the repertoire of molecular biologists who aim to manipulate crop plants for stress tolerance. Alternatively, novel genetic stocks, commonly available in wheat and maize, representing addition or deletion of chromosomes or their arms can be used for similar functionality assays. This is, however, complicated by the presence of a large number of genes (some of them even affecting the overall growth of the plant) on any particular chromosome or it's arms. Use of natural genetic variation appears to be the only amenable approach to resolve some of the complexities of this problem.

Crop plants like other organisms are known to induce genotype specific stress responsive proteins in several cases (Krishnan et al., 1989; King et al., 1992; Jorgensen et al., 1992). Among hundreds of protein spots visible on 2-D gel electrophoresis profiles after heat or water deficit stress, identification of the most important proteins which impart or are tightly linked to stress tolerance traits has proved to be the most difficult task. Genetic studies can be of a greater help in deciphering this problem. Two parents of contrasting stress tolerance behavior can be first identified from the diverse germplasm resources available in a crop species. Their F_2 segregating progeny, recombinant inbred lines, or doubled haploids can be used for the assessment of stress tolerance traits using physiological assays and 2-D profiles of stress proteins which assist the identification of individual proteins. A tight linkage between these two parameters may allow us to narrow down our choices to 10-15 proteins as compared to hundreds. Routine molecular biological techniques can then be employed for identification and characterization of cDNA clones encoding these stress proteins that are tightly associated with stress tolerance traits. The identification of 3' gene-specific probes from these cDNAs will assist in easy monitoring of the expression of these genes in stress tolerant and susceptible plants by simple presence and absence tests on northern blots. If identified proteins are shown to be tightly associated with a resistant phenotype, then these clones can be used for direct, function- based selection of stress resistant crops. There is also a possibility that the kinetics of the production and accumulation of these stress proteins over a period of stress regimes is responsible for acquired thermotolerance or water deficit stress tolerance. Production of specific antibodies against stress proteins may help in studying these changes in protein accumulation. High antigenic similarity between stress proteins from the same family is a well known phenomenon and isolation of monoclonal antibodies against a specific protein moiety will be advantageous. Isolation of gene-specific probes and specific antibodies for stress proteins can elucidate the genetic organization of stress proteins and unravel trends in their evolution across the plant kingdom.

When we know that a specific protein is important for stress tolerance, there are still many challenges left. Targeting genes at a specific location in the genome so that its expression is not be affected by positional effects or that will not interfere in the functions of other important genes is still in the experimental stages. There is no control over the number of copies introduced in a cell. Rather than overexpressing a gene under a constitutive promoter we need more specific experiments planned in the genetic manipulation of crops. A temporally regulated gene which is expressed in a specific tissue or even in a specific organelle might produce a better effect than a rampant overproduced protein disturbing the cellular environment or clogging the protein synthesis machinery. Modifying a gene at a time may not be a cost-effective way of introducing stress tolerance. We definitely know that heat shock responsive elements or ABA responsive elements in plant gene promoters control expression of a large number of HSPs or WSPs. More research is required to manipulate expression of these transcription factors which could be the key to manipulating stress responses in a more global way. Post-translational modifications of transcriptional factors are not new and a strategy must be developed regarding the temporal activation of these transcription factors. We should also know how these stresses are perceived by plant cells, tissues and organs through more studies into the mechanisms of signal transduction pathways.

The good news is right around the corner. Ho and colleagues have observed that the transfer of the barley lea III gene to tobacco plants has triggered most of the drought stress responses which are normally seen in plants growing in a water-limiting environment (Fitzpatrick, 1993). These include early flowering, short plant height and desiccation tolerance. This observation was surprising because drought stress response which is believed to be under multigenic control has been elicited by transfer of a single gene. Perhaps due to the high similarity between monocot- and dicot lea III proteins, the transgenic tobacco plant perceives overproduction of a drought stress induced protein as a drought response and turns on its own drought tolerance mechanism. Researchers in Dorothea Bartel's laboratory have transferred a few drought responsive genes from the resurrection plant to tobacco under constitutive promoters (Iturriaga et al., 1992). They have overexpressed three desiccation related proteins from resurrection plants into transgenic tobacco plants. These plants expressed mRNAs and proteins at high levels. However, these proteins were not sufficient to increase water deficit stress tolerance as measured by ion leakage assays. The resulting protein products were transported appropriately to the corresponding cellular compartments but this did not induce drought tolerance in transgenic plants. However, future experiments can be planned by using these observations. Work is in progress in Elizabeth Vierling's laboratory (personal communication) with overexpression of plastid-localized HSP21 at non-stress conditions. Their results show that the transgenic plants show normal phenotype although heat tolerance aspects are still under investigation.

Outside the area of heat and drought tolerance, two excellent experiments by Murata et al. (1992) and Tarczynski et al. (1993) are worth mentioning. These studies have indicated that stress tolerance in higher plants can be improved by manipulation of single gene. Murata et al (1992) have overexpressed cDNAs for glycerol-3-phosphate acyltransferases from squash and *Arabidopsis* to tobacco. Transgenic plants were more cold resistant than control plants. In another set of experiments, transgenic tobacco plants overexpressing a bacterial gene encoding mannitol 1-phosphate dehydrogenase showed synthesis and accumulation of mannitol. These plants have been shown to have an increased ability to tolerate high saline conditions (Tarczynski et al., 1993). These experiments have opened doors to the manipulation of sugar alcohols, proline and glycine-betaine concentrations in plant cells which, in turn, are likely to provide protection against drought stress. This does not mean we have finished our job, in fact, we have just began to unravel some of the mysteries of stress tolerance.

Genome mapping and genetic markers associated with heat and drought stress tolerance traits

Paterson et al. (1991) reviewed the advantages of using molecular marker based selection for crop improvement over selection based solely on plant phenotype. Little is known about molecular markers such as RFLPs (restriction fragment length polymorphisms) and RAPDs (random amplified polymorphic DNAs) that are tightly linked to the heat and water deficit stress tolerance traits. RFLP markers are inherited codominantly, are independent of tissue or environmental effects and rarely have epistatic or pleiotropic effects. Moreover, a large number of RFLP loci can be mapped which will aid in development of high resolution linkage maps. Potential uses of RFLP maps have been described in the literature (Paterson et al., 1991). Currently, good resolution RFLP linkage maps are available for several cereals including rice, barley, wheat, and maize. These maps are likely to assist in identifying molecular markers linked to traits of interest including stress tolerance traits. Several of these traits are expected to be polygenic and techniques of QTL (quantitative trait loci) analysis should be employed to investigate this approach. Two inbreds of contrasting stress tolerance phenotypes should be crossed and polymorphisms associated with stress tolerance should be determined in F_2 segregating and advanced populations. Recombinant inbred lines or doubled haploid lines are preferred since they allow extensive replicated evaluation of stress tolerance traits. The only report in this direction appears to be that of Ottaviano et al. (1991) who attempted to find linkage between cellular membrane stability (CMS) as a measurement of heat tolerance and RFLPs in recombinant inbred populations of maize. Regression analysis of CMS on RFLPs detected a minimum number of six QTLs accounting for 53% of the genetic variability. Work on tagging genes involved in drought resistance traits such as osmotic adjustment, root penetration ability and stay green traits is in progress in various laboratories (Ludlow M.M., personal communication; Nguyen H.T., Rosenow D.R., and Mullet J.E., personal communication). Preliminary report of the location of dehydrin (RFLP marker number UNC170-DHN) clone from corn to long arm of chromosome 6 should indicate the possibility of chromosomal assignments of different RFLP markers linked to heat and drought stress tolerance traits in future (Close et al., 1989).

RAPDs have recently become popular for the rapid detection of polymorphisms among individuals from a population using a single primer of arbitrary sequence and the PCR (polymerase chain reaction) mediated amplification of random genomic DNA fragments (Williams et al., 1990). Rapid analysis, avoidance of radioactive waste, requirement of small amounts of genomic DNAs, high reproducibility and easy data scoring as either the presence or absence of fragments from ethidium bromide stained agarose gels are some of the advantages of the RAPD technique over RFLPs used in the past. Moreover, Williams et al. (1990) have indicated that the RAPDs are not limited to single copy sequences and it is possible to amplify fragments from repetitive DNA fractions. There has been a great flurry of activity following the original reports of the RAPD technique and it has been applied in plants for construction of genetic linkage maps, estimation of genetic relationships, tagging disease resistance traits, identification of cultivars, parentage determinations and population genetics (see Joshi and Nguyen, 1993 for review). The dominant behavior of RAPDs in segregating populations may limit their value in similar analyses in F_2 generations. Suitable use of recombinant inbreds (Burr and Burr, 1991) have been shown to alleviate this problem and aid in construction of genetic linkage maps (Reiter et al., 1992). Although, simple traits such as disease resistance have been reported to be linked to RAPD markers, there are no reports of QTL tagging using RAPDs. Use of RFLPs and RAPDs for association with heat and water deficit stress tolerance traits still awaits a further exploration.

The advantages of these recent breakthroughs can be enormous. Specific molecular markers may be identified which arc tightly linked to stress tolerance traits and they may be further used in molecular marker-aided selection. By knowing the location of these markers on particular chromosomes, one can develop a more detailed map which will assist in the probable isolation of genes which are the cause of the stress protection and are not only stress responsive. Knowledge gained by locating these genes in relation to other genes in its vicinity will provide a clear picture of how these genes are organized in a crop genome. This information will be crucial in targeting modified genes at the correct position in a genome.

Conclusions

In this review, we have attempted to give an overview of the field of stress tolerance and how molecular biological advances are likely to supplement our efforts to produce stress tolerant plants. Biotechnology does hold great promise in the genetic improvement of stress tolerance in crop plants. However, much remains to be done and an active collaboration between plant breeders, geneticists, molecular biologists and plant physiologists is essential to make significant advances in this field of research. Plant molecular genetics as a discipline is just about ten years old and we may require more time to come to grips with the problems in stress tolerance. From a practical viewpoint, more research is required to understand the complexities of stress responses in crop plants in addition to continuing work with model plants with little or no economical importance. Recent public concerns on the application of genetic engineering techniques might hamper the progress in plant biotechnology. It is our responsibility to educate the public about the safety and potential benefits of this new technology in crop improvement.

Acknowledgments

We wish to thank Dr. Jeff Ray for critical reading of the manuscript and helpful suggestions.

Literature cited

Blum A (1988) Plant breeding for stress environments. CRC Press. Florida

Bozarth CS, Mullet JE, Boyer JS (1987) Cell wall proteins at low water potentials. Plant Physiol 85: 261-267

Burr B, Burr FA (1991) Recombinant inbreds for molecular mapping in maize. Trends Genet 7: 55-60

Close T, Kortt AA, Chandler PM (1989) A cDNA-based comparison of dehydration-induced proteins (dehydrins) in barley and corn. Plant Mol Biol 13:95-108

Creelman RA, Mason HS, Bensen RJ, Boyer JS, Mullet JE (1990) Water deficit and abscisic acid cause differential inhibition of shoot versus root growth in soybean seedlings. Plant Physiol 92: 205-214

Dure L, Crouch M, Harada J, Ho T-HD, Mundy J, Quatrano R, Thomas T, Sung Z (1989) Common amino acid sequence domain among the LEA proteins of higher plants. Plant Mol Biol 12: 475-486

Ellis RJ. (1991) Molecular chaperones: the plant connection. Science 250:954-959

Guerrero FD, Jones JT, Mullet JE (1990) Turgor-responsive gene transcription and RNA levels increase rapidly when pea shoots are wilted: Sequence and expression of three inducible genes. Plant Mol Biol 15: 11-26

Howarth CJ (1991) Molecular responses of plants to an increased incidence of heat shock. Plant Cell Environ 14: 831-841

Fitzpatrick T (1993) Pleiotropic gene found in barley plant. Genetic Eng News 13: 1

Itturiaga G, Schneider K, Salamini F, Bartels D (1992) Expression of desiccation related proteins from the resurrection plant in transgenic tobacco. Plant Mol Biol 20: 555-558

Jorgensen JA, Weng J, Ho T-HD, Nguyen HT (1992) Genotype-specific heat shock proteins in two maize inbreds. Plant Cell Rep 11: 576-580

Joshi CP, Nguyen HT (1993) Application of the random amplified polymorphic DNA technique for the detection of polymorphism among wild and cultivated tetraploid wheats. Genome 36: 602-609

Joshi CP, King SW, Nguyen HT (1992) Molecular cloning and characterization of a cDNA encoding water stress protein (WSP23) from wheat roots. Plant Sci 86:71-82

Kimpel JA, Key JL (1985) Presence of heat shock mRNAs in field grown soybeans. Plant Physiol 79: 672-678

King SW, Vierling RA, Nguyen HT (1992) Changes in mRNA species during drought stress in winter wheat. Crop Sci 32: 822-825

Krishnan M, Nguyen HT, Burke JJ (1989) Heat shock protein synthesis and thermal tolerance in wheat. Plant Physiol 90: 140-145

Ludlow MM (1993) Physiological mechanisms of drought resistance. In Mabry TJ, Nguyen HT, Dixon RA, eds, Proceedings of the International Symposium on Applications and Prospects of Biotechnology for Arid and Semi-arid Lands. IC^2 Institute, University of Texas at Austin, Texas. (in press)

McCue KF, Hanson AD (1990) Drought and salt tolerance: towards understanding and application. Trends Biotech 8: 358-362

Murata N, Ishizaki-Nishizawa O, Higashi S, Hayashi N, Tasaka Y and Nishida I (1992) Genetically engineered alteration in the chilling sensitivity of plants. Nature 356: 710-713

Nguyen HT, Hendershot KL, Joshi CP (1992) Molecular genetics of stress breeding: Heat shock proteins. In D. Buxton, ed, Proceedings of the First International Crop Science Congress. Crop Science Society of America, Wisconsin, USA (in press)

Nguyen HT, Joshi CP (1993a) Molecular strategies for the genetic dissection of water and high temperature stress adaptation in cereal crops. In G. Kuo, ed, Proceedings of International Symposium on Adaptation of Food Crops to Temperature and Water Stress. Asian Vegetable Research and Development center, Taipei, Taiwan (in press)

Nguyen HT, Joshi CP (1993b) Molecular and genetic analysis of heat tolerance in plants. In Mabry TJ , Nguyen HT, Dixon RA, eds, Proceedings of the International Symposium on Applications and Prospects of Biotechnology for Arid and Semi-arid Lands. IC^2 Institute, University of Texas at Austin, Texas. (in press)

Nover L, Neumann D, Scharf KD (1989) Heat shock and other stress response systems of plants. Springer-Verlag, Berlin

Ottaviano E, Gorla MS, Pe E, Frova C (1991) Molecular markers (RFLPs and HSPs) for the genetic dissection of thermotolerance in maize. Theor Appl Genet 81: 713-719

Paterson AH, Tanksley SD, Sorrells ME (1991) DNA markers in plant improvement. Advances in Agronomy 46: 39-90

Reiter RS, Coors JG, Sussman MR, Gabelman WH (1991) Genetic analysis of tolerance to low-phosphorous stress in maize using restriction fragment length polymorphisms. Theor Appl Genet 82: 561-568

Skriver K, Mundy J (1990) Gene expression in response to abscisic acid and osmotic stress. Plant Cell 2 : 503-512

Tarczynski MC, Jensen RG, Bohnert HJ (1993) Stress protection of transgenic tobacco by production of the osmolyte mannitol. Science 259: 508-510

Vierling E (1991) The roles of heat shock proteins in plants. Ann. Rev. Plant Physiol. and Plant Mol Biol 42: 579-620

Williams JGK, Kubelik AR, Livak KJ, Rafalski JA, Tingy SV (1990) DNA polymorphisms amplified by arbitrary primers are useful as genetic markers. Nucleic Acids Res 18: 6531-6535

EVOLUTION AND METABOLIC ENGINEERING OF OSMOPROTECTANT ACCUMULATION IN HIGHER PLANTS

Andrew D. Hanson and Michael Burnet
IRBV, Université de Montréal
4101 Rue Sherbrooke Est
Montreal, PQ HIX 2B2
Canada

Abstract   Glycine betaine can act an osmoprotectant in many organisms.   The capacity to accumulate glycine betaine in response to drought or salt stress is widespread in angiosperms, and probably appeared early in the evolution of this group. Despite the efficacy of glycine betaine, diverse angiosperm taxa have evolved a range of structurally analogous compounds which supplement or replace it.   These compounds include choline-O-sulfate, dimethylsulfoniopropionate, and the betaines of ß-alanine, proline and hydroxyproline.   This phytochemical diversity may reflect biosynthetic innovation driven by three types of selective forces: the need for effective osmoprotection, the interaction between osmotic and co-occurring physicochemical stresses, and the action of herbivores and pathogens.   The evolution of the biosynthetic pathways for alternative osmoprotectants to glycine betaine suggests new opportunities for the metabolic engineering of stress tolerance.

Introduction

Various zwitterionic quaternary ammonium and tertiary sulfonium compounds (QACs and TSCs) can act as compatible osmolytes or osmoprotectants in the cytoplasm of osmotically-stressed cells of bacteria (Csonka and Hanson, 1991), marine algae (Blunden and Gordon, 1986), animals (Somero, 1986), and higher plants (Rhodes and Hanson, 1993).   The biosynthetic pathways of such compounds are potential targets for the genetic engineering of osmotic stress tolerance in microorganisms and crop plants (LeRudulier et al., 1984b; McCue and Hanson, 1990).

NATO ASI Series, Vol. H 86
Biochemical and Cellular Mechanisms
of Stress Tolerance in Plants
Edited by J. H. Cherry
© Springer-Verlag Berlin Heidelberg 1994

Figure 1 shows the structures of the main QACs, and the only TSC, reported to accumulate in angiosperms (Rhodes and Hanson, 1993). The most common compound is glycine betaine; it occurs in at least 11 families, including mono- and dicotyledons. Proline betaine is also widespread (at least 7 families), often occurring with hydroxyproline betaine (of which there are several possible isomers). ß-Alanine betaine and choline-O-sulfate accumulation are known only in Plumbaginaceae, and accumulation of dimethyl-sulfoniopropionate (DMSP) only in Gramineae and Compositae.

Because glycine betaine is widespread among higher plants, and because comparative biochemical, immunological and nucleotide sequence data indicate that the glycine betaine biosynthesis pathway appeared early in angiosperm evolution (Weretilnyk et al., 1989), it can be viewed as the archetypal angiosperm betaine. The other compounds of Figure 1 can then be seen as evolutionarily novel substitutes which presumably confer some selective advantage(s) that glycine betaine does not. Such advantages are important to investigate in the context of genetic engineering of crop stress resistance. Accordingly, we will (a) summarize data on the biosynthetic pathways of the compounds of Figure 1, (b) consider the selective forces that may underlie the diversity in QAC and TSC osmoprotectants, and (c) discuss metabolic engineering of the accumulation of these compounds.

Biosynthetic Pathways

Glycine Betaine. For plants from several families, *in vivo* isotope tracer studies (e.g. Weretilnyk et al., 1989; Lerma et al., 1991) have confirmed that glycine betaine is synthesized by a two step oxidation of choline, via the intermediate betaine aldehyde. The enzymes mediating both steps have been studied in Chenopodiaceae, and that mediating the second step in Gramineae.

The first step is catalyzed by a stromal, ferredoxin-dependent choline monooxygenase (CMO) which introduces an atom of oxygen from O_2 into choline, giving the hydrate form of betaine aldehyde (Brouquisse et al., 1989; Lerma et al., 1988). The reduced ferredoxin is photosynthetically generated, and choline oxidation is promoted by light both *in vivo* (Lerma et al., 1988)

293

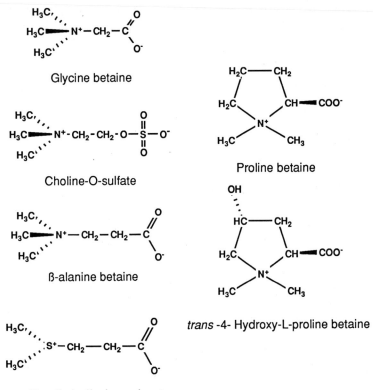

Figure 1.   Structures of zwitterionic quaternary ammonium and tertiary sulfonium compounds accumulated by angiosperms.

and in isolated chloroplasts (Weigel et al., 1988). Spinach CMO has been partially purified; it has a native M_r of 98,000, a pH optimum of approximately 8, and is stimulated by Mg^{2+} (Brouquisse et al., 1989).

The second step in glycine betaine synthesis is catalyzed by a pyridine nucleotide-dependent betaine aldehyde dehydrogenase (BADH) which has a strong preference for NAD^+ (Weretilnyk and Hanson, 1989; Arakawa et al., 1990). In spinach leaves the majority of the BADH activity is in the chloroplast stroma; the remainder apparently resides in a cytosolic isozyme (Weigel et al., 1986). Inheritance studies of BADH isozyme variants indicated that the stromal enzyme is a homodimer of a nuclear-encoded protein (Weretilnyk and Hanson, 1988). BADH has been purified to homogeneity from spinach (Arakawa et al., 1987; Weretilnyk and Hanson, 1989); in agreement with the genetic data, the purified enzyme is a dimer with subunits of M_r ~60,000. The spinach enzyme has a broad pH optimum around 8.5 and is substrate specific. BADH cDNA clones containing the entire protein coding sequence have been isolated from spinach (Weretilnyk and Hanson, 1990) and sugar beet (McCue and Hanson, 1992). The spinach and beet cDNA sequences have been expressed in tobacco, where they confer the capacity to oxidize exogenous betaine aldehyde to glycine betaine (Rathinasabapathi et al., 1993).

Choline-O-Sulfate. Choline-O-sulfate has been shown by *in vivo* radiotracer studies to be formed from choline by leaves and roots of *Limonium* species (Hanson et al., 1991) and by leaves of *Aegialitis annulata* (Benson and Atkinson, 1967). The enzyme responsible has been identified in *Limonium* species as a choline sulfotransferase which uses 3'-phosphoadenosine-5'-phosphosulfate as sulfate donor (Rivoal and Hanson, unpublished).

ß-Alanine Betaine. *In vivo* radiotracer studies of *Limonium* species have shown that ß-alanine betaine is synthesized in both leaves and roots by methylation of ß-alanine, with the methyl groups coming from methionine (Hanson et al., 1991; Larher, 1976). The *N*-methyltransferase(s) responsible have not yet been investigated. The origin of the ß-alanine moiety is not known.

The simplest possibility, aspartate decarboxylation, is unlikely because [^{14}C]aspartate does not act as a precursor of ß-alanine or ß-alanine betaine in *Limonium* (Hanson, unpublished).

Proline Betaine and Hydroxyproline Betaine. In alfalfa, proline betaine (stachydrine) has been shown by *in vivo* labeling to derive from methylation of proline (Essery et al., 1962). The hydroxylation steps in the synthesis of hydroxyproline betaines have not been elucidated. In only a few cases is it clear which isomer of hydroxproline betaine is present: *trans*-4-hydroxy-L-proline betaine (betonicine) has been found in Lamiaceae (Mauger and Witkop, 1966; Yuan et al., 1992), Myrtaceae (Jones et al., 1987), and Plumbaginaceae (Hanson and Zamir, unpublished); *trans*-3-hydroxy-L-proline betaine has been reported from Capparidaceae (Mauger and Witkop, 1966).

Dimethylsulfoniopropionate. DMSP biosynthesis has been studied in higher plants and marine algae by *in vivo* radiotracer methods. In the green alga *Ulva lactuca* the carbon skeleton, the sulfur and the methyl groups of DMSP are derived from methionine (Greene, 1962; Kahn, 1964); this is probably also so in the red alga *Chondria coerulescens* (Chillemi et al., 1990), in brown algae (Pokorny et al., 1970), and in *Wedelia biflora* (Hanson and Rivoal, unpublished). DMSP therefore presumably originates from methionine via deamination, decarboxylation, oxidation and methylation, although the order of these steps is not known. The enzymes involved have not been identified. Methylation seems unlikely to be the final step, because ß-methylthiopropionate was not a detectable metabolite of [^{14}C]methionine in DMSP-accumulating algae (Pokorny et al., 1970), and because in *W. biflora* ß-[^{14}C]methylthiopropionate was a poor precursor of DMSP (Hanson, unpublished). It is possible that methylation is the first step, because S-methylmethionine is a major metabolite of methionine in *W. biflora* (Hanson and Rivoal, unpublished) and in many other plants (Mudd and Datko, 1990; Pokorny et al., 1970). In algae, supplied [^{14}C]S-methylmethionine was not converted to DMSP (Greene, 1962; Chillemi et al., 1990); this may have been due to poor intracellular uptake, however.

Evolution of Chemical Diversity in QAC and TSC Osmoprotectants

Effectiveness of Osmoprotectants. Some compounds may be superior osmoprotectants in a physicochemical sense, i.e. less disruptive to cellular functions when present at high concentration. In bacteria, glycine betaine is a more potent alleviator of osmotic stress than most other osmoprotectants (Csonka and Hanson, 1991), and generally also ranks as most effective in studies with isolated enzymes (Mason and Blunden, 1989; Wyn Jones, 1984). There is evidence that proline betaine and hydroxyproline betaine are superior to proline (LeRudulier et al., 1984a; Hanson and Rathinasabapathi, unpublished). Dry or saline conditions elicit accumulation of proline in many angiosperms (Hanson and Hitz, 1982); certain Plumbaginaceae have evolved the ability to convert this proline pool to proline betaine and hydroxyproline betaine (Hanson and Rathinasabapathi, unpublished).

Interaction of Osmotic Stress and Other Abiotic Stresses. Some compounds may have an advantage over glycine betaine with respect to a second stress associated with dry or saline conditions. In the case of choline-O-sulfate, the second stress may be a high level of SO_4^{2-} within the plant. The Plumbaginaceae, the family in which choline-O-sulfate accumulation occurs, possess salt glands in their leaves (Faraday and Thompson, 1986). These glands can excrete large amounts of Na^+ and Cl^-, but not SO_4^{2-} (Hanson and Gage, 1991). This can produce internal accumulation of SO_4^{2-} during growth on saline soils, which often contain levels of SO_4^{2-} similar to those in seawater (30 mM). Because SO_4^{2-} ions are inhibitory to metabolism, conjugation with choline may represent a means of detoxifying part of the excess SO_4^{2-}.

For ß-alanine betaine, the second stress may be oxygen deficiency, which frequently accompanies salinity in agricultural and natural soils. The oxygenase (CMO) in the biosynthetic pathway of glycine betaine has an absolute requirement for O_2. There is no such direct O_2 requirement for ß-alanine betaine biosynthesis from ß-alanine (Hanson et al., 1991). Thus, ß-alanine betaine synthesis could proceed in anoxic roots, whereas glycine betaine synthesis could not. The habitats of glycine betaine- and ß-alanine betaine-accumulating species of Plumbaginacaeae are consistent with this: the former tend to occur in dry, sandy or rocky soils, while many of the latter colonize salt-marshes (Hanson et al., 1991). Another possible factor favoring ß-alanine betaine over glycine betaine in the Plumbaginaceae is elimination of the competition between choline-O-sulfate and glycine betaine for choline (Hanson et al., 1991).

For DMSP, the second stress may be low N availability. In both *Spartina* and *Wedelia*, DMSP co-occurs with glycine betaine, and in *Spartina,* field experiments have shown that N-fertilization increases the level of glycine betaine and decreases that of DMSP (Dacey et. al. 1987). Because DMSP contains S in place of N, and because glycine betaine can represent up to about 20% of total reduced plant N, substituting DMSP for glycine betaine can have a major sparing effect on the N budget, and hence make significantly more N available for growth. This has been proposed as the basis for the widespread occurrence of DMSP in marine algae (Grone and Kirst, 1992).

Interaction with Herbivores and Microrganisms. Osmoprotectants accumulate to several percent of dry weight in stressed plants. Osmoprotectants which contain N (proline and QACs) therefore represent a rich potential source of a limiting nutrient for phytophagous insects, other herbivores, and pathogenic or symbiotic microorganisms (White, 1984). Little is known about interactions between osmoprotectant accumulation and biotic factors, but the following examples suggest that they are important. (a) Plants with high levels of proline (Haglund, 1980) or glycine betaine (Zuñiga and Corcuera, 1987) are more susceptible to insect predators. (b) Growth and infection in the cereal pathogen *Fusarium graminearum* are promoted by glycine betaine (Strange et al., 1974). (c) *Rhizobium meliloti* has catabolic functions for *Medicago* betaines encoded on the symbiotic megaplasmid (Goldmann et al., 1991).

It is therefore possible that osmoprotectant evolution has involved selection for less available N-sources than glycine betaine or proline, and perhaps also for outright toxicity. The following cases are consistent with this possibility, and suggest that it is worth exploring more thoroughly. (a) DMSP, besides being of no value as an N source, readily decomposes to give acrylic acid which has antibacterial activity (Kodama and Ogata, 1983). (b) Proline betaine is probably less easily utilized by microorganisms and animals than proline itself (Goldmann et al., 1991; Chambers and Kunin, 1987). (c) The prevalent naturally occurring hydroxyproline betaine (*trans*-4-hydroxy-L-proline betaine) is a potent acetylcholine esterase inhibitor whereas other possible 4-hydroxy isomers are not (Friess et al., 1957).

Metabolic Engineering Considerations

All the compounds in Figure 1 are related to primary metabolites by short (one- to three-step) biosynthetic pathways. Assuming that only these steps would need to be inserted into a target species, then the genes for one to three enzymes would be involved. It is quite possible in principle to engineer biosynthetic sequences of this size, once the enzymes are known and the corresponding cDNA or genomic clones are available. The limitation to doing this lies at the level of biochemistry and cDNA cloning. As outlined above, only for glycine betaine and choline-O-sulfate are the biosynthetic enzymes known, and of these only BADH has been purified and used to isolate cDNA clones. None of the enzymes for the synthesis of ß-alanine betaine, proline betaine, hydroxyproline betaines and DMSP have been identified, and in fact the steps in the biosynthetic pathways of these compounds are mostly unclear.

It has been emphasized for some time that metabolic engineering of glycine betaine accumulation may have potential for increasing crop resistance to salt- and drought stress (LeRudulier et al., 1984b; McCue and Hanson, 1990). The value of engineering the other compounds in Figure 1 has had less attention. The preceding discussion of their possible selective advantages suggests that they could be more appropriate than glycine betaine for certain environments, and that they may have roles in plant defence as well as in osmotic regulation.

299

References

Arakawa K, Katayama M, Takabe T (1990) Levels of betaine and betaine aldehyde dehydrogenase in the green leaves, and etiolated leaves and roots of barley. Plant and Cell Physiology 31:797-803

Arakawa K, Takabe T, Sugiyama T, Akazawa T (1987) Purification of betaine-aldehyde dehydrogenase from spinach leaves and preparation of its antibody. Jounal of Biochemistry 101:1485-1488

Benson AA, Atkinson, MR (1967) Choline sulfate and phosphate in salt excreting plants. Federation Proceedings 26:394

Blunden G, Gordon SM (1986) Betaines and their sulphonio analogues in marine algae. Progress in Phycological Research 4:39-79

Brouquisse R, Weigel P, Rhodes D, Yocum CF, Hanson AD (1989) Evidence for a ferredoxin-dependent choline monooxygenase from spinach chloroplast stroma. Plant Physiology 90:322-329

Chambers ST, Kunin CM (1987) Isolation of glycine betaine and proline betaine from human urine. Journal of Clinical Investigation 79:731-737

Chillemi R, Patti A, Morrone R, Piattelli M, Sciuto S (1990) The role of methylsulfonium compounds in the biosynthesis of *N*-methylated metabolites in *Chondria coerulescens*. Journal of Natural Products 53:87-93

Csonka LN, Hanson AD (1991) Prokaryotic osmoregulation: genetics and physiology. Annual Review of Microbiology 45:569-606

Dacey JWH, King GM, Wakeham SG (1987) Factors controlling emission of dimethylsulphide from salt marshes. Nature 330:643-645

Essery JM, McCaldin DJ, Marion L (1962) The biogenesis of stachydrine. Phytochemistry 1:209-213

Faraday CD, Thomson WW (1986) Structural aspects of the salt glands of the Plumbaginaceae. Journal of Experimental Botany 37:461-470

Friess SL, Patchett AA, Witkop B (1957) The acetylcholinesterase surface. VII. Interference with surface binding as reflected by enzymatic responses to turicine, betonicine and related heterocycles. Journal of the American Chemical Society 79:459-462

Goldmann A, Boivin C, Fleury V, Message B, Lecoeur L, Maille M, Tepfer D (1991) Betaine use by rhizosphere bacteria: genes essential for trigonelline, stachydrine, and carnitine catabolism in *Rhizobium meliloti* are located on pSym in the symbiotic region. Molecular Plant-Microbe Interactions 4:571-578

Greene RC (1962) Biosynthesis of dimethyl-ß-propiothetin. Journal of Biological Chemistry 237:2251-2254

Grone T, Kirst GO (1992) The effect of nitrogen deficiency, methionine and inhibitors of methionine metabolism on the DMSP contents of *Tetraselmis subcordiformis* (Stein) Marine Biolgy 112:497-503

Haglund BM (1980) Proline and valine - cues which stimulate grasshopper herbivory during drought stress? Nature 288:697-698

Hanson AD, Gage D (1991) Identification and determination by fast atom bombardment mass spectrometry of the compatible solute choline-O-sulfate in *Limonium* species and other halophytes. Australian Journal of Plant Physiology 18:317-327

Hanson AD, Hitz WD (1982) Metabolic responses of mesophytes to plant water deficits. Annual Review of Plant Physiology 33:163-203

Hanson AD, Rathinasabapathi B, Chamberlin B, Gage DA (1991) Comparative physiological evidence that ß-alanine betaine and choline-O-sulfate act as compatible osmolytes in halophytic *Limonium* species. Plant Physiology 97:1199-1205

Jones GP, Naidu BP, Paleg LG, Tiekink ERT, Snow MR (1987) 4-hydroxy-*N*-methylproline analogues in *Melaleuca* spp. Phytochemistry 26:3343-3344

Kahn V (1964) Glycine as a methyl donor in dimethyl-ß-propio-thetin synthesis. Journal of Experimental Botany 15:225-231

Kodama M, Ogata T (1983) Acrylic acid, as an antibacterial substance in scallop. Bulletin of the Japanese Society of Scientific Fisheries 49:1103-1107

Larher F (1976) Sur quelques particularités du métabolisme azoté d'une halophyte: *Limonium vulgare* Mill. D.Sc. Thesis, Université de Rennes

Lerma C, Hanson AD, Rhodes D (1988) Oxygen-18 and deuterium labeling studies of choline oxidation by spinach and sugar beet. Plant Physiology 88:695-702

Lerma C, Rich PJ, Ju GC, Yang W-J, Hanson AD, Rhodes D (1991) Betaine deficiency in maize: complementation tests and metabolic basis. Plant Physiology 95:1113-1119

LeRudulier D, Bernard T, Goas G, Hamelin J (1984a) Osmoregulation in *Klebsiella pneumoniae*: enhancement of anaerobic growth and nitrogen fixation under stress by proline betaine, butyrobetaine, and other related compounds. Canadian Journal of Microbiology 30:299-305

LeRudulier D, Strom AR, Dandekar AM, Smith LT, Valentine RC (1984b) Molecular biology of osmoregulation. Science 224:1064-1068

Mason TG, Blunden G (1989) Quaternary ammonium and tertiary sulfonium compounds of algal origin as alleviators of osmotic stress. Botanica Marina 32:313-316

Mauger AB, Witkop B (1966) Analogs and homologs of proline and hydroxyproline. Chemical Reviews 66:47-86

McCue KF, Hanson AD (1990) Drought and salt tolerance: towards understanding and application. Trends in Biotechnology 8:358-362

McCue KF, Hanson AD (1992) Salt-inducible betaine aldehyde dehydrogenase from sugar beet: cDNA cloning and expression. Plant Molecular Biology 18:1-11

Mudd SH, Datko AH (1990) The *S*-methylmethionine cycle in *Lemna paucicostata*. Plant Physiology 93:623-630

Pokorny M, Marcenko E, Keglevic D (1970) Comparative studies of L- and D-methionine metabolism in lower and higher plants. Phytochemistry 9:2175-2188

Rathinasabapathi B, McCue KF, Gage DA, Hanson AD (1993) Metabolic engineering of glycine betaine synthesis: plant betaine aldehyde dehydrogenases lacking typical transit peptides are targeted to tobacco chloroplasts where they confer betaine aldehyde resistance. The Plant Journal (to be published)

Rhodes D, Hanson AD (1993) Quaternary ammonium and tertiary sulfonium compounds in higher plants. Annual Review of Plant Physiology and Plant Molecular Biology 44:357-384

Somero GN (1986) Protons, osmolytes, and fitness of internal milieu for protein function. American Journal of Physiology 251:R197-R213

Strange RN, Majer JR, Smith H (1974) The isolation and identification of choline and betaine as the two major components in anthers and wheat germ that stimulate *Fusarium graminearum in vitro*. Physiological Plant Pathology 4:277-290

Weigel P, Lerma C, Hanson AD (1988) Choline oxidation by intact spinach chloroplasts. Plant Physiology 86:54-60

Weigel P, Weretilnyk EA, Hanson AD (1986) Betaine aldehyde oxidation by spinach chloroplasts. Plant Physiology 82:753-759

Weretilnyk EA, Bednarek S, McCue KF, Rhodes D, Hanson AD (1989) Comparative biochemical and immunological studies of the glycine betaine synthesis pathway in diverse families of dicotyledons. Planta 178:342-352

Weretilnyk EA, Hanson AD (1988) Betaine aldehyde dehydrogenase polymorphism in spinach: genetic and biochemical characterization. Biochemical Genetics 26:143-151

Weretilnyk EA, Hanson AD (1989) Betaine aldehyde dehydrogenase from spinach leaves: purification, *in vitro* translation of the mRNA, and regulation by salinity. Archives of Biochemistry and Biophysics 271:56-63

Weretilnyk EA, Hanson AD (1990) Molecular cloning of a plant betaine-aldehyde dehydrogenase, an enzyme implicated in adaptation to salinity and drought. Proceedings of the National Academy of Sciences USA 87:2745-2749

White TCR (1984) The abundance of invertebrate herbivores in relation to the availability of nitrogen in stressed food plants. Oecologia 63:90-105

Wyn Jones RG (1984) Phytochemical aspects of osmotic adaptation. Recent Advances in Phytochemistry 18:55-78

Yuan Z-X, Patel A, Blunden G, Turner CH (1992) *Trans*-4-hydroxypipecolic acid betaine from *Lamium maculatum*. Phytochemistry 31:4351-4352

Zuñiga GE, Corcuera LJ (1987) Glycine-betaine accumulation influences susceptibility of water-stressed barley to the aphid *Schizaphis graminum*. Phytochemistry 26:367-369

REGULATION OF SHOOT GROWTH IN DRY SOILS BY ABSCISIC ACID AND BY ROOT MESSAGES

Rana Munns[1] and Robert E. Sharp[2]
[1]Cooperative Research Centre for Plant Science
CSIRO Division of Plant Industry
GPO Box 1600
Canberra ACT 2601
Australia

Abstract

Leaf expansion in drying soil is affected by growth substances produced by roots. The role of ABA in this root-to-shoot communication is unclear: the concentrations of free ABA in xylem sap seem too low to affect leaf expansion. Yet, ABA can inhibit shoot expansion in dry soil, as shown by the greater shoot growth when ABA synthesis is inhibited. A possible solution to this puzzle is that ABA is transported in xylem sap in a complexed form, or that another compound in xylem sap stimulates the synthesis or activity of ABA in leaves. Drying soil can stimulate the continued growth of a particular shoot tissue: the mesocotyl. The role of ABA and root messages in this unique response is being investigated.

Introduction

The factors most likely to limit productivity under drought are those that control leaf expansion. Root growth is generally much less inhibited than shoot growth in dry soils, and the root system biomass may even be greater in dry soils than in well-watered soils. Further, in most experiments in drying soils, leaf expansion is affected before stomatal conductance and photosynthesis (e.g. Passioura 1988, Saab and Sharp 1989; Gowing et al. 1990). Thus it is likely that biomass

---

[2] Department of Agronomy, University of Missouri, Columbia MO 65211, USA

NATO ASI Series, Vol. H 86
Biochemical and Cellular Mechanisms
of Stress Tolerance in Plants
Edited by J. H. Cherry
© Springer-Verlag Berlin Heidelberg 1994

production in dry soils is limited directly by the rate of leaf expansion rather than by the rate of photosynthesis.

Leaf growth can be significantly inhibited by drying soil without a change in leaf water status (Passioura 1988; Saab & Sharp 1989). As it is the roots that directly perceive the low soil water potential, some kind of signal must communicate the soil condition to the shoot. This suggests that roots may produce growth inhibitors, or decrease their production of growth promoters. The evidence for the control of leaf growth or stomatal conductance in dry soils by root signals has been reviewed by Davies and Zhang (1991). A particularly pertinent result comes from split root experiments on apple trees by Gowing, Davies and Jones (1990): excision of roots in the dry compartment relieved the inhibition of shoot growth, showing that growth had been restricted by an inhibitor coming from the roots, not just by a reduction in the supply of growth promoters.

Here we summarise some recent work on the identity of root messages controlling leaf expansion, and describe preliminary studies on a shoot tissue with a unique response to water stress: the mesocotyl.

Materials and methods

Collection of xylem sap. Wheat or barley seedlings were grown in pots designed to fit in pressure chambers, and the pots were either watered regularly or allowed to dry down. To collect sap, the pots were pressurised so that the xylem sap was on the verge of bleeding from the tip of the leaf. Half the lamina was removed, and sap flowed from the cut at the same rate as the leaf was previously transpiring. Sap was analysed for ABA by ELISA immunoassy (Munns and King 1988).

Bioassay for leaf elongation. This used whole shoots of well-watered 14 d wheat or barley seedlings detached below the crown (Munns 1992).

Bioassay for transpiration. This used expanded leaves detached from well-watered wheat seedlings (Munns and King 1988).
Measurement of mesocotyl elongation. Maize or rice seeds were germinated until the root was 2-5 mm long, transferred to tubes containing vermiculite that was fully hydrated with water or ABA solutions, or was partially watered to a water potential of -0.3 MPa (water content of about 25%), and grown in the dark at 29°C and near-saturation humidity. Measurements of mesocotyl and coleoptile length were made daily with a ruler.

Results and Discussion

Inhibition of leaf expansion by root messages, and the role of ABA

Xylem sap collected from wheat and barley in dry soil contains a substance that inhibits leaf expansion. This was shown by feeding shoots detached from well-watered plants and measuring leaf elongation rates over the subsequent 16 h (Munns 1992). The ABA concentration in the sap increased as the soil dried, but not to levels high enough to account for the inhibitory effect of the sap on the bioassay: synthetic ABA had no effect on leaf expansion when supplied at the concentrations found in xylem sap from water-stressed plants (Fig. 1). Leaf elongation rate was affected by ABA only at concentrations greater than 100 nM, yet the concentration of ABA in xylem sap in the plants whose growth was reduced by drying soil was only about 20 nM (Munns 1992).

Others have suggested that the ABA in the xylem causes the decreased rate of leaf expansion in drying soil. Zhang and Davies (1990a,b) found a relation between ABA in xylem exudates and leaf expansion rate in well-watered maize fed ABA through root tips that was quite similar to the relation found in maize in drying soil. However, both experiments were done at very low light levels, and leaf expansion started to decline only after stomatal conductance was significantly reduced, so

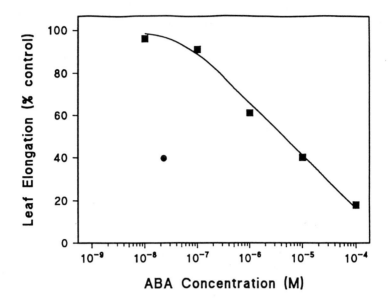

Figure 1. Effect of ABA on leaf elongation in shoots detached from wheat seedlings. Solutions were fed through the sub-crown internode. Squares show the increase in length of the emerging leaf between 8 and 16 h after feeding (during the period of maximum and steady inhibition. The isolated circle shows the relationship between the activity of xylem sap collected from wheat plants in drying soil and its ABA concentration. From Munns (1992).

carbon limitation may have caused the growth inhibition. Thus, there is no proof that the concentration of ABA in the xylem directly controls leaf expansion in dry soils.

Yet, it is surprising that ABA would not be involved in the inhibition of leaf elongation by root messages. That ABA inhibits shoot growth is shown by experiments manipulating endogenous ABA levels in water-stressed plants. Such evidence comes from Saab *et al.* (1990) in maize using fluridone which inhibits the synthesis of ABA. Seeds were germinated in well-watered vermiculite for 60 h, then seedlings were transferred to vermiculite at -0.3 MPa. Water stress decreased the rate of shoot elongation by more than 50%, but pre-treatment with fluridone prevented this to a large extent. Similarly, shoots of the vp5 mutant, which is deficient in ABA, grew faster than the wildtype at -0.3 MPa (Saab *et al.* 1990). These results indicate that the production of ABA at low water potential inhibits shoot growth.

It is possible that the bioassay used to detect root signals (Fig. 1) is not sufficiently sensitive to ABA. Well-watered plants are routinely used for the bioassay material but perhaps these are not as sensitive to ABA as water-stressed plants in which another root-produced substance may have altered the response of leaves to ABA coming in from the xylem. In this context, it is significant that ABA has a very different effect on roots growing in dry soil than in wet: it maintains root growth in dry soil while the same endogenous concentrations are inhibitory to root growth in wet soil (Saab *et al.* 1990; Sharp *et al.*, unpublished data). Unfortunately, attempts to use stressed plants as the source of material for the bioassay have been unsuccessful. When leaves were taken from plants growing slowly in either drying or saline soil, stomates opened within a few hours of feeding water, and the stomatal response to ABA was no different than in leaves from well-watered and non-saline plants (Munns and King 1988). ABA was also fed mixed with xylem sap, in case there was another compound or ion that made ABA more effective, but there was no synergism between xylem sap and synthetic ABA on stomatal conductance. Leaf expansion has not been measured in response to these treatments. It would be interesting to feed xylem sap collected from stressed plants to shoots from unstressed plants for some period of time before feeding ABA. The leaves might then be more responsive to the low levels of ABA in the xylem.

On the other hand, it is possible that xylem ABA does not control leaf expansion, even in stressed plants, and that the following alternatives are likely:
1. The compound might be a complexed or bound form of ABA and so be undetected by the usual methods of chemical or immunological analysis. ABA could be conjugated to a sugar or unknown compounds such as in the "ABA-adduct" discovered by Netting, Willows and Milborrow (1992). This possibility is

supported by the results of Bano *et al.* (1993) with xylem sap of rice in drying soil: they found five times more ABA after hydrolysis with alkali.

2. The compound could be chemically unrelated to ABA, but affect the accumulation of ABA in leaves, by stimulating its synthesis or repressing its metabolism. This seems quite likely, as we have found an increase in ABA in leaves fed xylem sap from stressed plants - an increase greater than expected from the small amount of ABA in the sap.

3. The compound could have an independent effect.

Identification of a new inhibitor in the xylem of wheat and barley

An inhibitor of transpiration was found in xylem sap from water-stressed plants, using a bioassay of detached leaves. This inhibitor was not ABA (Munns and King 1988). Subsequently it was found that the inhibitory activity of xylem sap increased with time of storage (Munns *et al.* 1993). The same phenomenon has been observed with the effect of xylem sap on leaf elongation of detached whole shoots (Munns, unpublished), suggesting that the same inhibitor affects both transpiration and leaf elongation. To identify the unknown inhibitor found in xylem sap, the transpiration bioassay rather than the leaf elongation bioassay was chosen because the transpiration bioassay uses less sap. The results are given in Munns *et al.* (1993), and the main findings are summarised here.

The inhibitory activity of fresh sap was small, and could be partly accounted for by the ABA acid content (about 20 nM). When fresh sap was stored at -20°C for several days, the activity increased. Maximum activity developed after a week. This increase in activity was due to a compound that increased in size with storage at -20°C. When sap already stored at -20°C for some time was fractionated with filters of different molecular size exclusion characteristics, activity was only in fractions containing compounds larger than 10 kD. However, when fresh sap was fractionated and the separated fractions stored at -20°C for a week, activity developed only in the fraction containing compounds smaller than 0.3 kD. These results suggest that storage at -20°C causes the aggregation or polymerisation of a small compound with low activity to form a large compound with high activity.

The increase in activity and in size did not occur with storage in liquid nitrogen (-196°C) or at -80°C. Sap collected from plants in soils of high water potential had little or no activity after storage at -20°C.

To confirm that the ABA in the sap was not involved in the development of the high molecular weight compound, ABA was measured in sap before and after passage through a filter excluding compounds greater than 0.3 kD. This was done both before and after sap was stored at -20°C. Regardless of the treatment, all the ABA passed through the filter, and none was in the retained fraction. This experiment shows that the high molecular weight compound is not a polymer of ABA, and that ABA does not aggregate after storage in sap at -20°C.

The increase in size with storage at -20°C could be due to aggregation or polymerisation of one or more small compounds. Although xylem sap contains enzymes that could catalyse polymerisation or aggregation, such as peroxidases, enzymes could not be responsible for the change as it occurred in fractions from which molecules greater than 1 kD were excluded. Solutes would increase to saturated concentrations as ice crystals formed during freezing, and the "monomer(s)" could concentrate enough to polymerise or aggregate spontaneously. We always took care to minimise such an event: the tubes containing sap were routinely snap-frozen before placing in a freezer at -20°C, to avoid the concentration of solutes in the middle of the tube as water froze slowly from the outside. However, even under these conditions, activity still developed during storage at -20°C. The fact that activity did not develop at -80°C or in liquid nitrogen suggests that slow movement of water molecules at -20°C to form microscopic ice crystals was enough to cause local high concentrations of solutes.

310

We are at present trying to identify this compound, making use of its strange behaviour during storage to purify it.

Effect of drying soil on mesocotyl expansion, and the role of ABA

While it is certain that drying soil reduces the rate of leaf expansion, not all plant tissues are inhibited. As mentioned above, there is relatively little effect on root growth, and sometimes root growth is actually stimulated (Jupp *et al.* 1987). A growth increase can also occur in another underground organ: the mesocotyl. This is the lowest stem internode in monocots, between the seed and the coleoptile. In

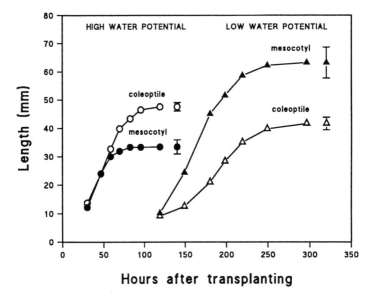

Figure 2. Effect of low water potential on the elongation of the coleoptile and mesocotyl of maize seedlings. Seedlings were imbibed for 36 h then transplanted to vermiculite that was well-watered (circles) or at a water potential of −0.3 MPa (triangles). Open symbols, coleoptile; closed symbols, mesocotyl. Bars show the average s.e.m.

some species (notably maize, rice and oats) the mesocotyl elongates in the absence of light and so raises the crown to the surface of the soil.

A dramatic response to drying soil was found in young rice seedlings (Terao and Inouye 1980): the mesocotyl grew much longer than in control conditions, while the coleoptile was shorter. We found a similar response with maize seedlings transplanted into vermiculite of low water potential (Fig. 2). Dry vermiculite greatly delayed the onset of both mesocotyl and coleoptile development, but the rate of mesocotyl elongation eventually reached that of the controls, and continued for longer, so that the final length of the mesocotyl was greater (Fig. 2). The rate of coleoptile elongation was much less than the controls, but continued for longer, and the final length was little affected.

We have evidence that ABA plays a major role in the inhibition of mesocotyl and coleoptile growth during the first 100 h after transfer to low water potential. First, the lag in growth was much less in seedlings pretreated with fluridone to inhibit ABA synthesis (Voetberg and Sharp, unpublished). In addition, Saab et al. (1992) showed that the initial reduction in the rate of mesocotyl elongation was largely prevented by fluridone. (This was done by germinating seedlings in wet vermiculite for 60 h, so that the shoot was elongating rapidly, then transferring to low water potential vermiculite and measuring the effect of fluridone pre-treatment on elongation rates at 5 h.)

The explanation for the recovery and eventual promotion of mesocotyl elongation at low water potential is unknown. Does the tissue become less susceptible to inhibition by ABA? Our initial attempts to address this question using fluridone were unsuccessful because there was no longer an effect of the fluridone treatment on shoot ABA levels by 100 h after transplanting (presumably because of low rates of fluridone transport to the shoot under the non-transpiring condition in which the experiments were conducted). Further experiments are planned.

Table 1.   Effect   of   water   stress   or   ABA   application   on
mesocotyl growth in maize and rice.   See Fig. 2 for
experimental details.

Treatment	Length of mesocotyl (mm)	
	Maize	Rice
Well-watered vermiculite	59 ± 2	2 ± 1
Dry vermiculite (-0.3 MPa)	73 ± 6	19 ± 2
ABA (10^{-5} M)	50 ± 2	9 ± 2

It is noteworthy that growth of the rice mesocotyl was
stimulated 4-fold by high concentrations of ABA (over 1 μM)
(Takahashi 1973). We also found this, although we could not
obtain the same effect in maize (Table 1). Lower concentrations
of ABA (1 nM) caused a small stimulation of growth of *Lemna*
fronds and of the mesocotyl and coleoptile of several cereals
(McWha and Jackson 1976). It should be noted, however, that in
all these examples of shoot growth promotion by ABA, the
experiments were done at high water potential. Whether ABA can
play a similar role at low water potential remains to be seen.

The mesocotyl should provide a useful tissue in which to
identify genes for drought resistance, particularly when
contrasted with coleoptile and leaf tissue. As growth of these
latter tissues is inhibited by water stress, many changes in
gene expression there may merely be secondary responses to the
reduced growth.

References

Bano A, Dörffling K, Bettin D, Hahn H   (1993)   Abscisic acid
    and cytokinins as possible root-to-shoot signals in xylem
    sap of rice plants in drying soil.   Aust J Plant Physiol
    20: 109-115.
Davies WJ, Zhang J   (1991)   Root signals and the regulation of
    growth and development of plants in drying soil. Ann Rev
    Plant Physiol Plant Mol Biol 42: 55-76

313

Gowing DJG, Davies WJ, Jones HG (1990) A positive root-sourced signal as an indicator of soil drying in apple, *Malus X domestica* Borkh. J Exp Bot 41: 1535-540

Jupp AP, Newman EI (1987) Morphological and anatomical effects of severe drought on the roots of *Lolium perenne* L. New Phytol 105: 393-402

McWha JA, Jackson DL (1976) Some growth promotive effects of abscisic acid. J Exp Bot 27: 1004-1008

Munns R (1992) A leaf elongation bioassay detects an unknown growth inhibitor in xylem sap from wheat and barley. Aust J Plant Physiol 19: 127-135

Munns R, King RW (1988) Abscisic acid is not the only stomatal inhibitor in the transpiration stream of wheat plants. Plant Physiol 88: 703-708

Munns R, Passioura JB, Milborrow BV, James RA, Close TJ (1993) Stored xylem sap from wheat and barley contains a transpiration inhibitor that is retained by high molecular weight exclusion filters. Plant Cell Environ (in press)

Netting AG, Willows RD, Milborrow BV (1992) The isolation and identification of the prosthetic group released from a bound form of abscisic acid. Plant Growth Regulation 11: 327-334

Passioura JB (1988) Root signals control leaf expansion in wheat seedlings growing in drying soil. Aust J Plant Physiol 15: 687-693

Saab IN, Sharp RE (1989) Non-hydraulic signals from maize roots in drying soil: inhibition of leaf elongation but not stomatal conductance. Planta 179: 466-474

Saab IN, Sharp RE, Pritchard J (1992) Effect of inhibition of abscisic acid accumulation on the spatial distribution of elongation in the primary root and mesocotyl of maize at low water potentials. Plant Physiol. 99:26-33

Saab IN, Sharp RE, Pritchard J, Voetberg GS (1990) Increased endogenous abscisic acid maintains primary root growth and inhibits shoot growth of maize seedlings at low water potentials. Plant Physiol 93: 1329-1336

Takahashi K (1973) Interaction between ethylene, abscisic acid and gibberellic acid in elongation of rice mesocotyl. Planta 109: 363-364

Tereo H, Inouye J (1980) Effect of low water potential of the culture medium on mesocotyl elongation of rice seedlings. Plant Cell Physiol 21: 1661-1666

Zhang J, Davies WJ (1990a) Changes in the concentration of ABA in xylem sap as a function of changing soil water status can account for changes in leaf conductance and growth Plant Cell Environ 13: 277-285

Zhang J, Davies WJ (1990b) Does ABA in the xylem control the rate of leaf growth in soil-dried maize and sunflower plants? J Exp Bot 41: 1125-1132

"STRESS TOLERANCE IN PLANTS": WHAT ARE WE LOOKING FOR?

Abraham Blum
Institute of Field Crops
The Volcani Center, ARO
P.O. Box 6, Bet Dagan
Israel

The huge progress made in molecular biology techniques is allowing us today to probe the plant at very low levels of biological organization and to reveal minute sub cellular and molecular modifications in response to stress. We have great expectations of these capabilities. At the same time we are largely incapable of understanding the meaning of such modifications at the higher levels of plant organization, largely because of the imbalance in emphasis between various plant research disciplines that address problems of stress tolerance. A typical and a disturbing symptom of this situation is the widely spread tendency to accept molecular modifications which are caused by stress as being "adaptive". This presentation is an offered clarification of some of the issues involved in whole-plant stress tolerance, being more of a reminder than a revelation.

In reality, stress is a most unstable phenomenon that varies in intensity and may occur at any plant developmental stage. The intensity of stress is probably the most important factor in eliciting plant responses to stress and it is a main reason for misinterpretation of responses. Any level of plant stress tolerance may be overcome by stress, depending on stress intensity. In terms used in phytopathology it can be said that there is no stable immunity against environmental stress. Therefore, the definition of stress tolerance is intimately linked with the definition of stresswhereas the latter determines what plant traits are important in conditioning tolerance.

What then may be considered as important plant traits in ascribing stress tolerance? In the biological domain, the discovery of a novel plant response to stress, say, a specific protein that accumulates in response to heat stress and found to protect the function of a certain enzyme, is most important. Novelty is the essence here. On the other hand, in the agronomic domain what is important is the function and the quantified impact of the biological discovery within the integrated plant and crop system. Depending on that, some discoveries may be important in the scientific domain and trivial in the agronomic domain, and vise versa. For example, whole plant phenology may be considered today as a biological trivia, but its role in crop productivity under drought stress is critical. Therefore, what is ultimately important is the ability to measure the impact of the discovery on the proper scale. Evidently, the biologist and the agronomist often use different scales. I shall try to translate some of the issues from one scale to the other.

NATO ASI Series, Vol. H 86
Biochemical and Cellular Mechanisms
of Stress Tolerance in Plants
Edited by J. H. Cherry
© Springer-Verlag Berlin Heidelberg 1994

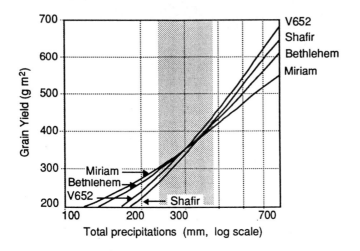

<u>Fig.1</u>. Results for 4 spring wheat cultivars tested over 16 locations in Israel. The linear regression of wheat cultivar yield on log total trial precipitation (+irrigation). The shaded area marks the range of yield 'crossover' (see text). All regressions were better than R2=0.70. From Blum and Pnuel (1990) with permission.

First and foremost, plant performance under stress may be affected by both adaptive and constitutive plant traits. It may be affected by traits which seemingly have nothing to do with the response to stress. Fig.1 describes the actual yield response of several wheat cultivars to conditions of an increasing drought stress. V652 represents a high yielding cultivar that performs best when seasonal precipitation are optimal (*ca.* 700mm). V652 is therefore a superior cultivar of high yield potential. However, it also performed better than all other cultivars under moderate drought stress (400-600mm of precipitation) merely because of its high productivity. When drought stress was severe enough to reduce mean yield by about 50%, other cultivars performed better than V652. This interaction, with a crossover at about 300mm (shaded area in Fig.1), indicated that while cv. V652 was drought susceptible cv's. Miriam and Bethlehem were drought resistant in that environment. They were resistant because of specific drought adaptive traits (*e.g.* osmotic adjustment) and constitutive traits (*e.g.* early flowering).

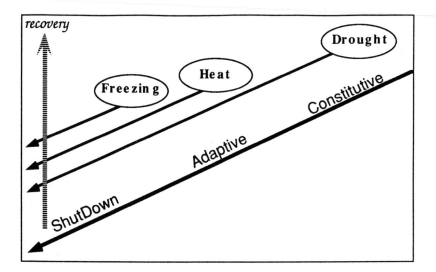

<u>Fig.2</u>. A conceptual schematic representation of the role of different plant traits and processes in supporting plant tolerance and productivity over a gradient of an increasing plant stress (see text).

Fig.2 is a conceptual and schematic representation of the role of different plant traits and processes in supporting plant productivity over a gradient of an increasing plant stress. As stress intensifies, different plant traits may assume an important role in supporting plant productivity. Under mild stress, several constitutive traits may carry a major impact. Yield potential, plant phenology and plant developmental characteristics are of primary importance in this respect. Stress responsive genes are not involved here. As stress intensifies and the respective genes respond, hardening (*syn.* acclimation) and adaptation are important, among which osmotic adjustment is a classical example. As stress further increases the plant ceases to grow it approaches what may be defined as a system-shutdown.

As can be seen in Fig.2, not all phases are relevant for all stresses. While they are all very relevant for drought stress, constitutive plant traits have only some impact on mineral deficiency tolerance and very little impact on heat and freezing tolerance. Freezing tolerance is almost exclusively a function of slow hardening which will determine the capacity for recovery. Freezing tolerance is rated solely by the plant ability to resurrect after the relief of stress.

Constitutive plant traits and plant production under stress.

Numerous constitutive morphological, developmental and phenological features of plants affect their performance under various stresses, without being considered as specifically adaptive. The more evident traits involve growth duration, leaf area, leaf surface properties, existence of awns in the cereal inflorescence and root developmental attributes. A unique constitutive phenological/developmental trait of maize was found to be the most effective, among many other traits studied, in ascribing better yields under drought stress (Bolanos and Edmeades, 1993). It is the time duration between anthesis and silking, defined as 'anthesis-to-silking interval' (ASI). Grain yield and kernel number per plant increased with shorter ASI of genotypes, irrespective of the environmental conditions. Selection for shorter ASI was very effective in improving yield under drought stress.

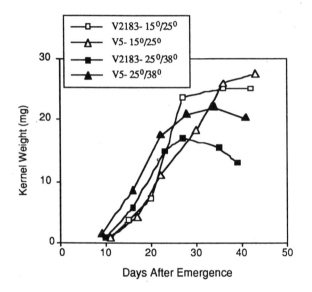

Fig.3. Kernel growth (by weight) in two wheat cultivars under low and high temperatures. Plants were grown at low (optimum) temperature up to anthesis.

In some respects a high potential plant productivity would support plant production under stress. In some other respects it may pose a disadvantage. There may be an inherent physiological advantage to the production and maintenance of a large carbon pool in the plant before entering a stress situation, which is typical of productive genotypes. A large reserve of carbon is important for plants entering a phase of chronic heat stress, where carbon starvation is a major limitation to growth and productivity. A large carbon pool may be important for osmotic adjustment (McCree, 1986) as plants enter drought, cold or salt stress. It may enable better carbon partitioning to the root when plants enter a soil afflicted stress (*e.g.* Nicolas *et al.*, 1985). It is quite fascinating that the importance of constitutive carbon partitioning to roots in relations to root osmotic adjustment and drought resistance of sorghum has already been suggested more than 60 years ago (Martin, 1930). Recently, it has been suggested that growth retardation by water stress could be partially caused by carbon starvation, since perturbations to productivity under water stress were partially eliminated by supplying carbon to the plant, at a low water potential (Boyle *et al.*, 1991).

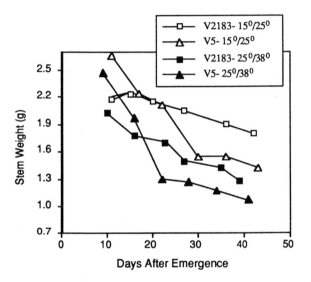

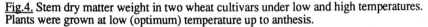

Fig.4. Stem dry matter weight in two wheat cultivars under low and high temperatures. Plants were grown at low (optimum) temperature up to anthesis.

A unique case of carbon storage capacity prior to stress as a mechanism for supporting plant productivity under stress is seen in stem reserve accumulation and its remobilization for use in kernel growth in the cereals. Two wheat cultivars of similar phenology and plant stature were found to differ in the extent of kernel shriveling under drought stress in the field. Cv. V5, which resulted from selection under plant chemical desiccation treatment (Blum et al., 1991), showed less shriveling while another (V2183) tended to shrivel more. When subjected to heat stress after anthesis, grain filling was reduced less in V5 than in V2183 (Fig.3). The reduction of kernel weight by heat stress was 26% and 48% in V5 and V2183, respectively. V5 therefore appeared to be more heat tolerant than V2183, in grain filling. Stems were heavier at anthesis in V5 than in V2183 (Fig.4; Table 1). The amount of dry matter lost from stems was greater in V5 than in V2183, irrespective of temperature (Table 1). V5 had a longer grain filling duration than V2183, irrespective of temperature (Fig.3). Therefore, V5 had a relatively better constitutive capacity to support grain filling from mobilized stem reserves. This capacity was a function of a large carbon storage in the stem and a longer duration of grain filling which allowed time for remobilization of reserves into the grain. In that sense V5 was not heat tolerant. Whenever the transient photosynthetic source during grain filling is inhibited by stress (*e.g.*, drought, heat or foliar diseases stress), V5 would be expected to show better grain filling than V2183.

Table 1. Stem dry matter weight at anthesis and the loss of stem dry matter (±SE) between anthesis and maturity in two wheat cultivars, under low and high temperatures during grain filling. Temperatures are for night/day, at 12h (Fig.3).

Cultivar	Temperatures (C^0)	Stem weight (g)	Stem weight loss (g)
V2183	$15^0/25^0$	2.22 ±0.064	0.39 ±0.047
V2183	$25^0/38^0$	2.06 ±0.103	0.74 ±0.039
V5	$15^0/25^0$	2.66 ±0.042	0.86 ±0.042
V5	$25^0/38^0$	2.34 ±063	1.36 ±0.096

Adaptation (hardening) to stress

Upon exposure to stress plants undergo various modifications. Some modifications are adaptive and allow continued plant functions, at lower than potential rates. Adaptation has been also referred to as "hardening" or "acclimation". The sometimes confusing terminology may be traced to the fact that adaptation requires time, and a prolonged process of adaptation is often referred to as hardening or acclimation. In many cases the normal progress of adaptation can be eliminated simply by not allowing sufficient time. Plant response to desiccation, heat or salinity will be totally different if subjected to stress abruptly or gradually. Offcourse, the time scale may differ for different processes. For example, osmotic adjustment requires much more time than the appearance of heat shock proteins. Thus, the rating of different genotypes for stress tolerance depends on their capacity to harden and the hardening protocol used in tests, as demonstrated for heat (*e.g.* Chaisompongpan *et al.*, 1990) and desiccation (*e.g.* Blum and Ebercon,1981) stresses.

Hardening against extreme temperatures is a well known phenomenon, affecting various plant life functions such as membrane thermostability (Blum and Ebercon, 1981), photosynthesis (Chaisompongpan et al., 1990), enzyme thermostability (Kinbacher, 1970), etc.' The consequence is often in the widening of the "thermo-kinetic window" (Burke et al., 1988) for plant growth and function, which may interact with the genotype. Thus, the capacity for heat hardening is under genetic control. In the case of F_1 hybrids of sorghum (Blum, 1989), maize (McWilliams and Griffing, 1965) and *Drosophila* (Ehiobu and Goddard, 1989), a wider thermo-kinetic window was a manifestation of heterosis, reminding us of the decades of discussion among population biologists of the homeostatic capacity of heterozygotes.

While temperature hardening widens the temperature response curve of plants, it generally tends to reduce the maximum rate at the optimum temperature (Blum, 1986) - which again is an indication of a trading off between maximum rates and wider stability of processes. This pertains to the well-grounded feeling among plant breeders that genotypic stability across environments is associated with mediocrity of performance in the better environments, which brings one to reflect on some wider issues of excellence in performance and the environment.

There may be a certain trading off between the capacity to harden and the potential growth rate or the potential size of the plant. This is in some contradiction to the case made above for high potential productivity in support of plant performance under stress. However, the existence of contrasting associations in plants is not a novelty. Ecotypes of natural vegetation in stress environments are generally characterized by small plants and small organs . This can be seen from the Alpine flora to the desert vegetation. The importance of plant or organ size under stress conditions has not been studied sufficiently. There is scattered evidence on possible negative genetic or phenotypic association between processes of stress adaptation (such as osmotic adjustment) and potential growth (*e.g.* Quisenberry *et al.*, 1984). Small organs are associated with small cells and small cells may have better capacity for osmotic adjustment (Cutler *et al.*, 1977) or turgor maintenance. It would be most interesting to learn what other cellular processes interact with cell size in response to stress. In both sorghum (Blum and Sullivan, 1986) and wheat (Blum and Sullivan, unpublished data) potentially smaller plants or slow growth rates were associated across genotypes with better osmotic adjustment and/or growth under drought stress.

The accumulation of cell compatible solutes have an important role in maintaining cellular hydration as cells dehydrate under the effect of drought, freezing and salt stresses. Apart from the effect on the maintenance of cell turgor, the accumulated solutes were implicated as protectants of several distinct cellular functions. Certain accumulated compounds were shown to serve as cryo-protectants (Grittith *et al.*, 1992) or to protect protein function and integrity under heat and salt stress (Paleg *et al.*, 1985; Shomerilan *et al.*, 1991).

The recent exponential increase in our capacity to identify the appearance and accumulation of specific compounds under the effect of various plant stresses (see this conference) raises the question of their role, which may not be necessarily adaptive.

A noted case is the accumulation of metal-binding cysteine-rich compounds in plants exposed to metal toxicity. The majority of this group, the phytochelatins, were long implicated in ascribing plant tolerance to metal toxicity. An interesting review on how this concept came about is presented in the introduction of the paper by Schat and Kalff (1992). However, in their own critical investigation Schat and Kalff provided evidence to show that phytochelatin production in root tips of copper sensitive and tolerant genotypes of *Silene vulgaris* challenged by copper was no indication of tolerance. Phytochelatin was merely an expression of strain caused by copper. They concluded that phytochelatin content of the root apex could be used as a quantitative tolerance-independent measurement of the degree of toxicity experienced by the plant.

Cross-hardening among different stresses was a popular topic in the literature some years ago (reviewed by Blum, 1988). The phenomenon was seen in the acquisition of tolerance to one stress after exposure to another. Examples are noted for the effect of water deficit on cold tolerance (Cloutier and Andrews, 1984) and more recently for the effect of mineral toxification on heat tolerance (Bonham-Smith *et al.*, 1987). Cross-hardening is an important topic not only for practical purpose but also because it may provide a lead towards understanding the transduction of hardening by the environment. At present there are numerous indications that ABA (abscisic acid) (e.g. Heino *et al.*, 1990; Ward and Lawlor, 1990) may serve as the transducer.

System Shutdown

The fact that severely stressed plants, which sometimes appear dead for all practical purposes, may recover after the relief of stress, implies that plants evolved mechanisms for the preservation of life in a dormant state. This is off course carried out to an extreme in the case of seed and resurrection plants, and it is common-place in the backyard of anyone who lives in the Northern latitudes. A system shutdown is the primary mechanism involved with freezing tolerance in buds (*e.g.* deciduous trees) or meristems (*e.g.* grasses) which may be achieved only in a dormant state, after all live plant tissues were shed or died. The ability to avoid death and to recover upon the relief of stress depends on a timely and an orderly protocol of system shutdown. I feel this is very much analogous to computer shutdown. A computer can be shut down by pulling the electrical cord off the wall, which may crash the system and cause serious problems in recovering the system upon power-up. The system should be shutdown by protocol, which allows it to proceed in an orderly fashion, exiting programs, closing files, scratching deleted files, rewriting the directory on the hard disk, etc.' - all of which will enable normal system recovery upon power-up. This, off course requires sufficient time - up to several weeks in winter wheat, by analogy.

The hard question is which of all the cellular responses to stress that we reveal and tag do support continued life and function under stress and which are simply components of a system shutdown. This distinction is extremely important for practical applications. In agricultural systems where continuous production despite an increasing stress is sought, early (in the course of stress) system shutdown is a drawback. There, tolerance is to be based largely on hardening and adaptation for continuous growth. Early system shutdown is can be essential where recovery after stress is an important agronomic option. For example, for maize production in the Northern Corn Belt of the USA where the season is short, the option of recovery from drought stress is impractical because by recovery time the crop will be out of season and into the autumn frost. On the other hand, in the Sahel region, any recovery from drought which may produce even a few heads of millet is critical for people subsistence.

Therefore, as much as it is of interest, the identification of stress responsive genes and their products must be interpreted in terms of physiological and agronomic significance before usefulness is implied.

References

Blum A (1986) The effect of heat stress on leaf and ear photosynthesis. J Exp Bot 37:111-118.

Blum A (1989) The temperature response of gas exchange in sorghum leaves and the effect of heterosis. J Exp Bot 40: 453-460.

Blum A, A Ebercon (198)1 Cell membrane stability as a measure of drought and heat tolerance in wheat. Crop Sci 21:43-47.

Blum A , Y Pnuel (1990) Physiological attributes associated with drought resistance of wheat cultivars in a Mediterranean environment. Aust J Agric Res 41:799-810.

Blum A, Shpiler L, Golan G, Mayer J, Sinmena B (1991) Mass selection of wheat for grain filling without transient photosynthesis. Euphytica 54:111-116.

Blum A, Sullivan CY (1986) The comparative drought resistance of landraces of sorghum and millet from dry and humid regions. Ann Bot 57:835-846.

Bolanos J, Edmeades GO (1993) 8 cycles of selection for drought tolerance in lowland tropical maize 2 responses in reproductive behavior. Field Crop Res 31:253-268.

Bonham-Smith PC, Kappor Manju, Bewley JD (1987) Establishment of thermotolerance in maize by exposure to stresses other than heat shock does not require heat shock protein synthesis. Plant Physiol 85:575-580.

Boyle MG, Boyer JS, Morgan PW (1991) Stem infusion of liquid culture medium prevents reproductive failure of maize at low water potential. Crop Sci 31:1246-1252.

Burke JJ, Mahan JR, Hatfield JL (1988) Crop-specific thermal kinetic window in relation to wheat, cotton biomass production. Agron J 80:553-560.

Chaisompongpan N, Li PH, Davis DW, Markhart AH (1990) Photosynthetic responses to heat stress in common bean genotypes differing in heat acclimation potential. Crop Sci 30:100-104.

Cloutier Y, Andrews CJ (1984) Efficiency of cold hardiness induction by desiccation stress in four winter cereals. Plant Physiol 76:595-598.

Cutler JM, Rains DW, Loomis RS (1977) The importance of cell size in the water relations of plants. Physiol Plant 40:255-260.

Ehiobu NG, Goddard ME (1989) Heterosis in crosses between lines of *Drosophila melanogaster* selected for adaptation to different environments. TAG 77:253-259.

Grittith M, Ala P, Yang DC, Hon WC, Moffatt BA (1992) Antifreeze protein produced endogenously in winter rye leaves. Plant Physiol 100:593-596.

Heino P, Sandman G, Lang V, Nordin K, Palva ET (1990) Abscisic acid deficiency prevents development of freezing tolerance in *Arabidopsis-Thaliana* (L) Heynh. TAG 79:801-806.

Kinbacher EJ (1970) Relative thermal stability of malic dehydrogenase from heat-hardened, unhardened *Phaseolus* sp. Crop Sci 10:181-186.

Martin JH (1930) The comparative drought resistance of sorghum and corn. Agron J 22:993-1003.

McCree KJ (1986) Whole-plant carbon balance during osmotic adjustment to drought and salinty stress. Aust J Plant Physiol 13:33-43.

McWilliams JR, Griffing B (1965) Temperature-dependant heterosis in maize. Aust J Biol Sci 18:569-583 1965.

Nicolas ME, Lambers H, Simpson RJ, Dalling MJ (1985) The effect of drought on metabolism and partitioning of carbon in two wheat varieties differing in drought-tolerance. Ann Bot 55:727-742.

Paleg LG, Stewart GR, Starr R (1985) The effect of compatible solutes on proteins. Plant a nd Soil 89:83-94.

Quisenberry JE, Cartwright GB, McMichael BL (1984) Genetic relationship between turgor maintenance and growth in cotton germplasm. Crop Sci 24:479-482.

Schat H, Kalff MMA (1992) Are phytochelatins involved in differential metal tolerance or do they merely reflect Metal-Imposed strain. Plant Physiol 99:1475-1480.

Shomerilan A, Jones GP, Paleg LG (1991) In vitro thermal and salt stability of pyruvate kinase are increased by proline analogues and trigonelline. Aust J Plant Physiol 18:279-286.

Ward DA, Lawlor DW (1990) Abscisic acid may mediate the rapid thermal acclimatization of photosynthesis in wheat. J Exp Bot 41:309-314.

PROTEOLYSIS AND PROTEOLYTIC ACTIVITIES IN THE ACCLIMATION TO STRESS: THE CASE OF SUGAR STARVATION IN MAIZE ROOT TIPS.

Philippe Raymond, R. Brouquisse, C. Chevalier, I. Couée, M. Dieuaide, F. James, D. Just, A. Pradet

Station de Physiologie Végétale
INRA Centre de Recherche de Bordeaux BP 81
33140 Villenave d'Ornon Cedex

The changes that stress induce in a plant may be either specific of the type of stress, or non–specific, ie, an effect of, or a response to, the stress situation itself. Several phenomena have been found to occur in a number of plant tissues under different stresses and are therefore of the non–specific type. This is the case of the production of toxic oxygen species (Foyer et al., 1991), the production of ethylene (Wang et al, 1990), increasing levels of polyamines (Flores, 1990), or the induction of heat–shock–cognate proteins (Cabané et al., 1993) in response to temperature, osmotic, mineral or wounding stresses. It is not always clear whether these changes are part of an acclimation mechanism which would improve stress tolerance, or are symptoms of the cellular degradations caused by the environmental constraints. These observations however suggest that in spite of the diversity of stresses, similar mechanisms may be involved in the response to stress. In addition, modifications by stresses in the pattern of protein synthesis, in carbon allocation among plant organs, in maintenance respiration, or in developmental programs and senescence induction (Amthor and McCree, 1990) indicate that, at the same time as specific responses are induced, the metabolic background is profoundly modified. We believe that proteolysis is part of the common response system.

Stress and proteolysis

Unfavourable growth conditions (osmotic stress, carbon starvation, absence of nitrate, presence of ABA) increase the rate of protein degradation and decrease the rate of protein synthesis in *Lemna minor* (Cooke *et al.*, 1979). Net protein degradation under

NATO ASI Series, Vol. H 86
Biochemical and Cellular Mechanisms
of Stress Tolerance in Plants
Edited by J. H. Cherry
© Springer-Verlag Berlin Heidelberg 1994

the effects of stress sometimes occurs without a detectable increase in proteolytic activities: for example, in needles of spruce (*Picea abies* L), the measured endoprotease activity decreased before any significant loss of proteins was detected (Pierre and Savouré, 1990). Increased proteolysis in this case could result from an activation of the proteases by either a change in pH, or the local Ca++ concentration. Alternatively protein loss may result from an increased susceptibility of proteins to degradation as may occur after denaturation by heat treatment (Caldwell, 1993a), by the photolysis of tryptophanyl residues (Caldwell, 1993b), by oxidative stress–induced cross–linking (Mehta et al;, 1992) or by ubiquitination as shown in animal and yeast (Hilt and Wolf, 1992), but not yet in plant cells (Genschick et al., 1992). In other cases, proteases have been shown to be induced, or activated by stresses: cysteine proteinases, identified by sequence analysis of the corresponding cDNA, are induced in response to cold treatment in tomato fruits (Schaffer and Fischer, 1988), and by wounding in tobacco leaves (Linthorst et al., 1992).

The function of stress–induced protein degradation in the acclimation phase may be essential since it provides amino acids for the synthesis of new proteins more suited for growth or survival under the modified conditions (Cooke et al., 1979) and also substrates to the energy metabolism (see below). Prolonged stress often leads to senescence as occurs for example in poplar leaves under ozone treatment (Landry and Pell, 1993). One function of proteolysis in senescing tissues is clearly to recycle the amino acid nitrogen for other parts of the plant (People and Dalling, 1988). Senescence related proteases in plants are better known than the stress–related ones (Feller, 1986; Huffaker, 1990). It is interesting to observe that this proteolysis activity can be activated by high salt concentration by two different mechanims: *de novo* synthesis of proteases and activation of the existing ones (Kang and Titus, 1989).

Sugar starvation and stress

Although carbohydrates are usually abundant in plants, and are indeed the major substrates of plant respiration, there is increasing evidence that the supply of carbohydrates may become limiting during normal plant life, for example after prolonged darkness or shading (Brouquisse *et al.*, 1991, and references therein). The study of the effects of sugar limitation on plant tissues is related to the response to stress for different reasons. First, the response to this constraint may be regarded as a particular stress response, and the mechanisms involved may be at least in part common to other stresses. Secondly, limiting sugar availability may also occur as a consequence of stresses which decrease the photosynthetic activity such as drought,

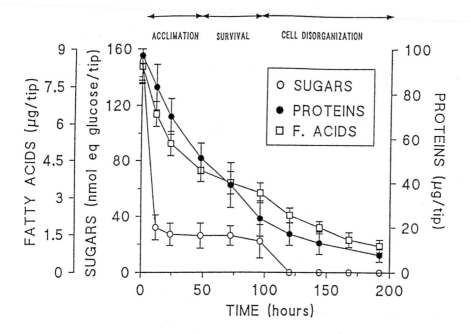

FIGURE 1 : Changes in total sugar, protein and fatty acid contents of excised maize root tips during glucose starvation.

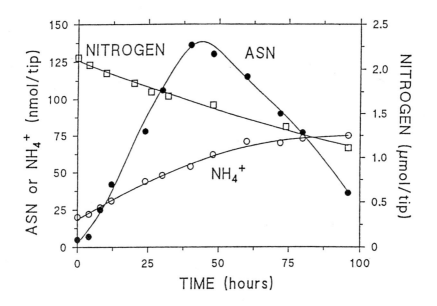

FIGURE 2 : Changes in the contents of asparagine (ASN), ammonium (NH4+) and total nitrogen in excised maize root tips submitted to sugar starvation.

328

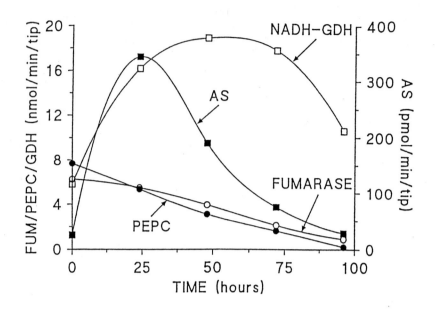

FIGURE 3 : Changes in some enzymic activities of excised maize root tips during sugar starvation. AS : asparagine synthetase; NADH-GDH : NADH-dependent glutamate dehydrogenase; PEPC : phosphoenol pyruvate carboxylase.

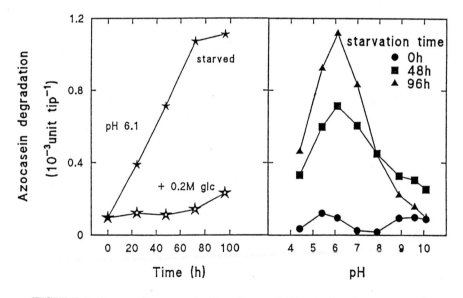

FIGURE 4 : Changes in azocasein degrading activities during glucose starvation in excised maize root tips. A : Time course of proteolytic activities in the presence or absence of 0.2 M glucose in the incubation medium. B : pH-dependent changes in proteolytic activities after 0, 48 and 96 hours of starvation.

levels after 24 h, whereas glutamine synthetase and glutamate synthase decrease more slowly, as did the alanine and aspartate amino transferases (Brouquisse et al., 1992). In radish leaves senescing in the dark (Kawakami and Watanabe, 1988), the change of the total glutamine synthetase (GS) activity was similar to that in maize, although the delay of ca. 24 h before decrease was more evident. In addition, Kawakami and Watanabe, (1988) showed that this change mainly reflects the decrease of the chloroplastic isoform whereas the cytosolic GS isoform increases. Similar data of selective loss of enzymic activities were obtained in *Lemna* submitted to different stresses (Cooke et al., 1979). As far as proteolysis is involved in these decreases of activity, these data indicate some specificity of proteolysis.

Interestingly, we found that in spite of the loss of total proteins, the activities of some enzymes or whole pathways increased. Increases were observed in the pathways of amino acid catabolism and asparagine metabolism, of fatty acid β-oxidation, and in proteolytic activities, in agreement with the increased fluxes which occur in these pathways as a consequence of the re-direction of the metabolism.

Increased activities of glutamate dehydrogenase and asparagine synthetase. In maize the glutamate dehydrogenase activity increases during 2 d, and asparagine synthetase activity peaks at day 1 (Brouquisse et al., 1992; Fig. 3). As for the cytosolic glutamine synthetase in radish leaves (Kawakami and Watanabe, 1988), we recently found that these changes are related to increased gene expression.

Increased activities of fatty acid β-oxidation. In plants, the β-oxidation of fatty acids appears to be localized in two cellular compartments: its presence in the peroxisomes of all cells is well established (Gerhardt, 1986), but it is also present, at least in some cases, in the mitochondria (Miernyk et al., 1991). We first compared (Dieuaide et al., 1992) the rate of CO_2 production from labelled fatty acids by control or sugar-starved root tips, and observed an increase after sugar starvation: this suggested that the β-oxidation activity could be increased. The peroxisomal β-oxidation activity, measured *in vitro*, was effectively increased by a similar factor of 2 to 5 according to root tip batches, after 1 d. This increase did not occur in root tips incubated for the same time in the presence of glucose, showing that the increase is specifically related to sugar starvation. Other peroxisomal enzyme activities, catalase, DTNB-insensitive citrate synthase and malate synthase also increased during sugar starvation, whereas the mitochondrial marker fumarase decreased. Malate synthase is one of the two enzymes which characterize the glyoxylate cycle, the pathway which allows the synthesis of 4-carbon compounds from the acetyl-CoA produced by β-oxidation, thus allowing the synthesis of carbohydrates from fatty acids. The other enzyme needed in this cycle, isocitrate lyase, was not detected in the root tip preparation, suggesting that gluconeogenesis does not occur in this tissue: we therefore suggested that the acetyl–CoA produced by β–oxidation could feed respiration directly, presumably after being incorporated into citrate by the peroxisomal citrate synthase (Dieuaide et al., 1992). More recently, we found β–oxidation activity in purified mitochondria from sugar-starved root tips (Dieuaide et al. unpublished). This activity was not detectable in mitochondria from control root tips. It was sensitive to respiration inhibitors and an acyl–CoA dehydrogenase activity was detected: these two results are characteristic of the mitochondrial β–oxidation from animal tissues. The function of β–oxidation is not only to oxidize fatty acids from lipids, but also the derivatives of the branched chain amino acids valine, leucine and isoleucine. These amino acids are first converted by transamination to 2–oxo acid derivatives which are then decarboxylated to the acyl–CoAs and converted to propionyl– and acetyl–CoA in peroxisomes (Gerbling and Gerhardt, 1988).

Increased proteolytic activities. The decrease in total protein which occurs in sugar starved maize root tips is associated with an increase in proteolytic activities (Fig. 4); however, the aminopeptidase activity decreased (James et al., 1993). Not only the total activity but also the pH profile and the sensitivity to inhibitors of both endoprotease and carboxypeptidase activities were modified in the course of starvation suggesting that *de novo* synthesis of these types of proteases was involved in the increase of their activity; moreover, the increase of endoprotease activity did not occur when inhibitors of transcription and translation (cordycepin and cycloheximide) were present. Interestingly, the induction of endoproteases did not occur in control root tips incubated in the presence of 0.2 M glucose; the rate of increase of this activity was directly related to the concentration of glucose between 0 and 0.2 M. The control of the proteolytic activity by glucose was efficient at any time during the 4 day period of starvation (James et al., 1993). The effect of protease inhibitors showed that the major proteinases were of thiol and serine types. The protease with the major activity at 4 d was purified. Immunodetection on Western–blots showed that the protein is nearly undetectable at time zero of starvation but is present after only 6 h and increases gradually during the following 4 d. The acidic optimum pH of the endoprotease and carboxypeptidase activities suggested that these proteases are located in the vacuoles. Lysosomes, which are the animal cell equivalent of vacuoles (Feller, 1986) are involved in different types of proteolyses (Marzella and Glaumann, 1987): Autophagy is a mechanism for the sequestration of cytoplasm into lysosome; large portions of cytoplasm, containing organelles such as mitochondria, are enclosed in macroautophagic vacuoles which then merge with lysosomes, whereas in microautophgy, the vesicles formed by invagination of the lysosomal membrane are smaller. In both cases, proteolysis is ATP dependent but non–specific: it is likely that degradation by such mechanisms are active during the senescence. Earlier in the response to stress, particularly during the acclimation phase, proteolysis would be expected to depend on more specific mechanisms. The specific import into lysosomes of some proteins containing characteristic sequences have been described (Chiang et al, 1989); the mechanism of recognition involves a hsp70 protein, and is activated when the animal cells are deprived of serum growth factors. An alternative mechanism for the selective degradation of intracellular proteins is the ubiquitin–mediated pathway which has been described in detail in a number of recent reviews (Hershko and Ciechanover, 1992; Hilt and Wolf, 1992). The system involves a set of ubiquitinating enzymes on the one hand, and a multicatalytic proteinase complex of high molecular weight called 26 S proteasome, which specifically degrades the ubiquitinated proteins; a component of this complex, the 20 S proteasome, is also active on non ubiquitinated proteins and peptides. Studies on mammalian cells or yeast cells showed that mutations on enzymes of the ubiquitin system or on proteasome subunits make the cells very sensitive to stress (Hershko and Ciechanover, 1992; Hilt and Wolf, 1992). In plants the expression of ubiquitin genes has been shown to be modified in response to a number of stresses (Christensen and Quail, 1989; Genshick et al., 1992). The ubiquitination of proteins was found to increase following chilling (Gindin and Borochov, 1992), and also in chloroplasts of senescing leaves (Veierskov and Ferguson, 1991); however the latter is controversial (Beers et al., 1992). The 20 S proteasome, but not the 26 S proteasome, has been characterized in some plant tissues (Skoda and Malek, 1992; Ozaki et al., 1992). Changes in the expression of ubiquitin genes and in the pattern of protein ubiquitination have also been observed during the early stages of sugar starvation in maize root tips, and we have recently characterized the 20 S proteasome in the root tips (unpublished results).

Two different proteolytic systems appear to be active in response to sugar starvation: we have obtained preliminary evidence for the involvement of the ubiquitin–mediated pathway in the selective proteolysis of the acclimation period (Brouquisse et al., 1992) but there may also be a contribution of the vacuolar/lysosomal pathway. It is not clear whether the protease which is induced during the first hours of sugar starvation and becomes the major one at day 4, is first involved in selective proteolysis or if it is the sign that senescence has begun already.

LITERATURE CITED

Amthor and McCree, 1990 Carbon balance of stressed plants *In* RG Alscher, JR Cummings, eds Stress response in plants: adaptation and acclimation mechanisms. Wiley–Liss, NY, pp 1–15

Beers EP, Moreno TN, Callis J (1992) Subcellular localization of ubiquitin and ubiquitinated proteins in *Arabidopsis thaliana*. J Biol Chem **267**: 15432–15439

Brouquisse, R., James, F., Raymond, P., Pradet, A. (1991) Study of glucose starvation in excised maize root tips. Plant Physiol 96:619–626.

Brouquisse, R., James, F., Pradet, A., Raymond, P. (1992) Asparagine metabolism and nitrogen distribution during protein degradation in sugar–starved maize root tips. Planta 188: 384–395

Cabané M, Calvet P, Vincens P, Boudet AM (1993) Characterization of chilling-acclimation–related proteins in soybean and identification of one as a member of the heat shock protein (HSP 70) family. Planta **190**:346–353

Caldwell CR (1993a) Estimation and analysis of cucumber (*Cucumis sativus* L.) leaf cellular heat sensitivity. Plant Physiol. **101** 939–945

Caldwell CR (1993b) Ultraviolet–induced photodegradation of cucumber (*Cucumis sativus* L.) microsomal and soluble protein tryptophanyl residues in vitro. Plant Physiol. **101** 947–953

Chiang H–L, Terlecky SR, Plant CP, Dice JF (1989) A role for a 70–kilodalton heat shock protein in lysosomal degradation of intracellular proteins. Science **246**: 382–385

Christensen AH, Quail PH (1989) Sequence analysis and transcriptional regulation by heat shock of polyubiquitin transcripts from maize. Plant Mol Biol **12**: 619–632

Cooke RJ, Oliver J, Davies DD (1979) Stress and protein turnover in *Lemna minor*. Plant Physiol **64** 1109–1113

Couée, I., Jan, M., Carde, J.-P., Brouquisse, R., Pradet, A., Raymond, P. (1992) Selective effects of glucose starvation on mitochondrial subpopulations in the meristematic and submeristematic regions of maize roots. Plant Physiol. 100 1891–1900

Dieuaide, M., Brouquisse, R., Pradet, A., Raymond, P. (1992) Increased fatty acid β–oxidation after glucose starvation in maize root tips. Plant Physiol. 99: 595–600.

Dorne A–J, Bligny R, Rébeillé F, Roby C, Douce R (1987) Fatty acid disappearance and phosphorylcholine accumulation in higher plant cells after a long period of sucrose deprivation. Plant Physiol Biochem **25**: 589–595

Feller U (1986) Proteolytic enzymes in relation to leaf senescence. *In* MJ Dalling ed, Plant proteolytic enzymes, Vol **2**, CRC Press, Boca Raton Fl, pp 49–68

Flores, HE (1990) Role of ethylene under stress. *In* RG Alscher, JR Cummings, eds Stress response in plants: adaptation and acclimation mechanisms Wiley–Liss, NY, pp 217–239

Foyer C, Lelandais M, Galap C, Kunert KJ (1991) Effect of elevated cytosolic glutathione reductase activity on the cellular glutathione pool and photosynthesis in leaves under normal and stress conditions. Plant Physiol 97: 863–872

Genix P, Bligny R, Martin JB, Douce R (1990) Transient accumulation of asparagine in sycamore cells after a long period of sucrose starvation. Plant Physiol 94: 717–722

Genschick P, Parmentier Y, Durr A, Marbach J, Criqui MC, Jamet E, Fleck J (1992) Ubiquitin genes are differentially regulated in protoplast–derived cultures of *Nicotiana sylvestris* and in response to various stresses. Plant Mol Biol 20: 897–910

Gerbling H, Gerhardt B (1989) Peroxisomal degradation of branched–chain 2–oxo acids. Plant Physiol 91: 1387–1392

Gerhardt B (1986) Basic metabolic function of the higher plant peroxisome. Physiol Veg 24: 397–410

Gindin E, Borochov A (1992) Ubiquitin conjugation to protein increases following chilling of Clerodendrum leaves. Plant Physiol. 100: 1392– 1395

Hilt W, Wolf DH (1992) Stress–induced proteolysis in yeast. Mol Microbiol 6 :2437–2442

Huffaker RC (1990) Proteolytic activities during senescence of plants. New Phytol 116:199–231

James F, Brouquisse R, Pradet A, Raymond P (1993) Changes in proteolytic activities in glucose–starved maize root tips – Regulation by sugars. Plant Physiol Biochem in press

Kang S–M, Titus JS (1989) Increased proteolysis of senescing rice leaves in the presence of NaCl and KCl. Plant Physiol 91: 1232–1237

Kawakami N, Watanabe A (1988) Senescence–specific increase in cytosolic glutamine synthetase and its mRNA in radish cotyledons. Plant Physiol 88: 1430–1434

Kerr PS, Rufty JrTW, Huber SC (1985) Changes in nonstructural carbohydrates in different parts of soybean (*Glycine max* [L] Merr.) plants during a light/dark cycle and in extended darkness. Plant Physiol 78: 576–581

King GA, Woollard DC, Irving DE, Borst WM (1990) Physiological changes in asparagus spear tips after harvest. Physiol Plant 80: 393–400

Landry LG, Pell EJ (1993) Modification of rubisco and altered proteolytic activity in O3–stressed hybrid poplar (Populus maximowizii x trichocarpa). Plant Physiol. 101: 1355–1362

Linthorst HJM, van der Does C, Brederode FT, Bol JF (1993) Circadian expression and induction by wounding of tobacco genes for cysteine proteinase. Plant Mol Biol 21: 685–694

Marzella L, Glaumann H (1987) Autophagy, microautophagy and crinophagy as mechanisms for protein degradation. *In* H Glaumann and FJ Ballard eds, Lysosomes: their role in protein breakdown. Academic Press, London pp 319–367

Mehta RA, Fawcett TW, Porath D, Mattoo AK (1992) Oxidative stress causes rapid membrane translocation and *in vivo* degradation of ribulose–1,5–bisphosphate carboxylase/oxygenase J Biol Chem **267**: 2810–2816

Miernyk JA, Thomas DR, Wood C (1991) Partial purification and characterization of the mitochondrial and peroxisomal isozymes of enoyl–Coenzyme A hydratase from germinating pea seedlings. Plant Physiol **95** 564–569

Ozaki M, Fujinami K, Tanaka K, Amemiya Y, Sato T, Ogura N, Nakagawa H (1992) Purification and initial characterization of the proteasome from the higher plant Spinacia oleracea. J Biol Chem **267**: 21678–21684

People MB, Dalling MJ (1988) The interplay between proteolysis and amino acid metabolism during senescence and nitrogen reallocation. *In* LD Noodén and AC Leopold eds, Senescence and aging in plants, Academic press Inc, London, pp 181–217

Pierre M, Savouré A (1990) Effects of water stress and SO_2 pollution on spruce endoproteases. Plant Physiol Biochem **28**: 95–104

Saglio PH, Pradet A (1980) Soluble sugars, respiration, and energy charge during aging of excised maize root tips. Plant Physiol **66**: 516–519.

Thomas H (1978) Enzymes of nitrogen mobilization in detached leaves of *Folium tementulum* during senescence. Planta **142**: 161–169.

Veierskov B, Ferguson IB (1991) Ubiquitin conjugating activities in leaves and isolated chloroplasts from *Avena sativa* L. during senescence. J Plant Physiol **138**: 608– 613

Wang SY, Wang CY, Wellburn AR (1990) Role of ethylene under stress. *In* RG Alscher, JR Cummings, eds Stress response in plants: adaptation and acclimation mechanisms. Wiley–Liss, NY, pp 147–173

REGULATION OF THE *rab17* GENE IN ABA-DEFICIENT AND ABA-INSENSITIVE *viviparous* MUTANTS OF MAIZE

Adela Goday, Maria Pla, Mª Françoise Niogret, Anders Jensen, Mercè Figueras and Montserrat Pagès

Departament de Genética Molecular
Centre d'Investigació i Desenvolupament
C.S.I.C.
Jorge Girona 18-26
08034 Barcelona
Spain.

INTRODUCTION

Previous studies on the regulation of gene expression during embryogenesis in *Zea mays* L. have shown that a group of stage-specific mRNAs and proteins which accumulate in late embryogenesis are induced precociously in immature embryos when abscisic acid (ABA) is supplied in the culture medium (Sanchez-Martinez et al. 1986, Goday et al 1988). Abscisic acid (ABA) is supposed to mediate a large number of physiological and developmental processes in plants such as embryo maturation, seed dormancy and the response of plants to osmotic stress (for a review see Zeevaart and Creelman 1988). The response of the plant cells to ABA include modulation of specific mRNAs and proteins in embryo and vegetative tissues (Quatrano et al. 1986, Gomez et al 1988, Mundy et al. 1988). Although the molecular basis of the effect of ABA on gene expression has been the subject of numerous studies, to date little progress has been made in the elucidation of the intermediates in the ABA signal transduction pathway.

Many genes responsive to ABA (*rab*) are controlled at the transcriptional level. Functional dissection of *rab* promoters using transient expression assays in protoplasts has revealed the existence of a specific sequence element (ABRE) which confers ABA responsiveness upon the reporter gene (Marcotte et al. 1989, Mundy et al. 1990, Vilardell et al. 1991). Moreover, leucine zipper-type proteins binding to the ABRE sequence have been identified (Guiltinan et al. 1990, Oeda et al. 1991). However, recent studies have shown that the expression of *rab* genes are

NATO ASI Series, Vol. H 86
Biochemical and Cellular Mechanisms
of Stress Tolerance in Plants
Edited by J. H. Cherry
© Springer-Verlag Berlin Heidelberg 1994

also controlled by a variety of other processes and that changes in ABA levels are not always correlated with changes in gene expression (Pla et al. 1991, Nordin et al. 1991, Espelund et al. 1992).

To further investigate the role of ABA in the control of *rab* gene expression we have initiated a genetic approach to characterize the transcriptional regulation of the maize (ABA-responsive) *rab* genes by using different mutants as genetic tools. In maize *viviparous* mutants, which are deficient or insensitive to ABA, have been isolated and characterized genetically by their ability to respond to ABA and by a variety of ABA-inducible responses throughout the life cycle of the plant (Robertson 1955, Robichaud et al 1980, Neill et al 1986). In the maize *viviparous* mutants the embryo fails to become dormant but precociously germinates in the mother plant without having first undergone the normal period of desiccation. Thus, these mutations provide an invaluable tool for the elucidation of the role of ABA in the induction of *rab* gene expression in response to developmental and physiological cues. Moreover they can help in the dissection of the ABA signal transduction pathway.

REGULATION OF *rab* GENES IN ABA-DEFICIENT *vp2* MUTANTS

The *vp2* mutant (Robertson 1955) is carotenoid deficient and contains subnormal levels of ABA. However, *vp2* embryos respond normally to exogenously applied ABA by inhibiting precocious germination (Robichaud et al 1980). *vp2* seedlings contain less ABA than do wild-type seedlings and the ABA concentration does not increase in response to water deficit (Neill et al 1986).

In earlier studies we analyzed the steady state levels of *rab* 17 (Vilardell et al. 1990) and *rab* 28 (Pla et al 1991) mRNAs in embryos and seedlings of wild type and *vp2* mutants (Pla et al 1989. Pla et al 1991). The kinetics of mRNA accumulation during embryogenesis of *vp2* mutants closely resembled that obtained for non-mutant embryos. No detectable levels of *rab* 17 and *rab* 28 mRNAs were found in young embryos, increasing levels of these mRNAs accumulated in embryos during maturation and they disappeared within the first hours of germination in both *vp2* and wild type embryos. *rab* expression was also induced in immature *vp2* embryos cultured with ABA.

These results indicated that *vp2* embryos are capable of regulating the expression of *rab* genes during development in a pattern similar to the one in the wild-type and that expression of *rab* mRNAs was positively regulated by ABA in the mutant.

However, both genes *rab* 17 and *rab* 28 showed a differential regulation in embryo and vegetative tissues of the *vp2* mutant. Accumulation of *rab* mRNAs was induced by water stress or exogenous ABA in wild type young seedlings. However,in seedlings of *vp2* mutants subjected to water stress no induction of *rab* mRNAs was detected, although normal induction with exogenously applied ABA was observed. The absence of detectable levels of *rab* mRNAs in desiccated *vp2* leaves indicates that *rab* expression is dependent upon an increase of ABA that occurs in response to dehydration in wild-type vegetative tissues. These results also suggested that different ABA response pathways to developmental and environmental cues operate on the regulation of *rab* genes in maize.

EFFECT OF ABA AND DESICCATION ON *rab*17 GENE EXPRESSION IN *vp2* EMBRYOS

We have extended our studies on developmental and stress regulation of *rab* genes by using wild type and *vp2* excised young embryos incubated with, or without, ABA and subsequently subjected to desiccation. Accumulation of *rab* 17 mRNA was assessed by northern blot analysis.

At early stages of embryogenesis freshly excised wild-type and *vp2* embryos show absence of detectable *rab* 17 mRNA. Figure 1 shows that when young embryos are excised and placed on moist filter paper without ABA, *rab* 17 mRNA is transiently expressed prematurely both in wild-type and *vp2* mutants. *rab* 17 mRNA increases during the first 3 days in *vp2* as in the wild-type, although with lower amounts in the mutant. However, when young embryos are incubated with exogenous ABA, *rab* 17 mRNA is expressed in a sustained manner at a very high level both in *vp2* and the corresponding wild-type. Subsequent desiccation for 1 hour of wild-type and *vp2* embryos treated with ABA for different periods of time (from 1 hour to 3 days) has no significant effect on *rab* 17 mRNA accumulation. However, parallel experiments with embryos cultivated in a medium without ABA causes a stimulatory effect on gene expression in wild type embryos, but not in *vp2* embryos where the level of *rab* 17 mRNA does not increase.

The *vp2* mutant has reduced but significant levels of ABA in embryo tissues (Neill et al. 1986), therefore developmental and stress induction of *rab* genes may be dependent on different thresholds for ABA responsiveness. Alternatively the lack of *rab* gene induction in *vp2* embryos after desiccation could be due to the absence in this mutant of specific factors involved in *rab* gene regulation during the stress response. These results also imply that different molecular mechanisms are involved in the response of the embryo to developmental and environmental signals.

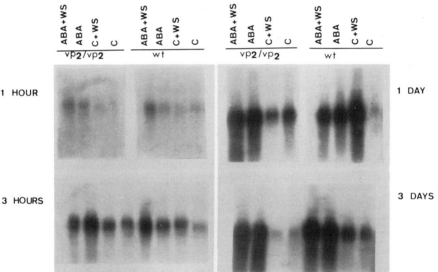

Figure 1 Northern blot analysis of RNA from wild-type and *vp2* mutant embryos. *rab* 17 cDNA was hybridized to total RNA isolated from wild-type and *vp2* embryos (20 days after pollination) incubated in water (C), or in 10 uM ABA (ABA) for the time indicated. w.s. desiccation for 1 hour.

REGULATION OF *rab* 17 IN ABA-INSENSITIVE *vp1* MUTANTS

The *vp1* mutant contains normal levels of ABA in embryo and seedlings (Neill et al. 1987), but reduced levels of anthocyanin pigments in the endosperm (Robertson 1955, Paz-Ares et al 1986). The ABA concentration of *vp1* seedlings increases in response to water deficit (Neill et al 1987). However *vp1* mutant is insensitive to exogenously applied ABA and is considered to be defective in its responsiveness to this hormone (Mc Daniel et al.1977, Robichaud et al. 1980).

The level of *rab* 17 mRNA in *vp1* mutants was assessed by northern blot analyses with total RNAs from different tissues and callus cells treated with or without ABA, and from young seedlings subjected to different stresses. Figure 2A shows that in young *vp1* embryos (20 d.a.p. *rab*17 mRNA is not present at detectable levels, and in mature *vp1* embryos (30 and 40 d.a.p.) *rab*17 transcripts do not accumulate at a significant level compared with the wild-type. However, when young *vp1* embryos were treated with 10 uM ABA for 24 hours there is a strong induction of *rab* 17 mRNA.

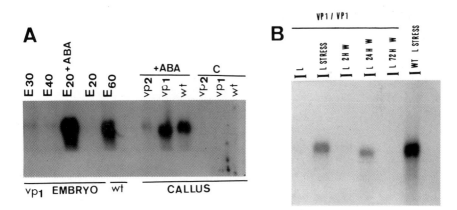

Figure 2 Northern blot analysis of RNA from embryos, callus and leaves of wild-type and mutants hybridized with *rab* 17 cDNA. E: embryos at 20, 40 and 60 days after pollination. +ABA: 24 hours in 10 uM ABA. L:leaves of 6 day old seedlings. +stress: desiccated for 3 hours. +W: wounded.

Synthesis of *rab*17 protein in the *vp1* mutant was assessed by two-dimensional electrophoresis of *in vivo* labelled proteins from *vp1* and corresponding wild-type embryos. In agreement with northern analysis Figure 3 shows that the *rab* 17 protein is synthesized at a very low level in the mutant.

To determine if expression of *rab* 17 could also be induced   in vegetative tissues, young seedlings (6 days old) of *vp1* were treated with 10 uM ABA, subjected to water stress or wounded. The results of northern blot analysis (Figure 2B) show the induction of *rab* 17 mRNA by ABA, desiccation and wounding in *vp1* plants. Callus of maize derived from wild-type, *vp1* and *vp2* young embryos when treated with ABA during 24 hours show the induction of *rab*17

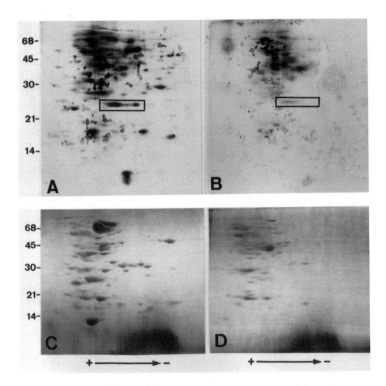

Figure 3 In vivo protein synthesis of wild-type and *vp1* mature embryos. Two dimensional electrophoresis of labelled proteins from wild-type(A) and *vp1* (B) embryos incubated for 2 hours with ^{35}S-methionine. Coomassie blue staining of total proteins from wild-type (C) and *vp1* (D).

mRNA (Figure 2A). The relative levels of induction are similar in wild type and *vp1* cells and significantly lower in *vp2* callus cells.

The *vp1* protein has structural features consistent with a transcription activator, and it has been suggested that it potentiates the hormone response in seed tissues by interacting with ABA regulated transcription factors at the level of the gene promoter (Mc Carthy et al 1991).

The low level of *rab* 17 message detected in mature *vp1* embryos may reflect a selective regulation of *rab* 17 by *vp1* at the level of gene expression during embryo development. However, the induction of *rab* 17 mRNA by exogenous ABA and water stress in *vp1* mutants seems to indicate that the product of the *vp1* gene (Mc Carthy et al 1989) is not involved in the complex signal transduction pathway of *rab* 17 gene induction by ABA.

These data strongly suggest that alternative pathways involving different molecular mechanisms may be responsible for developmental, and ABA/osmotic stress induction of *rab* gene expression. Indeed, preliminary evidence of specific factors involved in the regulation of *rab* genes has been recently obtained. Distinct regulatory protein factors with apparent similar affinity for the ABRE sequence have been detected in embryo and vegetative tissues of maize (Pla et al. 1993).

In conclusion, the results obtained on the expression of *rab*17 (here and Pla et al 1989), and on *rab* 28 (Pla et al 1991), in the ABA-insensitive and ABA-deficient *viviparous* mutants provide a starting point for addressing questions concerning the regulatory mechanisms that operate in the developmental and stress induction of ABA responsive genes in these mutants, and for investigating the cis-elements and protein factors involved in the expression of *rab* genes.

ACKNOWLEDGEMENTS. This work was supported by grant BIO91-0546 from CICYT.

REFERENCES

Espelund M Saeboe-Larssed S Hughes D.W. Galau G.A. Larsen F and Jakobsen K.S. (1992) Late embryogenesis-abundant genes encoding proteins with different numbers of hydrophilic repeats are regulated differentially by abscisic acid and osmotic stress. The Plant Journal 2 241-252.

Goday A, Sanchez-Martinez D, Gomez J, Puigdomenech P, Pages M (1988) Gene expression in developing Zea mays embryos : Regulation by abscisic acid of a highly phosphorylated 23- to 25-group of proteins. Plant Physiol. 88, 564-569.

Gomez J, Sanchez-Martinez D, Steifel V, Rigau J, Puigdomenech P, Pages M, (1988) A gene induced by the plant hormone abscisic acid in response to water stress encodes a glycine-rich protein. Nature 334, 262-264.

Guiltinan MJ, Marcotte WR, Quatrano RS (1990) A plant leucine zipper protein recognizes an abscisic acid response element. Science 250, 267-270.

Marcotte WR, Russell SH, Quatrano RS (1989) Abscisic acid responsive sequences from the Em gene of wheat. Plant Cell 1, 969-976.

Mc Carthy DR, Carson CB, Stinard PS, Robertson DS (1989) Molecular analysis of *viviparous-1*: an abscisic acid insensitive mutants of maize.The Plant Cell 1, 523-532.

Mc Carthy DR, Hattori T, Vasil V, Vasil IK (1991) A regulatoty hierarchy in seed development: Interaction of *viviparous 1* and abscisic acid in regulation of the C1 gene in maize. J Cell Biochem.(abstr.) 25, 047.

Mc Daniel S, Smith JD, Price HJ (1977) Response of viviparous mutants to abscisic acid in embryo culture. Maize Gen Coop Newsl 51, 85-86.

Mundy J, Chua NH (1988) Abscisic acid and water stress induce the expression of a novel rice gene. EMBO J. 7, 2279-2286.

Mundy J, Yamaguchi-Shinozaki K, Chua NH (1990) Nuclear proteins bind conserved elements in the abscisic acid-responsive promoter of a rice RAB gene. Proc. Natl. Acad. Sci. USA 87, 1406-1410.

Neill SJ, Horgan R, Parry AD (1986) The carotenoid and abscisic acid content of viviparous kernels and seedlings of Zea mays L. Planta 169, 87-96.

Neill SJ, Horgan R, Rees AF (1987) Seed development and vivipary in Zea mays L Planta 171 358-364.

Nordin, K. Heinno, P. Palva E.T. (1991). Separate signal pathways regulate the expression of a low-temperature-induced gene in Arabidopsis thaliana (L.) Heynh. Plant Mol. Biol. 16, 1061-1071.

Oeda, K. Salinas, J. Chua, N.H. (1991). A tobacco bZip transcription activator (TAF-1) binds to a G-box-like motif conserved in plant genes. EMBO J. 10, 1793-1802.

Paz-Ares J, Wienand U, Peterson PA, Saedler PA (1986) Molecular cloning of the *c1*locus of *Zea mays*:A locus regulating the anthocyanin pathway in maize. EMBO J.5:829-833.

Pla M, Goday A, Vilardell J, Gomez J, Pages M (1989) Differential regulation of ABA-induced 23-25 kDa proteins in embryo and vegetative tissues of the viviparous mutants of maize. Plant Mol. Biol.13, 385-394.

Pla M. Gomez, J. Goday, A. Pagès M. (1991). Regulation of the abscisic acid-responsive gene *rab 28* in maize *viviparous* mutants. Mol. Gen. Genet. 230, 394-400

Pla, M. Vilardell, J. Guiltinan M.J. Marcotte, W.R. Niogret, M.F. Quatrano, R.S. Pagès, M. (1993) The cis-regulatory element CCACGTGG is involved in ABA and water-stress responses of the maize gene *rab 28*. Plant Mol. Biol. 21, 259-266.

Quatrano RS (1986) Regulation of gene expression by abscisic acid during angiosperm embryo development. Oxford Surveys Plant Mol Cell Biol. 3, 467-477.

Robertson DS (1955) The genetics of vivipary in maize Genetics 40, 745-760.

Robichaud CS, Wong J, Sussex IM (1980) Control of in vitro growth of viviparous embryo mutants of maize by abscisic acid. Dev Genet. 1 325-330.

Sanchez-Martinez D, Puigdomenech P, Pages M (1986) Regulation of gene expression in developing *Zea mays* embryos. Protein synthesis during embryogenesis and early germination of maize. Plant Physiol. 82, 543-549.

Skriver K, Mundy J (1990) Gene expression in response to abscisic acid and osmotic stress. Plant Cell 2, 503-512.

Vilardell J, Goday A, Freire MA, Torrent M, Martinez MC, Torne JM, Pages M (1990) Gene sequence, developmental expression, and protein phosphorylation of RAB-17 in maize. Plant Mol. Biol.14, 423-432.

Vilardell, J. Mundy. J. Stilling, B. Leroux, B. Pla, M. Freyssinet, G. Pages, M. (1991). Regulation of the maize *rab* 17 gene promoter in transgenic heterologous systems. Plant Mol. Biol.17 985-993.

Zeevaart, J.A.D. Creelman, R.A. (1988). Metabolism and physiology of abscisic acid. Ann. Rev. Plant Physiol. Plant Mol. Biol. 39, 439-473.

A ROLE FOR SORBITOL IN DESICCATION TOLERANCE OF DEVELOPING MAIZE KERNELS: INFERENCE FROM THE PROPERTIES OF MAIZE SORBITOL DEHYDROGENASE[1]

R. D. Locy
Auburn University Cell Science Center and
Department of Botany and Microbiology
Auburn University
Auburn, AL 36849

INTRODUCTION

Some tissues of most higher plant species are capable of tolerating extremes of desiccation. Typically, these tissues are found within the developing seed (Kermode et al., 1986), but for some species a wider range of tissues may be desiccation tolerant (Bartels, this volume). Within the developing seed the ability to tolerate desiccation is typically acquired during seed development, and is typically lost upon germination (Bartels et al., 1991; Kermode et al., 1986). Thus, the developing seed provides a system which can be utilized to elucidate at the biochemical and molecular level those factors which are involved in the acquisition and loss of desiccation tolerance.

One aspect of desiccation tolerance involves the accumulation of osmoprotective compounds within desiccation tolerant tissues (see Bartels, this volume; Hanson, this volume; Csonka, this volume; for review). Such compounds are thought to protect intracellular components such as enzymes, membranes, and other macromolecules against damage due to desiccation (Ahmad et al., 1979; Crowe et al., 1984). Among the osmoprotective organic solutes which accumulate in higher plants the most extensively studied compounds include amino acids such as proline (see Paleg, Aspinall, 1981; Csonka, this volume; for review), quaternary ammonium compounds such as glycine betaine (see Hanson, this volume for review), carbohydrates including sucrose and various reducing sugars (Briens, Larher, 1982; Crowe et al., 1984), and 2 classes of polyhydric alcohols, the straight chain alditols (Ahmad et al., 1979; Briens, Larher, 1982; Briens, Larher, 1983; Hellebust, 1985; Lewis, 1984) and the cyclitols derived from inositol (Ford, 1984; Loewus, 1990; Nguyen, Lamant, 1988; Paul, Cockburn, 1989; Streeter, 1985).

---

[1]This is paper #   6-933595   in the Alabama Agricultural Experiment Station journal series.

NATO ASI Series, Vol. H 86
Biochemical and Cellular Mechanisms
of Stress Tolerance in Plants
Edited by J. H. Cherry
© Springer-Verlag Berlin Heidelberg 1994

While the accumulation of proline and quaternary ammonium compounds has been extensively studied in higher plants, the accumulation of polyhydric alcohols during stress has received much less attention. Despite the occurrence of sorbitol and mannitol as well as other alditols within numerous higher plant families (Bieleski, 1982; Lewis, 1984; Loescher, 1987) information relating the physiology of these compounds to osmotic stress tolerance is lacking in higher plant systems. This situation persists despite the fact that information relating alditol metabolism to osmotic stress has been prevalent in fungal, algal, and animal systems. For example the alditols, glycerol, arabitol, and mannitol, are the primary compounds which accumulate in response to low external water potentials in certain fungi (Brown, Edgley, 1980). Halotolerant algae such as *Dunaliella spp.* accumulate glycerol (Chitlaru, Pick, 1991) while other algal species accumulate mannitol or sorbitol (Hellebust, 1985). Mammalian tissues subjected to frequent osmotic stress, e.g. kidney tissues, accumulate sorbitol as an osmoregulatory compound (Cowley et al., 1990). Thus, the case can be made that alditols play a key osmoprotective/osmoregulatory function in some biological systems.

In higher plants only a single species, *Plantago maritima*, has been shown to accumulate an alditol (sorbitol) in response to osmotic (salt) stress (Ahmad et al., 1979; Briens, Larher, 1982; Briens, Larher, 1983). Apparently this species and other species of *Plantago* have the capacity to make and accumulate sorbitol in both stressed and unstressed conditions (Briens, Larher, 1983). Although the original report of sorbitol accumulation in response to salt stress in *Plantago maritima* is over a decade old, nothing is presently known about the pathway of biosynthesis or utilization of sorbitol within the Plantagenaceae nor have any of the enzymes involved in these pathways been studied within this family. Consequently, while sorbitol (Bieleski, 1982; Lewis, 1984; Loescher, 1987), mannitol (Flora, Madore, 1993; Loescher, 1987; Loescher et al., 1992), galactitol (Loescher, 1987), and other polyols (Beck, Hopf, 1990; Loescher, 1987) have significant metabolic roles in numerous higher plant species, a natural role for these compounds in the processes of osmoprotection and osmoregulation in higher plants has yet to be unequivocally established. The best evidence inferring that alditols are able to act as osmoprotective compounds in higher plants is derived from studies on transgenic tobacco plants (Bohnert, this volume; Tarczynski et al., 1992; Tarczynski et al., 1993), where it has been shown that tobacco plants transformed with a bacterial gene which imparts the ability to synthesize and accumulate mannitol to tobacco plants are more tolerant of salt stress than normal tobacco plants.

Sorbitol, mannitol, and galactitol are utilized as phloem translocated photoassimilates in selected genera of higher plants (Bieleski, 1982; Lewis, 1984; Loescher, 1987), and among the alditols, the metabolic pathways leading to the accumulation and utilization of sorbitol and mannitol are the most clearly established among the Rosaceae which translocate sorbitol (Bieleski, 1982; Loescher, 1987) and the Oleaceae (Flora, Madore, 1993) and celery (Umbeliferae) which translocate mannitol (Loescher, 1987; Bieleski, 1982).

Five enzymes have been reported to be involved in sorbitol metabolism in higher plants. The reactions catalyzed by these enzymes are given below:

$$\text{GLUCOSE-6-P} + \text{NADPH} \rightleftharpoons \text{NADP} + \text{SORBITOL-6-P} \tag{1}$$

$$\text{SORBITOL-6-P} \longrightarrow \text{SORBITOL} + P_i \tag{2}$$

$$\text{SORBITOL} + \text{NAD} \rightleftharpoons \text{NADH} + \text{FRUCTOSE} \tag{3}$$

$$\text{SORBITOL} + \text{NADP} \rightleftharpoons \text{NADPH} + \text{GLUCOSE} \tag{4}$$

$$\text{SORBITOL} + 1/2\ O_2 \longrightarrow H_2O + \text{GLUCOSE} \tag{5}$$

In sorbitol translocating species sorbitol is synthesized from recently fixed carbon (Redgwell, Bieleski, 1978), and an active aldose 6-phosphate reductase (sorbitol 6-phosphate dehydrogenase, reaction 1 above) can be demonstrated in leaf tissues actively synthesizing sorbitol (Hirai, 1981; Negm, Loescher, 1981). The enzyme has a molecular weight of 65,000 Daltons, and consists of a dimer of 33 kD subunits. Furthermore, the enzyme appears to show seasonal variation consistent with a role in the production of sorbitol for translocation (Hirai, 1983; Moriguchi et al., 1990; Yamaki, Moriguchi, 1986). The other enzyme which is thought to be involved in sorbitol metabolism in photosynthetic tissues of higher plants is sorbitol-6-phosphate phosphatase (reaction 2 above; Grant, Ap Rees, 1981). Beyond the fact that the enzyme is considered to be a specific phosphatase for the production of sorbitol from sorbitol-6-phosphate, little is known about this enzyme.

An NADPH-linked aldose reductase (NADP-linked sorbitol dehydrogenase; reaction 4 above) similar to the well characterized enzyme from animal systems (Nishimura et al., 1989), has also been reported in leaf tissues (Yamaki, 1984), but the activity of this enzyme is quite low in leaf and fruit tissues (Moriguchi et al., 1990; Yamaki, Moriguchi, 1989), and it is likely that the enzyme does not have a critical role in sorbitol metabolism in the rosaceae.

Apple leaf and fruit tissues also contain a sorbitol oxidase which converts sorbitol and 1/2 an O_2 into glucose and water (reaction 5 above; Yamaki, 1982). Sorbitol oxidase appears to be localized in the vacuole (Yamaki, 1982); and thus, it may be involved in the utilization of vacuole stored carbohydrate. This enzyme is also not abundant in sorbitol metabolizing higher plant tissues.

In sink tissues of sorbitol translocating species an active sorbitol dehydrogenase (reaction 3 above) can be demonstrated (Negm, Loescher, 1979) which appears to be the primary enzyme of sorbitol utilization in these tissues. The level of SDH activity appears to correlate well with the demand for carbohydrate within the tissue (Loescher, 1987; Moriguchi et al., 1990; Yamaki, Moriguchi, 1989); suggesting a role for SDH in the assimilation of sorbitol in sink tissues. Apple sorbitol dehydrogenase is remarkably similar to the well characterized animal sorbitol dehydrogenases (Jeffery, Jornvall, 1988; Maret, Auld, 1988; Walsall et al., 1978). The apple SDH has an absolute specificity for NAD, works with a wide range of polyols, and has a pH optimum of about 9.0 for sorbitol oxidation and 6.0 for fructose reduction. The maximum velocity of the enzyme is greater in the direction of sorbitol oxidation than for fructose reduction, and the substrate K_M's are similar to the animal enzyme.

Beside those rosaceous species where sorbitol is a translocation carbohydrate, developing maize kernels also contain a metabolically active pool of sorbitol (Carey et al., 1982; Shaw, Dickinson, 1984). This pool of sorbitol is rapidly labeled by ^{14}C-sucrose arriving at the kernel when [14]CO_2 is fed to leaves (Shaw, Dickinson, 1984). There is no evidence that ^{14}C-sorbitol is made in the leaf and translocated to the kernel. The only tissues within the maize plant which contain sorbitol appear to be the developing kernel tissues. However, the pathways and enzymes of sorbitol metabolism in developing maize kernels are largely uncharacterized and may be quite different from those found in the rosaceae.

At least two enzymes are present in maize kernel tissues which are capable of acting on the sorbitol pool. A highly active NAD-linked sorbitol dehydrogenase (SDH, reaction 3 above) has been purified and characterized from this tissue (Doehlert, 1987; Locy, this report). Also recent studies have shown that a low level of aldose reductase (reaction 4 above) is also present in maize kernel tissues (Locy, unpublished). In animal systems these two enzymes are also both present in tissues where sorbitol metabolism is important, and together they function in what has been referred to as the sorbitol pathway (Jeffery, Jornvall, 1983). The presence of both SDH and aldose reductase in maize kernel tissues might suggest that a sorbitol pathway analogous to the animal sorbitol pathway also functions in maize. This suggestion was made by Doehlert (1987) even in advance of the finding that aldose reductase was present in maize kernels. Doehlert further argued that the two reactions would work in opposite directions to those proposed for the animal enzymes in the sorbitol pathway, i.e SDH would function to make sorbitol from fructose and aldose reductase would function to make glucose from sorbitol. This would provide a means of assimilating fructose derived from sucrose breakdown via hexokinase rather than fructose kinase which is limiting in developing maize kernels (Doehlert, 1990a; Doehlert, 1990b). This suggestion seems unsupported since the levels of aldose reductase we have been able to detect are less than 1/20 the levels of SDH (on a per kernel or a specific activity basis) and no greater than fructose kinase activities whereas the levels of SDH are as great or greater than those of sucrose synthase the enzyme thought to be responsible for sucrose breakdown. Thus, it is difficult to envision such a sorbitol shuttle contributing greatly to the assimilation of fructose in the developing maize kernel.

An alternative hypothesis for the role of sorbitol metabolism within developing maize kernel tissues derives from the fact that the only enzyme thus far detected within the kernel in significant quantity to metabolize sorbitol is SDH. Thus, it is possible that SDH acts to move carbon from fructose to sorbitol and then either in a different tissue within the kernel or at a different time in kernel development the same enzyme or another isoform of the enzyme acts to remake fructose from sorbitol. The reduction status of the NAD pool would likely play a key role in modulating the direction in which the enzyme works according to this hypothesis. Perhaps the low level of aldose reductase found to date, acts in concert with SDH to reassimilate sorbitol as suggested by Doehlert (1987). The hypothesis presented here would also permit an osmoprotective role for sorbitol in the acquisition of desiccation tolerance within the kernel, and suggests that the regulation of SDH activity and/or isoforms may be critical to such a role.

That kernel sorbitol metabolism might be involved in the process of desiccation tolerance of the tissues within the cereal seed, has also been suggested from an entirely different perspective. Bartels et al. (1991) prepared cDNA libraries from developing barely embryos prior to and after the acquisition of desiccation tolerance. By differential screening of the desiccation tolerant library, several cDNA clones which appear to be associated with desiccation tolerance have been isolated. One clone which has been sequenced showed substantial sequence homology to animal aldose reductase. Antibodies raised against the fusion protein of this clone expressed in *E. coli* have been used to demonstrate the presence of aldose reductase in the immature barley embryo and to study its occurrence and possible regulation. From these studies it was inferred that a polyol, possibly sorbitol, may play a role in osmoprotection within the barley embryo during desiccation as the seed matures.

Thus, sorbitol metabolism has been shown to have a key role in carbon metabolism in several higher plant species, and sorbitol metabolism may play an osmoregulatory/osmoprotective function in other species. However, definitive work linking these two processes has yet to emerge. The work presented here is part of an ongoing effort attempting to describe the role of sorbitol metabolism in plant tissues. Specifically the studies reported are directed at characterizing the active sorbitol dehydrogenase activity present in developing maize kernels. Initial studies suggested that there were two sorbitol dehydrogenase activities present in kernel tissues. However, the studies described here suggest that the two activities which can be shown are likely the result of metal ion substitution in a single peptide chain, possibly at the active site. This substitution may be physiologically meaningful relating the function of this enzyme to processes of desiccation tolerance and osmoprotection in the developing seed.

MATERIALS AND METHODS

Materials - All chemicals used were reagent grade chemicals purchased from either Fisher Scientific or Sigma Chemical Co. DEAE Cellulose (DE52) was purchased from Whatman. Sephadex G-50 was purchased from Pharmacia Fine Chemicals, Inc, and Reactive Blue #4 agarose was purchased from Sigma Chemical Co (product R2507). Hydroxylapatite was prepared as described by Muench (1971). Prepacked Econo-Q columns were purchased from Biorad, Inc.

Enzyme Assays - All enzyme assays were performed by measuring the production or disappearance of NAD or NADH spectrophotometrically by continuously monitoring the absorbance at 340 nm using a Milton-Roy Spectronic 21 Spectrophotometer. Assays were performed either as sorbitol dehydrogenase assays by following the sorbitol dependent reduction of NAD, or as fructose reductase assays by following the fructose dependent oxidation of NADH. The sorbitol dehydrogenase assay contained in a 1 mL reaction volume, 20 mM N-2-hydroxyethylpiperazine-N'-2-ethanesulfonic acid (HEPES), pH 7.5; 100 mM sorbitol; 1 mM NAD; and enough enzyme to produce an absorbance change of no more than 0.2 per min. The fructose reductase assay contained in a 1 mL reaction volume 20 mM HEPES, pH 7.5; 400 mM fructose; 0.2 mM NADH; and enough enzyme to produce an absorbance change of no more than 0.2 per min. In some assays where pH dependent enzyme activity was determined and for determination of apparent kinetic constants various reaction components were varied from the standard assay conditions as described in the text, figure legends, and tables. In all cases one unit of enzyme activity is defined as that amount of enzyme which will produce 1 μmole of product per minute.

Enzyme Purification - NAD-linked Sorbitol dehydrogenase (SDH) was purified from developing maize kernels by two related methods. For both methods 600 G of kernel tissues were ground to a fine powder in liquid nitrogen. The powdered kernels were further homogenized in 1200 mL of 20 mM sodium phosphate, pH 6.7, containing 1 mM dithiothreitol using a Waring Blender. Following homogenization the filtrate was filtered through cheese cloth and Miracloth (Calbiochem, Inc.), and clarified by centrifugation at 12,000 x g for 15 min. The supernatant was made 30% saturated with ammonium sulfate and again centrifuged as above. This supernatant was then made 75% saturated with ammonium sulfate, and the precipitate was collected by centrifugation as above and dissolved in a minimum of homogenization buffer. The 30-75% ammonium sulfate preparation was then desalted by passing the solution through a 5 cm x 55 cm Sephadex G-50 equilibrated with homogenization buffer. The fraction containing the bulk of the protein eluting in the void volume of this column was collected and used as the starting material for the subsequent protocols outlined below.

Purification Method 1 - The desalted, 30-75% ammonium sulphate precipitated enzyme preparation from above was applied to a 2.5 cm x 30 cm DEAE Cellulose column equilibrated with 20 mM sodium phosphate buffer, pH 6.7 containing 0.1 mM dithiothreitol (DTT). The column was then washed with 3 column volumes of the same buffer, and eluted with a linear gradient of 0 to 1 M NaCl in 20 mM sodium phosphate, pH 6.7, and 0.1 mM DTT. The column flow rate was 3 mL per min., and 6 mL fractions were collected. The elution profile of a typical DEAE cellulose column is shown in Figure 1.

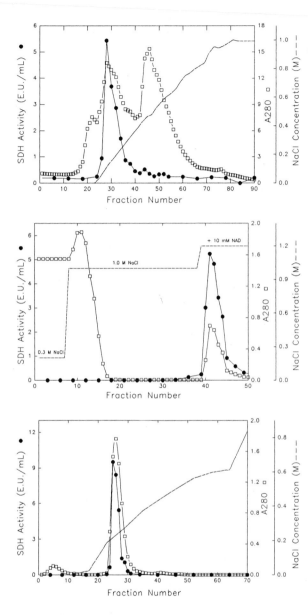

Figure 1. Ammonium sulphate precipitated, Sephadex G-50 desalted maize kernel SDH was chromatographed on a DEAE cellulose column as indicated in the Methods. The elution of SDH activity (●), protein as determined by absorbance at 280 nm (□), and the NaCl gradient (---) are shown.

Figure 2. DEAE cellulose column fractions showing SDH activity were pooled, directly eluted across a Reactive Blue #4 agarose dye affinity column, and chromatographed as described in Methods. The elution of SDH activity (●), protein as determined by absorbance at 280 nm (□), and the NaCl and NAD steps (---) are shown.

Figure 3. Reactive Blue #4 agarose purified SDH activity was eluted across an Econo-Q column, and chromatographed as described in Methods. The elution of SDH activity (●), protein as determined by absorbance at 280 nm (□), and the NaCl gradient (---) are shown.

is shown in Figure 1.

DEAE cellulose fractions containing SDH activity were pooled and eluted through a 1 cm x 15 cm Reactive Blue #4 agarose column. The column was eluted with 20 mM sodium phosphate, pH 6.7, 0.1 mM DTT, and 1 M NaCl until the A_{280} of the eluate was 0. SDH activity could then be eluted from the column with 20 mM sodium phosphate, pH 6.7, 0.1 mM DTT, 1 M NaCl, and 10 mM NAD. A flow rate of 1.5 mL per min. was used to elute the column and 1.5 mL fractions were collected. The elution profile of a typical Reactive Blue #4 agarose column is shown in Figure 2. The pooled fractions containing SDH activity were concentrated, and diafiltered by 2 consecutive 10 fold dilutions with 20 mM HEPES buffer, pH 7.5, and 0.1 mM DTT using an Amicon Ultrafiltration apparatus equipped with an Amicon YM-10 membrane.

The pooled diafiltered Reactive Blue #4 agarose fractions were eluted across a 5 mL Econo-Q column prepared according to the manufacturers specifications and equilibrated with 20 mM HEPES, pH 7.5 containing 0.1 mM DTT. The column was washed with 3 column volumes of the same buffer, and eluted with a linear 0 to 1 M NaCl gradient in the same buffer. One mL fractions were collected, and the column was eluted at a flow rate of 0.5 mL per min. The elution profile of a typical Econo-Q column is shown in Figure 3. Fractions containing SDH activity were pooled, concentrated and diafiltered as above using 2 consecutive 10 fold dilutions with 20 mM HEPES, pH 7.5 containing 0.1 mM DTT. This concentrated diafiltered enzyme was stored either at 4 C or at -20 C for further use.

Purification Method 2 - The desalted, 30-75% ammonium sulphate precipitated enzyme preparation from above was applied to a 2.5 cm x 30 cm DEAE Cellulose column equilibrated with 30 mM sodium phosphate buffer, pH 6.7 containing 0.1 mM DTT, 200 μM zinc sulfate, and 100 μM ethylenediaminetetraacetic acid (EDTA). The column was then washed with 3 column volumes of the same buffer, and eluted with a linear gradient of 30 to 300 mM sodium phosphate, pH 6.7, containing 0.1 mM DTT, 200 μM zinc sulfate, and 100 μM EDTA. The column flow rate was 3 mL per min., and 6 mL fractions were collected. The elution profile of a typical DEAE cellulose column is shown in Figure 4.

DEAE cellulose fractions containing SDH activity were pooled, diluted with distilled water to a phosphate concentration of 30 mM, and the diluted fractions were applied to a 1.5 cm x 20 cm hydroxyapatite column equilibrated with 30 mM sodium phosphate, pH 6.7 containing 0.1 mM DTT, 200 μM zinc sulfate, and 100 μM EDTA. The column was washed with 5 column volumes of the equilibrating buffer, and SDH activity was eluted from this column by a linear gradient from 30 to 500 mM sodium phosphate, pH 6.7 containing 0.1 mM DTT, 200 μM zinc sulfate, and 100 μM EDTA. The column was eluted at a flow rate of 1.5 mL per min. and 3 mL fractions were collected. A typical elution profile for a hydroxyapatite column is shown in Figure 5. Fractions containing SDH activity were pooled, concentrated, and diafiltered by 2 consecutive 10 fold dilutions with 10 mM tris (hydroxylmethyl) aminomethane (TRIS) buffer, pH 8.0, containing 0.1 mM DTT, 200 μM zinc sulfate, and 100 μM EDTA using an Amicon Ultrafiltration apparatus equipped with an Amicon YM-10 membrane.

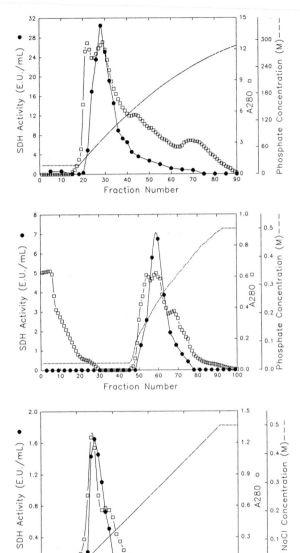

Figure 4. Ammonium sulphate precipitated, Sephadex G-50 desalted maize kernel SDH was chromatographed on a DEAE cellulose column as indicated in the Methods. The elution of SDH activity (●), protein as determined by absorbance at 280 nm (□), and the sodium phosphate gradient (---) are shown.

Figure 5. DEAE cellulose purified SDH activity was chromatographed on an hydroxyapatite column as described in Methods. The elution of SDH activity (●), protein as determined by absorbance at 280 nm (□), and the sodium phosphate gradient (---) are shown.

Figure 6. Hydroxyapatite purified SDH activity was eluted across an Econo-Q column, and the column was eluted as described in Methods. The elution of SDH activity (●), protein as determined by absorbance at 280 nm (□), and the NaCl gradient (---) are shown.

The pooled diafiltered hydroxyapatite fractions were eluted across a 5 mL Econo-Q column prepared according to the manufacturers specifications and equilibrated with 10 mM TRIS, pH 8.0 containing 0.1 mM DTT, 200 μM zinc sulfate, and 100 μM EDTA. The column was washed with 3 column volumes of the same buffer, and eluted with a linear 0 to 0.5 M NaCl gradient in the same buffer. One mL fractions were collected, and the column was eluted at 0.5 mL per min. The elution profile of a typical Econo-Q column is shown in Figure 6. Fractions containing SDH activity were pooled, concentrated and diafiltered as above using 2 consecutive 10 fold dilutions with 20 mM HEPES, pH 7.5 containing 0.1 mM DTT. This concentrated diafiltered enzyme was stored either at 4 C or at -20 C for further use.

Beside the obvious differences in column chromatographic steps, the two methods differed in that all buffers used in Method 2 also contained 200 μM $ZnSO_4$ and 100 μM EDTA. These addenda were found to be useful as results from one preparation to the next were less consistent when they were omitted. The choice of Zn^{2+} was made since all animal SDH's purified to date have been shown to contain 1 molecule of Zn^{2+} per subunit (Jeffery and Jornvall, 1988).

SDH prepared by Method 1 is essentially pure as determined by 1-D SDS-PAGE, giving only a single band at approximately 38 kD (data not shown). Method 2 enzyme preparations yield 3-4 bands in 1-D SDS PAGE showing a major band also at 38 kD (data not shown).

RESULTS

When sorbitol dehydrogenases (SDH's) prepared by the two methods were further characterized a number of striking differences between the preparations made by the two methods were noted. Figure 7 shows the pH versus specific activity profile in both the direction of sorbitol oxidation (SDH activity) and fructose reduction (FR activity) for SDH prepared by Method 1 (SDH-1). Figure 8 shows the same relationships for enzyme prepared by Method 2 (SDH-2).

The pH profile of SDH-1 (Figure 7) is similar to that previously reported for maize SDH (Shaw, Dickinson, 1984) where the pH optimum for the FR assay is near 7, and the maximum specific activity for the FR assay is 5 to 6 times higher than for the SDH assay. However, the pH profile of SDH-2 (Figure 8) is not at all like that of SDH-1. SDH-2 demonstrates a pH profile more similar to that reported for apple (Negm, Loescher, 1979), and for a number of animal SDH's (Jeffery, Jornvall, 1988). The maximum specific activity in the SDH assay is 2 to 4 fold higher than for the FR activity assay. In addition apparent K_m's for the 4 substrates for SDH-1 and SDH-2 reflect different substrate specificities for the two enzymes (see Table 1).

Thus, the two differing purification procedures yielded two apparently unique SDH enzymes. However, there was no evidence that SDH-1 was present in the Method 2 prepared enzyme and vice versa. Since purified SDH-2 either binds irreversibly to Reactive Blue #4 agarose dye affinity gel or loses activity when passed across the gel, it is not surprising that SDH-2 was not found in SDH-1 preparations since Reactive Blue #4 agarose was used as a purification step in Method 1. However,

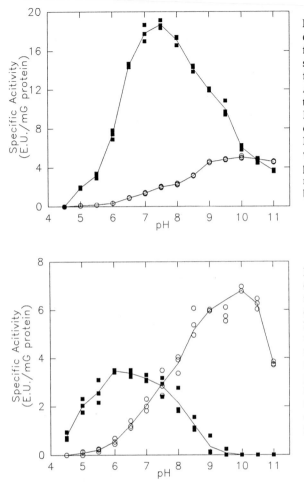

Figure 7. The pH versus specific enzyme activity curve in the fructose reductase assay (■) and sorbitol dehydrogenase assay (○) for SDH-1 from maize kernels. Assays were conducted according to the standard assay (see Methods) except that a combination of phosphate, HEPES, TRIS, and CAPS buffers were employed at appropriate pH's within 1 pH unit of the respective pK_a for each of the buffering ions.

Figure 8. The pH versus specific enzyme activity in the fructose reductase assay (■) and sorbitol dehydrogenase assay (○) for SDH-2 from maize kernels. Assays were conducted according to the standard assay (see Methods) except that a combination of phosphate, HEPES, TRIS, and CAPS buffers were employed at appropriate pH's within 1 pH unit of the respective pK_a for each of the buffering ions.

Table 1.

Apparent Michaelis Constants derived from Lineweaver-Burke plots of a single varied substrate at the concentration of the other substrate used in the standard assay.

Varied Substrate	SDH-1 Apparent K_m	SDH-2 Apparent K_m
Sorbitol	11.7 mM	1.45 mM
Fructose	128 mM	141 mM
NAD	840 μM	145 μM
NADH	17.2 μM	18.4 μM

extensive efforts to purify the two SDH forms from the same homogenate have failed. This prompted an investigation of possible factors used in Method 2 that would lead to a modification of the SDH-1 enzyme during purification.

Although addition of EDTA to the enzyme assay showed no immediate effect on SDH-1 or SDH-2 activity (data not shown), prolonged incubation of SDH-1 with EDTA in the absence of substrates at a series of temperatures led to a reversible decline in activity of the enzyme (Figures 9A and 9B). This decline took place after a modest initial, nearly instantaneous increase in activity when the assays were performed in the sorbitol dehydrogenase direction (Figure 9A), but the increase could be as much as 4 fold (at 30 C) when the assays were performed in the fructose reductase direction (Figure 9B, Table 2).

Both the extent of the increase and the subsequent rate of loss of enzyme activity were highly temperature dependent. Figure 10 demonstrates this temperature dependence for the rate of loss of activity when SDH-1 was incubated with EDTA. An Arrhenius plot of the logarithm of the exponential decay constant (slopes of the lines in Figure 9) versus the reciprocal of the absolute

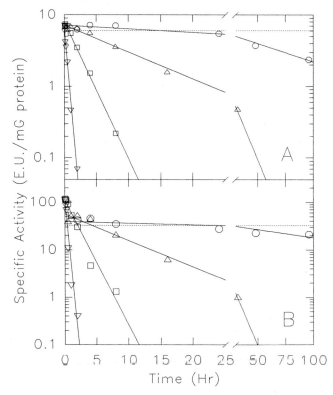

Figure 9. Purified SDH-1 was incubated for various periods of time with 2 mM EDTA at 0 C (o), 15 C (△), 30 C (□), or 45 C (▽). The specific activity of the incubated enzyme was then determined in the standard assay either in the direction of sorbitol oxidation (panel A) or in the direction of fructose reduction (panel B). The log of the specific activity is shown versus time. Each data point shown is the mean of 3 replications. For most data points standard error bars would fall within the symbol, and are therefore not shown. The dashed line through the figure represents the specific activity of the enzyme not incubated with EDTA.

357

Table 2.

Purified SDH-1 was incubated with 2 mM EDTA at varying temperatures as described in Figure 9. The maximum specific enzyme activity for sorbitol oxidation and fructose reduction as determined from the y-intercepts in Figure 9, and the ratio of these activities are shown at the incubation conditions used.

Incubation Conditions	Sorbitol Dehydrogenase Activity (EU/mG)	Fructose Reductase Activity (EU/mG)	FR/SDH
0 C + EDTA	7.47 ± 0.38	43.94 ± 2.74	5.88
15 C + EDTA	7.48 ± 0.20	64.69 ± 4.50	8.65
30 C + EDTA	6.81 ± 0.08	129.59 ± 2.87	19.03
45 C + EDTA	6.79 ± 0.32	105.41 ± 1.95	15.52
No Incubation	6.08 ± 0.32	31.48 ± 4.81	5.17
45 C No EDTA	5.66 ± 1.03	35.99 ± 1.04	6.36

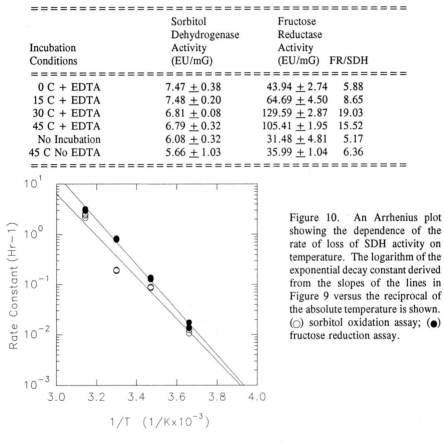

Figure 10. An Arrhenius plot showing the dependence of the rate of loss of SDH activity on temperature. The logarithm of the exponential decay constant derived from the slopes of the lines in Figure 9 versus the reciprocal of the absolute temperature is shown. (○) sorbitol oxidation assay; (●) fructose reduction assay.

temperature is shown. For both sorbitol oxidation and fructose reduction straight lines are obtained. The fact that the two lines have nearly identical slopes and intercepts indicates that both reactions are similarly inhibited by EDTA as a function of temperature. Table 2 shows degree of stimulation resulting from EDTA incubation. The maximum specific activity achieved at varying temperatures for each reaction (y-intercepts from the plots in Figure 9) when SDH-1 was incubated with EDTA.

Incubation of the enzyme in the absence of EDTA at temperatures up to 45 C for periods of several hours had almost no effect on the specific activity of the enzyme in either direction of assay (Figures 11).

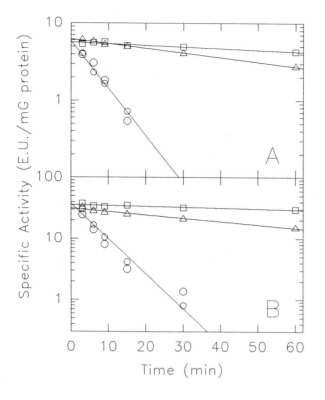

Figure 11. SDH-1 enzyme was incubated with 2 mM EDTA at 45 C for varying periods of time. Following the incubation period the enzyme was either assayed using standard assay conditions as described in Methods (without any metal ion added) (○), or assayed in standard assay conditions with 200 μM zinc sulfate added (△). For comparison SDH-1 activity incubated for varying periods of time in the absence of EDTA and assayed in the standard assay is shown (□). Panel A shows the results when sorbitol oxidation assays were performed, and panel B shows the results when fructose reduction assays were performed.

Table 3.

The ratio of fructose reductase activity to sorbitol dehydrogenase activity in SDH-1; SDH-2; EDTA treated SDH-1; and Zn^{2+}, Mg^{2+}, or Ca^{2+} substituted SDH-1 assayed in the standard reaction mixtures described in Table 1.

Enzyme Source	Fructose Reductase Activity (EU/mG)	Sorbitol Dehydrogenase Activity (EU/mG)	FR/SDH
SDH-1	31.8	6.05	5.26
SDH-2	2.81	3.54	0.79
EDTA treated SDH-1	114.2	7.37	15.49
Ca^{2+} substituted SDH-1	22.1	3.21	6.88
Mg^{2+} substituted SDH-1	28.7	5.22	5.50
Zn^{2+} substituted SDH-1	3.29	2.91	1.13

The effects of EDTA could be reversed by incubating the enzyme with a number of divalent metal ions, including Zn^{2+}, Ca^{2+}, and Mg^{2+} (Table 3, and Figure 11). This reversibility could be demonstrated at all times of incubation (Figure 11). Of most significance is the fact that when SDH-1 is incubated with EDTA and subsequently Zn^{2+} is added to the enzyme, the catalytic properties of the enzyme now appear to be similar to SDH-2 (Figure 12 and Table 3). The specific activity was higher

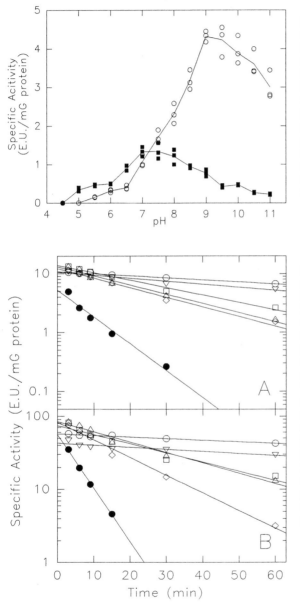

Figure 12. The pH versus specific enzyme activity curve in the fructose reductase assay (■) and sorbitol dehydrogenase assay (○) for zinc substituted SDH-1 from maize kernels. Assays were conducted according to the standard assay (see Methods) except that a combination of phosphate, HEPES, TRIS, and CAPS buffers were employed at appropriate pH's within 1 pH unit of the respective pK_a for each of the buffering ions.

Figure 13. The effect of various osmotic addenda on stabilizing SDH activity in the presence of EDTA is shown. Purified SDH-1 was incubated in the presence of 2 mM EDTA for varying periods of time at 45 C. Various solutes were added to the incubation medium all at 1 M concentration. The enzyme specific activity is shown as a function of time for each addendum. (●) No addendum, (○) 1 M NaCl, (▽) 1 M KCl, (□) 1 M mannitol, (△) 1 M sucrose, and (◇) 1 M glucose. Panel A shows the results when sorbitol oxidation assays were performed, and panel B shows the results when fructose reduction assays were performed.

for sorbitol oxidation than for fructose reduction. However, when either Ca^{2+} or Mg^{2+} were used to reactivate the enzyme SDH-1 type activity was recovered. Thus, it appears that metal ion substitution is capable of preferentially altering the pH profile and substrate specificity of SDH-1.

In order to establish that metal ion substitution may lead to physiologically meaningful regulation, two additional experiments were conducted. First, the effect of various osmotica on the EDTA activation and inactivation of SDH were investigated. Figures 13A and 13B show that the effect of EDTA on SDH-1 could be dramatically effected by including NaCl, KCl, mannitol, sucrose or glucose (at 1 M concentration) in the incubation medium along with EDTA and SDH. The two ionic osmotica (1 M NaCl or 1 M KCl) were both highly effective in slowing the rate of EDTA induced activity loss, while mannitol, sucrose, and glucose while effective in slowing the EDTA induced loss were less effective than the ionic osmotica.

Table 4.

Crude NAD-SDH activity in various kernel parts assayed either in the direction of sorbitol oxidation (SDH activity) or in the direction of fructose reduction (FR activity). The ratio of FR/SDH is also shown.

	SDH		FR		
Kernel Part	EU/ Kernel	EU/ Gfwt	EU/ Kernel	EU/ Gfwt	FR/SDH
Kernel Base	0.008	0.676	0.006	0.530	0.78
Embryo	0.004	0.292	0.003	0.274	0.94
Endosperm					
Basal	0.049	1.109	0.228	5.176	4.66
Middle	0.049	0.564	0.164	1.868	3.32
Apical	0.098	0.316	0.062	0.799	2.54
Pericarp	0.001	0.067	0.003	0.252	3.74

Table 4. shows that there is some evidence that modification of the FR/SDH ration may occur within various kernel tissues. Crude homogenates made from various kernel tissues were desalted by passing the crude homogenate across a 1 x 20 cm Sephadex G-50 column equilibrated with homogenization buffer (see Methods). The desalted crude homogenates were assayed for sorbitol dehydrogenase and fructose reductase activity as described in Methods. Embryos and kernel bases contain an SDH activity which has a fructose reductase to sorbitol dehydrogenase activity ratio more characteristic of SDH-2 than SDH-1 (Table 4). Whereas endosperm tissues and pericarp tissue contain an SDH with a FR/SDH ratio more characteristic of SDH-1 (Table 4).

DISCUSSION

The data presented above show that *in vitro* it is possible to dramatically modify the catalytic properties of maize SDH-1 by incubation with the chelating agent EDTA and subsequent substitution of the enzyme with specific divalent metal ions. By presumably removing endogenous metal ion with EDTA, and replacing that ion with Zn^{2+} it is possible to convert activity purified as SDH-1 into activity more similar to SDH-2. It is therefore logical to presume that SDH-2 may result from the inclusion of EDTA and Zn^{2+} in the Purification Method 2 buffers. The above data might indicate that SDH-2 is an artifact of preparation. However, the possibility also exists that metal ion substitution involving chelating agents and divalent metal ions might be capable of modulating SDH activity *in situ*. This suggestion is supported by the data presented which show that embryos and kernel bases appear to contain an SDH activity having a FR/SDH ratio more like SDH-2 whereas other kernel tissues have a FR/SDH ratio more like SDH-1. Also the fact that osmotica have the capacity to modulate the capacity of the enzyme for metal ion substitution is an additional piece of evidence that such modulation may be physiologically meaningful. It is highly plausible that sucrose, glucose, and fructose could accumulate to significant levels within a developing kernel actively importing photosynthate, and such regulation of SDH may be related to the accumulation of sorbitol within the kernel and/or to desiccation tolerance.

Thermodynamically the reaction catalyzed by SDH is readily reversible (Doehlert, 1987), and thus it is likely that the rate and direction in which SDH works *in vivo* is the result of the interaction of the intracellular concentrations of NAD, NADH, fructose, and sorbitol with the kinetic constants of the enzyme for each of its substrates. Based on these factors Doehlert (1987) proposed that SDH worked primarily in the direction of fructose conversion to sorbitol in the developing maize kernel. He further hypothesized (see Introduction) that the subsequent conversion of sorbitol to glucose by aldose reductase would allow the fructose generated from sucrose to be more readily assimilated since maize kernels did not contain appreciable quantities of fructose kinase (Doehlert et al., 1988; Doehlert, 1990a; Doehlert, 1990b). This hypothesis has proven difficult to support since neither Doehlert (1987) nor our efforts (Locy, unpublished) have demonstrated significant quantities of aldose reductase within the developing maize kernel. In fact the only enzyme which appears to be present within kernel tissues in sufficient quantity to metabolize fructose at a rate that would be proportionate to the rate at which sucrose synthase or invertase produce fructose is SDH (Doehlert et al., 1988; Locy, unpublished). Such negative evidence does not preclude that additional aldose reductase activity will be detected or that some other activity involved in sorbitol metabolism will be found in the developing kernel, and these possibilities are currently being examined. The fact that a low level of aldose reductase activity has been observed in developing maize kernels suggests that some carbon may flow from fructose to glucose through sorbitol as proposed by Doehlert. However, such a pathway would seem minor. Thus, we have considered that there may be more than one isoform of SDH. Separate isoforms of SDH with differing catalytic properties may be either temporally regulated or spatially compartmentalized within specific kernel tissues and may differentially function to first produce sorbitol from fructose and subsequently function to produce fructose again from sorbitol.

This hypothesis appeared to gain support when during the course of improving SDH purification protocols it was found that the enzyme prepared by two different methods had substantially different catalytic properties (refer to Table 1, and Figures 7 and 8). However, it was never possible to separate the two forms of the enzyme from each other by physical means. For example, two peaks of activity were never observed in any chromatography step during purification (Figures 1 through 6). In SDS-PAGE the single band observed in SDH-1 preparations was identical in size to the predominant band (which is also the one which enriches most during purification) in SDH-2 preparations. In native gel electrophoresis regardless of the conditions employed it was not possible to show that SDH-1 differed from SDH-2 when gels were stained for SDH activity. Furthermore, the demonstration that metal ion substitution can alter catalytic properties of SDH suggests that the only evidence found to date for two isoforms of SDH is highly questionable. Thus, it is clear that definitive resolution of whether isoforms of SDH exist within the maize plant must await resolution at the molecular level.

Maize SDH-1 is reversibly inhibited by the chelating agent EDTA. Since other SDH's are known to be Zn^{2+}-containing enzymes (Jeffery et al., 1984; Maret, Auld, 1988; Walsall et al., 1978), and since all SDH's thus far sequenced belong to the long chain alcohol dehydrogenase family of dehydrogenases all of which are Zn^{2+}-containing enzymes (Jeffery, Jornvall, 1988; Jornvall et al., 1984; Karlsson et al., 1989), this finding is not surprising. However, the inhibition by EDTA is a slow process which depending on temperature requires hours to days. The only rapid (nearly instantaneous) effect of EDTA on SDH appears to be an activation of the FR reaction which can be up to four fold depending on temperature. Both the extent of activation and the rate of inhibition are dependent on the temperature of incubation while the enzyme is exposed to EDTA. Furthermore, the reversibility of the phenomenon is dependent on the addition of a divalent metal ion back to the enzyme, and the properties of the reactivated enzyme are dependent on the metal ion used.

When EDTA-inactivated SDH is subsequently substituted by Zn^{2+} the properties of maize SDH-1 become more like those of SDH-2 and mammalian SDH's showing a low FR/SDH ratio similar to animal SDH's rather than a FR/SDH ratio typical of maize SDH-1. However, when Mg^{2+} or Ca_{2+} is substituted into the EDTA inhibited enzyme the properties of the enzyme remain like those of SDH-1. Thus, it appears that based on the data presented here maize SDH can be modified to preferentially catalyze the FR reaction or the SDH reaction based on metal ion substitution.

That metal ion substitution can alter preferentially the substrate specificity and catalytic properties of enzymes has been demonstrated for a number of metaloenzymes including collagenase (Angleton, Van Wart, 1988), carboxypeptidase (Davies et al., 1968; Auld, Holmquist, 1974), aminopeptidase (Bayliss, Prescott, 1986), carbonic anhydrase (Khalifah et al., 1987), and alcohol dehydrogenase (Maret, Makinen, 1991) among others. The present report is, however, the first demonstration that the kinetic constants and substrate specificity of sorbitol dehydrogenase is capable of modulation by metal ion substitution. The activity of *E. Coli* threonine dehydrogenase, also a member of the long chain alcohol dehydrogenase family, has been shown to be effected by metal ion removal and substitution (Aronson et al., 1989; Epperly, Dekker, 1991), but in this instance only the level of activity and not the substrate specificity was modulated by the metal ion.

Sorbitol dehydrogenases from a number of mammalian systems and *Bacillus subtilis* have been either directly sequenced or their sequences inferred from the sequence of cDNA clones (Jeffery, Jornvall, 1988; Jornvall et al., 1984; Karlsson et al., 1989; Ng et al., 1992). Comparisons of these sequences suggest that all SDH's whose sequence is thus far known belong to the long chain alcohol dehydrogenase (ADH) family of dehydrogenases which includes mammalian, yeast and higher plant ADH's as well as a number of bacterial ADH's (Eklund et al., 1985; Jeffery, Jornvall, 1988; Jornvall et al., 1984; Karlsson et al., 1989). Other proteins whose sequence is known which also belong to this family include *E. Coli* threonine dehydrogenase (Aronson et al., 1989; Epperly, Dekker, 1991) and mammalian lens crystallin (Borras et al., 1989). By analogy it is reasonable to expect that higher plant SDH's also belong to this family of dehydrogenases although the sequence of higher plant SDH's have not as yet been determined or inferred. The similarity of the enzymatic properties of apple SDH and the mammalian SDH's further reinforces this suggestion. However, maize kernel SDH has a number of unique properties from the other characterized SDH's (Doehlert, 1987).

In addition to the catalytic differences discussed above, it is noteworthy that all the mammalian SDH's are tetrameric enzymes containing one Zn^{2+} ion per subunit (Feiters, Jeffery, 1989; Jeffery et al., 1984), whereas *B. subtilis* SDH, like the mammalian ADH's and *E. coli* threonine dehydrogenase contain two Zn^{2+} ions per subunit (Aronson et al., 1989; Epperly, Dekker, 1991; Ng et al., 1992). For those members of the family whose sequence is known the second Zn^{2+} binding site has been shown to be located in the vicinity of residues 100-120 of the *B. subtilis* SDH sequence and mammalian ADH sequence (Eklund et al., 1985; Jornvall et al., 1987; Ng et al., 1992) in a region which contains 4 cysteine residues thought to be ligands for chelation of the Zn^{2+}. Since the complete tertiary structure of ADH has been solved (Eklund et al., 1985) it is known that this second so called structural Zn^{2+} binding domain is located in an external loop located on the outside of the ADH molecule. However, the function of this second Zn^{2+} binding domain is unknown for any of the enzymes of this family. Thus, it will be interesting to determine the number of Zn^{2+} ions bound per subunit of maize SDH and the amino acid sequence of the enzyme in the region analogous to the structural Zn^{2+} binding site of the ADH's.

SUMMARY

The accumulation of polyols in response to osmotic stresses is a widespread biological occurrence. Sorbitol in particular is found to accumulate in fungi, algae, higher plants, and animals in response to osmotic stresses. In some species of higher plants (selected genera of Rosaceae) sorbitol is made in the leaves from glucose-6-phosphate, and sorbitol is a form of carbon translocated throughout the plant to sink tissues. However, in these plants a role for sorbitol in stress metabolism has yet to be established. In maize, sorbitol is found only in kernel tissues, and the only enzyme thus far found in significant quantity in the kernel which utilizes sorbitol as a substrate is sorbitol dehydrogenase (SDH). Two sorbitol dehydrogenase activities have been purified from developing maize kernels designated SDH-1 and SDH-2. SDH-1 has a pH versus activity profile and kinetic constants similar to the SDH previously reported from maize endosperm. However, SDH-2 has a pH versus activity profile and kinetic constants more characteristic of SDH's purified from animal systems. Rather than being isoforms of SDH these two activities appear to result from metal ion substitution effects on a single polypeptide chain. While EDTA has little effect on SDH activity when included in the enzyme assay, incubation of SDH-1 with EDTA at varying temperatures results in an apparent loss of activity of the enzyme. This loss of activity follows an exponential decay, and the rate of decay increases with temperature from 0 to 45 C. The fructose reductase reaction catalyzed by maize SDH is stimulated up to 4 fold by incubation with EDTA prior to the decay of activity, while the sorbitol dehydrogenase activity of SDH shows only limited stimulation. The degree of stimulation of activity also increases with temperature of incubation. The loss of activity when SDH-1 is incubated with EDTA is completely reversible by the addition of divalent metal ions back to the enzyme, and the specific metal ion used can dramatically influence the nature of the activity obtained. When Zn^{2+} ion is added back to the enzyme, an activity having a fructose reductase to SDH ratio characteristic of SDH-2 is obtained. However, if Ca^{2+} or Mg^{2+} are added back the reconstituted enzyme retains the fructose reductase to sorbitol dehydrogenase ratio of SDH-1. Whether these phenomena are physiologically meaningful is currently being investigated, but the fact that the fructose reductase to sorbitol dehydrogenase ratio of SDH obtained from embryo and kernel bases is similar to SDH-2 while endosperm is more like SDH 1 suggests that these phenomena may be physiologically meaningful. Such modulation of SDH activity within the kernel may relate to the acquisition of tolerance to desiccation within the developing kernel.

ACKNOWLEDGEMENT

The author gratefully acknowledges the technical support provided in the work presented here by two Auburn University undergraduate students, Tami Andrews and Michelle Locy.

REFERENCES

Ahmad I, Lahrer F, Stewart GR (1979) Sorbitol, a compatible osmotic solute in *Plantago maritima*. New Phytol. **82**:671-678.

Angleton EL, Van Wart HE (1988) Preparation by direct metal exchange and kinetic study of active site metal substituted class I and class II *Clostridium histolyticum* collagenases. Biochem. **27**:7413-7418.

Aronson BD, Somerville RL, Epperly BR, Dekker EE (1989) The primary structure of *Escherichia coli* L-threonine dehydrogenase. J.Biol.Chem. **264**:5226-5232.

Auld DS, Holmquist B (1974) Carboxypeptidase A differences in the mechanisms of ester and peptide hydrolysis. Biochem. **13**:4355-4361.

Bartels D, Engelhardt K, Roncarati R, Schneider K, Rotter M, Salamini F (1991) An ABA and GA modulated gene expressed in the barley embryo encodes an aldose reductase related protein. EMBO J. **10**:1037-1043.

Bayliss ME, Prescott JM (1986) Modified activity of *Aeromonas* aminopeptidase: metal ion substitutions and role of substrates. Biochem. **25**:8113-8117.

Beck E, Hopf H (1990) Branched-chain sugars and sugar alcohols. *In* PM Dey, JB Harborne, eds. Methods in Plant Biochemistry, Vol. 2, Academic Press, Inc., New York, pp 235-289

Bieleski RL (1982) Sugar Alcohols. *In* FA Loewus, W Tanner, G Avigad, eds. Encyclopedia of Plant Physiology, Vol 13A, Intracellular Carbohydrates, Springer-Verlag, Berlin, pp 158-192

Borras T, Persson B, Jornvall H (1989) Eye lens z-crystallin relationships to the family of "long-chain" alcohol/polyol dehydrogenases. Protein trimming and conservation of stable parts. Biochemistry **28**:6133-6139.

Briens M, Larher F (1982) Osmoregulation in halophytic higher plants: a comparative study of soluble carbohydrates, polyols, betaines and free proline. Plant,Cell and Environment **5**:287-292.

Briens M, Larher F (1983) Sorbitol accumulation in plantaginaceae: further evidence for a function in stress tolerance. Z.Pflanzenphysiol. **110**:447-458.

Brown AD, Edgley M (1980) Osmoregulation in yeast. *In* DW Rains, C Valentine, A Hollaender, eds. Genetic Engineering of Osmoregulation, Plenum, New York, pp 75

Carey EE, Rhodes AM, Dickinson DB (1982) Postharvest levels of sugars and sorbitol in sugary enhancer (su se) and sugary (su Se) maize. HortSci. **17**:241-242.

Chitlaru E, Pick U (1991) Regulation of glycerol synthesis in response to osmotic changes in *Dunaliella*. Plant Physiol. **96**:50-60.

Cowley BD,Jr., Ferraris JD, Carper D, Burg MB (1990) *In vivo* osmoregulation of aldose reductase mRNA, protein, and sorbitol in renal medulla. Amer.J.Physiol. **258**:F154-F161.

Crowe JH, Crowe LM, Chapman D (1984) Preservation of membranes in anhydrobiotic organisms: the role of trehalose. Science **223**:701-703.

Davies RC, Riordan JF, Auld DS, Vallee BL (1968) Kinetics of carboxypeptidase A. I. Hydrolysis of carbobenzoxyglycyl-L-phenylalanine, Benzoylglycyl-L-phenylalanine, and hippuryl-DL-B-phenyllactic acid by metal-substituted and acetylated carboxypeptidases. Biochem. **7**:1090-1099.

Doehlert DC (1987) Ketose reductase activity in developing maize endosperm. Plant Physiol. **84**:830-834.

Doehlert DC (1990a) Distribution of enzyme activities within the developing maize (Zea mays) kernel in relation to starch, oil and protein accumulation. Physiol.Plant. **78**:560-567.

Doehlert DC (1990b) Fructokinases from developing maize kernels differ in their specificity for nucleoside triphosphates. Plant Physiol. **93**:353-355.

Doehlert DC, Kuo TM, Felker FC (1988) Enzymes of sucrose and hexose metabolism in developing kernels of two inbreds of maize. Plant Physiol. **86**:1013-1019.

Eklund H, Horjales E, Jornvall H, Branden CI, Jeffery J (1985) Molecular aspects of functional differences between alcohol and sorbitol dehydrogenases. Biochem. **24**:8005-8012.

Epperly BR, Dekker EE (1991) L-Threonine dehydrogenase from *Escherichia coli*: Identification of an active site cysteine residue and metal ion studies. J.Biol.Chem. **266**:6086-6092.

Feiters MC, Jeffery J (1989) Zinc environment in sheep liver sorbitol dehydrogenase. Biochem. **28**:7257-7262.

Flora LL, Madore MA (1993) Stachyose and mannitol transport in olive (*Olea europaea* L). Planta **189**:484-490.

Ford CW (1984) Accumulation of low molecular weight solutes in water-stressed tropical legumes. Phytochem. **23**:1007-1015.

Grant CR, Ap Rees T (1981) Sorbitol metabolism by apple seedlings. Phytochem. **20**:1505-1511.

Hellebust JA (1985) Mechanisms of response to salinity in halotolerant microalgae. Pl.& Soil **89**:69-81.

Hirai M (1981) Purification and characteristics of sorbitol 6-phosphate dehydrogenase from loquat leaves. Plant Physiol. **67**:221-224.

Hirai M (1983) Seasonal changes in sorbitol-6-phosphate dehydrogenase in loquat leaf. Plant & Cell Physiol. **24**:925-931.

Jeffery J, Chesters J, Mills C, Sadler PJ, Jornvall H (1984) Sorbitol dehydrogenase is a zinc enzyme. EMBO J. **3**:357-360.

Jeffery J, Jornvall H (1983) Enzyme relationships in a sorbitol pathway that bypasses glycolysis and pentose phosphates in glucose metabolism. Proc.Nat.Acad.Sci.(USA) **80**:901-905.

Jeffery J, Jornvall H (1988) Sorbitol dehydrogenase. Adv.Enzymol.Relat.Areas Mol.Biol. **61**:47-106.

Jornvall H, Bahr-Lindstrom V, Jeffery J (1984) Extensive variations and basic features in the alcohol dehydrogenase - sorbitol dehydrogenase family. Eur.J.Biochem. **140**:17-23.

Jornvall H, Persson B, Jeffery J (1987) Characteristics of alcohol/polyol dehydrogenases. The zinc-containing long-chain alcohol dehydrogenases. Eur.J.Biochem. **167**:195-201.

Karlsson C, Maret W, Auld DS, Hoog JO, Jornvall H (1989) Variability within mammalian sorbitol dehydrogenases. The primary structure of the human liver enzyme. Eur.J.Biochem. **186**:543-550.

Kermode AR, Bewley JD, Dasgupta J, Misra S (1986) The transition from seed development to germination: A Key role for desiccation. HortSci. **21**:1113-1118.

Khalifah RG, Rogers JI, Mukherjee J (1987) Interaction of the unique competitive inhibitor imidazole and related compounds with the active site metal of carbonic anhydrase: linkage between pH effects on the inhibitor binding affinity and pH effects on the visible spectra of inhibitor complexes with the cobalt-substituted enzyme. Biochem. **26**:7057-7063.

Lewis DH (1984) Physiology and metabolism of alditols. *In* DH Lewis, ed. Storage Carbohydrates in Vascular Plants, Cambridge Univ. Press, Cambridge, pp 157-179

Loescher WH (1987) Physiology and metabolism of sugar alcohols in higher plants. Physiol.Plant. **70**:553-557.

Loescher WH, Tyson RH, Everard JD (1992) Mannitol synthesis in higher plants. Evidence for the role and characterization of a NADPH-dependent mannose 6-phosphate reductase. Plant Physiol. **98**:1396-1402.

Loewus FA (1990) Cyclitols. *In* PM Dey, JB Harborne, eds. Methods in Plant Biochemistry, Vol. 2, Academic Press, Inc., New York, pp 219-233

Maret W, Auld DS (1988) Purification and characterization of human liver sorbitol dehydrogenase. Biochem. **27**:1622-1628.

Maret W, Makinen MW (1991) The pH variation of steady-state kinetic parameters of site- specific Co^{2+}-reconstituted liver alcohol dehydrogenase. A mechanistic probe for the assignment of metal-linked ionizations. J.Biol.Chem. **266**:20636-20644.

Moriguchi T, Sanada T, Yamaki S (1990) Seasonal fluctuations of some enzymes relating to sucrose and sorbitol metabolism in peach fruit. Journal of the American Society for Horticultural Science **115**:278-281.

Muench KH (1971) Fractionation of transfer RNA on columns of hydroxyl apatite. Proc. in Nucleic Acid Res. **2**:515-523.

Negm FB, Loescher WH (1979) Detection and characterization of sorbitol dehydrogenase from apple callus tissue. Plant Physiol. **64**:69-73.

Negm FB, Loescher WH (1981) Characterization and partial purification of Aldose-6-phosphate Reductase (Alditol-6-phosphate:NADP 1-Oxidoreductase) from apple leaves. Plant Physiol. **67**:139-142.

Ng K, Ye RQ, Wu XC, Wong SL (1992) Sorbitol dehydrogenase from *Bacillus subtilis* - purification, characterization, and gene cloning. J.Biol.Chem. **267**:24989-24994.

Nguyen A, Lamant A (1988) Pinitol and myo-inositol accumulation in water-stressed seedlings of maritime pine. Phytochem. **27**:3423-3427.

Nishimura C, Wistow G, Carper D (1989) Rat lens aldose reductase: Characterization of its primary structure, mRNA, and gene complexity. *In* H Weiner, TG Flynn, eds. Enzymology and Molecular Biology of Carbonyl Metabolism 2, Alan R. Liss, Inc., New York, pp 211-220

Paleg LG, Aspinall D (1981) Proline accumulation: physiological aspects. *In* LG Paleg, D Aspinall, eds. Physiology and Biochemistry of Drought Resistance, Academic Press, Inc., Sidney, pp 105-143

Paul MJ, Cockburn W (1989) Pinitol, a compatible solute in *Mesembryanthemum crystallinum* L. J.Exp.Bot. **40**:1093-1098.

Redgwell RJ, Bieleski RL (1978) Sorbitol 1-phosphate and sorbitol 6-phosphate in apricot leaves. Phytochem. **17**:407-409.

Shaw JR, Dickinson DB (1984) Studies of sugars and sorbitol in developing corn kernels. Plant Physiol. **75**:207-211.

Streeter JG (1985) Identification and distribution of ononitol in nodules of *Pisum sativum* and *Glycine max*. Phytochem. **24**:174-176.

Tarczynski MC, Jensen RC, Bohnert HJ (1992) Expression of a bacterial mtlD gene in transgenic tobacco leads to production and accumulation of mannitol. Proc.Nat.Acad.Sci.(USA) **89**:2600-2604.

Tarczynski MC, Jensen RG, Bohnert HJ (1993) Stress protection of transgenic tobacco by production of the osmolyte mannitol. Science **259**:508-510.

Walsall EP, Lyons SA, Metzer RP (1978) A Comparison of selected physical properties of hepatic sorbitol dehydrogenases [L-Iditol:Nad Oxidoreductases] from four mammalian species. Comp.Biochem.Physiol. **59B**:213-218.

Yamaki S (1982) Localization of sorbitol oxidase in vacuoles and other subcellular organelles in apple cotyledons. Plant & Cell Physiol. **23**:891-899.

Yamaki S (1984) NADP-dependent sorbitol dehydrogenase found in apple leaves. Plant & Cell Physiol. **25**:1323-1327.

Yamaki S, Moriguchi T (1989) Seasonal fluctuation of sorbitol-related enzyme/s and invertase activities accompanying maturation of Japanese pear (*Pyrus serotina* Rehder var. *culta* Rehder) fruit. Journal of the Japanese Society for Horticultural Science **57**:602-607.

Chapter 3

SALINITY STRESS

YEAST HALOTOLERANCE GENES: CRUCIAL ION TRANSPORT AND METABOLIC REACTIONS IN SALT TOLERANCE

Ramón Serrano
Departamento de Biotecnología
Universidad Politécnica de Valencia
Camino de Vera 14
46022 VALENCIA
SPAIN

The improvement of salt tolerance in crop plants is an urgent challenge for plant biotechnology. The technology for transfering genes to plants is readily available (Potrykus, 1991), but the major problem is the isolation of genes with the capability to improve salt tolerance, which I have called halotolerance genes (Serrano and Gaxiola, 1993). These genes could correspond to either stress-sensitive systems or to tolerance responses involving either ion transport or osmolyte synthesis (figure 1). Salt stress is a very complicated phenomenon, involving both ionic toxicity and osmotic stress. The cytoplasmic loading of Cl^- and Na^+ may be toxic to many cellular systems. On the other hand, water loss from the cells induced by external salt reduces turgor and may also affect cellular systems sensitive to water activity. Any of these forms of the stress trigger a signal transduction pathway which results in some kind of adaptation or tolerance. Both ion transport phenomena and synthesis of organic osmolytes are involved and regulation can occur by either changes in enzyme activy or gene regulation.

NATO ASI Series, Vol. H 86
Biochemical and Cellular Mechanisms
of Stress Tolerance in Plants
Edited by J. H. Cherry
© Springer-Verlag Berlin Heidelberg 1994

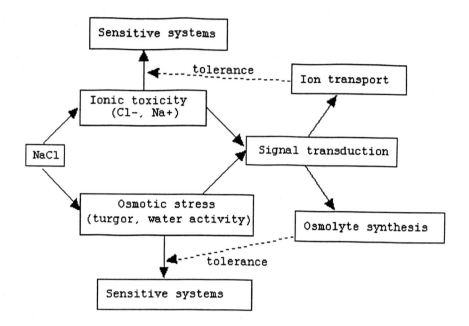

Figure 1. Phenomena involved in salt stress and tolerance.

The identification of the molecular basis for all the above phenomena may provide the clues for the biotechnological manipulation of salt tolerance. An straight-forward approach has been the isolation of salt-induced genes by differential screening (Skriver and Mundy, 1990). These normal components of adaptation could serve to improve salt tolerance by enhancing their expression. However, this approach may identify phenomena only peripheral to salt tolerance because plants are endowed with a general stress response of little specificity (Serrano and Gaxiola, 1993). Another approach is the direct identification of genes involved in known adaptation phenomena, such as ion transport at both the vacuole and plasma membrane and synthesis of organic osmolytes. In this case, although the relevance of these phenomena seems established by physiological studies, it is not clear what are the rate-limiting steps (Kacser and Burns, 1981) in salt tolerance. Only genes corresponding to such bottle-necks would be useful for improvement of salt tolerance. Finally, signal transduction events and cellular systems sensitive to the different forms of salt stres may be relevant to biotechnological manipulation, but much less is known about their identity.

Most of the current research on salt tolerance can be considered as "phenomenological" because it concentrates on phenomena occurring during salt adaptation (gene induction, ion transport, osmolyte synthesis). We have advocated (Serrano and Gaxiola, 1993) a novel "functional" approach based on the direct identification of genes by their capability to improve salt tolerance under overexpression. We have first implemented this approach to the model microorganism *Saccharomyces cerevisiae* (Gaxiola et al., 1992; Gläser et al., 1993) and our results identify potassium homeostasis and methionine biosynthesis as crucial reactions in salt tolerance.

Methodology

The yeast *Saccharomyces cerevisiae* serves as a convenient model system to isolate halotolerance genes and to uncover rate-limiting steps for growth under salt stress. The availability of multicopy plasmid vectors allows the selection of specific genes under conditions where their overexpression improves growth (Rine et al., 1983). Yeast cells were transformed with a genomic library in multicopy plasmid YEp24 (Carlson and Botstein, 1982). Transformed colonies were selected by the nutritional *URA3* marker in the plasmid and afterwards were tested for salt tolerance. Attempts to directly select for salt tolerance were not succesful because the yeast cells are treated with polyethylenglycol for plasmid uptake (Ito et al., 1983) and this results in adaptation to osmotic stress. Plasmid was extracted from colonies exhibiting improved

374

salt tolerance and the responsible genes characterized.
By following this approach we have isolated three
halotolerance genes (*HAL1*, *HAL2* and *HAL3*) which identify
different rate-limiting steps in salt tolerance.

The importance of potassium homeostasis in specific salt tolerance

Overexpression and underexpression of the yeast HAL1
gene specifically affect growth in media with NaCl and
is without effect in media with either sorbitol or KCl
(figure 2). In yeast NaCl is much more toxic than
equivalent concentrations of KCl and sorbitol.

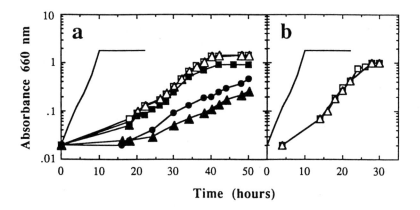

Figure 2. Growth inhibition by 1 M NaCl (closed
symbols), 1.5 M sorbitol (open symbols, part a) and 1 M
KCl (open symbols, part b). Wild type yeast: circles;
overexpression of *HAL1* in multicopy plasmid: squares;
gene disruption of *HAL1*: triangles. The continuous line
reflects growth in normal medium which is identical in
all strains. Reproduced from Gaxiola et al., 1992, with
permission.

These results indicate that *HAL1* is specifically involved in NaCl tolerance and not in non-specific osmotic stress. The sequence of the *HAL1* gene provides no clues about its mechanism: it encodes a cytoplasmic protein of 32 kDa without homologies in data banks (Gaxiola et al., 1992). Measurements of intracellular ion concentrations indicate that the *HAL1* protein participates in potassium homeostasis (figure 3).

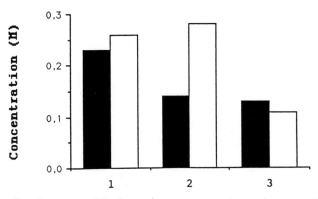

Figure 3. Intracellular ion concentrations of wild type yeast (black bars) and of yeast cells overexpressing the *HAL1* gene (white bars). 1: K^+ levels of cells growing in normal medium (7 mM K^+ and 2 mM Na^+). 2: K^+ levels of cells growing in medium with 7 mM K^+ and 1 M Na^+. 3: $Na+$ levels of cells growing in medium with 7 mM K^+ and 1 M Na^+.

In high-NaCl medium cells overexpressing *HAL1* have about twice the K^+ level than controls. In normal medium the effect is much less probably because the expression of the *HAL1* genes is induced by NaCl (Gaxiola et al., 1992). We have recently isolated another gene, *HAL3*, which is expressed constitutively and which also increases the K^+ level. In this case the effect is also observed in normal growth medium (A. Ferrando and R. Serrano, unpublished). A small decrease in Na^+ is also observed with overexpression of these two genes but it seems that the increase in K^+ is the primary effect.

The mechanisms which regulate intracellular potassium concentrations in fungal and plant cells remain largely unknown (Serrano, 1985). In addition to a balance between uptake and efflux, one important factor is a feed-back inhibition of uptake by intracellular K^+ (Glass, 1983). The molecular basis of K^+ transport are beginning to be elucidated. In yeast two genes, TRK1 and TRK2, have been identified by complementing K^+-uptake mutants and show no homology to animal K^+ channels (Gaber, 1992). Surprisingly, the same yeast mutants were complemented by two Arabidopsis thaliana cDNAs encoding prteins with homology to animal K^+ channels (Anderson et al., 1992; Sentenac et al., 1992). The physiological role of these plant genes, however, remains to be established by mutational analysis.

The mechanism by which overexpression of HAL1 and HAL3 increase intracellular potassium levels can only be especulated. It seems that sodium toxicity can be counteracted by increased potassium concentrations (Greenway and Munns, 1980). Interestingly, salt tolerance correlates with enhanced potassium transport capability in different experimental systems (Rush and Epstein, 1981; Hajibagheri et al., 1989; Muralitharan et al., 1990; Watad et al., 1991). Our working hypothesis is that the proteins encoded by the HAL1 and HAL3 genes interact with the potassium homeostatic system and determine a different "set-point" for the intracellular potassium level. This regulatory system may include the transporters themselves, regulatory factors such as protein kinases, second messengers, potassium sensors, turgor sensors etc.

Methionine biosynthesis as a sensitive step in salt toxicity

Another gene which by overexpression improves salt tolerance in yeast, HAL2, is identical to the methionine biosynthetic gene MET22 (Gläser et al., 1993). This suggests that methionine biosynthesis is a salt-sensitive metabolic pathway. Accordingly, methionine supplementation improves salt tolerance (figure 4). As observed with *HAL1* and *HAL3*, the effect of *HAL2* overexpression and of methionine supplementation is specific for Na$^+$ and Li$^+$ toxicities, no improvement being observed with the less inhibitory solutes KCl and sorbitol.

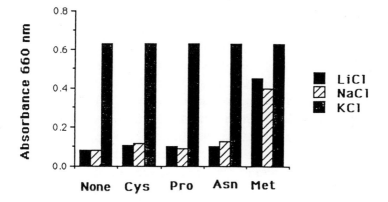

<u>Figure 4</u>. Methionine supplementation improves salt tolerance. Yeast cells were grown in medium supplemented with the indicated amino acids and either 0.2 M LiCl, 1 M NaCl or 1 M KCl as indicated. From Gläser et al., 1993, with permission.

The *HAL2* (*MET22*) gene is required for the assimilation of sulfate into homocysteine (Thomas et al., 1992). It does correspond, however, neither to any of the enzymes involved nor to a regulatory factor controlling the expression of these enzymes (Thomas et al., 1992). The *HAL2* protein has homology to animal inositol phosphatases and to bacterial and fungal regulatory proteins which probably have phosphatase activity (Neuwald et al., 1991; Bone et al., 1992). One of these proteins, encoded by the *cysQ* gene, is required for sulfate assimilation in *Escherichia coli* and it has been proposed to hydrolyze the toxic intermediate 3'-phosphoadenosine-5'-phosphosulfate (Neuwald et al., 1992). The yeast homolog could perform a similar role. Animal inositol phosphatases are sensitive to lithium salts and our working hypothesis is that the product of the *HAL2* (*MET22*) gene is a salt-sensitive phosphatase needed for sulfate assimilation.

Yeast as model system for salt tolerance in plants

The above results illustrate the utilization of the yeast *Saccharomyces cerevisiae* in studies of salt tolerance. Both an important defense reaction (potassium transport regulated by *HAL1* and *HAL3*) and a sensitive metabolic pathway (sulfate assimilation regulated by *HAL2*) have been identified. The relevance to plants of these findings is that: a) plant genes with homology to the yeast halotolerance genes can be identified by Southern analysis (Gaxiola et al., 1992); b) potassium transport has also been correlated with salt tolerance in plants (see above); c) methionine supplementation also improves salt tolerance in plants (U. Gläser and R. Serrano, unpublished).

There are, however, important differences between the yeast and plant systems. Yeast cells are much more sensitive to Na^+ and Li^+ than to non-specific osmotic stress. The situation in plants is not so clear (Serrano and Gaxiola, 1992). It has recently been reported that transgenic tobacco plants expressing bacterial mannitol-1-phosphate dehydrogenase accumulate mannitol and exhibit improved salt tolerance (Tarczynski et al., 1993). This result has two interpretation: a) osmotic stress is the major component of salt toxicity (figure 1); b) mannitol can protect cellular proteins affected by ionic toxicity. Another difference between yeast and plants is that in the last organisms chloride is as toxic as sodium while in yeast sodium is clearly the toxic component of salt.

Despite these reservations, I believe that the yeast system may provide novel genes relevant to salt tolerance in plants.

Acknowledgements. I want to thak the support of both the European Molecular Biology Laboratory, Heidelberg, Germany, and the spanish CICYT for financial support. The work from my laboratory discussed in this paper was made possible by the effort of R. Gaxiola, U. Gläser, I. F. de Larrinoa, F. Montrichard and A. Ferrando.

References
Anderson, J.A., Huprikar, S.S., Kochian, L.V., Lucas,W.J., Gaber, R.F. (1992) Functional expression of a probable *Arabidopsis thaliana* potassium channel in *Saccharomyces cerevisiae*. Proc. Natl. Acad. Sci. USA 89, 3736-3740.
Bone, R., Springer, J.P., Atack, J.R. (1992) Structure of inositol monophosphatase, the putative target of lithium therapy. Proc. Natl. Acad. Sci. USA 89, 10031-10035.
Carlson, M., Botstein, D. (1982) Two differentially regulated mRNAs with different 5' ends encode secreted and intracellular forms of yeast invertase. Cell 28, 145-154.
Gaber, R.F. (1992) Molecular genetics of yeast ion transport. Int. Rev. Cytol. 137A, 299-353.
Glass, A.D. (1983) Regulation of ion transport. Annu. Rev. Plant Physiol. 34, 311-326.
Greenway, H. Munns, R. (1980) Mechanisms of salt tolerance in nonhalophytes. Annu. Rev. Plant Physiol. 31, 149-190.
Hajibagheri, M.A., Yeo, A.R., Flowers, T.J., Collins, J.C. (1989) Salinity resisyance in *Zea mays*: fluxes of potassium, sodium and chloride, cytoplasmic concentrations and microsomal membrane lipids. Plant Cell Environ. 12, 753-757.
Ito, H., Fukuda, Y., Murata, K., Kimura, A. (1983) Transformation of intact yeast cells with alkali cations. J. Bacteriol. 153, 163-168.
Muralitharan, M.S., Van Steveninck, R.F.M., Chandler, S.F. (1990) Growth characteristics and ion contents of non-selected and salt selected callus lines of highbush blueberry (*Vaccinium corymbosum*) cultivars Blue Crop and Denise Blue. Plant Cell Rep. 9, 151-155.

Neuwald, A.F., York, J.D., Majerus, P.W. (1991) Diverse proteins homologous to inositol monophosphatase. FEBS Lett. 294, 16-18.

Neuwald, A.F., Krishnan, B.R., Brikun, I., Kulakauskas, S., Suziedelis, N., Tomcsanyi, T., Leyh, T.S., Berg, D.E. (1992) *cysQ*, a gene needed for cysteine synthesis in *Escherichia coli* K-12 only during aerobic growth. J. Bacteriol. 174, 415-425.

Potrykus, I. (1991) Gene transfer to plants: assessment of published approaches and results. Annu. Rev. Plant Physiol. Plant Mol. Biol. 42, 205-225.

Rine, J., Hansen, W., Hardeman, E., Davis, R.W. (1983) Targeted selection of recombinant clones through gene dosage effects. Proc. Natl. Acad. Sci. USA 80, 6750-6754.

Rush, D.W., Epstein, E. (1981) Comparative studies on the sodium, potassium, and chloride relations of a wild halophytic and domestic salt-sensitive tomato species. Plant Physiol. 68, 1308-1313.

Sentenac, H., Bonneaud, N., Minet, M., Lacroute, F., Salmon, J.M., Gaymard, F., Grignon, C. (1992) Cloning and expression in yeast of a plant potassium ion transport system. Science 256, 663-665.

Serrano, R. (1985) Plasma Membrane ATPase of Plants and Fungi. CRC Press, Boca Raton, Florida.

Serrano, R., Gaxiola, R. (1993) Microbial models and salt stress tolerance in plants. CRC Crit. Rev. Plant Sci. in press.

Skriver, K., Mundy, J. (1990) Gene expression in response to abscisic acid and osmotic stress. Plant Cell 2, 503-512.

Thomas, D., Barbey, R., Henry, D., Surdin-Kerjan, Y. (1992) J. Gen. Microbiol. 138, 2021-2028.

Watad, A.E.A., Reuveni, M., Bressan, R.A., Hasegawa, P.M. (1991) Enhanced net K^+ uptake capacity of NaCl-adapted cells. Plant Physiol. 95, 1265-1269.

STRUCTURE, REGULATION AND FUNCTION OF THE OSMOTIN GENE

Andrzej K. Kononowicz, Kashchandra G. Raghothama, Ana M. Casas,
Donald E. Nelson, Dong Liu, Meena L. Narasimhan, P. Christopher
LaRosa, Narendra K. Singh, Ray A. Bressan, Paul M. Hasegawa

Center for Plant Environmental
Stress Physiology
Department of Horticulture
Purdue University
West Lafayette
IN 47907-1165, USA

INTRODUCTION

Over the past years several genes have been reported to be
osmotically regulated (Storey and Storey, 1981; Holtum and
Winter, 1982; Singh et al., 1985; Singh et al., 1987a; Bedford et
al., 1987; Close et al., 1989; Cushman et al., 1989; Singh et
al., 1989a; 1989b; Delauney and Verma, 1990; Perez-Prat et al.,
1990; Skriver and Mundy, 1990; Bartels et al., 1991; Dhindsa,
1991, Estragarcia et al., 1991; Narasimhan et al., 1991;
Perez-Prat et al., 1992; Kononowicz et al., 1992; Nelson et al.,
1992; Niu et al., 1993; Zhu et al., 1993b). These studies have
been rationalized on the assumption that amongst these genes are
molecular determinants of osmotic tolerance. Although the
products of many of these genes still remain unidentified there
are a number that have been well characterized and are of
interest of several laboratories. One of these genes is osmotin
(Singh et al., 1987a; LaRosa et al., 1989, Meeks-Wagner et al.,
1989; Grosset et al., 1990; Roberts and Selitrennikoff, 1990;
Stintzi et al., 1991; Woloshuk et al., 1991; Casas et al., 1992;
LaRosa et al., 1992, Kononowicz et al., 1993).

Department of Plant Cytology and Cytochemistry, University of
Lodz, Banacha 12/16, 90-237 Lodz, Poland (A.K.K.)
Department of Biochemistry, University of Arizona, Biological
Sciences, West Tucson, AZ 85721-0001 (D.E.N.)
USDA, Agricultural Research Service, Beltsville, MA 20705
(P.C.L.)
Department of Botany and Microbiology, Auburn University, Auburn,
AL 36849-5407 (N.K.S.)

NATO ASI Series, Vol. H 86
Biochemical and Cellular Mechanisms
of Stress Tolerance in Plants
Edited by J. H. Cherry
© Springer-Verlag Berlin Heidelberg 1994

Osmotin was originally identified by SDS-PAGE as the most abundant polypeptide present in tobacco (Nicotiana tabacum L. cv. Wisconsin 38) cells after in vitro adaptation to NaCl (Singh et al., 1985). Interestingly, this polypeptide accumulated in both NaCl and PEG adapted cells but was not prevalent in cells under osmotic shock. The accumulation of the protein was found to be correlated with osmotic adaptation (up to 12% of total protein) hence the name - osmotin. In tobacco, there are at least 3 isoforms of osmotin all of which are cationic (LaRosa et al., 1992).

The open reading frame of tobacco osmotin cDNA encodes a deduced polypeptide of Mr 26.38 kD containing a 2.5 kD signal sequence (Singh et al., 1989a). The Mr of the mature osmotin was found to be 23.88 kD. The amino acid composition, a portion of which was determined from purified protein, was deduced by analysis of the nucleotide sequence of the isolated cDNA clone (Singh et al., 1989a). The leader sequence has a N-terminal methionine and two positively charged amino acids at position 3 and 5 (Appendix). The remainder of the amino acids in the leader sequence are hydrophobic (Singh et al., 1989a). In tobacco cells, osmotin has been localized to vacuolar inclusion bodies. Furthermore, it was found to be associated with the plasma membrane and the tonoplast and extracellular matrix fractions (Singh et al., 1987a).

Two osmotin-like proteins of homologs of Mr 50 and 24 kDhave been identified in cell cultures of the euhalophyte Atriplex nummularia (Casas et al., 1992); a peptide of Mr 50 kD detcted in protein extracts of cells and the other of Mr 24 kD identified in culture media after cell growth. Since the two unique cDNA clones, pA8 and pA9, encode polypeptides of 23.808 kD and 23.827 kD respectively, it is possible that the 50 kD protein represents a dimer while the 24 kD protein is a secreted form that is the result of processing of the dimer before release or it is a protected from dimerization form transportedto the extracellular matrix.

Comparative analysis of the primary structure of tobacco osmotin with several proteins, including the sweet protein thaumatin [Edens et al., 1982], tobacco minor form of PR-R protein [Cornelissen et al., 1986], maize -amylase/trypsin inhibitor [Richardson et al., 1987], and a tomato salt induced NP-24 protein [King et al., 1988], reveales several interesting conserved features of these proteins that this protein shares very high level of sequence homology with them (Appendix). The site of cleavage of the leader sequence at the N-terminal alanine and the sixteen cysteine residues involved in disulfide-bridges of thaumatin (De Vos et al., 1985) were found to be perfectly conserved among most of the osmotin-like proteins reported thus far (Appendix).the five proteins as well as several other osmotin-like proteins not listed here. On the basis of sequence similarity and characteristics of inducibility and expression pattern of the gene, osmotin has been classified as a pathogenesis-related (PR) type 5 protein of tobacco (Cutt and Klessig, 1992).

Selected results from our laboratories on studies of osmotin gene expression and osmotin protein function will be presented here:
(A) biochemical and histochemical analyses of osmotin gene expression regulation in tobacco in response to developmental cues and during adaptation to moderate (171 mM) NaCl stress; (B) molecular and histochemical analyses of osmotin promoter regulation by selected abiotic inducers performed on transgenic tobacco plants containing different chimeric gene constructs consisting of osmotin promoter::ß-glucuronidase reporter gene (OSM::GUS) fusions; and (C) function analyses of osmotin using CaMV 35S promoter::osmotin constructs that resulted in overproduction of the protein in transgenic tobacco and potato plants.

DEVELOPMENTAL REGULATION OF THE OSMOTIN PROMOTER

Histochemical localization of GUS activity in transgenic tobacco plants carrying a chimeric gene construct, consisting of the osmotin promoter fused upstream of the translation initiation codon of a ß-glucuronidase gene, allowed us to determine a detailed pattern of temporal and spatial regulation of the osmotin promoter both during normal plant development and after adaptation to a moderate concentration of NaCl (Kononowicz et al., 1992). A 1.8 kb fragment (-1595 to +45) of the OSM promoter was fused to a GUS reporter gene and introduced into N. tabacum L. cv. W38 by Agrobacterium-mediated leaf disc transformation. Regenerated plants were examined for the number and structure of integrated genes. R1 plants and seeds obtained from three independent transformants (GUS1A, GUS1B and GUS1C) were used for experiments (Kononowicz et al., 1992; Nelson et al., 1992).

Osmotin promoter activity in vegetative organs of tobacco

Little expression of the osmotin promoter was detected in tissues of most plant organs of tobacco plants grown hydroponically. In vegetative organs the exceptions were the epidermis the root elongation zone and trichomes and epidermis in stem and fully expanded leaves. No GUS activity was detected either in meristems or parenchyma or vascular tissues of any vegetative organs unless plants were grown in either soil or vermiculite. In these plants, presumable due to undetectable and thus difficult to avoid osmotic fluctuation,, GUS activity was localized to most tissues of the root, stem and mature leaves, with the highest level in xylem parenchyma. The only exceptions were young, expanding leaves (Kononowicz et al., 1992). This observation was verified and entirely confirmed by Northern biochemicalanalyses of osmotin message and SDS-PAGE of protein accumulation (LaRosa et al., 1992).

Osmotin promoter activity in flower organs and tissues

Important information concerning the regulation of osmotin promoter was provided by the analysis of OSM::GUS gene expression during formation and development of flowers and fruits in transgenic tobacco plants (Kononowicz et al., 1992). In both hydroponically and soil grown plants very high GUS activity was detected in mature pollen grains during anther dehiscence, in corolla tissues at the onset of flower senescence and in pericarp tissue (Fig. 1A) at the final stages of fruit maturation. All of these tissues/organs, at the above indicated stages of development, are under severe (for two of them - terminal) osmotic stress imposed by desiccation. While in pollen grains and fruits, up-regulation of osmotin promoter activity appeared to be strictly correlated with changes in water potential and/or other physiological changes accompanying desiccation (e.g. increase in endogenous ABA concentration), induction of osmotin promoter in corolla tissues seems to be closely related to flower senescence (K. G. Raghothama, unpublished data).

In flowers, no detectable GUS activity was detected until the final stages of development. At anthesis, relatively high OSM::GUS gene expression was detected in tissues of the top regions of corolla and calyx and in vascular bundles of the basal part of gynoecium. Induction of osmotin promoter and increase in GUS activity apparently coincided with the onset of flower senescence and accompanied the increase in endogenous ethylene level (Kononowicz et al., 1992; K. G. Raghothama, unpublished data).

Unlike several anther and pollen specific genes reported by other investigators (McCormick et al., 1987; Smith et al., 1987; Goldberg, 1988; Mascarenhas, 1988; Drews and Goldberg, 1989; Hanson et al., 1989; Koltunow et al., 1990; Twell et al., 1990, 1991), osmotin promoter activity was not detected either in sporophytic tissues of the anthers, at any developmental stage, or in immature pollen, including during microsporogenesis and pollen maturation (Kononowicz et al., 1992). Sudden induction of the osmotin promoter in pollen occurred during anther dehiscence and pollen release. High GUS activity in mature pollen grains, that decreased rapidly soon after pollen hydration and during early stages of pollen germination and pollen tube growth, suggests suppression of osmotin promoter activity occurs at this time.

Osmotin promoter activity in fruits, seeds and seedlings

In the fruit, changes in GUS activity that reflect alterations in osmotin promoter activity, followed the spatial and temporal pattern of osmotic stress exerted on fruit tissues

and created by desiccation. Barely detectable GUS activity
appeared first in the pericarp tissues of the upper portion of
the fruit and then "moved" down toward the stalk increasing in
intensity and gradually comprising the entire fruit (Fig.1A)
(Kononowicz et al., 1992).

Surprisingly, during seed development (including
megasporogenesis, fertilization, embryogenesis, seed maturation
and desiccation) no osmotin promoter activity was remains silent
and no GUS activity was detected either in embryo or seed coat
tissues (Fig. 1A). (Kononowicz et al., 1992; LaRosa et al., 1992;
A. K. Kononowicz, unpublished data). During seed germination (a
resumption of embryo growth after imbibition) and seedling
development, the OSM::GUS gene was expressed at very low level;
GUS activity was detected only in the root/hypocotyl transition
region and in seed coat tissue at the micropyle (Fig. 2) (A. K.
Kononowicz, unpublished data). Since the osmotin gene appears to
be wound responsive (Neale et al., 1990; Nelson et al., 1992),
the expression of the chimeric OSM::GUS gene described above
seems to be due to a wounding effect that occurs during seed coat
splitting and radicle emergence.

**MODIFICATION OF OSMOTIN PROMOTER ACTIVITY IN PLANTA DURING NaCl
ADAPTATION**

Transgenic OSM::GUS tobacco plants were also utilized to
verify results of our earlier experiments that indicated
transcriptional activation of the osmotin promoter occurs during
adaptation of tobacco cells to osmotic stress evoked by NaCl or
PEG (Singh et al., 1989b). Histological localization of GUS
activity allowed us to follow changes in osmotin promoter
activity during adaptation to NaCl in a developmental and organ-
and tissue-specific context (Kononowicz et al., 1992, 1993).

The root is the plant organ that is first exposed to water
deficit and ion toxicityosmotic stress imposed by high NaCl. In
roots of hydroponically grown OSM::GUS transgenic tobacco plants,
activation of the osmotin promoter was detected a few days after
the NaCl treatment was initiated. Three distinct phases of
osmotin promoter activation in root tissues could be

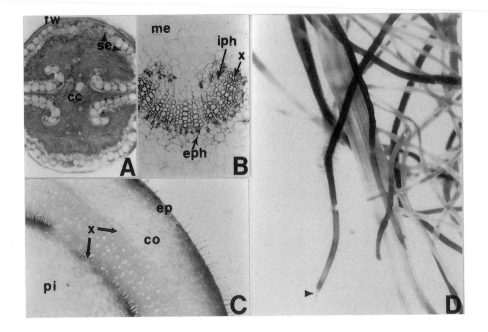

Figure 1. Tissue- and organ-specific localization of OSM promoter activity in transgenic tobacco plants. (A) Cross-section through mature OSM::GUS tobacco fruit. At the final stages of fruit maturation very high GUS activity was observed in fruit wall (fw), in the pericarp tissue (pc) and central column (cc). No GUS activity was detected in seeds (ss), at any stage of seed development. (B) Cross-section illustrating OSM promoter activity in the petiole of OSM::GUS tobacco leaf. High GUS activity was detected in the xylem parenchyma cells of the vascular bundle. Relatively high GUS activity was observed also in epidermis and mesophyll both in the leaf blade (not shown) and in the petiole of mature leaves. (C) OSM promoter activity in stem tissues. In stem, GUS expression was detected in epidermis and in the inner portion of xylem parenchyma. (D) Pattern of GUS activity in the root of the OSM::GUS tobacco plants adapted to 171 mM NaCl. After prolonged (3 months) NaCl treatment very high GUS expression was detected in root elongation zone and root hair zone. Induction of OSM::GUS gene did not occur in the root meristem (arrowhead), whereas in differentiated zone, OSM promoter activity was low and limited to the vascular cylinder.

demonstrated. Initially OSM::GUS gene expression was located only in the epidermis of the root elongation zone. Two weeks later, extremely high GUS activity was detected in the region of the elongation zone, next to the root tip meristem, including epidermis, cortex parenchyma and the differentiating vascular cylinder. After prolonged NaCl treatment, the region of high GUS activity had enlarged to comprise a several-millimeter-long fragment of root elongation zone (Fig. 1C). Induction of the osmotin promoter did not occur in the root meristem, even after prolonged (3 months) NaCl treatment (Kononowicz et al., 1992). Two other NaCl up-regulated genes studied in our laboratories, peroxidase and plasma membrane H+ ATPase, exhibited a very similar pattern of expression to osmotin in roots of Atriplex and tobacco plants, respectively (M. Botella, unpublished data; X. Niu, unpublished data).

In stems, activity of the osmotin promoter induced by adaptation to NaCl was limited to the epidermis and the xylem parenchyma. Detectable GUS activity was only occasionally observed in cortex parenchyma (Fig. 1B) (Kononowicz et al., 1992). Young, developing leaves, did not show a significant activity of the osmotin promoter regardless of the stage of salt adaptation. However, in fully expanded leaves, the level of promoter activity was related to the salt adaptation stage and an increase in GUS activity was detected in the epidermis, mesophyll and xylem parenchyma, especially in the petiole (Kononowicz et al., 1992). No changes in osmotin promoter activity were found either in developing flowers or fruits. One can suggest that defense mechanisms activated by NaCl and operating in the plant effectively protect these organs from stress imposed by high salt (Kononowicz et al., 1992).

Interesting results on NaCl inducibility of osmotin promoter were provided by experiments utilizing R1 seedlings of transgenic OSM::GUS plants (Nelson et al., 1992; A. K. Kononowicz, unpublished data; K. G. Raghothama, unpublished data). In OSM::GUS seedlings, 24 h NaCl treatments induced GUS activity. However, NaCl inducibility was tissue-specific and dependent on the growth stage of the seedlings. Maximum response of the osmotin promoter to NaCl shock was detected in 4- and 5-day-old seedlings and decreased gradually with the age of seedlings (Fig. 2). In seedlings at allgrowth and development stages (up to 14 day after germination), the greatest increase in the level of GUS activity was detected in the root elongation zone but not in root tip meristem, resembling the pattern observed in roots of mature tobacco plants adapted to NaCl (Fig. 2). A very low level and no significant changes inof osmotin promoter activity was found in primary leaves of developing seedlings treated with NaCl. These data confirmed the tissue-specific inducibility of osmotin promoter by NaCl.

STRUCTURAL ANALYSIS OF THE OSMOTIN PROMOTER

Regulation of osmotin and osmotin-like gene expression has been extensively studied in ours and in numerous other laboratories. In tobacco, expression of this gene is regulated by, besides salinity, several other environmental signals, including tobacco mosaic virus infection, water deficit, UV light, wounding and fungal infection (LaRosa et al., 1987; Singh et al., 1987b; Neale et al., 1990; Brederode et al., 1991; Stintzi et al., 1991; Nelson et al., 1992). Biochemical studies utilizing tobacco cell suspensions, tobacco plants and seedlings indicated that response of the osmotin promoter to environmental signals may be mediated by changes in the endogenous concentration of ABA and ethylene (LaRosa et al., 1987; Singh et al., 1987b; Nelson et al., 1992). This suggests the presence of specific, cis-acting elements in the osmotin promoter that interact with various developmentally, hormonally or environmentally activated trans-acting factors. To ascertain this possibility molecular and histochemical approaches have been utilized to analyze the signal responsive characteristics of the upstream region of osmotin gene (K G. Raghothama, unpublished data; A. K. Kononowicz, unpublished data).

Nuclear factors that interact with the osmotin promoter DNA

Gel retardation assays demonstrated two promoter fragments that interacted with nuclear proteins that are abundant in NaCl-adapted tobacco cells and ABA and ethylene treated tobacco leaves. One of these regions was localized near the TATA-box (-108 to +45) (Fig. 3) and another upstream from TATA-box (Cla I - Cla I fragment, -762 to -642). The latter one was chosen for further analyses. DNAase I footprinting assays revealed that

CONTROL

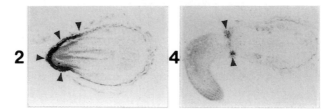

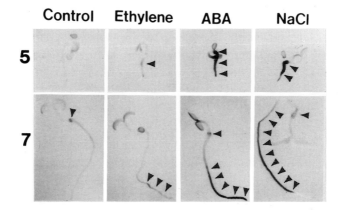

Figure 2. Expression of OSM::GUS chimeric gene in GUS1 tobacco seedlings. GUS1B seeds were germinated under sterile conditions and samples of seedlings were treated for 24 h with 5 ppm ethylene, 100 μM ABA or 171 mM NaCl on day 5 and 7. Non-treated seedlings were used as a control. In addition control 2 and 4-day-old seedlings are shown in the top panel of the figure. Histochemical GUS assays were performed according to Kononowicz et al. (1992). Regions of GUS activity are pointed out by arrowheads. On day 2, high GUS expression was detected in seed coat tissues at the micropyle. No differences in distribution and intensity of GUS activity were detected either in treated or nontreated seeds. On day 4, GUS activity was observed at the root/hypocotyl transition zone and surrounding seed coat tissue. On days 5 and 7, each of the treatments produced specific tissue/organ-specific patterns of GUS expression. In control seedlings GUS expression was detected only in the root/hypocotyl transition zone and barely detectable GUS activity was observed in the epidermis of the root elongation zone. No GUS activity was found in primary leaves. Ethylene slightly stimulated GUS activity in epidermis of the root elongation zone as compared to control seedlings. Furthermore, ethylene induced OSM::GUS gene expression in the primary leaf tips. Both, ABA and NaCl highly induced GUS activity in the root tissues. ABA treatment induced the OSM::GUS gene in the root tip meristem and elongation zone. In response to NaCl treatment the highest increase in the GUS activity was detected in the root elongation zone but not in the root tip meristem. Furthermore, very low GUS activity was found also in primary leaves of developing seedlings.

nuclear factors protect a 35 bp fragment interacting with promoter region from -757 to -723 bp (Fig. 3). This was further confirmed in retardation studies by using two complementary oligonucleotides containing the 35 bp region (Kononowicz et al., 1993; K. G. Raghothama et al., in press).

ANALYSIS OF OSMOTIN PROMOTER DELETION FRAGMENTS BY TRANSIENT AND STABLE TRANSFORMATION

Transient expression

Several deletions were made from 5'-end of the osmotin promoter and the resulting fragments were fused upstream of the translation initiation codon of a reporter GUS gene (Fig. 3). End-points of ten selected deletions corresponded to positions:

Deletion number	5' end nucleotide	Deletion number	5' end nucleotide
5:	-1382	16:	-496
8:	-1196	18:	-336
10:	-1052	22:	-248
11:	-842	23:	-108
13:	-653	24:	-49

Analyses of a series of 5'-end deletions of the osmotin promoter by transient expression in tobacco leaves after microprojectile bombardment demonstrated that the minimum length of the promoter required to transcribe the reporter gene was -248 to +45 (Fig. 4). This promoter region appeared to be the region required for NaCl, wounding, desiccation, ABA and ethylene induced activation of the reporter gene. The expression of GUS reached the maximum with promoter fragment of -1052 to +45 (Fig. 4). This promoter region contains the 35 bp sequence that shows specific interaction with nuclear factors. Further increase in the promoter size resulted in decreased GUS activity (Fig. 4) suggesting the presence of negative or silencer-like element(s) in the upstream region of promoter. This general pattern of

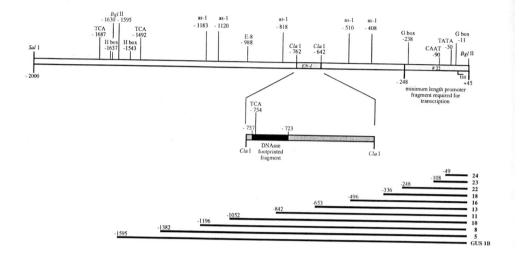

Figure 3. Map of the osmotin promoter showing the location of different cis-regulatory elements, restriction enzyme cleavage sites, and deletion constructs used in promoter structure and regulation studies. Close-up of ClaI-ClaI fragment shows localization of DNAase footprinted fragment. The sequences corresponding to G-box, as-1, E-8, TCA and H-box are centered at the positions indicated in the figure. Positions of the two possible translation start sites (ATG) are marked in the figure as tss.

osmotin promoter activity was observed for all of inducers tested, including NaCl, wounding, desiccation, ABA, and ethylene (Kononowicz et al., 1993; K. G. Raghothama et al., in press; Y. Xu, unpublished data).

Transgenic tobacco seedlings

All of the deletion gene constructs were stably integrated into the genome of tobacco plants by Agrobacterium-mediated transformation. Transgenic plants exhibiting normal morphology, high seed fertility and kanamycin resistance (KanR) phenotype were analyzed for the number and structure of the integrated genes.

Further analysis of 5'-deletion fragments of the osmotin promoter performed on 5-day-old transgenic R1 seedlings confirmed

the OSM::GUS gene expression pattern revealed by transient
expression experiments (Fig. 5). Both, the maximum GUS activity
in non-treated seedlings and the maximum response to each of the
inducers reported here (NaCl, ABA and ethylene) have been
demonstrated for promoter fragments with 5' end-points in
positions -1052 and -842 (K. G. Raghothama, unpublished data).
However, histological analysis revealed distinct, tissue-specific
differences in response to each of the inducers tested (A. K.
Kononowicz, unpublished data). After both NaCl and ethylene
treatment, GUS activity was present in root elongation zone and
root/stem transition region. In root tip meristem, the osmotin
promoter did not respond either to 24 h NaCl shock or to ethylene
treatment (Fig. 2). Whereas after ethylene treatment, increase in
GUS activity was detected only in the epidermis, NaCl treatment
induced the osmotin promoter also in cortex and vascular cylinder
tissues. In 4 to 7-day-old seedlings, ABA treatment resulted in a
rapid increase in GUS activity in both tissues of the root
elongation zone and the root tip meristem. While maximum ethylene
and ABA inducibility of the osmotin promoter was observed between
the 4th and 7th day of seedling development, NaCl treatment
induced the osmotin promoter even on day 12. It must be
emphasized that sensitivity of seedlings to all tested inducers
is growth stage dependent indicating strong developmental control
over sensitivity of the osmotin promoter to hormonal inducers.

Transgenic tobacco plants

Transgenic tobacco plants contain the GUS reporter gene
cloned to a series of 5' deletions of the osmotin promoter have
been utilized to further characterize developmental regulation of
osmotin promoter (Kononowicz et al., 1993; A. K. Kononowicz,
unpublished data; K. G. Raghothama, unpublished data).
Quantitative and histochemical analysis of GUS expression in
different plant tissues and organs at different stages of plant
development confirmed the general pattern established earlier on
seedlings. The minimum length of promoter required for reporter
gene transcription was that of -248 to +45 bp. Increase in
promoter fragment size up to -1052 bp resulted in gradual

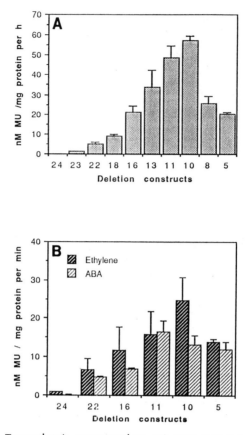

Figure 4. (A) Transient expression of GUS reporter gene fused to
the osmotin promoter. The pBI201 plasmids carrying the osmotin
promoter fragments of different lengths were fused to
GUS-reporter gene and introduced into tobacco leaves utilizing a
Du Pont Biolistic Gun. Plasmids carrying the luciferase gene
driven by CaMV 35S promoter were delivered simultaneously with
OSM::GUS constructs. Based on the luciferase activity, a
multiplication factor was obtained, and this factor was used to
normalize the level of GUS expression (nmol/mg protein/h) shown
in the histogram. The vertical bars represent standard errors.
(B) Effect of ethylene and ABA on transient expression of the
GUS-reporter gene. Plasmids carrying different length fragments
of the osmotin promoter along with CaMV 35S-luciferase chimeric
gene were precipitated onto gold particles and introduced to
tobacco leaves utilizing a Du Pont Biolistic Gun. Following
bombardment the leaves were treated for 24 h in the presence of
100 μM ABA or 5 ppm ethylene. The leaves without any treatment
represented controls. The control values were subtracted and the
net increase in GUS activity in response to ethylene or ABA
treatment is presented in the histogram. The vertical bars
represents standard error values.

increase in GUS activity but after that (deletion fragments of -1382 to +45 and -1595 to +45) activity decreased substantially. The histological pattern of GUS activity location characteristic of OSM::GUS plants did not change Couple sentences on osmotin promoter induction in flowers needed. It should be emphasized, that opposite patterns of inducibility of OSM promoter deletion fragments by ABA and ethylene were observed in the roots and in the corolla of transgenic tobacco flowers. While ABA induced the osmotin promoter in the roots but not in flowers, ethylene strongly induced this promoter in flowers and only slightly in roots (Fig. 6).

The presence of consensus elements in the osmotin promoter

Comparison of the upstream region of the osmotin gene with other ABA regulated genes revealed the presence of sequences with a high degree of identity to G-box elements (Marcotte et al., 1989; Pla et al., 1993) (Fig. 3). One of the G-box like sequences was localized close to the transcription start site and the TATA-box (-14 to -8) and another upstream of TATA-box, between positions -242 and -235. The transient expression studies indicated that the promoter deletion containing the TATA-box and the CAAT-box like sequences is not enough to activate the reporter gene. It appears that the osmotin promoter may contain ABA-responsive elements, which are distinct from G-box like sequences. Examination of the promoter revealed the presence of other sequences with identity to previously reported cis-acting regulatory sequences including five as-1 sequences (Lam et al., 1989) (centered at -1183, -1120, -818, -510 and -408), an E-8 sequence (Cordes et al., 1989) (-988), two H-box sequence motifs (Loake et al., 1992) (centered at -1637 and -1543), and three TCA motifs (Goldsbrough et al., 1993) (centered at -1687, -1492 and -754) (Fig. 3). Several interesting direct and indirect repeats were also noted (Kononowicz et al., 1993; K. G. Raghothama et al., in press).

OSMOTIN CAN FUNCTION IN DEFENSE AGAINST PLANT PATHOGENS

When attacked by pathogens, plants react in many ways that affect their ability to resist infection, colonization and

396

CONTROL   ABA

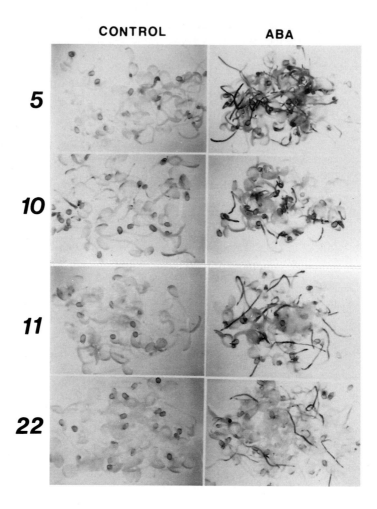

Figure 5. ABA inducibility of the GUS reporter gene driven by OSM
promoter deletion fragments of different sizes. The minimum
length of the promoter fragment that responds to ABA treatment
was -248 to +45. Increase in the promoter size resulted in
increased GUS activity but did not affect the spatial pattern of
GUS (compare with Fig. 2).

ultimate injury. One important and well-studied response of
plants to pathogens is the inductions of PR-proteins. Initially,
the PR-proteins were defined as proteins that are induced by TMV

Ethylene Control ABA

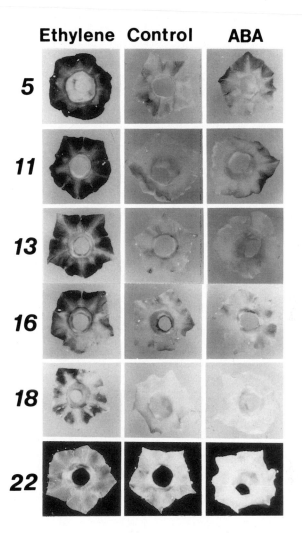

Figure 6. Induction of GUS reporter gene driven by OSM promoter
deletion fragments by ABA and ethylene in the corolla of
transgenic tobacco flowers at anthesis. Mature flowers of
transgenic tobacco plants were treated with 100 μM ABA or 5 ppm
ethylene for 24 h. Histochemical GUS assay was conducted
according to Kononowicz et al. (1992), however incubation time (3
h) was chosen to show high inducibility of osmotin promoter by
ethylene. The minimum size of the promoter to transcribe the
reporter gene was -248 to +45 (# 22). This promoter fragment
appeared to be the shortest one that responds to ethylene
treatment in corolla tissues. Increase in the promoter size
resulted in increased GUS activity and in increased responsivenes
to ethylene. ABA did not affect activity of any of OSM promoter
deletion fragments in the corolla.

398

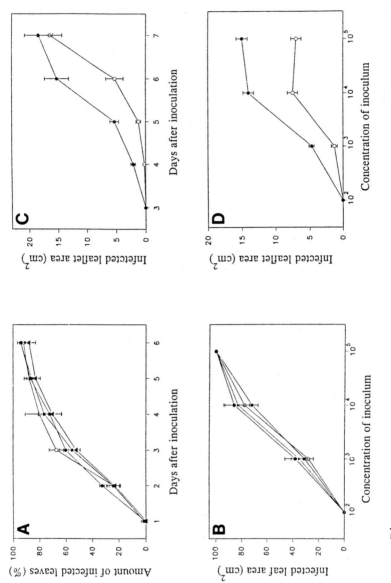

Figure 7.

Figure 7. Disease response of transgenic tobacco (A and B) and potato (C and D) plants to fungal infection of. (A) Plant leaves were inoculated with fungal (P. parasitica) mycelium. The number of infected leaves was determined every day after inoculation and expressed as a percentage of total inoculated leaf number. (B) Plant leaf discs were inoculated with fungal zoospore suspensions and total area of infected leaf tissue was determined at day 3 after inoculation. o: nontransformant; o,   ,   : transformed plant lines, TL11, TL14, and TL19. (C) Potato plant leaflets were inoculated with P. infestans sporangia suspension at a concentration of 105 sporangia/ml and infected leaflet areas were determined every day after inoculation. (D) Potato leaflets were inoculated with sporangia suspensions of different concentrations and infected leaflet areas were determined at a day 6 after inoculation. o: nontransformed plant; o: transformed plant PL11.

infections in tobacco plants. Moreover, it has been found that the expression of the genes encoding these PR-proteins can be induced not only by virus infection but also by other biotic or abiotic factors, such as bacteria, fungi, ethylene, ABA wounding, etc. We have indicated in the past that there is a connection between the inducibility of PR-genes by osmotic stress and their function in defense against pathogens (Kononowicz et al., 1993). We have pointed out that many plant pathogens themselves induce symptoms of osmotic stress and therefore defense genes may have evolved osmotic responsiveness to increase their efficacy against pathogen invasion.

Although a great deal of work has been performed to characterize PR-proteins and their genes, few reports exist that provide definitive evidence that these genes can actually function in plant defense. Broglie et al. (1991) showed that transgenic tobacco plants that over-express the bean chitinase gene exhibited enhanced resistance to the root pathogen Rhizoctonia solani. Recently Alexander et al. (1993) provided evidence that transgenic plants that over-produce PR-1 protein exhibit increased tolerance to two oomycete pathogens, Peronospora tabacina and P. parasitica var. nicotianae. We have recently found that constitutive accumulation of osmotin in transformed plants led to enhanced fungal resistance only in the case of potato and not in tobacco (Fig. 7) (Liu et al., in press). Such experiments indicate that some PR-proteins can play a defensive role during fungal infection when they are over-expressed in a heterologous system. It is interesting to note that there is other evidence that an early or high level of accumulation of PR-protein in transgenic plants of the same species from which the PR-gene was obtained, does not have an obvious anti-pathogen effect. For example, Linthorst et al. (1989) made transgenic tobacco plants that over-expressed the tobacco PR-1, GRP and PR-S genes, and Cutt et al. (1989) produced transgenic tobacco that constitutively produced tobacco PR-1b protein, and these investigators did not find any alteration of resistance to TMV infection in the transgenic plants. In addition Neuhaus et al. produced transgenic plants of tobacco (Nicotiana sylvestris) where tobacco chitinase accumulation was enhanced (Neuhaus et al., 1991), or where the tobacco ß-glucanase was blocked (Neuhaus et al., 1992), and these plants did not have altered susceptibility to the tobacco fungal pathogen Cercospora nicotianae. Taken together, these results reflect a specificity of particular PR-gene products for activity against pathogens that are not pathogenic against the plant species that served as the source of the PR-gene. This could be the result of pathogen/host co-evolution, resulting in PR-genes with diminished effects on co-evolving pathogens as proposed by Lamb et al. (1992). However, it still should be considered that over-expression of one tobacco PR-protein is not sufficient to confer the plant with a high level of resistance, because it must work with other defense factors to cause strong resistance or especially immunity. In addition, the resistance of the transgenic plants has not been tested under the variable conditions in the field with large scale experiments where other environmental factors may confound the genetic differences seen here.

The first evidence that osmotin and other PR family-5 proteins might have antifungal properties emerged from the discovery by Vigers et al. (1991) that the N-terminal sequence of zeamatin, an antifungal protein from corn, was very similar to

osmotin. Also Woloshuk et al. (1991), using a spore germination
bioassay with Phytophtora infestans, identified from tobacco and
tomato an anti-fungal protein that by amino acid sequence
comparison was found to be osmotin. Recently, more osmotin-like
proteins or cDNAs that encode these proteins have been identified
in many plant species, such as tomato (Woloshuk et al., 1991;
King et al., 1988; Rodrigo et al., 1991), potato (Pierpoint et
al., 1990), Atriplex (Casas et al., 1992), Arabidopsis (Uknes et
al., 1992), rice (Reimmann and Dudler, 1993), etc. Some of them
have been demonstrated in in vitro assays to be antifungal
proteins. Our results, using CaMV 35S promoter::osmotin
transgenic potato plants, are consistent with the inhibitory
effects of purified osmotin protein on P. infestans mycelia
growth found by ourselves and inhibition of sporangia germination
by others (Woloshuk et al., 1991). The fact that the
over-expression of osmotin can only delay the development of
disease symptoms indicates that even greater over-expression or a
redirected targeting of over-expressed osmotin from the vacuole
to the extracellular matrix may be needed to further enhance
resistance. Furthermore, as more antifungal gene products are
identified, transgenic experiments using combinations of these
genes will be of great interest.

The mechanism of the inhibitory effect of osmotin on fungal
growth is not completely understood. Woloshuk's experiments
indicate that osmotin can cause sporangia lysis of P. infestans
(Woloshuk et al., 1991). A high concentration of osmotin also can
cause the lysis of hyphae tips (Woloshuk et al., 1991; Vigers et
al., 1991). Moreover, many hyphae ruptured following zeamatin
treatment apparently releasing cytoplasm, and Vigers et al.
(1991) have proposed that osmotin belongs to a large class of
proteins that induce fungal cell permeability and are termed
permatins. Recently we have presented evidence that osmotin can
cause dissipation of the plasma membrane pH gradient. of fungal
species whose growth is inhibited by osmotin. Fungal species
whose growth is not inhibited by osmotin do not exhibit
sensitivity of the plasma membrane pH gradient to osmotin

(Reuveni et al., 1993). Based on these data, and on sequence identity to thaumatin, which has antifungal activity (Vigers et al., 1991) and whose 3-dimensional structure has been determined (de Vos et al., 1985), osmotin may contain structural domains that are involved in membrane receptor binding and membrane pore formation.

SUMMARY AND CONCLUSIONS

The osmotin gene expression is regulated at both transcriptional and post-transcriptional levels.

Osmotin promoter shows unique, organ- and tissue; specific pattern of activity - activity of osmotin promoter is under control of developmental cues.

Distinct expression patterns of osmotin promoter have been demonstrated for desiccated and germinating pollen grains and seeds.

NaCl adaptation modulate osmotin promoter activity in transgenic tobacco.

A specific cis-acting elements of osmotin promoter may be responsive to several environmental and hormonal signals.

Responsive elements of osmotin promoter show both enhancer and suppressor activities.

Osmotin protein has anti-fungal activity against different classes of phytopathogenic fungi.

ACKNOWLEDGMENTS

We are grateful to Jean Clithero and Glenda McClatchey to their excellent technical assistance.

LITERATURE CITED

Alexander D, Goodman RM, Gut-Rella M, Glascock C, Weymann K, Friedrich L, Maddox D, Ahl-Goy P, Luntz T, Ward E, Ryals J (1993) Increased tolerance to two oomycete pathogens in transgenic tobacco expressing pathogenesis-related protein 1a. Proc Natl Acad Sci USA 90: 7327-7331

Bartels D, Engelhardt K, Roncarati R, Schneider K, Rotter M, Salamini F (1991) An ABA and GA modulated gene expressed in the barley embryo encodes an aldose reductase related protein. EMBO J 10: 1037-1043

Bedford J, Bagnasco S, Kador P, Harris W, Burg W (1987) Characterization and purification of a mammalian osmoregulatory protein, aldose reductase, induced in renal medullary cells by high extracellular NaCl. J Biol Chem 222: 14255-14259

Brederode FT, Linthorst HJM, Bol JF (1991) Differential induction of acquired resistance and PR gene expression in tobacco by virus infection, ethephon treatment, UV light and wounding. Plant Mol Biol 17: 1117-1125

Broglie K, Chet I, Holliday M, Cressman R, Biddle P, Knowlton S, Mauvais CJ, Broglie R (1991) Transgenic plants with enhanced resistance to the fungal pathogen Rhizoctonia solani. Science 254: 1194-1197

Casas AM, Nelson DE, Raghothama KG, Paino D'Urzo M, Singh NK, Bressan RA, Hasegawa PM (1992) Expression of osmotin-like genes in the halophyte Atriplex nummularia L. Plant Physiol 99: 329-33

Close TJ, Kortt AA, Chandler PM (1989) A cDNA-based comparison of dehydration-induced proteins (dehydrins) in barley and corn. Plant Mol Biol 13: 95-108

Cordes S, Deikman J, Margossian LJ, Fisher RL (1989) Interaction of a developmentally regulated DNA-binding factor with sites flanking two different fruit-ripening genes from tomato. Plant Cell 1: 1025-1034

Cornelissen BJC, Hooft van Hievjsduijnen RAM, Bol JF (1986) A tobacco mosaic virus-induced tobacco protein is homologous to the sweet-tasting protein thaumatin. Nature 321: 531-532

Cushman JC, Meyer G, Michalowski CB, Schmitt JM, Bohnert HJ (1989) Salt stress leads to differential expression of two isogenes of phosphoenolpyruvate carboxylase during Crassulacean acid metabolism induction in the common ice plant. Plant Cell 1: 715-725

Cutt JR, Harpster MH, Dixon DC, Carr JP, Dunsmuir P, Klessig DF (1989) Disease response to tobacco mosaic virus in transgenic tobacco plants that constitutively express the pathogenesis related PR-1b gene. Virology 173: 89-97

Cutt JR, Klessig DF (1992) Pathogenesis-related proteins. In T Boller, F Meins, eds, Genes Involved in Plant Defense. Springer-Verlag, New York, pp 209-243

Delauney A, Verma D (1990) A soybean 1-pyrroline-5-carboxylate reductase gene was isolated by functional complementation in Escherichia coli and is found to be osmoregulated. Mol Gen Genet 221: 299-305

de Vos AM, Hatada M, Van del Wel H, Krabbendam H, Peerdeman AF, Kim S-H (1985) Three-dimensional structure of thaumatin I, an intensely sweet protein. Proc Natl Acad Sci USA 82: 1406-1409

Dhindsa RS (1991) Drought stress, enzymes of glutathione metabolism, oxidation injury, and protein synthesis in Tortula ruralis. Plant Physiol 95: 648-651

Drews GN, Goldberg RB (1989) Genetic control of flower development. TIG 5: 256-261.

Edelbaum O, Ilan N, Grafi G, Sher N, Stram Y, Novick D, Tal N, Sela I, Rubinstein M (1990) Two antiviral proteins from tobacco: Purification and characterization by monoclonal antibodies to human ß-interferon. Proc Natl Acad Sci USA 87: 588-592

Edens L, Heslinga L, Klok R, Ledeboer AM, Maat J, Toonen MY, Visser C, Verrips CT (1982) Cloning of cDNA encoding the sweet-tasting plant protein thaumatin and its expression in Escherichia coli. Gene 18: 1-12

Estragarcia T, Craxton A, Kirk C, Michell R (1991) A salt-activated inositol 1,3,4,5-tetrakisphosphate 3-phosphatase at the inner surface of the human erythrocyte membrane. Proc Royal Soc London Series B 244: 63-68

Frendo P, Didierjean L, Passelegue E, Burkard G (1992) Abiotic stresses induce a thaumatin-like protein in maize; cDNA isolation and sequence analysis. Plant Sci 85: 61-69

Goldberg RB (1988) Plants: Novel developmental processes. Science 240: 1460-1467

Goldsbrough AP, Albrecht H, Stratford R (1993) Salicylic acid-inducible binding of a tobacco nuclear protein to a 10 bp sequence which is highly conserved amongst stress-inducible genes. Plant J 3: 563-571

Gomez J, Sanchez-Martinez D, Stiefel V, Rigau J, Puigdomenech P, Pages M (1988) A gene induced by the plant hormone abscisic acid in response to water stress encodes a glycine-rich protein. Nature 334: 262-264

Graham JS, Burkhart W, Xiong J, Gillikin JW (1992) Complete amino acid sequence of soybean leaf P21. Plant Physiol 98: 163-165

Grosset J, Meyer Y, Chartier Y, Kauffmann S, Legrand M, Fritig B (1990) Tobacco mesophyll protoplasts synthesize 1,3-ß-glucanase, chitinase and "osmotins" during in vitro culture. Plant Physiol 92: 520-52

Grumet R, Hanson A, (1986) Genetic evidence for an osmoregulatory function of glycinbetaine accumulation in barley. Aust J Plant Physiol 13: 353-364

Hahn M, Lehnackers H, Knogge W (1991) Sequence. H. vulgare mRNA for a pathogenesis-related protein (Hv-1b). EMBL Data Library, unpublished.

Hanson DD, Hamilton DA, Travis JL, Bashe DM, Mascarenhas JP (1989) Characterization of a pollen-specific cDNA clone from Zea mays and its expression. Plant Cell 1: 173-179

Hasegawa PM, Bressan RA, Handa AK (1986) Cellular mechanisms of salinity tolerance. HortScience 21: 1317-1324

Hejgaard J, Jacobsen S, Svendsen I (1991) Two antifungal thaumatin-like proteins from barley grain. FEBS Lett 291: 127-131

Holtum J, Winter K (1982) Activity of enzymes of carbon metabolism during the induction of Crassulacean acid metabolism in Mesembryanthemum crystallinum L. Planta 155: 8-16

Kauffmann S, Legrand M, Fritig B (1990) Isolation and characterization of six pathogenesis-related (PR) proteins of Samsun NN tobacco. Plant Mol Biol 14: 381-390

King GJ, Turner VA, Hussey CE, Wurtele ES, Lee SM (1988) Isolation and characterization of a tomato cDNA cline which codes for a salt-induced protein. Plant Mol Biol 10: 401-412

Koltunow AM, Triettner J, Cox KH, Wallroth M, Goldberg RB (1990) Different temporal and spatial gene expression patterns occur during anther development. Plant Cell 2: 1201-1224

Kononowicz AK, Nelson DE, Singh NK, Hasegawa PM, Bressan RA (1992) Regulation of the osmotin gene promoter. Plant Cell 4: 513-52

Kononowicz AK, Raghothama KG, Casas AM, Reuveni M, Watad A-EA, Liu D, Bressan RA, Hasegawa PM (1993) Osmotin: Regulation of gene expression and function. In TJ Close, EA Bray, eds, Plant Responses to Cellular Dehydration during Environmental Stress, The American Society of Plant Physiologists, Rockville, MD, pp 144-158

Kumar V, Spencer ME (1992) Nucleotide sequence of an osmotin cDNA from the Nicotiana tabacum cv. White Burley generated by the polymerase chain reaction. Plant Mol Biol 18: 621-622

Lam E, Benfey PN, Gilmartin PM, Rong-Xiang F, Chua N-H (1989) Site-specific mutations alter in vitro factor binding and change promoter expression pattern in transgenic plants. Proc Natl Acad Sci USA 86: 7890-7894

Lamb CJ, Ryals JA, Ward ER, Dixon RA (1992) Emerging strategies for enhancing crop resistance to microbial pathogens. Bio/Technology 10: 1436-1445

LaRosa CP, Hasegawa PM, Rhodes D, Clithero JM, Watad A-EA, Bressan RA (1987) Abscisic acid stimulated osmotic adjustment and its involvement in adaptation of tobacco cells to NaCl. Plant Physiol 85: 174-185

LaRosa PC, Singh NK, Hasegawa PM, Bressan RA (1989) Stable NaCl tolerance of tobacco cells is associated with enhanced accumulation of osmotin. Plant Physiol 91: 855-861

LaRosa PC, Chen Z, Nelson DE, Singh NK, Hasegawa PM, Bressan RA (1992) Osmotin gene expression is posttranscriptionally regulated. Plant Physiol 100: 409-415

Linthorst HJM, Meuwissen RLJ, Kauffmann S, Bol JF (1989) Constitutive expression of pathogenesis-related proteins PR-1, GRP, and PR-S in tobacco has no effect on virus infection. Plant Cell 1: 285-291

Liu D, Raghothama KG, Hasegawa PM, Bressan RA (1993) Osmotin overexpression in potato delays development of disease symptoms. Proc Natl Acad Sci USA, in press

Loake GJ, Faktor O, Lamb CJ, Dixon RA (1992) Combination of H-box [CCTACC(N)7CT] and G-box (CACGTG) cis elements is necessary for feed-forward stimulation of a chalcone synthase promoter by the phenylpropanoid-pathway intermediate p-coumaric acid. Proc Natl Acad Sci USA 89: 9230-9234

Marcotte WR Jr, Russel SH, Quatrano RS (1989) Abscisuc acid-responsive sequence from the EM gene of wheat. Plant Cell 1: 969-976

Mascarenhas JP (1988) Anther- and pollen-expressed genes. In DPS Verma, RB Goldberg, eds, Temporal and Spatial Regulation of Plant Genes. Springer-Verlag, New York, pp 97-115

McCormick S, Smith A, Gasser C, Sachs K, Hinchee M, Horsch R, Fraley R (1987) The identification of genes specifically expressed in reproductive organs of tomato. In D Nevins, R Jones, eds, Tomato Biotechnology. Alan Liss, New York, pp 255-265

Meeks-Wagner DR, Dennis ES, Tran Thanh Van K, Peacock WJ (1989) Tobacco genes expressed during in vitro floral initiation and their expression during normal plant development. Plant Cell 1: 25-35.

Narasimhan ML, Binzel ML, Perez-Prat E, Chen Z, Nelson DE, Singh NK, Bressan RA, Hasegawa PM (1991) NaCl regulation of tonoplast ATPase 70-kilodalton subunit mRNA in tobacco cells. Plant Physiol 97: 562-568

Neale AD, Wahleithner JA, Lund M, Bonnett HT, Kelly A, Meeks-Wagner DR, Peacock WJ, Dennis ES (1990) Chitinase, ß-1,3-glucanase, osmotin, and extensin are expressed in tobacco explants during flower formation. Plant Cell 2: 673-684

Nelson DE, Raghothama KG, Singh NK, Hasegawa PM, Bressan RA (1992) Analysis of structure and transcriptional activation of an osmotin gene. Plant Mol Biol 19: 577-588

Neuhaus J-M, Ahl-Goy P, Hinz U, Flores S, Meins F Jr (1991) High-level expression of a tobacco chitinase gene in Nicotiana sylvestris. Susceptibility of transgenic plants to Cercospora nicotianae infection. Plant Mol Biol 16: 141-151

Neuhaus J-M, Flores S, Keefe D, Ahl-Goy P, Meins F Jr (1991) The function of vacuolar ß-1,3-glucanase investigated by antisense transformation. Susceptibility of transgenic Nicotiana sylvestris plants to Cercospora nicotianae infection. Plant Mol Biol 19: 803-813

Niu X, Zhu J-K, Narasimhan ML, Bressan RA, Hasegawa PM (1993) Plasma-membrane H+-ATPase gene expression is regulated by NaCl in cells of the halophyte Atriplex nummularia L. Planta 190: 433-438

Payne G, Middlesteadt W, Williams S, Desai N, Parks TD, Dincher S, Carnes M, Ryals J (1988) Isolation and nucleotide sequence of a novel cDNA clone encoding the major form of pathogenesis-related protein R. Plant Mol Biol 11: 223-224

Perez-Prat E, Narasimhan ML, Binzel ML, Botella MA, Chen Z, Valpuesta V, Bressan RA, Hasegawa PM (1992) Induction of a putative Ca2+-ATPase mRNA in NaCl-adapted cells. Plant Physiol 100: 1471-1478

Pierpoint WS, Tatham AS, Pappin DJC (1987) Identification of the virus-induced protein of tobacco leaves that resembles the sweet-protein thaumatin. Physiol Mol Plant Pathol 31: 291-298

Pierpoint WS, Jackson PJ, Evans RM (1990) The presence of a thaumatin-like protein, a chitinase and glucanase among the pathogenesis-related proteins of potato (Solanum tuberosum). Physiol Mol Plant Pathol 36: 325-338

Pla M, Vilardell J, Guiltinan MJ, Marcotte WR, Niogret MF, Quatrano RS, Pages M (1993) The cis-regulatory element CCACGTGG is involved in ABA and water-stress responses of the maize gene rab28. Plant Mol Biol 21: 259-266

Raghothama KG, Liu D, Nelson DE, Hasegawa PM, Bressan RA (1993) Analysis of an osmotically-regulated pathogenesis-related osmotin gene promoter. Plant Mol Biol, in press

Rebmann G, Mauch F, Dudler R (1991) Sequence of a wheat cDNA encoding a pathogen-induced thaumatin-like protein. Plant Mol Biol 17: 283-285

Reimmann C, Dudler R (1993) cDNA cloning and sequence analysis of a pathogen-induced thaumatin-like protein from rice (Oryza sativa). Plant Physiol 101: 1113-1114

Reuveni M, Liu D, Singh NK, Chen Z, Narasimhan ML, Hasegawa PM, Bressan RA (1993) Osmotin acts to permebilize cells of fungal pathogens. (Abstract No. 633) Plant Physiol 102: S-112

Richardson M, Valdes-Rodriguez S, Blanco-Labia A (1987) A possible function for thaumatin and a TMV- induced protein suggested by homology to a maize inhibitor. Nature 327: 432-434

Roberts WK, Selitrennikoff CP (1990) Zeamatin, an antifungal protein from maize with membrane-permeabilizing activity. J Gen Microbiol 136: 1771-1778

Rodrigo I, Vera P, Frank R, Conejero V (1991) Identification of the viroid-induced tomato pathogenesis-related (PR) protein P23 as a thaumatin-like tomato protein NP24 associated with osmotic stress. Plant Mol Biol 16: 931-934

Rodrigo I, Vera P, Tornero P, Hernandez-Yago J, Conejero V (1993) cDNA cloning of viroid-induced tomato pathogenesis-related protein P23. Characterization as a vacuolar antifungal factor. Plant Physiol 102: 939-945

Ruiz-Medrano R, Jimenez-Moraila B, Herrera-Estrella L, Rivera-Bustamante RF (1992) Nucleotide sequence of an osmotin-like cDNA induced in tomato during viroid infection. Plant Mol Biol 20: 1199-1202

Singh NK, Handa AK, Hasegawa PM, Bressan RA (1985) Proteins associated with adaptation of cultured tobacco cells to NaCl. Plant Physiol 79: 126-137

Singh NK, Bracker CA, Hasegawa PM, Handa AK, Buckel S, Hermodson MA, Pfankoch E, Regnier FE, Bressan RA (1987a) Characterization of osmotin. Plant Physiol 85: 529-536

Singh NK, LaRosa PC, Handa AK, Hasegawa PM, Bressan RA (1987b) Hormonal regulation of protein synthesis associated with salt tolerance in plant cells. Proc Natl Acad Sci USA 84: 739-743

Singh NK, Nelson DE, Kuhn D, Hasegawa PM, Bressan RA (1989a) Molecular cloning of osmotin and regulation of its expression by ABA and adaptation to low water potential. Plant Physiol 90: 1096-1101

Singh NK, Nelson DE, LaRosa PC, Bracker CE, Handa AK, Hasegawa PM, Bressan RA (1989b) Osmotin: a protein associated with osmotic stress adaptation in plant cells. In JH Cherry, ed, Environmental Stress in Plants, NATO ASI Series, vol G19. Springer-Verlag, Berlin-Heidelberg, pp 67-87

Skriver K, Mundy J (1990) Gene expression in response to abscisic acid and osmotic stress Plant Cell 2: 503-512 Smith, A., Hinchee, M.A., and Horsch, R. (1987). Cell and tissue specific expression localized by in situ RNA hybridization in floral tissues. Plant Mol. Biol. Rep. 5, 237-241.

Smith A, Hinchee MA, Horsch R (1987) Cell and tissue specific expression localized by in situ RNA hybridization in floral tissues. Plant Mol Biol Rep 5: 237-241

Stintzi A, Heitz T, Kauffmann S, Legrand M, Fritig B (1991) Identification of a basic pathogenesis-related thaumatin-like protein of virus-infected tobacco as osmotin. Physiol Mol Plant Pathol 38: 137-146

Storey K, Storey J (1981) Biochemical strategies of overwintering in the gall fly larva, Eurosta solidaginis: control of cryoprotectant polyol synthesis. Insect Biochem 12: 501-505

Takeda S, Sato F, Ida K, Yamada Y (1991) Nucleotide sequence of a cDNA for osmotin-like protein from cultured tobacco cells. Plant Physiol 97: 844-846

Twell D, Yamaguchi J, McCormick S (1990) Pollen-specific gene expression in transgenic plants: Coordinate regulation of two different tomato gene promoters during microsporogenesis. Development 109: 705-713

Twell D, Yamaguchi J, Wing RA, Ushiba J, McCormick S (1991) Promoter analysis of genes that are coordinately expressed during pollen development reveals pollen-specific enhancer sequences and shared regulatory elements. Gen Dev 5: 496-507

Uknes S, Mauch-Mani B, Moyer M, Potter S, Williams S, Dincher S, Chandler D, Slusarenko A, Ward E, Ryals J (1992) Acquired resistance in Arabidopsis. Plant Cell 911 4: 645-656

van Kan AL, van de Rhee MD, Zuidema D, Cornelissen BJC, Bol JF (1989) Structure of tobacco genes encoding thaumatin-like proteins. Plant Mol Biol 12: 153-155

van Loon LC, Gerritsen YAM, Ritter CE (1987) Identification, purification, and characterization of pathogenesis-related proteins from virus-infected Samsun NN tobacco leaves. Plant Mol Biol 9: 593-609

Vigers AJ, Wiedemann S, Roberts WK, Legrand M, Selitrennikoff CP, Fritig B (1992) Thaumatin-like pathogenesis-related proteins are antifungal. Plant Sci 83: 155-161

Vigers AJ, Roberts WK, Selitrennikoff CP (1991) A new family of plant antifungal proteins. Mol Plant-Microb Interact 4: 315-323

Weretilnyk E, Hanson A (1990) Molecular cloning of a plant betaine-aldehyde dehydrogenase, an enzyme implicated in adaptation to salinity and drought. Proc Natl Acad Sci USA 87: 2745-2759

Woloshuk CP, Meulenhoff JS, Sela-Buurlage M, van den Elzen PJM, Cornelissen BJC (1991) Pathogen-induced proteins with inhibitory activity toward Phytophthora infestans. Plant Cell 3: 619-628

Zhu B. Chen THH, Li PH (1993a) Expression of an ABA-responsive osmotin-like gene during the induction of freezing tolerance in Solanum commersonii. Plant Mol Biol 21:729-735

Zhu J-K, Shi J, Bressan RA, Hasegawa PM (1993b) Expression of an Atriplex nummularia gene encoding protein homologous to the bacterial molecular chaperone DnaJ. Plant Cell 5: 341-349

Appendix

Comparison of amino acid sequences of osmotin and related proteins. Source of the amino acid sequence: **1** - cDNA; **2** - protein. Completely conserved amino acids are in boldface type. The positions of the conserved cysteines are marked with: *; † and • represent 80 to 99% identity and conservative change (amino acid has been replaced by synonym amino acid having functionally similar R groups) of the amino acids, respectively.

Osmotin (Singh et al, 1989; Nelson et al., 1992); White Burley (Kumar and Spencer, 1992); Samsun NN (Takeda et al., 1991); AP24 (Woloshuk et al., 1991; Melchers et al., 1993); gp22 (Edelbaum et al., 1990); PR-R major (Payne et al, 1988); PR-R minor (Cornelissen et al., 1986); NP24 (King et al., 1988); TPM1 (Ruiz-Medrano et al., 1992); P23 (Rodrigo et al., 1991; 1993); Potato C (Pierpoint et al., 1990); pA13 (Zhu et at., 1993a); pA8-pA9 (Casas et al., 1992); Thaumatin (Edens et al., 1982); P21 (Graham et al., 1992); PR-5 (Uknes et al., 1992); Mai (Richardson et al., 1987); Zeamatin, Sormatin, Avematin, Trimatin (Vigers et al., 1991); CHEM4 (Frendo et al., 1992); Hv-1b (Bryngelsson and Green, 1989; Hahn et al., 1991); BP-R, BP-S (Hejgaard et al., 1991); pWIR2 (Rebman et al., 1991); pPIR2 (Reimann and Dudler, 1993);

```
OSMOTIN        1,2 (Nicotiana tabacum)            MGNLRSSFVFFLLALVTYTYA
White Burley1           "                         MGNLRSSFVFFLLALVTYTYA
Samsun NN      1        "                         MSHLTTFLVFFLLAFVFVTYTYA
AP24           1,2      "                         MSNNMGNLRSSFVFFLLALVTYTYA
gp22           2        "
PR-R maj       1        "                         MNFLKSFPFFAFLYFGQYFVAVTHA
PR-R min       1        "                         MNFLKSFPFYAFLCFGQYFVAVTHA
Np24           1   (Lycopersicon esculentum)           .VLFFLLCVTYTYA
TPM1           1        "                               .FFFLLAFVTYTYA
AP24           2        "
P23            1,2      "                                     .AFVTYTYA
Potato C       2   (Solanum tuberosum)
pA13           1   (S. commersonii)              MAYLRSSFVFFLLAFVTYTYA
pA8            1   (Atriplex nummularia)         MSPLSASLVTFLLSIVTITHA
pA9            1        "                    MNSSLMKSLLISSLLLMIASLITTPTHA
Thaumatin      1   (Thaumatococcus daniellii)     MAATTCFFFLFPFLLLLTLSRA
P21            2   (Glycine max)
PR-5           1   (Arabidopsis thaliana)        MANISSIHILFLVFITSGIAVMA
Mai            2   (Zea mays)
Zeamatin       2        "
CHEM4          1        "                        MAAASSSSVLLLLLAAALAGMSANA
Sormatin       2   (Sorghum bicolor)
Avematin       2   (Avena sativa)
Hv-1b          1   (Hordeum vulgare)             MATSPVLFLLLAVFAAGASA
BP-R           2        "
BP-S           2        "
pWIR2          1        "                        ..ATSPVLFLLLAVFAAGASA
Trimatin       2   (Triticum aestivum)
pPIR2          1   (Oryza sativa)                MASPATSSAVLVVVLVATLAAGGANA
```

```
                                    †    *  ••†             •••                        •         †
OSM           ATIE.VRNNCPYTVWAASTPI......GGGRRLDRGQTWVINAPRGTKMARV
W.B.          ATIE.VRNNCPYTVWAASTPI......GGGRRLDRGQTWVINAPRGTNMARV
Sam.NN        SGVFEVHNNCPYTVWAA.TPV......GGGRRLERGQSWWFWAPPGTKMARI
AP24          ATIE.VRNNCPYTVWAASTPI......GGGRRLDRGQTWVINAPRGTKMARV
gp22          SGVFEVHNNAPYTVWAAAWPV......GGGKRLERSQS
PR-R maj      ATFD.IVNKCTYTVWAAASP.......GGGRRLDSGQSWSINVNPGTVQARI
PR-R min      ATFD.IVNQCTYTVWAAASP.......GGGRQLNSGQSWSINVNPGTVQARI
NP24          ATIE.VRNNCPYTVWAASTPI......GGGRRLNRGQTWVINAPRGTKMARI
TPM1          ATFE.VRNNCPYTVWAASTPI......GGGRRLDRGQTWVINAPRGTKMARI
AP24          ATFE.VRNNCPYTVWAASTPI......GGGRRLD
P23           ATFE.VRNNCPYTVWAASTPI......GGGRRLDRGQTWVINAPRGTKMARI
Potato C      ATFD.ITNRCTYPVWAAASP.......GGGRRLDSGQTWNLNVNPGTIQARI
pA13          ATIE.VRNNCPYTVWAASTPI......GGGRRLDRGQTWVINAPRGTKMARI
pA8           TTIT.VVNNCNFPVWPPRPIGPTAIN.AQQRKLP.....CLKPTRRTKMARI
pA9           TTFT.VKNNCGYTVWGARHRWAKELP.ANAHGLSMSLP........ARAGV
Thaumatin     ATFE.IVNRCSYTVWAAASKGDAALD.AGGRQLNSGESWTINVEPGTNGGKI
P21           ARFE.ITNRCTYTVWAASVPV......GGGVQLNPGQSWSVDVPAGTKGARV
PR-5          TDFT.LRNNCPTTVWAGTLAGQGPKLGDGGFELTPGASRQLTAPAGWSGRF.
Mai           AVFT.VVNQCPFTVWAASVPV......GGGRQLNRGESWRITAPAGTTAARI
Zeamatin      AVFT.VVNQCPFTVWAASVPV......GGGRQLN
CHEM4         ATFT.ITNNCGFTVWPAATPV......GGGTQLNPGGTWTVNVPAGTSSGRV
Sormatin      AVFT.VVNRCPYTVWAASVPV......GG
Avematin      TTIT.VVNKCSYTVWPGALP......GGGVVLD
Hv-1b         ATFN.IKNNCGSTIWPAGIPV......GGGFELGSSGQTSSINVPAGTQAGRI
BP-R          ATIT.VVNRCSYTVWPGALP.......GGGVRLDPGQRWALNMPAGTAGAAV
BP-S          ATFT.VINKCQYTVWAAAVPA......GGGQKLDAGQTWSIXXP
pWIR2         ATFN.IKNNCGFTIWPAGIPV......GGGFALGSSGQTSSINVPAGTQAGRI
Trimatin      ATIT.VVNRCSYTVWPGALP.......GGGA
pPIR2         ATFT.ITNRCSFTVWPAATPV......GGGVQLSPGQTWTINVPAGTSSGRV
```

```
                               *  •     • •      *     •*••      *            •    †
OSM           WGRTNCNFNAAGRGT..CQTGDCGGVL.QCTGWGKPPNTLAEYALDQFSG
W.B.          WGRTNCNFNAAGRGT..CQTGDCGGVL.QCTGWGKPPNTLAEYALDQFSG
Sam.NN        WGRTNCNFDGAGRGW..CQTGDCGGVL.ECKGWGKPPNTLAEYALNQFSN
AP24          WGRTNCNFNAAGRGT..CQTGDCGGVL.QCTGWGKPPNTLAEYALDQFSG
gp22
PR maj        WGRTNCNFDGSGRGN..CETGDCNGML.ECQGYGKAPNTLAEFALNQ.PN
PR min        WGRTNCNFDGSGRGN..CETGDCNGML.ECQGYGKPPNTLAEFALNQ.PN
NP24          WGRTGCNFNAAGRGT..CQTGDCGGVL.QCTGWGKPPNTLAEYALDQFSN
TPM1          WGRTNCNFDGDGRGS..CQTGDCGGVL.QCTGWGKPPNTLAEYALDQFSN
AP24
P23           WGRTNCNFDGAGRGS..CQTGDCGGVL.QCTGWGKPPNTLAEYALDQFSN
Pot C         WGRTNCNFDGSGRG
pA13          WGRTNCNFDGAGRGS..CQTGDCGGVL.QCTGWGKPPNTLAEYALDQFSN
pA8           WGRTGCNFDGNGNGNGRCQTGGCGKLVCDPGNWGEIPKTLFEYTLAQPNN
pA9           WGRTGC..TSNGGNNLQCTTGGCGTLFDCGMNSGAPPLTIAEYTL...TN
Thaumatin     WARTDCYFDDSGSGI..CKTGDCGGLL.RCKRFGRPPTTLAEFSLNQYGK
P21           WARTGCNFDGSGRGG..CQTGDCGGVLD.CKAYGAPPNTLAEYGLNGFNN
PR-5          WARTGCNFDASGNGR..CVTGDCGGLR..CNGGGVPPVTLAEFTLVGDGG
Mai           WARTGCQFDASGRGS..CRTGDCGGVV.QCTGYGRAPNTLAEYALKQFNN
Zeamatin
CHEM4         WGRTGCSFNG.NSGS..CQTGDCGGAL.ACTLSGQPPLTLAEFTIGGS..
Sormatin
Avematin
Hv-1b         WARTGCSFNG.GSGS..CQTGDCGGQL.SCSLSGQPPATLAEFTIGGGST
BP-R
BP-S
pWIR2         WARTGCSFNG.GSGS..CQTGDCGGQL.SCSLSGRPPATLAEYTIGGGST
Trimatin
pPIR2         WGRTGCSFDGSGRGS..CATGDCAGAL.SCTLSGQKPLTLAEFTICGS..
```

```
                           †  † • † †    •              *       *        *
OSM        L.DFWDISLVDGFNIPMTFAPTNPSG...GKCHAIHCTANINGECPRE
W.B.       L.DFWDISLVDGFNIPMTFAPTNPSG...GKCHAIHCTANIRRMSR.E
Sam.NN     L.DFWDISVIDGFNIPMSFGPTKPGP...GKCHGIQCTANINGECPGS
AP24       L.DFWDISLVDGFNIPMTFAPTNPSG...GKCHAIHCTANINGECPRE
gp22                         MSFGPTKPGP...GKYHVIQA
PR maj     Q.DFVDISLVDGFNIPMEFSPTNGG......CRNLRCTAPINEQCPAQ
PR min     Q.DFVDISLVDGFNIPMEFSPTNGG......CRNLRCTAPINEQCPAQ
NP24       L.DFWDISLVDGFNIPMTFAPTKPSG...GKCHAIHCTANINGECPRA
TPM1       L.DFWDISLVDGFNIPMTFAPTNPSG...GKCHAIHCTANINGECPGS
AP24
P23        L.DFWDISLVDGFNIPMTFAPTNPSG...GKCHAIHCTANINGECPGS
Potato C
pA13       L.DFWDISLVDGFNIPMTFAPTNPSG...GKCHAIHCTANINGECPGS
pA8        PTDTIDISLIEGFNLPISFTPTSNAGALNGKCRTISCTADINGQCVTG
pA9        TLDTIDISLVDGFNVPMSFGGCPNSP.........SCASNILDSCPSD
Thaumatin  ..DYIDISNIKGFNVPMNFSPTTRG......CRGVRCAADIVGQCPAK
P21        L.DFFDISLVDGFNVPMDFSPTSNG....CTRGISCTADINGQCPSE
PR-5       K.DFYDVSLVDGYNVKLGIRPSGGS....GDCKYAGCVSDLNAACPDM
Mai        L.DFFDISILDGFNVPYSFLPDGGSG...CSRGPRCAVDVNARCPAE
Zeamatin
CHEM4      Q.DFYDISVIDGYNLAMAFSCSTGVRLV...CTDPGC...........
Sormatin
Avematin
Hv-1b      Q.DFYDISVIDGFNLAMDFSCSTGDALQ...CRDPSCPP.........
BP-R
BP-S
pWIR2      Q.DFYDISVIDGFNLAMDFSCSTGDALQ...CRDPSCPP.........
Trimatin
pPIR2      Q.DFYDLSVIDGYNVAMSFSCSSGVTVT...CRDSRC...........
```

```
           •          ••*  •*         **           * •             *
OSM        LRVP.....GGCNNPCTTFGGQQYCCTQRP.....CGPTFFSKFFKQRC
W.B.       LRVP.....GGCNNPCTTFGGQQYCCTQGP.....CGPTFFSKFFKQRC
Sam.NN     LRVP.....GGCNNPCTTFGGQQYCCTQGP.....CGPTELSRWFKQRC
AP24       LRVP.....GGCNNPCTTFGGQQYCCTQGP.....CGPTFFSKFFKQRC
gp22
PR maj     LKTQ.....GGCNNPCTVIKTNEYCCTNGPG...SCGPTDLSRFFKERC
PR min     LKTQ.....GGCNNPCTVIKTNEFCCTNGPG...SCGPTDLSRFFKARC
NP24       LKVP.....GGCNNPCTTFGGQQYCCTQGP.....CGPTELSKFFKKRC
TPM1       LRVP.....GGCNNPCTTFGGQQYCCTQGP.....CGPTDLSRFFKQRC
AP24
P23        LRVP.....GGCNNPCTTFGGQQYCCTQGP.....CGPTDLSRFFKQRC
Potato C       ....            EFCCTNGP...GSCGPTDLSRFF
pA13       LRVP.....GGCNNPCTTFGGQQYCCTQGP.....CGPTDLSRFFKQRC
pA8        KYAG....VTTLHEPGT................QSCGANGDSKFFKQRC
pA9        LKVN.....GGCLSACNKYSTDEYCCRGQ..YEKNCPPNKYSMIFKGLC
Thaumatin  LKAP....GGGCNDACTVFQTSEYCCTTG.....KCGPTEYSRFFKRLC
P21        LKTQ.....GGCNNPCTVFKTDQYCCNSG.....SCGPTDYSRFFKQRC
PR-5       LKVMDQNNVVACKSACERFNTDQYCCRGANDKPETCPPTDYSRIFKNAC
Mai        LRQD.....GVCNNACPVFKKDEYCCVGS..AANNCHPTNYSRYFKGQC
Zeamatin
CHEM4      ............................................
Sormatin
Avematin
Hv-1b      ............................................
BP-R
BP-S
pWIR2      ............................................
Trimatin
pPIR2      ............................................
```

```
                  ·  ·        ··    *     ··  †   *
OSM      PDAYSYPQDDPTSTFTCPGGSTNYRVIFCPNGQAHPNFPLEMPGS....DEVAK
W.B.     PDAYSYPQDDPTSTFTCPGGSTNYRVIFCPNGQAHPNFPLEMPGS....DEVAK
Sam.NN   PDAYSYPQDDPTSTFTCTSWTTDYKVMFCPYGSAHNETTNFPLEMPTSTHEVAK
AP24     PDAYSYPQDDPTSTFTCPGGSTNYRVIFCPNGQAHPNFPLEMPGS....DEVAK
gp22
PR maj   PDAYSYPQDDPTSLFTCPSG.TNYRVVFCP
PR min   PDAYSYQ.DDPPSLFTCPPG.TNYRVVFCP
NP24     PDAYSYPQDDPTSTFTCPGGSTNYRVVFCPNGVADPNFPLEMPAST...DEVAK
TPM1     PDAYSYPQDDPTSTFTCPSGSTNYRVVFCPNGVTSPNFPLEMPSS....DEEAK
AP24
P23      PDAYSYPQDDPTSTFTCPSGSTNYRVVFCPNGVTSPNFPLEMPSS....DEEAK
Potato C
pA13     PDAYSYPQDDPTSTFTCPSGSTNYRVVFCPNGVTSPNFPLEMPAS....DEEAK
pA8      PDAYSFPKDDQSSTFSCPPG.TNYKLTFCP
pA9      PQAYSYAKDDQSSTFTCPSG.TNYVVTFCP
Thaumatin PDAFSYVLDKPT.TVTCP.GSSNYRVTFCPTALELEDE
P21      PDAYSYPKDDPPSTFTCNGG.TDYRVVFCP
PR-5     PDAYSYAYDDETSTFTC.TGA.NYEITFCP
Mai      PDAYSYPKDDATSTFTCPAG.TNYKVVFCP
Zeamatin
CHEM4    PDAYHNPPDMK..THACGGNS.NYQVTFCP
Sormatin
Avematin
Hv-1b    PQAYQHPNDVA..THACSGNN.NYQITFCP
BP-R
BP-S
pWIR2    PQAYQHPNDVA..THACSGNN.NYQITFCP
Trimatin
pPIR2    PDAYLFPEDNTK.THACSGNS.NYQVVFCP
```

RESPONSES TO SALT STRESS IN THE HALOPHYTE *MESEMBRYANTHEMUM CRYSTALLINUM*

Hans J. Bohnert, John C. Thomas, E. Jay DeRocher[1], Christine B. Michalowski, Heimo Breiteneder[2], Dan M. Vernon[3], Wei Deng, Shigehiro Yamada, and Richard G. Jensen
Departments of Biochemistry, Molecular and Cellular Biology, and Plant Sciences,
The University of Arizona
Biosciences West
Tucson, AZ 85721, U.S.A.

ABSTRACT

The halophyte *Mesembryanthemum crystallinum* (common ice plant) has emerged as a model system suitable to advance our understanding of mechanisms that have evolved in a species adapted to harsh environments, characterized by drought and salt stress conditions. Here, we include a discussion of distinguishing factors of cellular and organismic responses to salt stress, the role of growth regulators, and a discussion of molecular and metabolic mechanisms employed by this plant for survival. Several mechanisms appear to be important, including acceleration of ontogeny, protection of the photosynthetic apparatus, the rapid induction of presumably osmoprotective substances, and the long-term establishment of Crassulacean Acid Metabolism (CAM). Our understanding of the causes for tolerance and resistance towards abiotic environmental stresses, salt stress in particular, has been extended to include the transfer of functionally characterized genes from *M. crystallinum* into other plants. The use of transgenic plants will be an essential component of future work directed at the molecular dissection of mechanisms of transcription control, mRNA-stability and developmental competence in establishing salt stress tolerance.

INTRODUCTION

Studies on the effects of water stresses on plant life and performance are many. The stresses most commonly associated with water deficits are drought, high salinity and low temperature. All of these stresses include other factors as well, such as desiccation and the concentration of

----

[1]present address: DOE-Plant Research Laboratory, MSU, East Lansing, MI.
[2]present address: Institut for General and Experimental Pathology, University of Vienna, Vienna, Austria.
[3]present address: Department of Botany, Oklahoma State University, Stillwater, OK.

NATO ASI Series, Vol. H 86
Biochemical and Cellular Mechanisms
of Stress Tolerance in Plants
Edited by J. H. Cherry
© Springer-Verlag Berlin Heidelberg 1994

metabolites, or excess sodium, or a deceleration of enzymatic processes. These three stress factors have, however, the lack of water in common, and may be considered similar in this respect. A number of studies have been conducted in several model systems, in both halophytes, such as *M. crystallinum* or *Atriplex nummularia*, and glycophytes, such as tobacco or tomato, that have provided basic knowledge about processes that are associated with salt stress. Suggestions have been made about the relative importance of these particular responses to tolerance acquisition under salt stress. While the tolerance phenotype is clearly the result of many genes, some metabolic pathways appear to be uniquely important, such as the biosynthesis and accumulation of osmoprotective substances, salt partitioning, and changes in development and growth habit. We find it now possible to offer a mechanistic explanation of the salt resistance phenotype through the analysis of transgenic glycophytic plants that have been engineered to express characters identified as essential in halophytes.

Table 1: Phenotypic, Physiological, and Metabolic Effects of Salt Stress on Plants.

Observations	Cause and Putative Function	Selected References
Wilting	Osmotic imbalance/ Signaling	numerous.
Increased oxygen consumption in roots	Mitochondrial respiration Energy provision	Hanson et al., 1991. Schwarz et al., 1991
Sugar accumulation	Altered carbon storage	Loescher, 1987.
Accumulation of metabolites	Osmotic adjustment Osmoprotection (polyols, proline, glycine betaine, inorganic salt) Radical scavenging by polyols Compatible solutes	Fougere et al., 1991 Yancey et al., 1982: Vernon & Bohnert, 1992; McCue & Hanson, 1990 Smirnoff & Cumbes, 1989. Schobert, 1977
"Stress-protein" synthesis	Chaperones (some LEA proteins) Pathogenesis responses (osmotin)	Skriver & Mundy, 1990; Claes et al., 1990 Singh et al., 1989; Casas et al., 1992.
Growth	Growth maintenance/ adaptation CAM-induction	Casas et al., 1991; Brugnoli & Lauteri, 1991; Cushman & Bohnert, 1993; Waterborg et al., 1989.
Growth regulators change	Growth maintenance/ senescence	Chu et al., 1990; Thomas et al., 1992b, 1993.
Chloroplast performance	Photosynthesis maintenance (photoinhibition)	Michalowski et al., 1992; Brugnoli & Lauteri, 1991; Breiteneder et al., 1993 Demmig & Winter, 1986; Bartholomew et al., 1991.
Proline increases	Nitrogen metabolism impairment/ maintenance	Rhodes, 1987; McCue & Hanson, 1992; Thomas et al., 1992a.
Acceleration of ontogeny	Induction of flowering	Winter & Gademann, 1991; Cushman & Bohnert, 1993; Chu et al., 1990.
Reduced cell expansion	Reversible adaptation to salt before growth is resumed	Bressan et al., 1990.

Table 1 with selected references summarizes processes that have been identified in various species, including the ice plant, as typical reactions having a presumably stress-relieving function. With this list of functions as a yardstick it may be possible to arrive at an organized strategy to investigate and pinpoint mechanisms, to test whether these are concomitant stress reactions, possibly reflections of stress-induced pathological processes, or whether the reactions, and the mechanisms that lead to their manifestation, are at the basis of stress tolerance or resistance. In the following paper we will discuss some of these correlations in the context of how these phenotypic and metabolic effects of stress on plants are reflected in changes in gene expression of a halophytic plant. Our first attempts at including known stress-relieving mechanisms into a glycophytic plant will be discussed.

HOW ARE CELLULAR AND ORGANISMIC RESPONSES TO SALT STRESS RECONCILED?

In the ice plant, many stress responses require organized tissue (Thomas et al., 1992a; Adams et al., 1992; Andolfatto et al., 1993). Suspension cultures and hairy-root cultures of the ice plant do not reproduce the full spectrum of reactions to salt stress that characterize whole plants. Irrespective of the failure of these cultures or organs to mount gene expression responses identical to those observed in plants, such cultures are useful to study and understand cellular responses to high salinity. The accumulation of presumably osmoprotective compounds is a common response to salt stress in many organisms (Flowers et al., 1977; Yancey et al., 1982; Binzel et al., 1988) and in suspension cells of glycophytes and halophytes (Greenway & Munns, 1980; Wyn Jones et al., 1983). Accumulation of these compounds during salt stress indicates osmoregulation as a cellular trait (Csonka 1989; Treichel, 1975) which possibly does not require organized tissues. In *M. crystallinum*, NaCl leads to proline accumulating from 5% to 40 to 50% of total amino acids in suspension cells and in plants (Treichel, 1975; Thomas et al., 1992a), depending on the length of stress and plant age. A second putative osmoprotectant, D-pinitol, accumulates in salt-stressed plants and may also play a role in the maintenance of osmotic balance (Paul & Cockburn, 1989; Vernon & Bohnert, 1992). In specialized cells of the ice plant, epidermal bladder cells, the increases in NaCl and D-pinitol are most dramatic, indicating the crucial role in maintaining turgor and osmotic balance played by these cells (Adams et al., 1992).

However, non-adapted suspension cells of the ice plant do not accumulate D-pinitol (Adams et al., 1992). Salt stress leads to slower growth of root- and leaf-derived cells, while cells established from hypocotyl tissue are growth-arrested under stress (Thomas et al., 1992a). PEPCase protein levels in cell suspensions appear to be constitutive, while NaCl induces the synthesis of a CAM-specific PEPCase isoform in plants (Cushman et al., 1989). We discuss several explanations, including differences between cells and plants in the concentrations of growth regulator, the lack of organized tissue, absence of leaves that may function as NaCl-sinks, and the size of the vacuolar compartment. Reduced growth is commonly observed even when salt-adapted, tolerant suspension culture cells of halophytic or glycophytic origin are treated with NaCl (Greenway & Munns, 1980; Hasegawa et al., 1986).

STRESS RESPONSES AND PLANT DEVELOPMENT

The ice plant undergoes dramatic changes in habitus and physiology as the plants age in the absence of obvious stress (Table 2). During the juvenile phase of growth, the developmental program establishes a plant with a fixed number of usually six primary leaves. At a certain age the adult growth phase begins which is characterized by termination of growth of the

Table 2: Characteristics of juvenile and adult growth phases of *M. crystallinum*.[*]

Markers of ontogeny	juvenile growth phase	adult growth phase
Leaf shape	obovate with entire margins; net veined; symmetrical; nearly sessile; only slightly succulent; may be large (> 10 cm^2) declining in size from leaf pair three on.	bilaterally symmetrical; smaller; succulent; bearing anthocyanic mucro at apex when young, more pronounced with age; leaves becoming smaller with age.
Leaf position	along primary axis; primary leaves always in pairs; opposite; six to seven pairs in total.	along secondary axes; oldest secondary leaves in pairs and opposite; becoming alternate.
Growth habit	orthotropic; primary axis stunted.	plagiotropic; terminal flower at primary meristem.
Branching	equal to number of primary leaves or less.	unlimited side branch formation with terminal flowers.
Anthocyanin	absent.	strong in youngest leaves and in fruits and seed pods.
Flowers	absent.	present.
Bladder cells	present, but empty, appressed.	present, turgid, protruding; salt storage function.
Ploidy levels[**]	2 to 32 N; cell-specific.	increased ploidy in many, but not in all, cell types.
Pinitol/ Ononitol	absent or low amounts.	high and increasing with age; strongly induced by salt stress.
CAM-inducibility	incomplete under stress conditions	readily inducible by salt stress

[*]Modified after Cushman & Bohnert, 1993; [**]DeRocher et al., 1990. The presence of CAM-characters, however, is not a suitable distinction between juvenile and adult growth phases, because CAM-induction is incomplete in old plants without stress.

419

primary axis, development of axillary side-shoot, changes in leaf morphology, and, later, by flowering and fruit and seed production. These developmental processes are somewhat flexible and may be influenced by day-length, temperature and nutrient availability; adverse environmental conditions may accelerate timing of the transition to the adult phase of growth, thus determining the size of the plants and productivity.

MOLECULAR CONTROLS

Induction of osmoprotection. As reported earlier (Demmig & Winter, 1986; Paul and Cockburn, 1989), the ice plant accumulates low molecular weight substances that lead to either osmotic adjustment or osmoprotection. Accumulation of proline and of the cyclic sugar alcohol, D-pinitol, have been observed. One cDNA, *Imt1*, encoding an enzyme whose activity leads to the accumulation of the cyclitol, D-ononitol, has recently been detected among salt stress-induced transcripts in the ice plant (Vernon & Bohnert, 1992). Its induction is at the level of transcription. Dissection of the control elements that are involved in the transcriptional activation of this gene should provide information about the mechanism of induction. *Imt1* appears to be strictly stress-induced and is little influenced by age of the plants, since we could elicit maximal induction at any age of the plants (Vernon & Bohnert, 1992b). The transcript is inducible even when seeds are germinated in the presence of 250 mM NaCl (Figure 1).

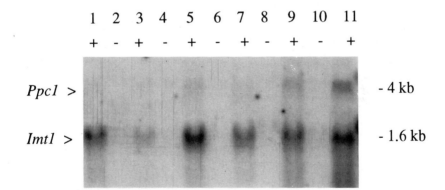

Figure 1: Northern hybridization of radiolabelled *Imt1*-probe to total RNA from plants of different ages.

Lanes 1 and 2 - germinating seedlings in soil with (+) or without (-) 250 mM NaCl, approximately one week old; lane 3 - germinating seeds grown on a paper surface in the presence of 200 mM NaCl; lanes 4 and 5 - plants two weeks old; lanes 6 and 7 - three weeks old; lanes 8 and 9 - four weeks of age; lanes 10 and 11 - plants five weeks old. Lanes 5, 7, 9 and 11 - plants stressed with 500 mM NaCl in soil. The signal produced by *Imt1* is indicated. The filter was subsequently hybridized with a *Ppc1* probe, indicating the slow accumulation of this transcript in young plants.

Photosynthesis-related mechanisms. Over the last several years we have concentrated on describing gene expression as the ice plant responds to high salinity stress. The changes in gene expression, measured initially as changes in steady-state amounts of mRNAs of genes encoding CAM-enzymes, in most instances were later found to be due to increased transcription of the responding genes (Cushman et al., 1989). This fact by itself underscores what we see as stress-adaptation, not a pathological effect, but that gene-induction follows an active mechanism. Furthermore, many genes are actively down-regulated, often by cessation or decline of transcription and, finally, transcription activity of the majority of genes is not altered under salt stress (Meyer et al., 1990). While this statement on gene expression is certainly true for the general behaviour of several classes of genes, the regulation of expression for any gene is more complex in detail. Either up-regulation or down-regulation, or continued presence of transcripts, may be brought about by different control mechanisms. For example, the genes *Fnr1* (encoding ferredoxin-NADP+ oxidoreductase) and *Prk1* (encoding phosphoribulokinase) are largely unaffected by the stress, but in the case of *Prk1* transcription declines, mRNA amounts decline subtly, but protein amounts are not reduced. In the case of *Fnr1,* transcription continues, while the mRNA amount declines marginally (Michalowski et al., 1992). In contrast, *RbcS* mRNAs decline (with the four transcribed members of the multigene family reacting differentially; DeRocher et al., 1993), while transcription of all *RbcS* gene family members actually increases during stress (DeRocher & Bohnert, 1993). Finally, for *Ppc1*, an example for a CAM-related gene, the mRNA amounts increase by approximately 40-fold, while transcriptional activity is elevated only by a factor of five to eight (Cushman et al., 1989).

The molecular mechanisms underlying these changes are unknown, but they can be deduced from the often opposing effects of stress on transcription of the genes and amount, i.e. stability, of the mRNAs. For one of the examples, *RbcS*, where salt stress induces transcription while transcript levels drop precipitously, the opposing mechanisms are particularly obvious when stressed plants are "de-stressed". This can be accomplished by reducing NaCl-levels rapidly through rinsing with nutrient solution. *RbcS* transcription declines rapidly and mRNA amounts recover to pre-stress levels (DeRocher & Bohnert, 1993). The result of such fine-tuned, reciprocal control *via* the stabilization of either transcription or mRNA half-life, obviously, provides for the maintenance of expression of subunit genes for Rubisco and maintenance of a functional chloroplast.

The mechanisms by which photosynthesis-related functions appear to be protected against adverse environmental factors become also clear when stress-mediated expression changes of another nuclear gene, termed *Crbp*, for a chloroplast protein are considered. We have recently isolated a protein which, upon partial sequencing, indicated that it might be a chloroplast RNA-binding protein (CRBP). The sequence of the gene, encoding a transit peptide sequence for chloroplast import and being highly sequence-identical to identified chloroplast RNA-binding proteins, supported that functional assignment (Breiteneder et al., 1993). The gene shows a developmental decline of transcription activity, with high transcription rates as functional chloroplasts are established and with lower maintenance transcription thereafter. The gene and protein are under transcription control. This changes when the plants are stressed. *Crbp* expression declines to barely above background under salt stress, while its mRNA increases, the lengths of the transcripts increase and the protein amount increases slightly (Breiteneder et al., 1993). We suggest that the control elements for mRNA-stability are located in the 3'-UTR of the mRNAs that accumulate under stress, in particular in a series of conserved elements that are found in other *Crbp* transcripts, although not all other *Crbp*-genes encoded all of these elements in their genes (Breitender et al., 1993).

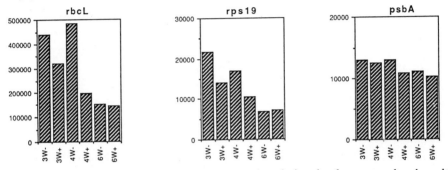

Figure 2: Expression of selected chloroplast transcripts during development and under salt stress of the ice plant.
Probes used: *RbcL*, encoding large subunit of Rubisco; *Rps19*, encoding chloroplast ribosomal protein S19 ribosomal protein ; *PsbA*, encoding 32kD PSII protein D1; Plants of age 3, 4, and 6 weeks were used either without (-) or with (+) 500 mM NaCl in soil for five days. Total plant RNA, 5 µg, were loaded in each lane. Hybridizations were quantitated using a β-scanner. Radioactivity (cpm) reflects specific activity of the probes used, not mRNA amount.

Included in Figure 2 are examples describing the behaviour of chloroplast mRNAs, products of the chloroplast transcription machinery. Stress during early plant development (3 and 4 weeks) does not severely affect the relative amounts of transcripts encoding *RbcL* and *Rps19*. At 6 weeks when the plants have initiated the reproductive growth phase, the mRNA amounts for *RbcL* and *Rps19* are much lower, but salt stress does not lead to a further reduction of transcript levels. The two examples chosen are examples of coding regions for relatively stable proteins. A slightly different result is observed for the mRNA encoding PsbA (D1), a protein that is the rapidly turned-over in functional chloroplasts. The mRNA for this protein is not significantly affected by stress during any time of development and may decrease only slightly as the plants age, underscoring the protection of the chloroplast compartment against environmental stress. We think that this behaviour reflects one of the mechanisms by which the ice plant is enabled to reach maturity under adverse environmental conditions. As important as it may be, we view protection of chloroplasts as an ancillary mechanism which is likely not sufficient and which is not the primary mechanism that establishes salt tolerance in plants.

CAM-induction. The inducibility of CAM is under developmental control, initiated approximately at the transition from juvenile to adult life. CAM-induction, however, is under environmental control. CAM-genes are not fully induced in young plants (Cushman et al., 1990; Vernon et al., 1992b). Plants that have already entered adulthood, will express CAM-genes in the absence of stress, although Crassulacean Acid Metabolism is slow to develop (Winter & Gademann, 1991). Salt stress, for example, during the transition to adult growth will transcriptionally induce CAM-genes, old plants contain mRNAs for CAM-genes, but little protein of CAM-related enzymes, although stress at this later stage will rapidly lead to the full transition to CAM. We consider this behaviour similar to what has been observed in the developmental progression of leaves of different age going through a gradual transition from C3 photosynthesis to CAM in *Peperomia scandens* (Holthe et al., 1987).

There are two aspects of these observations that may be pointed out as particularly interesting and worth further studies. First, the incomplete transcriptional induction in young plants, which is replaced by transcription competence at the time of transition to adult growth (with continuing transcription thereafter) may be accomplished by two mechanisms. Transcription may be repressed in young plants and the repressor protein(s) may be turned over more rapidly (or disappear) at a certain stage in development. Alternatively, transcription could be initiated by the synthesis of activators at the phase transition point. The second aspect views a different point in development, when transcripts of CAM-genes are already synthesized. The lack of appreciable amounts of CAM-enzymes may be due to control of transcript splicing, capping, or control over transport of mature mRNAs to the cytoplasm, mRNA stability, or translation control. Both aspects should be addressed in future experiments.

Genetic mechanisms. Little is known about genetic mechanisms characteristic of salt stress-tolerant plants. Multigenicity is indicated for drought tolerance (Martin et al., 1990) through the mapping of quantitative trait loci. Meyer et al. (1990) estimated a large number of genes to be altered in their expression after salt stress in the ice plant, based on the analysis of a fraction, 1 %, of the ice plant genome. It will be necessary to establish genetic studies in a halophytic species, because the well-studied, classical models are probably not well suited. Analysis of mutants, e.g. from *Arabidopsis thaliana* (Saleki et al., 1993), appear to indicate that loci that contribute incrementally to higher salt tolerance may be characterized. Compared with the marginal progress through genetic studies, the understanding of biochemical and molecular reactions to salt stress that has been gained through the work on halophytic models is much more advanced. It would appear beneficial if one could develop genetic studies in a halophyte for the study of salt stress.

Genetic studies, which would also have to include plant transformation and regeneration, may be possible with *Atriplex nummularia* (RA Bressan, P Hasegawa, personal communication) and in *Mesembryanthemum crystallinum* which has been shown to be amenable to regeneration (Meiners et al., 1991). The relatively small genome size of the ice plant, approximately 390 million bp (DeRocher et al., 1990), lends the plant to genetic studies. Through initial studies we know that the ice plant is relatively resilient to *Agrobacterium*-based transformation (JC Thomas, A Bornhouser, personal communication), although transformation is possible. We are, in addition, in the process of establishing lines of mutated ice plants with deletions and rearrangements generated by fast neutron bombardment (HJ Bohnert, unpublished).

METABOLIC MECHANISMS OF SALINITY TOLERANCE

Molecular mechanisms set the stage for metabolic adjustments to stressful environmental conditions. What are, then, the metabolic pathways that need adjustment ? Water stress leads to general ABA increases and the number of proteins induced by this growth regulator, although not all of these proteins are functionally characterized, may provide a list of necessary functions. Other studies emphasized a number of phenotypic indicators and reaction pathways which are summarized by Table 1. In principle, these functions should be recognizable as increasingly more genes become characterized as transcriptionally salt stress-induced or whose transcripts and encoded proteins are modulated by stress.

Our knowledge about alterations in gene expression in *Mesembryanthemum* suggests that transcripts for some of the functions listed in Table 1 are among the transcripts that accumulate and the genes whose transcription is activated. For example, a stress-induced protease, *Sep1*, of the thiol-protease family has been detected which we consider to be involved in the large-scale turnover of proteins that accompany stress and the transition from C3-photosynthesis to CAM (Ostrem JA & Vernon DM, unpublished). As we may infer from the relatively small changes in transcript abundance for proteins of photosynthesis-related pathways, as discussed above, the photosynthetic apparatus appears to be actively protected in young plants with lower maintenance protection at later stages in development. Experiments reported by Demmig & Winter (1986) and Köster & Anderson (1989) pointed towards protection of the plastids under salt stress. Reactions such as CAM-induction (Cushman et al.,

1990), filling of the epidermal bladder cell compartment (Adams et al., 1992), and morphological changes reflect ice plant-specific mechanisms which have exemplary significance, but which can not be modeled in other species. The establishment of functional CAM is an indication for enhancement of water-use-efficiency.

One ice plant-specific pathway, however, may be both at the basis of this plant's tolerance and may be useful for the transfer into other species. The transcript of gene *Imt1*, as described before, codes for a myo-inositol *O*-methyl transferase. *Imt1* is induced by salt stress and low temperature (Vernon & Bohnert, 1992; Vernon et al., 1993a). The increase of the encoded enzyme leads to the biosynthesis of D-ononitol which is epimerized to D-pinitol in the ice plant (Paul & Cockburn, 1989). The epimerase activity, as well as the methyl transferase activity, have not been detected in the glycophyte, tobacco (Vernon et al., 1993b). One question arising from these studies with a halophyte is whether halophytes include genetic material that is not found in glycophytes. The stress-related increase of Imt1-activity and the increase of D-pinitol may provide osmotically balancing substances and, at least in part, may prevent the influx of sodium. In the epidermal bladder cells located around flowers and seed pods, D-pinitol is by far the most prevalent sugar alcohol, approaching or even exceeding molar amounts (under the assumption that the polyols are mainly located in the cytosol) (Adams et al., 1992). In such concentrations osmotic adjustment is likely the mechanism through which D-pinitol acts. In addition, osmoprotection *via* a mechanism termed compatible solute action (Schobert, 1977; Vernon et al., 1993b) is also possible, especially in younger plants which do not accumulate as much D-pinitol as is found in epidermal bladder cells.

In as far as other mechanisms are involved in making the ice plant salt-tolerant, we have little information. We suggest, however, that several other pathways are important. We envision the control over water uptake and/ or partial sodium exclusion in the root system as one likely candidate for a stress-relieving pathway. Additionally, the increase in ABA under stress may be a factor overriding the cytokinin-stimulated effects that lead to continued growth. We have shown that, although ABA increases when the plants experience stress, cytokinin, which accelerates growth, also stimulates the build-up of CAM (Thomas et al., 1992b; Thomas & Bohnert, 1993). ABA alone was, however, not sufficient and not required to elicit PEPcase induction (Thomas et al., 1992b), although ABA can lead to the accumulation of CAM-related mRNAs. Thus, CAM-induction may be viewed as a mechanism that accompanies accelerated growth in the ice plant, rather than as a senescence-related mechanism (as is usually implied by its ABA-inducibility). Overriding stress-imposed growth restrictions by cytokinin may provide an important control leading to stress-tolerance in this plant.

How exactly cytokinin and ABA function will have to be studied. We have only recently begun to focus our attention on the developmental programs that have a determining influence on the timing and the magnitude of stress-responses in the ice plant. It may be that plant growth regulators act similarly on the expression of *Ppc1* and other CAM-related characters; cytokinin by accelerating development, and ABA *via* its specific mechanism in stress perception and/ or stress signalling. At this point it must also be indicated that the three stresses that effect water status of the plant are not equal in all aspects. Low temperature, drought and salt stress overlap only in some molecular responses, while each stress factor also has a specific component in how the stress affects gene expression (Vernon et al., 1993a). It appears that different stresses are transmitted in partially overlapping signal transduction pathways which involve various plant growth regulator ratios.

GENE TRANSFER AND ANALYSIS OF TRANSGENIC PLANTS

One way to prove that polyol biosynthesis in the halophyte, *M. crystallinum*, is important or essential for growth under salt stress is to transfer the pathway, or a possibly rate limiting enzyme of the pathway, into glycophytes. We have used *Agrobacterium tumefaciens* vectors and suitable strains to transform *Imt1* into tobacco and *Arabidopsis thaliana* (Vernon et al., 1993b). Transgenic tobacco plants have also been generated which express the bacterial gene, *mtlD*, encoding mannitol 1-phosphate dehydrogenase (Tarczynski et al., 1992), leading to the accumulation of mannitol. We have already investigated the reactions of tobacco containing mannitol to salt stress. Plants with mannitol could be exposed to NaCl at higher concentrations (250 mM NaCl) than control plants before the plants died (Tarczynski et al., 1993), while the presence of mannitol in these plants did not result in any phenotype in the absence of stress. Several aspects of stress of these plants require, however, more study. Mannitol-producing plants are not protected against NaCl-stress at the seedling stage, and the concentrations of mannitol in the plants are not high enough to be osmotically balancing the amount of NaCl found in the plants, possibly indicating that mannitol may act as a compatible solute (see Table 1), protecting cellular structures, enzymes or enzyme complexes. Ononitol-producing plants are being studied at present. From the preliminary data is appears that the presence of D-ononitol acts similar to that of mannitol, but it also appears that the presence of this cyclic sugar alcohol may provide even better protection than the presence of the acyclic polyol, mannitol. From these studies it is clear that sugar alcohols and their synthesis (which utilizes cytosolic [NADH]) in stressed plants is not a pathological event, but may have a function in stress-protection. Further studies and the use of transgenic plants to test mechanisms, such as those outlined in the Table 1, are necessary to understand how halophytes are protected from the stress imposed by high salinity in the soil.

The presence of the *Imt1*-gene leads to the appearance of the product, D-ononitol, without any phenotypically obvious detrimental effects on plant growth. D-ononitol is not found in tobacco, nor has Imt1-activity been observed unless the plants are transformed

(Vernon et al., 1993b). Stress experiments with transformed plants that express *Imt1* are in progress.

These results must be considered a strong indicator for the importance of polyols in salt stress protection. One might expect that other physiological characteristics may likewise prove important. We think that the control over sodium-uptake and salt partitioning and continued growth (which might continuously provide additional vacuoles as a sodium sink) are important characters as indicated by metabolic phenotypes, such as increased root respiration and accelerated ontogeny.

ACKNOWLEDGMENTS. Our work has been supported by USDA-NRI/ CRGP (Plant Responses to the Environment) and, in part, by NSF (Biochemistry Program), DOE (Biological Energy Program) and the Arizona Agricultural Experiment Station. H.B. and S.Y. acknowledge support by Max-Kade-Foundation and Japan Tobacco Inc., respectively.

REFERENCES

Adams P, Thomas JC, Vernon DM, Bohnert HJ, Jensen RG (1992) Distinct cellular and organismic responses to salt stress. Plant Cell Physiol 33: 1215-1223

Andolfatto P, Bornhouser A, Bohnert HJ, Thomas JC (1993) Transformed hairy-roots of *Mesembryanthemum crystallinum*: gene expression patterns upon salt stress. Submitted

Bartholomew DM, Bartley GE, Scolnik PA (1991) Abscisic acid control of *rbcS* and *cab* transcription in tomato leaves. Plant Physiol 96: 291-296

Binzel ML, Hess FD, Bressan RA, Hasegawa PM (1988) Intracellular compartmentation of ions in salt-adapted tobacco cells. Plant Physiol 86: 607-614

Bohnert HJ, Vernon DM, DeRocher EJ, Michalowski CB, Cushman JC (1992) Biochemistry and molecular biology of CAM. I: Wray JL (ed) Inducible plant proteins. Cambridge University Press, Cambridge, UK, pp. 113-137

Breiteneder H, Michalowski CB, Bohnert HJ (1993) Environmental stress-mediated differential 3'-end formation of chloroplast RNA-binding protein transcripts. Submitted

Bressan RA, Nelson DE, Iraki NM, LaRosa PC, Singh NK, Hasegawa PM, Carpita NC (1990) Reduced cell expansion and changes in cell walls of plant cells adapted to NaCl. In Katterman F (ed) Environmental injury to plants. Academic Press, San Diego, pp. 137-171

Brugnoli E, Lauteri M (1991) Effects of salinity on stomatal conductance, photosynthetic capacity, and carbon isotope discrimination of salt-tolerant (*Gossypium hirsutum* L.) and salt-sensitive (*Phaseolus vulgaris* L.) C3 non-halophytes. Plant Physiol 95: 628-635

Casas AM, Bressan RA, Hasegawa PM (1991) Cell growth and water relations of the halophyte, *Atriplex nummularia* L., in response to NaCl. Plant Cell Rep 10: 81-84

Casas AM, Nelson DE, Ragothama KG, D'Urzo MP, Singh NK, Bressan RA, Hasegawa PM (1992) Expression of osmotin-like genes in the halophyte *Atriplex nummularia* L. Plant Physiol 99: 329-337

Chu C, Ziyu D, Ku MSB, Edwards GE (1990) Induction of crassulacean acid metabolism in the facultative halophyte *Mesembryanthemum crystallinum*. Plant Physiol 93: 1253-1260

Claes B, Dekeyser R, Villaroel R, van den Bulcke M, Bauw G, van Montagu M, Caplan A (1990) Characterization of a rice gene showing organ-specific expression in response to salt stress and drought. The Plant Cell 2: 19-27

Csonka LN (1989) Physiological and genetic responses of bacteria to osmotic stress. Microbiol Rev 53: 121-147

Cushman JC (1993) Molecular cloning and expression of chloroplast NADP-malate dehydrogenase during Crassulacean acid metabolism induction by salt stress. Photosyn Res 35: 15-27

Cushman JC, Bohnert HJ (1993) Transcriptional activation of CAM-genes during development and environmental stress. In: Winter K, Smith AP, Smith JAC (eds) Crassulacean Acid Metabolism: Biochemistry, Ecophysiology and Evolution, Springer Verlag, Heidelberg, in press

Cushman JC, Michalowski CB, Bohnert HJ (1990) Developmental control of crassulacean acid metabolism inducibility by salt stress in the common ice plant. Plant Physiol 94: 1137-1142

Cushman JC, Meyer G, Michalowski CB, Schmitt JM, Bohnert HJ (1989) Salt stress leads to differential expression of two isogenes of PEPcarboxylase during crassulacean acid metabolism induction in the common ice plant. The Plant Cell 1: 715-725

Demmig B, Winter K (1986) Sodium, potassium, chloride and proline concentrations of chloroplasts isolated from a halophyte, *Mesembryanthemum crystallinum*. Planta 168: 421-426

DeRocher EJ, Harkins K, Galbraith DW, Bohnert HJ (1990) Developmentally regulated systemic endopolyploidy in succulents with small genomes. Science 250: 99-101

DeRocher EJ, Quigley F, Mache R, Bohnert HJ (1993) The six genes of the Rubisco small subunit multigene family from *Mesembryanthemum crystallinum*, a facultative CAM plant. Mol Gen Genet, In press

DeRocher EJ, Bohnert HJ (1993) Expression of the *RbcS* multigene family in *Mesembryanthemum crystallinum* during development and under environmental stress. Submitted

Flowers TJ, Troke PF, Yeo AR (1977) The mechanism of salt tolerance in halophytes. Annu Rev Plant Physiol 28: 89-121

Fougere F, Le RD, Streeter JG (1991) Effects of salt stress on amino acids, organic acid, and carbohydrate composition of roots, bacteroids, and cytosol of alfalfa (*Medicago sativa* L.). Plant Physiol 96: 1228-1236

Greenway H, Munns R (1980) Mechanisms of salt tolerance in non halophytes. Annu Rev Plant Physiol 31: 149-190

Hanson AD, Rathinasabapathi B, Chamberlin B, Gage DA (1991) Comparative physiological evidence that beta-alanine betaine and choline-O-sulfate act as compatible osmolytes in halophytic *Limonium* species. Plant Physiol 97: 1199-1205

Hasegawa PM, Bressan RA, Handa AK (1986) Cellular mechanisms of salinity tolerance. Hort Sci 21: 1317-1323

Holthe PA, Sternberg L, Ting IP (1987) Developmental control of CAM in *Perperomia scandens*. Plant Physiol 84: 640-642

Kononowicz AK, Hasegawa PM, Bressan RA (1992) Cell cycle duration in tobacco cells adapted to NaCl. Environ Exp Bot 32: 1-9

Köster S, Anderson JA (1989) The photosynthetic apparatus of C3 and CAM-induced *Mesembryanthemum crystallinum*. Photosyn Res 19: 251-264

Loescher WH (1987) Physiology and metabolism of sugar alcohols in higher plants. Plant Physiol 70: 553-557

Martin B, Nienhuis J, King G, Schaefer A (1989) Restriction fragment length polymorphisms associated with water use efficiency in tomato. Science 243: 1725-1728

McCue K, Hanson A (1990) Drought and salt tolerance: towards understanding and application. TIBTech 8: 358-362

Meiners M, Thomas JC, Bohnert HJ, Cushman JC (1991) Regeneration of multiple shoots and plants from *Mesembryanthemum crystallinum*. Plant Cell Rep 9: 563-566

Meyer G, Schmitt JM, Bohnert HJ (1990) Direct screening of a small genome: estimation of the magnitude of plant gene expression changes during adaptation to high salt. Mol Gen Genet 224: 347-356

Michalowski CB, Olson SW, Piepenbrock M, Schmitt JM, Bohnert HJ (1989) Time course of mRNA induction elicited by salt stress in the common ice plant. Plant Physiol 89: 811-816

Michalowski CB, DeRocher EJ, Bohnert HJ, Salvucci ME (1992) Phosphoribulokinase from ice plant: transcription, transcripts and protein expression during environmental stress. Photosyn Res 31: 127-138

Ostrem JA, Olson SW, Schmitt JM, Bohnert HJ (1987) Salt stress increases the level of translatable mRNA for PEPcarboxylase in *M. crystallinum*. Plant Physiol 84: 1270-1275

Paul MJ, Cockburn W (1989) Pinitol, a compatible solute in *Mesembryanthemum crystallinum* L? J Exp Bot 40: 1093-1098

Rhodes D (1987) Metabolic responses to stress. In: Stumpf P, Conn E (eds) Biochemistry of plants, vol 12, Academic Press, New York, pp. 201-241

Saleki R, Young PG, Lefebvre DD (1993) Mutants of *Arabidopsis thaliana* capable of germination under saline conditions. Plant Physiol 101: 839-845

Schobert B (1977) Is there an osmotic regulatory mechanism in algae and higher plants? J Theor Biol 68, 17-26

Schwarz M, Lerner HR, Reinhold L (1991) Mitochondria isolated from NaCl-adapted tobacco cell lines (*N. tabacum/ gossii*) maintain their phosphorylative capacity in highly saline media. Plant Physiol 96: 69-76

Singh NK, Nelson DE, Kuhn D, Hasegawa PM, Bressan RA (1989) Molecular cloning of osmotin and regulation of its expression by ABA and adaptation to low water potentials. Plant Physiol 90: 1096-1101

Skriver K, Mundy J (1990) Gene expression in response to abscisic acid and osmotic stress. Plant Cell 2: 503-512

Smirnoff N, Cumber QJ (1989) Hydroxyl-radical scavenging activity of compatible solutes. Phytochem 28: 1057-1060

Tarczynski MC, Jensen RG, Bohnert HJ (1992) Expression of a bacterial *mtlD* gene in transgenic tobacco leads to production and accumulation of mannitol. Proc Natl Acad Sci USA 89: 2600-2604

Tarczynski MC, Jensen RG, Bohnert HJ (1993) Stress protection in transgenic tobacco producing a putative osmoprotectant, mannitol. Science 259: 508-510

Thomas JC, deArmond R, Bohnert HJ (1992a) Influence of NaCl on growth, proline, and phosphoenolpyruvate carboxylase levels in *Mesembryanthemum crystallinum* suspension cultures. Plant Physiol 98: 626-631

Thomas JC, McElwain EF, Bohnert HJ (1992b) Convergent induction of osmotic stress responses. Abscisic acid, cytokinin and the effects of NaCl. Plant Physiol 100: 416-423

Thomas CJ, Bohnert HJ (1993) Linking salt stress perception and plant growth regulators in the halophyte, *Mesembryanthemum crystallinum*. Submitted

Treichel S (1975) The effect of NaCl on the concentration of proline in different halophytes. Z Pflanzenphysiol 76: 56-68

Vernon DM, Bohnert HJ (1992a) A novel methyl transferase induced by osmotic stress in the facultative halophyte *Mesembryanthemum crystallinum*. EMBO J 11: 2077-2085

Vernon DM, Bohnert HJ (1992b) Increased expression of a myo-inositol methyl transferase in *Mesembryanthemum crystallinum* is part of a stress response distinct from crassulacean acid metabolism induction. Plant Physiol 99: 1695-1698

Vernon DM, Ostrem JA, Bohnert HJ (1993) Stress perception and response in a facultative halophyte: the regulation of salinity-induced genes in *Mesembryanthemum crystallinum*. Plant Cell Environ 16: In press

Vernon DM, Tarczynski MC, Jensen RG, Bohnert HJ (1993) Cyclitol production in transgenic tobacco. Plant J 4: in press

Waterborg JH, Harrington RE, Winicov I (1989) Differential histone acetylation in alfalfa (*Medicago sativa*) due to growth in NaCl: responses in salt-stressed and salt-tolerant callus cultures. Plant Physiol 90: 237-245

Winter K, Gademann R (1991) Daily changes in CO_2 and water vapor exchange, chlorophyll fluorescence, and leaf water relations in the facultative halophyte *Mesembryanthemum crystallinum* during the induction of crassulacean acid metabolism in response to high NaCl salinity. Plant Physiol 95: 768-776

Wyn Jones RG, Gorham J (1983) Osmoregulation. In: Encyclopedia of Plant Physiology, New Series, vol 12C, Lange OL, Nobel PS, Osmond CB, Ziegler H (eds) Springer Verlag, Berlin, pp. 35-58

Yancey PH, Clark ME, Hand SC, Bowlus RD, Romero GN (1982) Living with water stress: evolution of osmolyte systems. Science 217: 1214-1222

ALTERATIONS IN H⁺-ATPASE GENE EXPRESSION IN RESPONSE TO SALT

Marla L. Binzel
Texas A&M Agricultural Research Center
1380 A&M Circle
El Paso, TX   79927

ABSTRACT

The accumulation of messages for the plasma membrane H⁺-ATPase and the 70 kD subunit of the tonoplast H⁺-ATPase was studied in tomato plants (*Lycopersicon esculentum* cv "Large Cherry Red") responding to salt and drought stress. There was an increased accumulation of plasma membrane H⁺-ATPase message in roots and expanded leaves of tomato plants treated with 400 mM NaCl. Message for the 70 kD subunit of the tonoplast H⁺-ATPase only increased in the expanded leaves of salt treated plants; no increased accumulation was observed in either roots or unexpanded leaves. In response to drought there was no increase in message accumulation for the 70 kD subunit of the tonoplast H⁺-ATPase in any of the tissues examined. Only a minimal increase in plasma membrane H⁺-ATPase accumulation was observed in the roots of drought stressed plants, and no increases were observed in the leaves.

INTRODUCTION

The ability to survive and grow in a saline environment requires that plants cope with a reduced availability of water, and an abundance of ions which are potentially toxic and may interfere with nutrient acquisition (Flowers 1977; Greenway & Munns, 1980; Wyn Jones, 1981; Poljakoff-Mayber, 1982). Plants need to effectively regulate solute uptake and accumulation in order to cope with these demands imposed by salinity. Naturally salt tolerant species (halophytes) typically utilize Na⁺ and Cl⁻

NATO ASI Series, Vol. H 86
Biochemical and Cellular Mechanisms
of Stress Tolerance in Plants
Edited by J. H. Cherry
© Springer-Verlag Berlin Heidelberg 1994

as principal sources of osmotica (Flowers et al., 1977; Poljakoff-Mayber, 1982; Wyn Jones and Gorham, 1983). The inability of metabolic enzymes to function in high ionic environments led to the theory that these ions might be sequestered in the vacuole (Flowers et al., 1977; Poljakoff-Mayber, 1982; Wyn Jones and Gorham, 1983). Since then, there have been a number of studies establishing the compartmentalization of Na^+ and Cl^- in the vacuoles of salt accumulating cells (Pitman et al., 1981; Yeo, 1981; Storey et al., 1983; Harvey, 1985; Harvey and Thorpe, 1986; Hajibagheri et al, 1987; Binzel et al., 1988; Hajibagheri et al., 1989). Even in salt sensitive species (glycophytes) considerable accumulation of Na^+ and Cl^- occurs in response to salt (Wyn Jones and Gorham, 1981; Sacher et al., 1982; Yeo and Flowers, 1984; Hajibagheri et al., 1989). Therefore the ability to effectively compartmentalize ions is likely to be important even in glycophytic species, and may in part explain examples where tolerant and susceptible genotypes accumulate equivalent amounts of Na^+ and/or Cl^-.

The ability to exclude ions and maintain the ion concentration gradients associated with vacuolar compartmentation under salinity requires efficient control of ion flux. In salt tolerant plants, this may involve alterations of specific proteins which act as ion channels or carriers, changes in passive permeabilities, and alterations in the processes which provide the driving forces for solute movement. H^+-translocating ATPases located on the plasma membrane and tonoplast, along with a pyrophosphatase on the tonoplast are considered to be the principle sources of the proton motive force (pmf) which provides the driving force for solute transport and osmotic adjustment (Sze, 1985; Braun et al., 1986; Reinhold et al., 1988; Rea 1993). Solute regulation at the tonoplast may be of particular importance for osmotic adjustment since the vacuole may occupy as much as 90% of the cell volume. Consequently as much as 90% of the cell osmoticum may require transport across the tonoplast (Daie, 1988). Sodium is thought to be evacuated from the cytosol by Na^+/H^+ antiporters which have been identified on both the plasma membrane and tonoplast (Blumwald and Poole, 1985; Braun et al., 1988; Garbarino and DuPont, 1989; Matoh et al., 1989). Enhanced capacity to generate and maintain electrochemical gradients, either from altered enzymes or elevated expression, potentially could be an integral component of the ability of cells to osmotically adjust.

Some of the earliest evidence supporting a possible role for H^+-ATPases in salinity tolerance came from the work of Bremberger et al. (1988) and Braun et al. (1986). When a shift from C3 to CAM metabolism was induced by NaCl in *Mesembryanthemum crystallinum,* the activity of the tonoplast H^+-ATPase increased approximately three fold (Bremberger et al., 1988). Similarly, *Atriplex nummularia* plants treated with NaCl exhibited a two fold increase in plasma membrane H^+-ATPase activity (Braun et al., 1986). Since then, there have been additional reports of increased H^+-ATPase activity associated with exposure to salinity for both the plasma membrane and the tonoplast H^+-ATPases. A two fold or more increase in the specific activity of tonoplast proton transport was observed in roots of barley plants treated with NaCl (Matusumoto and Chung, 1988; Dupont, 1992) and in salt adapted tobacco cells (Reuveni et al. 1990). Using cytochemical techniques, Sanchez-Aquayo et al. (1991) observed an increase in tonoplast ATPase activity in tomato roots treated with salt.

The cloning of genes encoding the plasma membrane and tonoplast H^+-ATPases have made it possible to begin examining whether or not exposure to salinity alters the genetic expression H^+-ATPase genes. An increased accumulation of message for the plasma membrane H^+-ATPase (Perez-Prat et al. 1993, Niu et al., 1993a) and the 70 kD subunit of the tonoplast H^+-ATPase (Narasimhan et al., 1991) occurred in response to NaCl in cultured cells. However no increase in message accumulation was detected in response to NaCl for the plasma membrane H⁺-ATPase in sunflower roots (Roldan et al., 1991) or for the tonoplast H⁺-ATPAse in barley roots (Dupont, 1992).

The present study was undertaken to determine whether NaCl induction of genes encoding H⁺-ATPases is limited to cells in culture, or perhaps occurs in tissues other than roots. Accumulation of message for the plasma membrane H⁺-ATPase, and the 70 kD subunit of the tonoplast H⁺-ATPase was examined in unexpanded and expanded leaves, and roots, of tomato plants responding to salt (400 mM NaCl) and drought stress. The results presented herein indicate that there is an increased accumulation of message for both the plasma membrane H⁺-ATPase and the 70 kD subunit of the tonoplast H⁺-ATPase, and that this increase is tissue specific. Furthermore, the pattern of enhanced accumulation in response to NaCl differs between the two H⁺-ATPases.

MATERIALS AND METHODS

Plant Material and Tissue Sampling

Seeds of *Lycopersicon esculentum* cv "Large Cherry Red" were germinated in vermiculite. For the salinization experiment, seedlings were transferred to a hydroponic system containing half strength Hoagland's nutrient solution. Water was replenished daily, and the nutrient solutions changed on a weekly basis. Once the plants reached the 5th leaf stage, NaCl was added to the solutions in 50 mM/day increments until a final treatment concentration of 400 mM NaCl was attained. Young unexpanded leaves (along with the apical meristem), mature expanded leaves, and roots were harvested 24 hrs later. Harvested tissues were immediately ground in liquid N_2 and stored at ⁻80°C for further analyses.

For the drought stress experiment, seedlings were transplanted into 1 L pots containing vermiculite and placed on a water saturated capillary mat. This constant source of moisture ensured that plants were maintained in a well watered status. Three times a week the plants were watered with half strength Hoagland's nutrient solution. When the plants reached the fifth leaf stage, fertilization was discontinued and half the plants were removed from the capillary mat. Water was withheld from these plants until the plants showed signs of permanent wilting. Tissues were harvested and stored as described for the salinization experiments.

Extraction of RNA and DNA

Total RNA was extracted from frozen ground tissue using an acid quanidinium thiocyanate phenol chloroform method (Chomiczinski and Sacchi, 1987). DNA was also extracted from frozen ground samples, using the method of Dellaporta et al. (1983). The DNA was further purified by cetyl trimethyl ammonium bromide (CTAB) extraction and digestion with RNAase A. DNA and RNA were both quantitated by spectrophotometry.

cDNA Probes

The probe used for analyses of the tonoplast H^+-ATPase was a 1.4 kb insert from a cDNA clone to the 70 kD subunit of the tobacco (*Nicotiana tabacum*) tonoplast H^+-ATPase (Narasimhan et al., 1991). The cDNA used to probe the plasma membrane H^+-ATPase was a 1.8 kb insert from the 3.4 kb LHA1 cDNA clone of the tomato plasma membrane H^+-ATPase (Ewing et al., 1990) which was generously provided by A. Bennett, U.C., Davis. The cDNA probes were labeled with ^{32}P-dCTP (Amersham) using a random prime DNA labeling kit (Promega).

Northern and Southern Analyses

Total RNA (20 μg/lane) was fractionated by electrophoresis through formaldehyde agarose gels and blotted onto nitrocellulose (Maniatis et al., 1989). RNA in the gels was visualized by staining with ethidium bromide prior to and following transfer to nitrocellulose to ensure that equivalent amounts of RNA had been loaded in each lane and that complete transfer to the membrane had occurred. DNA (20 μg/lane) was electrophoresed and transferred to nitrocellulose as described by Narasimhan et al. (1991).

Hybridizations were performed in 50% formamide, 6X SSPE, 5X Denhardt's, 0.1% SDS and 100 μg/ml herring sperm DNA at 47°C for 15 hours. The blots were washed as described in Narasimhan et al. (1991) with the addition of a final brief rinse at room temperature in 0.1X SSC (without SDS) prior to autoradiography at ⁻80°C.

RESULTS AND DISCUSSION

In the expanded leaves of tomato plants treated with NaCl, there is a large increase in the accumulation of message for the 70 kD subunit of the tonoplast H⁺-ATPase (Fig. 1A), 24 hours after a concentration of 400 mM was attained (NaCl was added at a rate of 50 mM/day). No increases were observed in this message level in either the unexpanded leaves, or the roots. Dupont (1992) obtained a similar result with barley; roots of plants treated with salt did not show any increase in tonoplast H⁺-ATPase message accumulation.

Examination of the accumulation of message for the plasma membrane H⁺-ATPase also reveals an increase in response to salt (Fig. 1B), however the pattern differs from that of the tonoplast H⁺-ATPase (Fig. 1A). Increased message accumulation for the plasma membrane H⁺-ATPase occurs in both roots and expanded leaves, however the increase in the roots is much larger. An increase in plasma membrane H⁺-ATPase message accumulation in the roots of salt treated plants has also been observed by Niu et al. (1993b), in studies involving both a glycophyte (tobacco) and a halophyte (*Atriplex nummularia*).

The largest increase in message accumulation for both the plasma membrane H⁺-ATPase and the 70 kD subunit of the tonoplast H⁺-ATPase occurred in tissues where the level of these messages

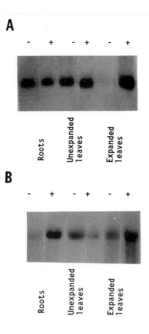

Fig. 1   *Northern analyses of accumulation of message for the 70 kD subunit of the tonoplast H^+-ATPase (A) and plasma membrane H^+-ATPase (B) in control (-) and salinized (+) tomato plants.* Total RNA (20 μg/lane) from roots (lanes 1-2), unexpanded leaves (lanes 3-4) and expanded leaves (lanes 5-6) were probed with the cDNA inserts described in "Material and Methods". Single bands corresponding to message sizes of 2.74 kb (A) and 4.24 kb (B) were detected. NaCl was added in increments of 50 mM/day until a treatment concentration of 400 mM was attained; tissue samples were harvested 24 hours later.

in the control tissue was relatively low (Fig. 1A,B). In the case of the tonoplast H^+-ATPase, low levels of message in the expanded leaves of control plants may reflect the relatively low demand for H^+-transport capacity. The net flux of solutes across the tonoplast of these cells can be envisioned to be minimal. When exposed to NaCl, the vacuoles of these expanded leaves may serve as a storage site for the Na^+ and Cl^- accumulated by the plant. Increased Na^+/H^+ antiport activity to load Na^+ into the vacuole would increase the demand for H^+ transport capacity and may be the reason for the increased accumulation of tonoplast H^+-ATPase message in this tissue after exposure to NaCl. The higher accumulation of 70 kD subunit message in roots and young leaves of control plants may reflect a greater steady-state demand for tonoplast H^+-transport capacity. The lack of increase in response to salinity is perhaps because there is sufficient expression of the tonoplast H^+-ATPase in these tissues to meet demands imposed by salinity. Relatively more accumulation of Na^+ was found to occur in the older, expanded leaves of plants treated with salt (Yeo and Flowers, 1984). This may suggest that a relatively low amount of Na^+ accumulation in the roots and young leaves, coupled with an intrinsically higher level of expression, results in little need for increased expression of the tonoplast H^+-ATPase in these tissues.

Based upon cytochemical localizations of ATPase activity by a method producing lead precipitate at the site of ATP hydrolysis, Sanchez-Aguayo et al. (1991) concluded that salinity causes an increase in root tonoplast ATPase activity and a decrease in root plasma membrane ATPase activity. Roldan et al. (1991) reported a lack of induction in response to salt in sunflower of either plasma membrane ATPase activity or message accumulation. These apparent contradictions may be due to the developmental differences between seedlings and more mature plants. The data reported here, and by Niu et al. (1993b) were obtained with plants that were several weeks old and had fully developed leaves and stems. Both Sanchez-Aguayo et al. (1991) and Roldan et al. (1991) utilized seedlings. The capacity to store accumulated Na^+ in the leaves is not present in seedlings. Thus the increased tonoplast H^+-ATPase activity in seedling roots (not observed in roots of older plants) may reflect a

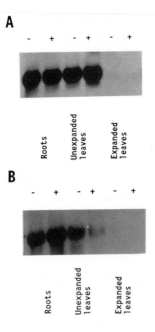

Fig. 2   *Northern analyses of accumulation of message for the*
          70 kD subunit of the tonoplast H^+-ATPase (A) and
          plasma membrane H^+-ATPase in control (-) and drought
          stressed (+) tomato plants. Lanes 1,2: roots; lanes
          3,4: unexpanded leaves; and lanes 5,6: expanded
          leaves. Total RNA, 20 µg/lane, cDNA probes used and
          size of detected bands were the same as in Fig. 1.

necessity to store Na^+ in the root cells.   If the increased
plasma membrane H^+-ATPase expression (Niu et al., 1993b and data
reported herein) is related to a role in symplastic Na^+
transport to aerial plant parts, then the absence of this
increase in seedlings may again reflect intrinsic developmental
differences between seedlings and young plants.

        No increase in message accumulation for the 70 kD subunit
of the tonoplast H^+-ATPase was observed in expanded or
unexpanded leaves, or roots, when plants were drought stressed.
There may be a very minimal increase in accumulation of message
for the plasma membrane H^+-ATPase in roots of drought stressed

plants (Fig. 2B). Plasma membrane H⁺-ATPase message levels were
unchanged in expanded leaves, and decreased in unexpanded
leaves, in response to drought stress (Fig. 2B).   H⁺-ATPases

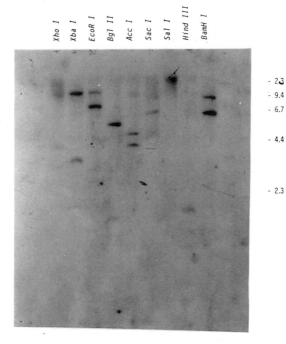

Fig. 3   *Southern analysis of tomato genomic DNA.*   Ten
          micrograms of genomic DNA was digested with the
          enzymes indicated and probed with a cDNA probe to the
          70 kD subunit of the tonoplast H⁺-ATPase.   The
          positions of molecular weight standards are indicated.

seem to play a more significant role in response to salt than
drought, based upon strong induction of both the plasma membrane
and tonoplast H⁺-ATPases by NaCl but not drought. The
accumulation of solutes for osmotic adjustment is common to
plants responding to drought and salinity (Hellebust, 1976;

Greenway and Munns, 1980). One notable difference however, is that compartmentalization of ions is likely to occur with salinity (Pitman et al., 1981; Storey et al., 1983; Binzel et al., 1988), whereas under drought stress there is probably relatively little partitioning of solutes between the vacuole and cytoplasm.

Unlike the plasma membrane H^+-ATPase, for which multiple genes have been identified, the 70 kD subunit of the tonoplast H^+-ATPase appears to be encoded by a small gene family, perhaps as few as 2 genes (Fig. 3). This data from tomato is consistent with results from Southern analysis of carrot (L. Taiz, personal communication). Based on 2D-PAGE, two protein isoforms of the 70 kD subunit have been identified in barley (Dupont et al., 1988) and tobacco (Binzel, unpublished data) based on immunological cross reactivity with antibodies to the corresponding subunit from red beet.

Although salinity or stress induced isoforms of neither the plasma membrane H^+-ATPase nor the tonoplast H^+-ATPase have yet to be identified, the function of one or more of these isoforms might be to respond to environmental stimuli. An ABA like responsive element was identified (Narasimhan et al., 1991) on a promoter sequence of the 70 kD subunit of the tonoplast H^+-ATPase from carrot (Struve et al., 1990). ABA increases in many plants in response to environmental stresses such as salinity and drought, and has been shown to regulate the expression of a number of genes (Skriver and Mundy, 1990). Perhaps ABA plays a role in the signal transduction pathway between exposure to salinity and enhanced H^+-ATPase gene expression. As isoform specific probes and expression constructs are developed, it will become possible to evaluate the influence of NaCl (and possibly ABA as well) on the expression of individual isoforms.

While the expression of both the plasma membrane H^+-ATPase and 70 kD subunit of the tonoplast H^+-ATPase respond to salinity, they do so in a non-coordinated manner. The study of the effect of NaCl on H^+-ATPase genes will provide a useful system to begin elucidating the hierarchy of developmental, cell type, and environmental control of gene expression.

Literature Cited

Binzel ML, Hess FD, Bressan, RA, Hasegawa PM (1988) Intracellular compartmentation of ions in salt adapted tobacco cells. Plant Physiol 86:607-614

Blumwald E, Poole RJ (1985) Na^+/H^+ antiport isolated tonoplast vesicles from storage tissue of *Beta vulgaris*. Plant Physiol 78:163-167

Braun Y, Hassidim M, Lerner HR, Reinhold L (1986) Studies on H^+-translocating ATPases in plants of varying resistance to salinity. Plant Physiol 81:1050-1056

Braun Y, Hassidim M, Lerner HR, and Reinhold L (1988) Evidence for a Na^+/H^+ antiporter in membrane vesicles isolated from roots of the halophyte *Atriplex nummularia*. Plant Physiol 87:104-108

Bremberger C, Haschke HP, Luttge U (1988) Separation and purification of the tonoplast ATPase and pyrophosphatase from plants with constitutive and inducible Crassulacean acid metabolism. Planta 175:465-470

Chomiczinski P, Sacchi N (1987) Single step method of RNA isolation by acid guanidinium thiocyanate-phenol-chloroform extraction. Anal Biochem 162:156-159

Daie J (1988) Mechanism of drought-induced alterations in assimilate partitioning and transport in crops. Crit Rev In Plant Sci 7:117-137

Dellaporta SL, Wood J, Hicks JB (1983) A plant DNA minipreparation: Version II. Plant Mol Biol Rep 1:19-21

Dupont FM, Tanaka CK, Hurkman WJ (1988) Separation and immunological characterization of membrane fractions from barley roots. Plant Physiol 86:717-724

DuPont FM (1992) Salt-Induced Changes in Ion Transport: Regulation of Primary Pumps and Secondary Transporters. *In* DT Cooke and DT Clarkson, eds, Transport and Receptor Proteins of Plant Membranes. Plenum Press, NY, pp 91-100

Ewing NM, Wimmers LE, Meher DJ, Chetelet RT, Bennet AB (1990) Molecular cloning of tomato plasma membrane H^+-ATPase. Plant Physiol 94:1874-1881

Flowers TJ, Troke PF, Yeo AR (1977) The mechanism of salt tolerance in halophytes. Annu Rev Plant Physiol 28:89-191

Garbarino J, DuPont FM (1988) NaCl induces a Na^+/H^+ antiport in tonoplast vesicles from barley roots. Plant Physiol 86:0231-0236

Greenway H, Munns R (1980) Mechanisms of salt tolerance in nonhalophytes. Annu Rev Plant Physiol 31:149-190

Hajibagheri MA, Harvey DMR, Flowers TJ (1987) Quantitative ion distribution within root cells of salt-sensitive and salt-tolerant maize varieties. New Phytol 105:367-379

Hajibagheri MA, Yeo AR, Flowers TJ, Collins JC (1989) Salinity resistance in *Zea mays*: fluxes of potassium, sodium and chloride, cytoplasmic concentrations and microsomal membrane lipids. Plant Cell and Environ 12:753-757

Harper JF, Manney L, Dewitt ND, Yoo MH, Sussman MR (1990) The *Arabidopsis thaliana* plasma membrane H⁺-ATPAse multigene family. Genomic sequence and expression of a third isoform. J Biol Chem 265:13601-13608

Harvey DMR (1985) The effects of salinity on ion concentrations within the root cells of *Zea mays* L. Planta 165:242-248

Harvey DMR, Thorpe Jr (1986) Some observations on the effects of salinity on ion distributions and cell ultrastructure in wheat leaf mesophyll cells. J Exp Bot 37:1-7

Matoh T, Ishikawa T, Takahashi E (1989) Collapse of ATP-induced pH gradient by sodium ions in microsomal membrane vesicles prepared from *Atriplex gmelini* leaves. Plant Physiol 89:180-183.

Matsumoto H, Chung GC (1988) Increase in proton-transport activity of tonoplast vesicles as an adaptive response of barley roots to NaCl stress. Plant Cell Physiology 29:1133-1140

Narasimhan ML, Binzel ML, Perez-Pratt E, Chen Z, Nelson DE, Singh NK, Bressan RA, Hasegawa PM (1991) NaCl regulation of tonoplast ATPase 70-kd subunit mRNA in tobacco cells. Plant Physiol 97:562-568

Niu X, Zhu JK, Narasimhan ML, Bressan RA, Hasegawa PM (1993a) Plasma membrane H⁺-ATPase gene expression is regulated by NaCl in halophyte (*Atriplex nummularia* L.) cell cultures. Planta In Press

Perez-Pratt E, Narasimhan ML, Niu X, Botella MA, Bressan RA, Valpuesta B, Hasegawa PM, Binzel ML (1993) NaCl induction of plasma membrane H⁺-ATPase mRNA accumulation in tobacco cells. Plant Cell and Environment (In press)

Pitman MG, Lauchli A, Stelzer R (1981) Ion distribution in roots of barley seedlings measured by electron probe x-ray microanalysis. Plant Physiol 68:673-679.

Poljakoff-Mayber A. (1982) Biochemical and physiological responses of higher plants to salinity stress. *In* A San Pietro, ed, Biosaline Research, A Look to the Future. Plenum Press, NY, pp 245-269

Rea PA, Poole RJ (1993) Vacuolar H⁺-translocating pyrophosphatase. Annu Rev Plant Physiol Mol Biol 44:53-76

Reinhold L, Braun Y, Hassidim M, Lerner HR (1988) The possible role of various membrane transport mechanisms in adaptation to salinity. *In* JH Cherry, ed, Biochemical & Physiological mechanisms associated with environmental stress tolerance. Springer-Verlag, Berlin, pp 121-130

Reuveni M, Bennett AB, Bressan RA, Hasegawa PM (1990) Enhanced H⁺-transport capacity and ATP hydrolysis activity of the tonoplast H⁺-ATPase after NaCl adaptation. Plant Physiol 94:524-530

Roldan M, Donaire JP, Pardo JM, Serrano R (1991) Regulation of root plasma membrane H⁺-ATPase in sunflower seedlings. Plant Science, 79:163-172

Sacher RF, Staples RC, Robinson RW (1982) Saline tolerance in hybrids of *Lycopersicon esculentum* x *Solanum penellii* and selected breeding lines. *In* A San Pietro, ed, Biosaline Research: A Look to the Future. Plenum Press, NY, pp 325-336

442

Sangez-Aguayo I, Gonzalez-Utor AL, Medina A (1991) Cytochemical localization of ATPase activity in salt-treated and salt-free grown *Lycopersicon esculentum roots*. Plant Physiol 96:153-158

Sambrook J, Fritsch Ef, Maniatis F (eds) (1989) Molecular cloning, a laboratory manual. Cold Spring Harbor Laboratory Press, NY

Skriver K, Mundy J (1990) Gene expression in response to abscisic acid and osmotic stress. Plant Cell 2:503-512

Storey R, Pitman MG, Stelzer R, Carter C (1983) X-ray microanalysis of cells and cell compartments of *Atriplex spongiosa*. J Exp Bot 34:778-794

Struve I, Rausch T, Bernasconi P, Taiz L (1990) Structure and function of the promoter of the carrot V-type H^+-ATPase catalytic subunit gene. J Biol Chem 265:7927-7932

Sze H (1985) H^+-translocating ATPases. Advances using membrane vesicles. Annu Rev Plant Physiol 36:175-208

Wyn Jones RG, Gorham J (1983) Aspects of salt and drought tolerance in higher plants. *In* T Kosuge, CP Meredith, A Hollaender, eds, Genetic Engineering of Plants, An Agricultural Perspective, Plenum Press, NY, pp 355-370

Yeo AR (1981) Salt tolerance in the halophyte *Suaeda maritima* L. Dum: intracellular compartmentation of ions. J Exp Bot 32:487-497

Yeo AR, Flowers TJ (1984) Mechanisms of salinity resistance in rice and their role as physiological criteria in plant breeding. *In* RC Staples, GH Toenniessen, eds, Salinity tolerance in plants. Strategies for crop improvement. J Wiley and Sons, NY, pp 151-170

Yeo AR, Lauchli A, Kramer D (1977) Ion measurements by x-ray microanalysis in unfixed, frozen, hydrated plant cells of species differing in salt tolerance. Planta 134:35-38

SOLUTE REGULATION BY CALCIUM IN SALT-STRESSED PLANTS

André Läuchli, Timothy D. Colmer, Teresa W-M. Fan and Richard M Higashi
Department of Land, Air and Water Resources, University of California, Davis, CA 95616, U.S.A.

ABSTRACT

An overview is presented on the documented interactions that Ca has with salt stress effects on plants. Significant interactions at the cell level occur in the cell wall, at membranes and membrane transport processes, cytosolic Ca^{2+} activity, and intracellular solute regulation, all contributing to effects on cell elongation. Recent research links the dynamics of cell growth quantitatively to the spatial distribution of solutes, with Ca playing a major regulatory role.

Salt stress interfered with Ca allocation to the growth zone of a young sorghum leaf, but supplemental Ca partly mitigated this effect. Regulation of organic solutes was investigated by means of ^{1}H-NMR spectroscopy and GLC. Proline and asparagine accumulated in the young sorghum leaf, more so for plants supplied with low Ca where growth was curtailed by salt. Salt stress also had major effects on photosynthetic and carotenoid pigments, as determined by HPLC. Chlorophyll a and b and violaxanthin levels decreased but zeaxanthin increased, and high Ca supply amplified these responses to salt. The depletion of violaxanthin may be related to enhanced ABA synthesis.

OVERVIEW AND INTRODUCTION

Many crop plants are sensitive to saline conditions in their root environment. Among the various growth and developmental processes, leaf elongation is recognized as the most salt-sensitive growth process (reviews: Munns and Termaat, 1986; Läuchli and Epstein, 1990). Salinization of the root medium may bring about cessation of leaf elongation within minutes (Thiel et al., 1988; Yeo et al., 1991), probably due to a limitation of water transport to the shoot. Longer-term effects on growth are related to salt accumulation in the leaves and disturbed mineral nutrition (Läuchli and Epstein, 1990). Calcium supply plays an important role in the responses of plants to salinity. Calcium-salinity

T.D.C. gratefully acknowledges support by the Hackett Studentship from the University of Western Australia. We owe thanks to Dr. Andrew Lane for his expert advice on NMR-spectroscopy, and to Dr. Susan Ustin for the use of her HPLC detector.
Corresponding author: André Läuchli, FAX 916-752-1552.

NATO ASI Series, Vol. H 86
Biochemical and Cellular Mechanisms
of Stress Tolerance in Plants
Edited by J. H. Cherry
© Springer-Verlag Berlin Heidelberg 1994

interactions were recognized by LaHaye and Epstein (1971) who found that elevated Ca^{2+}

concentrations in the medium (up to 10 mM) protected bean plants from the adverse effects of NaCl.

Since then, numerous examples of similar calcium-salinity interactions have been documented in the

literature, and NaCl stress has been shown to induce Ca deficiency in leaves in a number of plants

(e.g. Maas and Grieve, 1987; Grieve and Maas, 1988; Ehret et al., 1990). The mechanisms of

calcium-salinity interactions in plants have also received broad attention in recent years; pertinent

reviews are by Läuchli and Schubert, 1989; Läuchli, 1990; Rengel, 1992). Table I lists possible

levels of interactions of Ca with cellular responses to salt stress.

Table I. Possible levels of interactions of Ca with cellular responses to salt stress

Cell walls
Membranes and solute fluxes
Solute compartmentation
Organic solutes
Enzymes and biosynthetic pathways
Cytosolic Ca^{2+} activity
Cell growth and development

In plant cells the cell wall compartment is directly exposed to salt stress. Oertli (1968) first

postulated that for plants exposed to salt stress, ions could accumulate in leaf cell walls resulting in

cellular dehydration. Using X-ray microanalysis, Flowers et al. (1991) recently confirmed this

hypothesis for rice leaves. Only few studies, however, addressed the mechanical and biochemical

effects that may occur in cell walls following salt accumulation. Cramer (1992) found that salt stress

reduced cell wall extensibility in maize leaves, and earlier Iraki et al. (1989) determined alterations in

both physical and chemical structure of the primary cell wall following cell adaptation to osmotic

stress. Only very recently Zhong and Läuchli (1993) showed chemical changes in cell walls induced

by salt stress that were fully mitigated by supplemental Ca, i.e., decreased cellulose content and

inhibition of polysaccharide degradation. Clearly more research is needed on calcium-salinity

interactions with the cell wall.

More information is available on calcium-salinity interactions at the level of cell membranes and membrane transport, and the reader is referred to several recent reviews (Läuchli and Schubert, 1989; Läuchli, 1990; Rengel, 1992). The mechanism by which supplemental Ca protects the plasma membrane structure and function is unknown. However, the protective effect of Ca results in maintenance of K/Na-selectivity (Läuchli, 1989; Zhong, 1991) and prevention of ion and organic solute leakage through the membrane (Whittington and Smith, 1992; Colmer, Fan, Higashi and Läuchli, unpublished results). The latter effect appears to be rather unspecific and qualitatively similar to membrane leakage observed during anoxia (Zhang et al., 1992). Such membrane effects must influence solute compartmentation in plant cells, but experimental evidence regarding the putative role of Ca in this feature is lacking. Also lacking is information on the possible role of Ca in the accumulation of organic solutes and enzyme activities in salt stressed plant cells; first data on organic solutes will be presented in the experimental part of this paper.

Central to our understanding of the cellular mechanisms of plant response to salt stress is the finding that high concentrations of NaCl elicit a dramatic rise in cytosolic Ca^{2+} activity within minutes (short-term response) (Bittisnich et al., 1989; Lynch et al., 1989). Changes in cytosolic Ca^{2+} activity mediate specific types of signal transduction in plant cells (Marme, 1989; Johannes et al., 1991). This may trigger alterations in gene expression, and in metabolism, growth and development (Guilfoyle, 1989). However, further research is required to determine the cellular processes which are regulated by these changes in cytosolic Ca^{2+} activities.

Recent quantitative approaches by W.K.Silk on the spatial distribution of growth and solutes in developing organs (Meiri et al., 1992; Table II) led us to a better mechanistic understanding of the role of Ca and other solutes in growth responses to salt stress (Zhong, 1991; Bernstein, 1992). In sorghum leaves, 100 mM NaCl shortened the length of the growth zone from 30 to about 24 mm and reduced the maximal relative elemental growth rate (REGrate) (Bernstein et al., 1993b). Increased Ca supply from 1 to 10 mM restored the length of the growing zone of both emerged and unemerged

Table II.   Experimental approaches used to study plant responses to salt stress in our laboratory
and that of W.K. Silk (see Silk, 1984; Meiri et al., 1992).

X-ray microanalysis	——	elemental localization
^{31}P NMR spectroscopy	——	regulation of cellular pH and energy metabolism
^{1}H NMR spectroscopy	——	organic solutes and metabolites
GLC	——	organic solutes and metabolites
HPLC	——	plant pigments
Growth kinematics (W.K. Silk)	——	spatial distribution of growth and ion depositon rates

salinized leaves and increased the peak value of the REGrate (Bernstein et al., 1993a). Salt stress induced a decrease in K concentration in the basal part of the leaf growth zone and a dramatic decrease in Ca concentration that could be responsible for leaf growth inhibition (Bernstein, 1992). Earlier, Lynch et al. (1988) found that reduced Ca availability in expanding leaves contributes to growth reduction in salt stressed barley seedlings. Sodium did not accumulate in the region of decreased leaf growth, and hence a high concentration of Na in the tissue does not appear to be the cause of the observed pattern of growth inhibition (Bernstein, 1992). The latter conclusion is in agreement with earlier observations and conclusions reached by Termaat et al. (1985) and Cheeseman(1988). Interestingly, the K concentration was found to be greatly elevated by salt stress in the mature zone of the youngest leaf (Bernstein et al., 1992). Potassium may serve here as a transient osmolyte in response to salt stress.

Our laboratory, partly in collaboration with that of W.K.Silk, is using a variety of experimental techniques to unravel the plant cell responses to salt stress and its interaction with Ca (Table II), the focus being on solute regulation by Ca in plants. In addition to the methods developed by Silk, as highlighted above, elemental localization by X-ray microanalysis can contribute much needed information on the cellular and tissue distribution of salt and nutrient elements in leaf and other tissues under salt stress (Lazof and Läuchli, 1991; Leigh and Storey, 1993). We are also pursuing complementary information on organic solutes and biochemistry of salt stress, using NMR spectroscopy and chromatographic techniques. These approaches hold much promise, and recent results that support this premise are presented in this paper.

MATERIALS AND METHODS

Plant growth

Seeds of *Sorghum bicolor* cv Hegari were germinated at 28°C in the dark on a mesh screen over aerated 10x diluted modified Hoagland solution (pH 5.7) (Epstein, 1972), with 50 μM Fe EDTA and 20 μM $Fe(NH_4SO_4)_2$. After 2d, seedlings were transferred to aerated 4x diluted modified Hoagland solution with $Ca(NO_3)_2.4H_2O$ at 0.5 mM and $NaNO_3$ at 1mM to replace the NO_3^- removed by the lower Ca supply. At this time, the shoots of the seedlings were exposed to full sunlight ($\sim$ 14 h/d) in a greenhouse at 28.5/19.0°C day/night. When seedlings were 3d old (2nd leaf emerging), the Ca and NaCl treatments were imposed in fresh solutions renewed daily. The 0.5 mM Ca treatment received no supplemental $CaSO_4$, while the 5.0 mM Ca treatment received 4.5 mM $CaSO_4$. The youngest leaves (leaf #2) were harvested from seedlings after 2d of exposure to the Ca and NaCl treatments.

Analysis of inorganic ions

Leaf tissues were oven dried at 65 °C for 2d, followed by extraction in 0.5 M HCl for 2d. Flame atomic emission spectrophotometry was used to measure K^+ and Na^+, while flame atomic absorption spectrophotometry was used to measure Ca^{2+}. Each treatment was replicated three times, with ten plants per replicate.

Perchloric acid extraction

Sorghum tissues were harvested, lyophilized, and pulverized using a Braun Mikro-Dismembrator II to yield 1-2 μm particles. Water-soluble low molecular weight solutes were extracted from the tissue powders with 5% (w/v) perchloric acid (PCA) at 4°C, as described

previously (Fan et al., 1986; Fan et al., in press). Briefly, the PCA extracts were adjusted to pH 3 to

3.5 with K_2CO_3 to precipitate perchlorate as the insoluble potassium salt. An aliquot of the extract

was lyophilized, silylated with MTBSTFA:acetonitrile (1:1, v/v), sonicated at 60°C for 3 h, and

allowed to stand overnight before GLC analysis. The remaining extract was titrated to pH 7 with

KOH, followed by passage through Chelex 100 resin (200-400 mesh, BioRad, Richmond, CA, USA),

lyophilization, and redissolution in ^2H_2O for ^1H NMR spectroscopy.

Gas liquid chromatography

GLC-FID analysis of the silylated samples was performed using a Varian 3300 gas

chromatograph (Varian Instruments, Palo Alto, CA, USA) outfitted with a 0.18 mm i.d. x 40 m DB-

1 (0.4 μm coat) open tubular column (J&W Scientific, Folsom, CA, USA) with temperature of the

column held at 60°C for 2 min, followed by ramping to 150°C at 20°C/min, then to 300°C at

6°C/min, injector and flame ionization detector temperatures at 260 and 320°C, respectively, and H_2

carrier gas velocity set at 40 cm/s.

^1H NMR spectroscopy

^1H NMR spectra were acquired from PCA extracts of *S. bicolor* tissues at 600 MHz using a

Varian Unity-600 NMR spectrometer. Acquisition parameters included a 90° observe pulse, a 1 s

pulse delay, a 1.5 s acquisition time, a 6000 Hz spectral width, 18000 points, and 256 transients.

The spectral assignment of various components was based on our previous data (Fan et al., 1986)

and confirmed by two-dimensional total correlation spectroscopy (TOCSY) of the same extracts

according to a previous procedure (Fan et al., in press).

HPLC

Leaf tissues were harvested, lyophilized, and pulverized using a Braun Mikro-Dismembrator

II to yield 1-2 μm particles. All the following procedures were performed under red light or in the

dark. Tissue powder was extracted by adding 1mL isopropanol per 20-25 mg of tissue, mixed, let stand for 1 h, centrifuged to pellet the majority of particles, and the supernatant filtered through a

0.22 μm Teflon filter to yield the final extract. This procedure typically yields an extract that exhibits

a stable pigment profile for > 1 month when stored at -20°C; nevertheless, all *Sorghum* samples were

stored at -70 °C and run within 24 h of extraction. HPLC consisted of a C_{18} Adsorbosphere 5 μm

particle, 4.6 mm x 250 mm column (Alltech Associates, Deerfield, IL, USA) fitted to a ISCO ternary

gradient HPLC system (ISCO, Lincoln, NB, USA) which employed a ternary gradient program

identical to that of Wright et al. (1991), except for a 2.5 min delay of the gradient onset. Detection

was by spectrophotometry on a Dionex DSA-1 instrument (Dionex Inc., San Jose, CA, USA)

scanning from 370-800 nm (approximately 0.67 sec/spectrum). Pigments were identified and

quantified by comparison to authentic standards for Chl a and Chl b (Sigma Chemical Co., St. Louis,

MO, USA), by relative retention, collection of peaks, and spectral comparison to literature data for

violaxanthin, zeaxanthin, and lutein (Davies, 1976; Wright et al., 1991), or by relative retention plus

spectral comparison to literature data (Rüdiger and Schoch, 1988; Wright et al., 1991) for the rest.

RESULTS AND DISCUSSION

Impact of Ca on salt stress responses of young S. bicolor leaves

1. Growth

Supplemental Ca has previously been shown to enhance the growth of non-halophytes exposed

to NaCl (LaHaye and Epstein, 1971). In the present experiments, root growth of *Sorghum* seedlings

exposed to 150 mM NaCl for 2d was reduced by 80% for plants supplied with 0.5 mM Ca, but only

by 18% for plants supplied with 5.0 mM Ca. On the other hand, 150 mM NaCl reduced the shoot

growth of *Sorghum* seedlings by about 45% after 2 d regardless of the external Ca concentration.

Upon longer exposure to 5.0 mM Ca, the salt-induced reduction of shoot growth was partially

ameliorated. These results indicated that for plants exposed to NaCl, the benefical effect of Ca on

shoot growth may be delayed relative to that on root growth.

2. Inorganic solutes

Na concentrations were 33% lower in the youngest leaf of *Sorghum* seedlings exposed to 150 mM NaCl with 5.0 mM Ca, compared to 0.5 mM Ca (Table III). Hence, supplemental Ca supply enhanced the ability for Na exclusion from the leaves of *Sorghum*, as has been previously reported for other non-halophytes (e.g. LaHaye and Epstein, 1971). Furthermore, leaf K concentrations of seedlings supplied with 0.5 mM Ca and exposed to 150 mM NaCl were reduced by 36%; whereas seedlings supplied with 5.0 mM Ca were able to maintain their youngest leaf K concentrations at levels observed in the control plants (Table III). The enhanced exclusion of Na and maintenance of K concentrations in the youngest leaf of plants supplied with 5.0 mM Ca is presumably related to the longer term protective effect of Ca on the growth of non-halophytes exposed to NaCl. Exposure to 150 mM NaCl for 2 d resulted in large decreases in the Ca concentrations of the youngest leaf of seedlings supplied with 0.5 (84% decrease) or 5.0 mM Ca (66% decrease), which may have resulted in Ca deficiency in the NaCl treated plants supplied with 0.5mM Ca (Table III). However, it is noteworthy that high Ca and salt treatment led only to a slight Ca decrease in the leaf as compared to the control treatment.

3. Organic solutes

It is clear from Figures 1 and 2 and Table III that salt treatment elicited an accumulation of Pro and Asn in young Sorghum leaves; these two metabolites have been implicated as organic solutes in both vascular and non-vascular plants (Ullrich, 1956; Flowers et al., 1977; Yancey et al., 1982; Munns et al., 1983; Wyn Jones and Gorham, 1983). Pro has been demonstrated to accumulate 80-fold during salt adaptation in suspension culture of tobacco (Rhodes and Handa, 1989). It was also shown to accumulate in sorghum leaves under salt Stress (Weimberg et al., 1982). However, in our

Table III. Effect of Ca on selected salt-induced responses (inorganic and organic solutes; chloroplastic pigments) in young *Sorghum* leaves.

| | Treatment | | | | | | |
	#1 Low Ca & Low NaCl[a]	#2 Low Ca & High NaCl[a]	#3 High Ca & Low NaCl[a]	#4 High Ca & High NaCl[a]	#5 % change #2/#1	#6 % change #2/#1	#7 Interaction of Ca with salt stress[d]
Solutes							
K_b	78.7 (0.3)	50.4 (1.8)	83.3 (1.2)	81.7 (0.8)	-36.0	-1.9	1.53
Na_b	0.9 (0.3)	104.9 (1.1)	0.8 (0.1)	70.2 (5.6)	11555.6	8675.0	0.75
Ca_b	4.5 (0.2)	0.7 (0.1)	10.7 (0.4)	3.6 (0.1)	-84.4	-66.4	2.16
Pro_b	0.7 (0.3)	7.8 (2.7)	0.4 (0.5)	4.3 (1.1)	1014.3	975.0	0.96
Asn_b	9.0 (1.8)	40.8 (4.9)	5.7 (3.1)	23.5 (0.5)	353.3	312.3	0.91
Glc_b	36.4 (2.6)	14.5 (1.5)	23.1 (11.6)	23.2 (8.4)	-60.2	0.4	2.52
Pigments							
Chl a[c]	143.7	88.0	168.9	66.3	-38.8	60.7	0.64
Chl b[c]	44.7	38.3	48.6	36.0	-14.3	-25.9	0.86
Chl a/Chl b	3.21	2.30	3.47	1.84	-28.3	-47.0	0.74
violaxanthin[c]	4.3	1.6	6.4	0.9	-62.8	-85.9	0.38
zeaxanthin[c]	1.1	1.4	1.0	3.0	27.3	200.00	2.36
lutein	23.0	25.4	24.5	25.8	10.4	5.3	0.95
phaeophythin a[c]	17.1	22.8	12.9	31.8	33.3	146.5	1.85

[a]Low Ca and low NaCl are 0.5 mM Ca and 1 mM Na, respectively while high Ca and high NaCl are 5 mM Ca and 150 mM NaCl, respectively.

[b]Values reported are in $\mu mol\ gm^{-1}$ fresh weight, and each is an average of three measurements; numbers in parenthesis are standard errors.

[c]Values reported are in $\mu g\ g^{-1}$ fresh weight, and each represents a single measurement.

[d]Calculated as the ratios of #4/#3 to #2/#1.

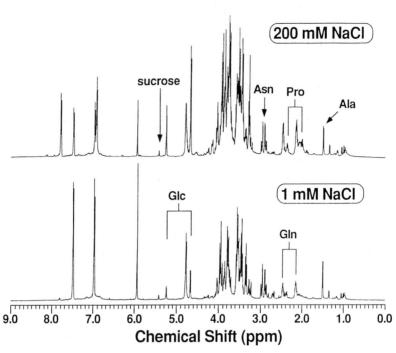

Figure 1. ¹H NMR spectra of PCA extracts of *S. bicolor* leaves subjected to NaCl treatments.

The treatment of *Sorghum* seedlings, PCA extraction, and ¹H NMR spectroscopy were as described in Materials and Methods. The two spectra shown illustrate the changes of water-soluble solute pattern in young *Sorghum* leaves induced by 200 mM NaCl treatment. The principal changes elicited by salt treatment included an increase in [Pro] and [Gln] (see also Table III).

study (Table III), the concentration of both Pro and Asn in the youngest leaf under salt stress was considerably lower at high Ca in comparison with low Ca. Thus, the accumulation of these two amino acids in salt-stressed leaves at low Ca may be important for maintenance rather than continued growth, as was suggested earlier for Pro accumulation by Greenway and Munns (1980). It should also be noted that salt caused a considerable decrease in Glc concentration at low Ca (Table III). This may be related to a diversion of carbon to Pro and Asn synthesis or an inhibition of dark carbon fixation. When superimposing Ca on the salt treatment, high Ca mitigated the salt effect on [Glc].

453

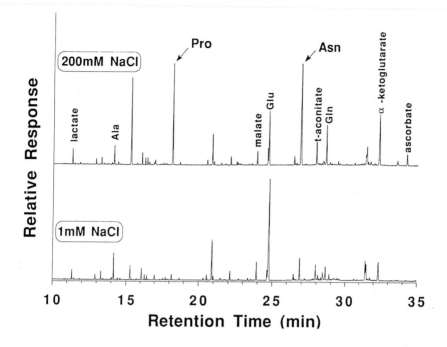

Figure 2. Gas Chromatograms of PCA leaf extracts of *S. bicolor* plants exposed to NaCl treatments.

Plant growth, PCA extraction, derivatization and gas chromatography procedures were as described in Materials and Methods. The two gas chromatograms show the resolution capacity of this technique for soluble metabolites such as amino acids, organic acids and other primary metabolites. Comparison of the 1 and 200 mM NaCl treatments demonstrates an accumulation of Pro and Asn in salt treated *Sorghum* leaves.

These data indicate that Ca may play a role in photosynthetic carbon fixation in addition to the

uptake/transport of inorganic solutes under salt stress.

4. Plant pigments

Table III and Figure 3 show the combined impact of Ca and salt on major photosynthetic and

carotenoid pigments. Both [Chl a] and [Chl b] decreased while [phaeophytin a] increased with 150

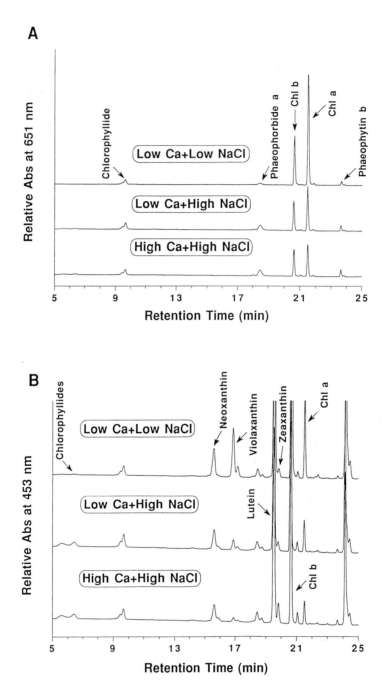

Figure 3. HPLC chromatograms of *S. bicolor* leaf isopropanol extracts under different salt and Ca treatments.

The extraction and HPLC procedures were as described in Materials and Methods. Panel A illustrates the 651 nm-based chromatograms for the three combinations of Ca and salt treatments while Panel B shows the 453 nm-based chromatograms from the same HPLC runs; 651 nm and 453 nm were chosen for the best average depiction of chlorophylls and carotenoids, respectively. Low and high Ca refer to 0.5 and 5 mM $CaCl_2$ while low and high NaCl represent 1 and 150 mM NaCl, respectively. Major changes in levels of violaxanthin, zeaxanthin, Chl a, Chl b, chlorophyllides, phaeophytin b, and phaeophorbide a in response to salt and Ca treatments are evident.

mM NaCl treatment. These salt effects could result from a disturbance of the photosynthetic apparatus. However, these changes may represent more than a simple breakdown of Chl to phaeophytin since the decrease in [Chl a] was not balanced by the increase in [phaeophytin a] and the Chl a to Chl b ratio was also affected. In addition, Ca appeared to enhance the salt effect on these photosynthetic pigments, which further complicates the interpretation. One possible explanation is that the Ca enhancement of [phaeophytin a] is associated with some alteration of the photosynthetic apparatus (rather than strictly a result of Chl a breakdown) since phaeophytin appears to be a functional component of photosystem II (Cogdell, 1988). This may be related to the requirement for bound Ca in electron flow of photosystem II (Crane and Barr, 1985).

As for carotenoid pigments, salt treatment had opposite effects on two major xanthophylls, violaxanthin and zeaxanthin (Table III). Moreover, high Ca treatment amplified the effect of salt on these two xanthophylls, while the level of the predominant xanthophyll, lutein, was unaffected by either salt or Ca treatment. Violaxanthin and zeaxanthin are the stoichiometrically interconverted end-products of the "xanthophyll cycle", which has been recently proposed to play a photoprotective role in vascular plants (Demmig-Adams and Adams III, 1992) and is related to changes in phytochemical efficiency (*c.f.* Björkman, 1987). Interestingly, the decrease in [violaxanthin] was not compensated by the increase in [zeaxanthin] in the salt treatments, as would be expected if the xanthophyll cycle was the only reason for the changes. It should also be noted that changes in both [violaxanthin] and [zeaxanthin] were further enhanced by the high Ca and high salt treatment. Since violaxanthin is reported to be a precursor of ABA synthesis (Zeevaart, in press), it is possible that its excess depletion under salt and Ca treatments is associated with enhanced ABA synthesis. Further studies are needed to clarify the interplay of Ca, ABA, the xanthophyll cycle, and changes in the photosynthetic apparatus in response to salt stress, including measurement of tissue ABA levels with the various treatments.

LITERATURE CITED

Bernstein N (1992) Distribution of growth and mineral element deposition in sorghum leaves: effects of salinity and calcium. PhD thesis. University of California, Davis

Bernstein N, Läuchli A, Silk WK (1992) Spatial and temporal distribution of inorganic solutes in salt affected leaves. Current Topics in Plant Biochemistry and Physiology, Vol 11. University of Missouri, Columbia, p 305

Bernstein N, Läuchli A, Silk WK (1993a) Kinematics and dynamics of sorghum leaf elongation at various Na/Ca salinities: distribution of expansive growth. Plant Physiol (in press)

Bernstein N, Silk WK, Läuchli A (1993b) Growth and development of sorghum leaves under conditions of NaCl stress: I. Spatial and temporal aspects of leaf growth inhibition. Planta (in press)

Bittisnich D, Robinson D, Whitecross M (1989) Membrane-associated and intracellular free calcium levels in root cells under NaCl stress. *In* J Dainty, MI DeMichelis, E Marré, F Rasi-Caldogno, eds, Plant Membrane Transport: The Current Position. Elsevier, Amsterdam-New York, pp 681-682

Björkman O (1987) High-irradiance stress in higher plants and interaction with other stress factors. *In* J. Biggins, ed., Progress in Photosynthesis Research, Vol. IV, Martinus Nijhoff Publishers, Dordrecht, pp 11-18

Cheeseman JM (1988) Mechanisms of salinity tolerance in plants. Plant Physiol 87:547-550

Cogdell, R. (1988) The function of pigments in chloroplasts. *In* T.W. Goodwin, ed., Plant Pigments, Academic Press, London, pp 183-230

Cramer GR (1992) Kinetics of maize leaf elongation. II. Responses of a Na-excluding cultivar and a Na-including cultivar to varying Na/Ca salinities. J Exp Bot **43**:857-864

Crane FL Barr R (1985) Function of calcium-calmodulin in chloroplasts. *In* A.J. Trewavas, ed., Molecular and Cellular Aspects of Calcium in Plant Development, Plenum, New York, pp 269-276

Davies BH (1976) Carotenoids. *In* T.W. Goodwin, ed., Chemistry and Biochemistry of Plant Pigments, Academic Press, London, pp 108-124, 150-153

Demmig-Adams B, Adams III WW (1992) Photoprotection and other responses of plants to high light stress. Annu. Rev. Plant Physiol. Plant Mol. Biol. **43**: 599-626

Ehret DL, Redman RE, Harvey BL, Cipywnyk A (1990) Salinity-induced calcium deficiencies in wheat and barley. Plant and Soil **128**:143-151

Epstein E (1972). Mineral Nutrition of Plants: Principles and Perspectives, John Wiley and Sons, New York

Fan T W-M, Higashi RM, Lane AN, Jardetzky O (1986) Combined use of ^{1}H-NMR and GC-MS for metabolite monitoring and in vivo ^{1}H-NMR assignments. Biochim. Biophys. Acta **882**: 154-167

Fan T W-M, Lane AN, Higashi RM Energy and Fermentation Metabolism in Hypoxic Rice Coleoptiles – A Multinuclear NMR Approach", *In* NATO Advanced Research Workshop "Interacting Stresses on Plants in a Changing Climate", Springer, Berlin, in press

Flowers TJ, Hajibagheri MA, Yeo AR (1991) Ion accumulation in the cell walls of rice plants growing under saline conditions: evidence for the Oertli hypothesis. Plant, Cell and Environment **14**:319-325

Flowers TJ, Troke PF, and Yeo AR (1977) The mechanism of salt tolerance in halophytes. *Ann. Rev. Plant. Physiol.* **29**: 89-121

Greenway H, Munns R (1980) Mechanisms of salt tolerance in nonhalophytes. Ann Rev Plant Physiol **31**:149-190

Grieve CM, Maas EV (1988) Differential effects of sodium/calcium ratios on sorghum genotypes. Crop Sci **28**:659-665

Guilfoyle TJ (1989) Second messengers and gene expression. *In* WF Boss, DJ Morre, eds, Second Messengers in Plant Growth and Development, Alan R Liss, Inc, New York, pp 315-326

Iraki NM, Bressan RA, Hasegawa PM, Carpita N (1989) Alteration of physical and chemical structure of the primary cell wall of growth-limited plant cells adapted to osmotic stress. Plant Physiol **91**:39-47

Johannes E, Brosnan JM, Sanders D (1991) Calcium channels and signal transduction in plant cells. BioEssays **13**:331-336

LaHaye PA and Epstein E (1971). Calcium and salt toleration by bean plants. Physiol. Plant. **25**: 213-218

Läuchli A (1989) Selectivity and energy-coupling of cation uptake. *In* Proc. 21st Colloquium, International Potash Institute, Bern, Switzerland, pp 13-26

Läuchli A (1990) Calcium, salinity and the plasma membrane. *In* RT Leonard, PK Hepler, eds, Calcium in Plant Growth and Development. The American Society of Plant Physiologists Symposium Series, Vol 4, Rockville, Maryland, pp 26-35

Läuchli A, Epstein E (1990) Plant responses to saline and sodic conditions. *In* KK Tanji ed, Agricultural Salinity Assessment and Management. American Society of Civil Engineers, New York, pp 113-137

459

Läuchli A, Schubert S (1989) The role of calcium in the regulation of membrane and cellular growth processes under salt stress. *In* JH Cherry, ed, Environmental Stress in Plants. Biochemical and Physiological Mechanisms. NATO ASI Series, Vol G19. Springer-Verlag, Berlin, pp 131-138

Lazof D, Läuchli A (1991) The nutritional status of the apical meristem of *Lactuca sativa* as affected by NaCl salinization: an electron-probe microanalytic study. Planta **184**: 334-342

Leigh RA, Storey R (1993) Intercellular compartmentation of ions in barley leaves in relation to potassium nutrition and salinity. J Exp Bot **44**:755-762

Lynch J, Polito VS, Läuchli A (1989) Salinity stress increases cytoplasmic Ca activity in maize root protoplasts. Plant Physiol **90**:1271-1274

Lynch J, Thiel G, Läuchli, A (1988) Effects of salinity on the extensibility and Ca availability in the expanding region of growing barley leaves. Botanica Acta **101**:355-361

Maas EV, Grieve CM (1987) Sodium-induced calcium deficiency in salt stressed corn. Plant, Cell and Environment **10**:559-564

Marmé D (1989) The role of calcium and calmodulin in signal transduction. *In* WF Boss, DJ Morré, eds, Second Messengers in Plant Growth and Development. Alan R Liss, Inc, New York, pp 57-80

Meiri A, Silk WK, Läuchli A (1992) Growth and deposition of inorganic nutrient elements in developing leaves of *Zea mays* L. Plant Physiol **99**:972-978

Munns R, Greenway H, Kirst GO (1983) Halotolerant eukaryotes. *In* O.L. Lange, P.S. Nobel, C.G. Osmond, and H. Ziegler, Eds., Physiological Plant Ecology III., Encyclopedia of Plant Physiology, Vol. 12C, Springer-Verlag, Berlin, pp 59-136

Munns R, Termaat A (1986) Whole plant responses to salinity. Aust J Plant Physiol **13**: 143-160

Oertli JJ (1968) Extracellular salt accumulation, a possible mechanism of salt injury in plants. Agrochimica **12**: 461-469

Rengel Z (1992) The role of calcium in salt toxicity. Plant, Cell and Environment **15**: 625-632

Rhodes D, Handa S (1989) Amino acid metabolism in relation to osmotic adjustment in plant cells. *In* JH Cherry, ed, Environmental Stress in Plants. Biochemical and Physiological Mechanisms. NATO ASI Series, Vol G19. Springer-Verlag, Berlin, pp 41-62

Rüdiger W, Schoch S (1988) Chlorophylls. *In* T.W. Goodwin, ed., Plant Pigments, Academic Press, London, p.8

Silk WK (1984) Quantitative description of development. Ann Rev Plant Physiol 35:479-518

Termaat A, Passioura JB, Munns R (1985) Shoot turgor does not limit shoot growth of NaCl affected wheat and barley. Plant Physiol 77:869-872

Thiel G, Lynch J, Läuchli A (1988) Short-term effects of salinity stress on the turgor and elongation of growing barley leaves. J Plant Physiol 132:38-44

Ullrich KJ (1956) *Pflügers Arch. Gesamte Physiol.* **262**: 551-561

Weimberg RW, Lerner HR, Poljakoff-Mayber A (1982) A relationship between potassium and proline accumulation in salt-stressed *Sorghum bicolor*. Physiol Plant 55:5-10

Whittington J, Smith FA (1992) Calcium-salinity interactions affect ion transport in *Chara corallina*. Plant, Cell and Environment 15:727-733

Wright SW, Jeffrey SW, Mantoura RFC, Llewellyn CA, Bjørnland T, Repeta D, Welschmeyer N (1991) Improved HPLC method for the analysis of chlorophylls and carotenoids from marine phytoplankton. Mar. Ecol. Prog. Ser. **77**: 183-196

Wyn Jones RG, Gorham J (1983) Osmoregulation. *In* O.L. Lange, P.S. Nobel, C.B. Osmond, H. Ziegler, Eds., Physiological Plant Ecology III., Encyclopedia of Plant Physiology, Vol. 12C, Springer-Verlag, Berlin, pp.35-38

Yancey PH, Clark ME, Hand SC, Bowlus RD, Somero, GN (1982) Living with water stress: evolution of osmolyte systems. Science **217**: 1214-1222

Yeo AR, Lee K-S, Izard P, Boursier BJ, Flowers TJ (1991) Short- and long-term effects of salinity on leaf growth in rice (*Oryza sativa* L.). J Exp Bot **42**:881-889

Zeevaart JAD Stress-enhanced metabolism of abscisic acid. *In* NATO Advanced Research Workshop "Interacting Stresses on Plants in a Changing Climate", Springer-Verlag, Berlin, in press.

Zhang Q, Läuchli A, Greenway H (1992) Effects of anoxia on solute loss from beetroot storage tissue. J. Exp Bot **43**:897-905

Zhong H (1991) Growth kinematics and cell wall metabolism in root tops of salt-stressed cotton seedlings (*Gossypium hirsutum* L.). PhD thesis. University of California, Davis

Zhong H. Läuchli A (1993) Changes of cell wall composition and polymer size in primary roots of cotton seedlings under high salinity. J Exp Bot **44**:773-778

THE RESPONSE OF PLANTS TO SALINITY: A WORKING HYPOTHESIS

H.R. Lerner and G.N.Amzallag
Department of Botany
The Hebrew University of Jerusalem
Jerusalem 91904
Israel

INTRODUCTION

The contrast between the sensitivity of some plants and the tolerance of others, towards salinity, has been difficult to explain. Some authors have ascribed the inhibitory effects of salt to Na^+ and Cl^- toxicity and to an osmotic effect (decrease in turgor). They have further suggested that compartmentation of the toxic ions in the vacuole and accumulation of compatible solutes in the cytoplasm, as well as the presence of genes for salt tolerance, bestow salt resistance. The demonstration that cultured cells from salt tolerant plants were not more tolerant than cells from salt sensitive plants (McCoy, 1987) as well as adaptation of cultured cells from a salt sensitive plant, tobacco, enabling them to grow in medium containing very high salt concentrations at a rate comparable to that of cells in medium containing no salt (Binzel et al., 1985, Watad et al., 1985) represent results not quite compatible with the concept of salt toxicity.

Some workers in the field expressed the opinion that the mechanisms responsible for the sensitivity of plants to salinity are not yet understood (Cheesmann, 1988; Munns,

NATO ASI Series, Vol. H 86
Biochemical and Cellular Mechanisms
of Stress Tolerance in Plants
Edited by J. H. Cherry
© Springer-Verlag Berlin Heidelberg 1994

1988). Other authors, such as Itai et al. (1968) and Vaadia
(1976) have suggested that the response of plants to salinity
derived from a supracellular regulatory mechanism. They
proposed that stress-induced changes in the quantity of
phytohormones originating from the root and transferred to the
shoot affect plant growth and development. Indeed, induction
of callus formation by appropriate exogenous auxin and
cytokinin (CK) concentrations, or regeneration and
determination of organ formation by modification of these
concentrations (Skoog and Miller, 1957), demonstrate the
central role played by hormones in integrated whole plant
development as well as in controlling differentiation of
cells, tissues and organs.

The intimate relation between hormones and photosynthate
partitioning has been demonstrated by many authors. For
example, fruit sink activity was dependent on auxin (Nitsch,
1950), while GA-stimulated internode elongation was correlated
with an increase in acid invertase activity in the vicinity of
the elongating cells, allowing enhanced sugar unloading
(Morris and Arthur, 1985). In a very elegent demonstration
Jahnke et al. (1989) followed GA-induced changes in sink
activities in a whole pea plant by monitoring movement of
^{11}C-photoassimilates while it was occurring.

The report of McDaniel et al. (1991) illustrates hormone
mediated plant development. Shoot apices of *Lolium
temulentum*, excised after exposure of the whole plant to long
day (LD), undergo floral development *in vitro* in appropriate
medium, whereas apices excised from plants exposed to short
day (SD) remain vegetative. If gibberellic acid (GA) is
omitted from the *in vitro* medium the apices of plants
exposed to LD remain vegetative. Floral development requires
both the cue from the environment, LD, as well as the hormone,
GA.

It seems that changes in hormonal balance, together with
development of competence to respond, is a central phenomenon
guiding plant development (differentiation, photosynthate

partition), see: Osborne (1987), Guern (1987) and Trewavas (1991) for a complete discussion.

From the data in the literature it is possible to develop an hypothesis describing the response of plants to salinity as deriving from environmentally-induced modifications in the hormonal balance of the plant. In the following we shall describe some of these data. For simplicity the expression "hormone" is used for plant growth regulator or phytohormone.

THE ROOT AS A SENSING ORGAN OF THE SOIL ENVIRONMENT

During the past 25 years, several authors have suggested that information concerning parameters of the soil environment are signaled to the shoot through transport of hormones from root to shoot. Indeed saline stress causes a decrease in CK and an increase in ABA transported from the root to the shoot (Itai et al., 1968; Vaadia, 1976; Downton and Loveys, 1981). Other stressing soil parameters affect endogenous hormones as well as shoot growth in a similar fashion. Mineral nutrient stress (either too little or too much minerals), results in decrease in CK, GA and IAA (for references see Table 1 in Amzallag et al., 1992).

The work of Davies and coworkers has been particularly useful in developing the concept that the root is a sensing organ. They showed that, even in absence of decreased shoot turgor, water-stressing of only part of the root system increases ABA transported to the shoot resulting in decreased shoot growth. While the part of the root which is well-watered supplies the shoot with sufficient water, the part of the root which is water-stressed sends the ABA signal. Inhibition of shoot growth results from a change in its hormonal balance (see Davies and Zhang (1991) for a review of

the literature). In a similar way low oxygen concentration at
the level of the root (flooding) causes increase in 1-
aminocyclopronane-1-carboxylic acid (ACC) or ABA, depending on
the species, which is transported to the shoot via the xylem
(Jackson, 1987). By maintaining the root of corn plants at
different temperatures, while keeping shoot temperature
constant, Atkin et al. (1973) showed that the amount of CK, GA
and ABA exported from the root is a function of root
temperature. Xylem exudate showed maximum ABA concentration
at $18^{0}C$ and maximum CK and GA concentrations at $28^{0}C$.
Lately it was shown that root penetration through compressed
media is enhanced by ethylene; transport to the shoot has not
been demonstrated yet (Zacarias and Reid, 1992).

It seems that stressing soil parameters such as salinity,
mineral nutrients, drought, lack of aeration or temperature,
all result in similar changes in hormones transported from
root to shoot. These stresses inhibit growth through their
effects on the hormonal balance of the plant.

The change in the root:shoot (R:S) ratio in stressed
plants results from the change in the plant's hormonal
balance. This can be demonstrated by exposing plants to

Table 1. *Root:Shoot ratio of* Agropyron elongatum *exposed to various NaCl concentrations*

NaCl (mM):	Ø	48	96	192	288	384
potential[a]:	Ø	-Ø.2	-Ø.4	-Ø.8	-1.2	-1.6
R:S ratio[b]:	Ø.77	Ø.86	Ø.98	Ø.63	Ø.84	Ø.4Ø

[a] change of osmotic potential, in MP, due to NaCl addition.
[b] R:S ratio calculated fron data of Weimberg (1986).

exogenous hormones (Arteca et al. 1991). Exposure to salinity
may either result in an increase or a decrease in the R:S
ratio depending on the effect of salinity on the hormonal
balance of the plant. The data from Weimberg (1986) on
Agropyron elongatum shows that exposure to 48, 96 or 288 mM
resulted in an increased, while exposure to 192 or 384 mM NaCl
resulted in decrease in the R:S ratio (Table 1).

Wareing et al. (1968) showed that CK and GA of root
origin affected photosynthesis (increse in CK or GA enhances,
while a decrease inhibits).

From these data, we can conclude that the hormonal
balance controls at the same time source and sink activities.
In plants, source and sink activities are coordinated through
several mechanisms, one of which is the hormonal balance.
Furthermore, accumulation of photosynthates for which there is
no sink results at first in inhibition of photosynthesis
(Stitt, 1991), then, in their oxidation by an overflow
mechanism, cyanide-resistant respiration (Lambers, 1982).

CHANGE OF PHYSIOLOGICAL MODE: GLYCOMODE - HALOMODE

Amzallag et al. (1990, 1993), growing plants in aerated
liquid medium, demonstrated that exposure of *Sorghum bicolor*
to salinity may result in one of two responses. Either the
plant copes with the stress using its mechanism of salt
resistance existing prior to exposure to stress, which was
labelled "pre-existing resistance", or the plant develops
mechanisms of increased resistance to salinity which was
labelled "adaptation". Adaptated plants grow and set seeds in
NaCl concentrations which are lethal for non-adapted plants.
For the plant to adapt, the exposure to salinity must be
initiated 5 to 10 days following germination (while a

developmental window for adaptation to environmental conditions is opened), and salinity has to be maintained at a stable NaCl concentration during the process of adaptation. If these requirements are not met, the plant responds to salinity through its pre-existing resistance mechanisms. If the *Sorghum* plants are sprayed, daily, with an ABA solution during the first 1Ø days of the adaptation process (exposure to salinity) the plants are adapted within 1Ø days instead of the 2Ø days required in absence of exogenous ABA. While the ABA treatment inhibits growth of control plants it enhances growth of the salt treated plants (Amzallag et al., 199Øa). If CK is added to the medium during the saline treatment the plants do not adapt; exogenous GA perturbs the adaptation process, but differently from CK (Amzallag et al., 1992). Evidently the hormonal balance of the plant plays an important role in adaptation, ABA enhances, CK inhibits, while GA perturbs the process. The environmental cue for adaptation is Na^+ since exposure to $CaCl_2$ or polyethylene glycol (PEG) does not result in adaptation to NaCl (Amzallag et al. 199Øb).

Adaptation is a change in the physiological mode of the plant. *Sorghum* is a glycophite. The maximum NaCl concentration in which it can survive is about 225 mM. However, at such salinity its growth is completely inhibited; we called this physiological mode "glycomode". Salt-adapted *Sorghum* grows at 15Ø mM NaCl at the same mean relative growth rate ($\overline{RGR}$) as in absence of salinity. It can even grow in medium containing 3ØØ mM NaCl, at the same $\overline{RGR}$ as in absence of NaCl, if the Hoagland solution is increased from half-strenght to full-strength, or if CK+GA (both at 1 nM) are added to the medium, we called this physiological mode "halomode".

This effect of CK+GA on growth of salt-adapted *Sorghum* allowed Liat Gonen to determine the effect of growth (sink activity) on photosynthate concentration. She measured reducing sugars, sucrose and starch at 6h, 14h, 2Øh and 2h in leaves of salt-adapted *Surghum* exposed to 3ØØ mM NaCl and

Hoagland's solution and CK+GA concentrations as indicated in Table 1 (unpublished results). Average amount of carbohydrates in mg per gFW leaf was calculated. From Table 1 it can be seen that while $\overline{RGR}$ was increased 4-fold by addition of either nutrients or CK+GA, total average carbohydrates were not decreased by the increase in growth (in fact there was a 3Ø percent increase).

Table 1. $\overline{RGR}$ *and carbohydrate of shoots of salt-adapted* Sorghum *exposed to 3ØØ mM NaCl and half-strength Hoagland, full-strength Hoagland or half-strenth Hoagland plus CK+GA*

M e d i u m			$\overline{RGR}$	Carbohydrates (mg/gFW leaf)					
NaCl (mM)	Hoag. sol.	CK+GA (nM)	(g/g·d)		6 AM	2 PM	8 PM	2 AM	Ave
3ØØ	½	Ø	Ø.Ø17	RS:	8	7	6	7	7
				Suc:	16	14	17	12	15
				Sta:	2	1	3	3	2
								Total:	24
3ØØ	1	Ø	Ø.Ø67	RS:	7	8	6	6	7
				Suc:	18	27	22	17	21
				Sta:	4	5	5	5	5
								Total:	33
3ØØ	½	1+1	Ø.Ø69	RS:	11	8	8	6	8
				Suc:	9	19	21	18	17
				Sta:	3	3	6	5	4
								Total:	29

RS, reducing sugars; Suc, sucrose; Sta, starch; Ave, average.

Thus demonstrating that CK+GA enhanced in parallel growth (increased sink activity) as well as photosynthesis activity. Using a completely different approach, these results confirm those of Wareing et al. (1968).

Adaptation to salinity of cultured plant cells has been demonstrated in many laboratories using cells derived from different species. Watad et al. (1985) adapted tobacco cells to 500 mM NaCl and Binzel et al. (1985) to 600 mM. We were also able to adapt tobacco cells to salinity by increasing NaCl concentration by 35 mM every 6 generations. Under such conditions growth is practically not affected by the increase in salinity, the bulk of the cells become tolerant to the increasing salt concentration, thus demonstrating that the phenomenon does not depend on the selection of rare cells having unusual properties, but rather that it is a property of the bulk of the cells. Several aspects of adaptation of plant cells show similarities to that of *Sorghum*. It seems that the developmental window is open at least once every cell generation allowing the cells to undergo the adaptation process. Although this may not necessarily be the opinion of the authors, in our opinion, LaRosa et al. (1987) demonstrated that ABA also accelerates the adaptation of cultured tobacco cells to salinity. In conclusion, adaptation of plant cells is possible because it is a property of plants. The fact that cells of different origin have been adapted shows that it is a general property.

Somaclonal variation received its name from the fact that regeneration of clones of cultured somatic cells yeilded plants with new traits (Lee and Phillips 1988). We have obtained, by raising the offsprings of NaCl-adapted *Sorghum*, plants which display new characteristics (Amzallag et al. unpublished results). These results are very similar to those obtained by somaclonal variation. It seems that just as adaptation to salinity is obtained with cultured plant cells because it is a property of plants, similarly somaclonal variation is obtained because it is a property of plants. The so called "somaclonal variation" is simply one of the consequences, of the response of plants to stress, when the developmental window is open. The idea that somaclonal variation is a property of the whole plant under stress has previously been suggested by Cullis (1990) since RFLPs arising in cultured cells of soybean and of flax were also found to exist in natural populations of these species.

There are several advantages in developing variants through whole plant adaptation rather than through regeneration of adapted cultured cells. As McCoy (1987) showed, there is no correlation between tolerance at the cellular level and that at the whole plant level. In the whole plant tolerance is a property of the organism in which each cell plays it role, each cell is integrated in the response. Adaptation of cultured cells provides tolerance for the individual cell but does make it capable to play its role in the integrated community of cells which is a plant. Furthermore, cells in culture develop aberrations (Bayliss, 1980), and the frequency and types of aberrations are very much enhanced by salinity (McCoy, 1987a). Within the culture there is a selection of the fast growing cells. These phenomenona are probably reasons why plants regenerated from salt adapted cultured cells are usually not very vigorous.

ROLE OF HORMONES AND REPETITIVE DNA IN MODIFICATION OF THE PHYSIOLOGICAL MODE OF THE PLANT

The structure and conformation of the chromatin DNA-protein complex have been shown to affect gene expression (Stirdivant et al., 1985; Felsenfeld, 1992). Furthermore, rapid modifications of the genome, resulting in changes in genome expression, have been observed under stress conditions.

Several mechanisms, such as quantitative modification of repetitive DNA, DNA (cytosine) methylation, movement of transposable elements, gene amplification or deletion have been identified. Bassi (199Ø, 1991) and Cullis (199Ø) have reviewed the literature. Some of these changes, such as modification of repetitive DNA and cytosine methylation, are known to occur during plant development (Chen and Osborne, 197Ø; Kobayashi et al., 199Ø) as well as during enviromental stress.

From our present knowledge it is reasonable to suggest that during adaptation to salinity, the salt-induced changes in hormonal balance induce changes in chromatin structure and conformation which result in modification of genome expression. It seems that repetitive DNA is not "parasite", "junk" or "selfish", as was thought by some authors in the past, but that together with stress-induced changes in the hormonal balance plays an important role in the adaptation of plants to environmental stress by modifying structure and expression of the genome.

CONCLUSIONS

The data available at present in the literature indicate that hormonal balance, together with competence, play a central role in integrating plant development (cell and organ growth and differentiation), in coordinating source and sink activities, in controlling the response of plants to environmental parameters, and in changing the physiological mode of the plant.

The root, among it various functions, is also an organ sensing soil parameters. It transforms this information into hormonal signals, which modify the hormonal balance, resulting in accommodation of the plant to the environmental condition of the soil.

Plants have a capacity for changing their physiological mode (to adapt themselves to the environment). This can only occur when the plant is competent to do so (while the developmental window for adaptation is opened) and when the appropriate signal is perceived. Competence changes with time. It is modified as a function of plant development as well as by external signals.

Genome expression is a function of DNA structure and conformation. It seems that these parameters can be modified as a function of plant development as well as by external factors.

In conclusion, the response of plants to saline environments is not a function of Na^+ or Cl^- toxicity, it is a consequence of the effect of salt on the hormonal balance and a function of the physiological mode of the plant.

ACKNOWLEDGEMENTS

This work was supported by The U.S.-Israel Binational Agricultural Research and Development (BARD) Foundation Project No. US-1869-9ØR.

REFERENCES

Amzallag GN, Lerner HR, Poljakoff-Mayber (199Ø) Induction of increased salt tolerance in *Sorghum bicolor* by NaCl pretreatment. J. Exp. Bot. 41:29-34

Amzallag GN, Lerner HR, Poljakoff-Mayber (1990a) Exogenous ABA
  as a modulator of the response of *Sorghum* to high
  salinity. J Exp Bot 41:29-34.
Amzallag GN, Lerner HR, Poljakoff-Mayber (1990b) Na⁺ is
  necessary but not sufficient to induce salt adaptation.
  Supp Plant Physiol 93:156 (abs 907)
Amzallag GN, Lerner HR, Poljakoff-Mayber (1992) Interaction
  between mineral nutrients, cytokinin and gibberellic acid
  during growth of *Sorghum* at high NaCl salinity. J Exp
  Bot 43:81-87
Amzallag GN, Seligmann H, Lerner HR (1993) A developmental
  window for salt-adaptation in *Sorghum bicolor*. J Exp Bot
  44:645-652
Arteca RN, Schlagnhaufer CD, Arteca JM (1991) Root application
  of gibberellic acid enhance growth of seven *Pelargonium*
  cultivars. HortScience 26:555-556
Atkin RK, Barton GE, Robinson DK (1973) Effect of root-growing
  temperature on growth substances in xylem exudate of *Zea
  mays*. J Exp Bot 24:475-487
Bassi P (1990) Quantitative variation of nuclear DNA during
  plant development: A critical analysis. Biol Rev 65:185-
  225
Bassi P (1991) Repetitive non-coding DNA: A possible link
  between environment and gene expression in plants? Biol
  Zent 110:1-13
Bayliss MW (1980) Chromosomal variation in plant tissues in
  culture. Int Rev Cytol Supp 11A:113-144
Binzell ML, Hasegawa PM, Handa AK, Bressan RA (1985)
  Adaptation of tobacco cells to NaCl. Plant Physiol 79:118-
  125
Cheesemann JM (1988) Mechanisms of salinity tolerance in
  plants. Plant Physiol 87:547-550
Chen D, Osborne DJ (1970) Ribosomal gene and DNA replication
  in germinating wheat embryos. Nature 225:336-342
Cullis CA (1990) DNA rearrangements in response to
  environmental stress. Adv Genet 28:73-97
Davies WJ, Zhang J (1991) Root signals and the regulation of
  growth and development of plants in drying soil. Annu Rev
  Plant Physiol Plant Mol Biol 42:55-76
Downton WJS, Loveys BR (1981) Abscisic acid content and
  osmotic relations of salt-stressed grapvine leaves. Aust J
  Plant Physiol 8:443-453
Felsenfeld G (1992) Chromatin as an essential part of the
  transcriptional mechanism. Nature 355:219-224
Guern J (1987) Regulation from within: the hormone dilema.
  Ann Bot 60 supp. 4:75-102
Itai C, Richmond A, Vaadia Y (1968) The role of root
  cytokinins during water and salinity stress. Isr J Bot
  17:187-195
Jackson MB (1987) A structured evaluation of the involvement
  of ethylene and abscisic acid in plant responses to
  aeration stress. *In* GV Hoad, JR Lenton, MB Jackson, RK
  Atkin, eds, Hormone Action in Plant Development,
  Butterworths, London, pp 189-199

Jahnke S, Bier D, Estruch JJ, Beltran JP (1989) Distribution of photoassimilates in the pea plant: chronology of events in non-fertilized ovaries and effects of gibberellic acid. Planta 18Ø:53-6Ø

Kobayashi H, Ngernprasirtsiri J, Akazawa T (199Ø) Transcriptional regulation and DNA methylation in plastids during transitional conversion of chloroplasts to chromoplasts. EMBO J 9:3Ø7-313

Lambers H (1982) Cyanide-resistant respiration: A non-phosphorylating electron transport pathway acting as an energy overflow. Physiol Plant 55:478-485

LaRosa PC, Hasegawa PM, Rhodes D, Clither JM Watad AA, Bressan RA (1987) Abscisic acid stimulated osmotic adjustment and its involvment in adaptation of tobacco cells to NaCl. Plant Physiol 85:174-181

Lee M, Phillips RL (1988) The chromosomal basis of somaclonal variation. Ann Rev Plant Physiol Plant Mol Biol 39:413-437

McCoy TJ (1987) Tissue culture evaluation of NaCl tolerance in *Medicago* species: Cellular versus whole plant response. Plant Cll Rep 6:31-34

McCoy TJ (1987a) Charactertiszation of alfalfa (*Medicago sativa* L.) plants regenerated from selected NaCl tolerant cell lines. Plant Cell Rep 6:417-422

McDaniel CN, King RW, Evans LT (1991) Floral determination and in vitro floral differentiation in isolated shoot apices of *Lolium temulentum* L. Planta 185:9-16

Morris DA, Arthur ED (1985) Effects of gibberellic acid on patterns of carbohydrate distribution and acid invertase activity in Phaseolus vulgaris. Physiol Plant 65:257-262

Munns R (1988) Why measure osmotic adjustment? Aust J Plant Physiol 15:717-726

Nitsch JP (195Ø) Growth and morphogenesis of strawberry as related to auxin. Am J Bot 37:211-215

Osborne DJ (1987) Physiological considerations in developmental studies: hormones and target cells. *In* GV Hoad, JR Lenton, MB Jackson, RK Atkin, eds, Hormone Action in Plant Development, Butterworths, London, pp 265-274

Skoog F, Miller CO (1957) Chemical regulation of growth and organ formation in plant tissues culture in *vitro*. Sym Soc Exp Biol 11:118-13Ø

Stirdivant SM, Crossland LD, Bogorad L (1985) DNA supercoiling affects *in vitro* transcription in two maize chloroplast gene differently. Proc Natl Acad Sci USA 82:4886-489Ø

Stitt M (1991) Rising CO_2 levels and their potential significance for carbon flow in photosynthetic cells. Plant Cell Environ 14:741-762

Trewavas A (1991) How do plant growth substances work? II. Plant Cell Envir 14:1-12

Vaadia Y (1976) Plant hormones and water stress. Phil Trans R Soc Lond B 273:513-522

Wareing PF, Khalifa MM, Treharne KJ (1968) Rate-limiting processes in photosynthesis at saturating light intensities. Nature 22Ø:453-457

Watad AA, Lerner HR, Reinhold L (1985) Stability of salt-resistance character in *Nicotiana* cell lines adapted to grow in high NaCl concentrations. Physiol Veg 23:887-894

Weimberg R (1986) Growth and solute accumulation in 3-week-old seedlings of *Agropyron elongatum* stressed with sodium and potassium salts. Physiol Plant 67:129-135

Zacarias L, Reid MS (1992) Inhibition of ethylene action prevents root penetration through compressed media in tomato (*Lycopersicon esculentum*) seedlings. Physiol Plant 86:3Ø1-3Ø7

Chapter 4

LOW TEMPERATURE STRESS

CAPS, *CORS*, *DEHYDRINS*, AND MOLECULAR CHAPERONES: THEIR RELATIONSHIP WITH LOW TEMPERATURE RESPONSES IN SPINACH

Charles L. Guy, James V. Anderson, Dale W. Haskell, and Qin-Bao Li
Department of Environmental Horticulture
Institute of Food and Agricultural Sciences
University of Florida
Gainesville, Florida 32611
United States

ABSTRACT

When exposed to low non-freezing temperatures (0-10°C), the freezing tolerance of spinach increases in a time dependent fashion from an LT_{50} of -4°C to -14°C. Associated with the low temperature induction of freezing tolerance is the increased expression of at least 20 proteins. Three of these low temperature responsive proteins were selected for study and found to belong to two families of stress proteins, the *LEA/dehydrin/Rab* and the heat shock 70s. Their possible role in low temperature tolerance mechanisms and a model for the role of ATP in peptide binding activities of the 70 kDa heat shock molecular chaperones are discussed.

INTRODUCTION

Until recently, it was generally accepted that plants and other organisms suddenly exposed to low temperatures were not able over short periods to substantially alter metabolic functions. While long-term metabolic modifications in response to cold were known, it was not expected that most organisms could rapidly adjust to a sudden low temperature exposure. This was considered so because of the seemingly slow rate of metabolic processes at lower temperatures (Levitt, 1980). Earlier attempts by radioisotopic methods to study various metabolic processes, such as photosynthesis, respiration, nucleic acid and protein synthesis, consistently suggested that the activities of these processes were but a small fraction of that at temperatures conducive for growth (Guy et al., 1981). That certain responses to cold could indeed be quite swift, although somewhat unexpected, has now been clearly established. Significant changes in membrane composition, within hours of a reduction in temperature, is a notable example (Dickens and Thompson, 1981). However, the most striking display of fast responses to cold appertains to changes in gene expression. A number of studies have shown that within a few hours, low temperature exposure can result in an elevated expression of a variety of genes (Hajela et al., 1990; Kondo and Inouye, 1991; Nordin et al., 1991; Kondo et al., 1992; Sato, 1992; Willimsky et al., 1992). Also, somewhat unexpected, was the diversity of genes displaying increased expression at low temperature. Table 1 presents a partial list

NATO ASI Series, Vol. H 86
Biochemical and Cellular Mechanisms
of Stress Tolerance in Plants
Edited by J. H. Cherry
© Springer-Verlag Berlin Heidelberg 1994

of the growing number of genes responsive to low temperature. Many of the genes and proteins showing increased expression at low temperature have been termed *cors*[1] for cold responsive (Hajela et al., 1990), *Iti* for low temperature induced (Nordin et al., 1991) or CAPs for cold acclimation proteins (Guy and Haskell, 1987). Inspection of the listings in Table 1 shows that many of the genes responsive to low temperature encode enzymes of intermediary metabolism, while several others could be classified as stress proteins of unknown function. This fact reinforces the concept that for hardy plants, exposure to low temperature elicits two basic responses; the adjustment of metabolism to the kinetic constraints imposed by low temperature, and the induction of freezing tolerance (Guy, 1990).

Table I. Partial listing of low temperature responsive genes.

PROCESS/GENE	ORGANISM
RESPIRATION	
Aldolase, GA3PDH, PDC, ADH	Corn
Alternative Oxidase	Corn
ADH	Arabidopsis
CARBOHYDRATE METABOLISM	
Sucrose Synthase, α-Amylase	Wheat, *Anabaena*
LIPID METABOLISM	
Desaturase, Lipid Transfer Protein	*Synechocystsis*, Barley
PHENYLPROPANOID METABOLISM	
Chalcone Synthase, PAL, R1, C1, A1, A2, Bz1	Corn
ANTIOXIDANT METABOLISM	
Cu/Zn Superoxide Dismutase	*Nicotiana*
Mn & Cu/Zn Superoxide Dismutase, Catalase	Corn, Barley
OTHER ENZYMES OR REGULATORY PROTEINS	
Ethylene Forming Enzyme	Tomato
cspA, hns, recA, nusA, infB, pnp	*Escherichia coli*
DNA gyrase	*Escherichia coli*
HSC70s	Spinach
Peptidyl Prolyl *Cis-trans* Isomerase	Chlorella
Ribosomal Protein L2	Chlorella
EF-1α	Barley
Nucleolin (Like) Protein	Yeast
Cathepsin D	Bromegrass
Thiol Protease	Tomato
NONENZYMATIC OR FUNCTION UNKNOWN	
Antifreeze proteins	Cold Water Fishes
LMW Heat Shock Protein	Potato
Germin	Bromegrass
Late Embryogenesis Abundant Homologs	Arabidopsis, Wheat, Barley, Spinach, Potato
Glycine Rich Proteins	Alfalfa, Chlorella, Wheat, Barley
Glycine/Alanine Rich Protein	Arabidopsis, Canola
Alanine Rich Protein	Arabidopsis, Canola
Alanine/Serine Rich Protein	Yeast

In recent years, our laboratory has taken a broad approach focussing on both metabolic adjustments to low temperature and freezing tolerance. Work on the former is concerned with the importance of molecular chaperones in protein biogenesis at low temperature (Neven et al., 1992), and sucrose metabolism (Guy et al., 1992a). However, a significant part of our effort is directed at understanding the mechanisms responsible for enhanced freezing tolerance (Guy et al., 1992b; Neven et al., 1993). While in many respects the two processes of acclimation; metabolic adjustment and freezing tolerance go hand in hand, the responses of many of the enzymes of intermediary metabolism will be easiest

---

[1]Abbreviations: CAPs, cold acclimation proteins; *cors*, cold responsive; LEA, late embryogenesis abundant; *Iti*, low temperature induced; HS70, includes entire family of 70 kDa heat shock members; HSC70, 70 kDa heat shock cognate; HSP70, 70 kDa heat shock protein; *Rab*, responsive to abscisic acid.

to reconcile and understand. As the listing in Table 1 grows, it will become increasingly important to divine which metabolic responses are essential for the development of freezing tolerance and which are not. A good example of a metabolic response that may be important to both processes of cold acclimation is the accumulation of sucrose and increased activity of sucrose phosphate synthase (Guy et al., 1992a). Modification of sucrose metabolism might be necessary for continued flux of carbon and source/sink processes, but it is also possible that the accumulation of sucrose is important in freezing tolerance because of its powerful cryoprotectant properties. However, the more difficult task confronting researchers working on cold tolerance will be establishing the role of an increasing number of cold stress proteins with no known function (Table 1).

Although the concept that cold acclimation in plants involved altered gene expression during the induction of freezing tolerance (Weiser, 1970) is more than 20 years old, it has only been in the past few years that most of the information on the influence of low temperature on gene expression has become available. Studies with cold water fishes (Lin and Gross, 1981), bacteria (Goldstein et al., 1991), and yeast (Kondo and Inouye, 1991), complement the growing literature of low temperature regulated genes in plants (see Table 1). Interestingly, a significant portion of the genes demonstrated to be regulated by low temperature in plants are novel, and as of yet without known function. Some, based on sequence homologies, belong to a grouping of proteins collectively known as LEA/dehydrin/Rabs (Hahn and Walbot, 1989; Gilmour et al., 1992; Guo et al., 1992; Houde et al., 1992; Lång and Palva, 1992; Luo et al., 1992; Sutton et al., 1992). Others have properties in common with LEA/dehydrin/Rabs like boiling solubility, regulation by ABA or responsiveness to drought (Kurkela and Franck, 1990; Orr et al; 1992; Lin and Thomashow, 1992; Nordin et al., 1991; Hofig et al., 1993). The significance of the fact that proteins linked with water stress also appear to be associated with cold stress has not gone unnoticed. It is widely known that freezing is as much a desiccative stress as a physical stress (Pearce, 1988). Thus, the idea that many of these LEA/dehydrin/Rab-like proteins might function during osmotic stress and afford some manner of protection is not only very appealing, but plausible. Our own work with spinach cold stress proteins shows a striking concordance with these findings (Neven et al., 1993; Hofig et al., 1993).

The mechanism(s) by which plants become cold tolerant is a major issue yet to be resolved. Undoubtedly, the process is one of great complexity (Levitt, 1980; Sakai and Larcher, 1987; Steponkus, 1984). The finding that many cold stress proteins are related to proteins linked to water stress or anhydrous conditions is pivotal because it points to a specific facet of freezing stress that can be more rigorously examined. Also, valuable information can be gleaned from drought and desiccation tolerance research (this volume) which may be meaningful in understanding freezing stress tolerance. Unfortunately, an understanding of drought tolerance mechanisms remains incomplete just as it does with freezing tolerance mechanisms.

Protein denaturation has long been implicated as a potential feature of low temperature stress injury (Levitt, 1962). As originally envisioned by Levitt, freezing would lead to the unfolding of proteins. During the close approach of unfolded polypeptide chains in the freeze concentrated cytoplasm, intermolecular disulfide bonds would be more or less irreversibly formed leading to aggregation. The initial unfolding of globular proteins was postulated to occur as a result of the weakening of hydrophobic interactions at low temperature. Since its proposal, the sulfhydryl hypothesis has largely fallen out of favor and given way to the "membrane" hypothesis (Steponkus, 1984). Yet, the sulfhydryl hypothesis in its day was forward-thinking in that it tried to provide a molecular explanation for the basis of stress injury. While even the most ardent supporters doubt its validity as a stress injury mechanism, it may merit some consideration in a modified form.

MATERIALS AND METHODS

Plant Material

Seedlings of all species used were grown from seed in a controlled environment as previously described (Guy and Haskell, 1987). Cold acclimation and deacclimation treatments were conducted as previously described (Guy and Haskell, 1987). Plants were grown at 20°C at ambient or 100% relative humidity. For the acclimation induction analysis, the temperature was lowered at 5°C intervals beginning at 20°C and held for two days at each temperature until 5°C was reached. For the deacclimation repression analysis, plants were first acclimated at 5°C for seven days, then the temperature was raised at 5°C intervals to 20°C holding at each temperature for two days. Samples for RNA extraction were taken after two days at the indicated temperature. Plants for the heat shock treatment were grown at 20°C then shifted to 37°C for 2 hr. Water stress was imposed by withholding water, and tissue samples were prepared when the plants began to lose turgor.

RNA Purification and Northern Analyses

RNA was extracted from spinach leaves using the phenol/chloroform extraction method (Ausubel et al., 1989). Equal amounts of RNA were separated on 1.2% formaldehyde agarose gels. Gels were stained with ethidium bromide, photographed, and pressure blotted onto Hybond-N (Amersham) nylon membrane. The RNA was fixed and hybridized with random primed (^{32}P) dCTP labeled probes as indicated.

Protein Blot Analyses

Fresh leaf tissue was weighed and homogenized in buffer (62 mM Tris-HCl, pH 6.8, 5% glycerol, 2.5% mercaptoethanol, 1.0% sodium dodecyl sulfate) in a w/v ratio of 1:3. Homogenates were boiled for two min and cell debris pelleted by centrifugation at 15,000 g for 10 min at 4°C. Protein content of the supernatant was determined on diluted samples by the dye-binding method (Bradford 1976). Leaf proteins (10 μg) were fractionated by SDS-PAGE in 7.5% acrylamide gels, and electroblotted to Immobilon (Millipore, Bedford, MA). CAPs 160 and 85 were detected with mouse monoclonal antibodies raised against each; 2H8 for CAP160 and 5A10 for CAP85 (Guy et al., 1992). CAP79 (HSC70s) were detected with a mouse antiserum (Neven et al., 1992). Mouse antibodies bound to the CAP proteins were detected by alkaline phosphatase conjugated anti-mouse IgG and color development with NBT and BCIP. Monoclonal antibodies reactive against the various proteins were prepared as previously described (Guy et al., 1992; Neven et al. 1992). The anti-GroEL antisera was a generous gift from J. McCarty.

RESULTS

Figure 1 shows the influence of environmental conditions on the steady-state expression levels of CAPs 85, 160, and two HSC70s. For the set of experiments shown in Figure 1A, the plants were grown at 100% relative humidity to minimize the influence of altered water relations on the expression of any of the genes.

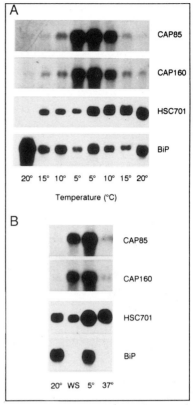

Figure 1. RNA blot analyses of spinach genes associated with the cold acclimation processes. A, influence of decreasing and increasing temperature on expression level at 100% relative humidity; B, responses to temperature and water stress.

As the temperature was lowered from 20° to 10°C, the mRNA steady-state levels only slightly increased for CAPs 85 and 160. At 5°C a dramatic increase occurred, and this level of expression was maintained for the seven day duration at 5°C. As the temperature was raised, the steady-state mRNA levels declined to the point where they became undetectable again at 20°C. However, the decline in CAPs 85 and 160 mRNA level during deacclimation compared to the increase during cold acclimation was asymmetric in that at 10°C deacclimation mRNA levels were higher than during acclimation. Figure 1B shows that both CAPs 85 and 160 are also responsive to water stress conditions, but not heat shock. These results are in keeping with the sequence homology which indicates that CAP85 belongs to the *LEA/dehydrin/Rab* superfamily of proteins (Neven et al., 1993). Although other properties of CAP160 link it to the *LEA/dehydrin/Rab* group its amino acid sequence does not (Hofig et al., 1993).

Compare the patterns for two HSC70s with CAPs 85 and 160 during an acclimation/deacclimation cycle. HSC701, a cytosolic form, was expressed at a low level at 20°C but showed increased expression at 15° to 5°C (Fig. 1A). After seven days at 5°C HSC701 showed greater expression than at 2 days at 5°C. During deacclimation the expression level remained high. In contrast, the ER-lumenal HSC70, BiP, showed very high expression at 20°C in this experiment. As the temperature was decreased to 15°C, the mRNA level decreased significantly. At lower temperatures the levels began to increase again followed by a further increase upon deacclimation. In contrast to CAPs 85 and 160, neither of the HSC70s mRNAs accumulate in response to water stress (Fig. 1B). The mRNA level for HSC701 remains unchanged during water stress, while the BiP mRNA becomes undetectable. The response to heat shock is also different. HSC701 is upregulated after 2 hrs at 37°C, while the BiP mRNA is again undetectable.

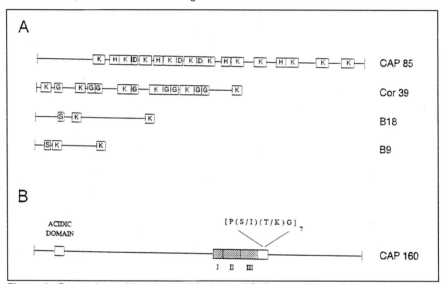

Figure 2. Comparison of the structural features of CAPs 85 and 160. Repeat designations K, D, H, and G are lysine, aspartic, histidine and glycine rich respectively. S designates a serine motif. CAPs 85 and 160 are not drawn to the same scale.

Analyses of the primary structure of the two CAPs, shows that CAP85 has some unusual characteristics (Neven et al., 1993). It has 11 lysine-rich repeats and two other repeats that occur 3 and 4 times respectively (Fig. 2). A comparison of its structure relative to a *cor* from wheat (Guo et al., 1992) and two barley dehydrins (Close et al., 1989) is shown. CAP85 is one of the largest members of this superfamily and this is mirrored by the number of repeats present. The predicted secondary structure is largely α-helical (60-80%). CAP85 contains a large number of charged residues (38%). Examination of the positions of the charged residues within the sequence indicates that many occur such that i + 3 or i + 4 arrangements would occur (Table 2). In contrast, CAP160 has less predicted α-helix (40-50%) and charged residue content (26%), but has a strongly acidic nature. The number of potential intrahelical salt linkages are also fewer than for CAP85.

Protein blots of extracts from 20° and 5°C grown spinach leaf tissue reacted with a polyclonal antibody (Neven et al., 1992) largely selective for cytosolic HSC70s or an antisera raised against *E. coli* GroEL show very little change in the steady-state protein levels (Fig. 3A).   In   contrast,   GroEL crossreactive proteins are elevated in the chilling sensitive species bean and tomato (Fig. 3B). Similar elevated levels of HSC70s following 5°C exposure have been observed for both bean and tomato tissues (Neven et al., 1993).

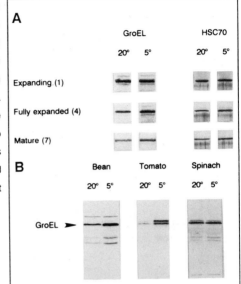

Figure 3. Protein blot analyses of the influence of low temperature on the abundance of GroEL and/or cytosolic HSC70s in spinach leaf tissue (A) or chilling sensitive plants (B).

DISCUSSION

We have found gene products for two hydrophilic high molecular weight proteins in spinach that are expressed in response to either low temperature or drought (Neven et al., 1993; Hofig et al., 1993). When these proteins are accumulated in leaf tissue at either low temperature or during desiccation, freezing tolerance is greatly enhanced. It is likely that these proteins play an adaptive role during periods of desiccation either during drought or as a result of freezing.

The amino acid sequence for CAP85 reveals similarity to the Group 2 LEA family (Neven et al., 1993).   The deduced protein has a typical Group 2-like 11 residue lysine-rich repeat, **LDKIKDKLPGQ**, contained within a larger 22 amino acid sequence that was repeated 11 times within the clone (Fig. 2). The repeats begin at around 90 residues from the N-terminus and continue to near the C-terminus. The repeats are not contiguous but are separated irregularly by spans of 9 to 28 amino acids. Database searches indicated the highest homology was to *Rab17* (Mundy and Chua, 1988). Two other imperfect repeating motifs are also found in the primary structure of the deduced protein (Fig. 2A).   Both were interspersed between the lysine repeats. The longer repeat - HTQQLYPASDHNYNTH- was present four times, while the shorter repeat -EDKKNDYH- was present three times. Structurally, CAP85 most closely resembles COR39 from wheat (Guo et al., 1992; Houde et al., 1992). At present, it is one of the largest members of the Group 2 LEA family although a larger protein may exist in wheat (Ouellet et al., 1993). The greater number of repeating units in CAP85 differs from the fewer number of repeats in most Group 2 *LEA*s. COR39 of wheat is most similar to CAP85 in the number of repeats and the absence of a serine cluster (Guo et al., 1992). Like CAP85, CAP160 has repeating motifs (Fig. 2B). However, the pattern and sequence of the repeating elements are very different. Both repeating motifs found in CAP160 occur as tandem arrays. Outside of a repeating motif feature, there is virtually no significant sequence similarity at the amino acid level between CAP160 and CAP85. CAP160 has one other interesting feature, a highly acidic domain of 26 residues of which 18 are acidic.

Secondary structure predictions suggest that both CAP85 and 160 may have high α-helical content. Inspection of the sequences reveals that CAP85 may contain an unusual number of intrahelical salt bridge linkages (Table 2). The lysine repeat has the potential to form 5 salt linkages if it assumes an α-helical structure. Fortuitously, the majority of the potential salt bridges occur in regions predicted to be α-helical further supporting the secondary structure predictions. Preliminary circular dichroism studies using synthetic peptides based on the lysine-rich repeat show about

Table II. Putative intrahelical salt bridges in CAPs 85 and 160.

	$i + 3$	$i + 4$
D/E--K/R		
CAP85	31	9
CAP160	6	9
K/R--D/E		
CAP85	20	4
CAP160	7	11

25% helical content in high concentrations of the helix stabilizing solvent trifluoroethanol further supporting the notion that CAP85 may indeed be mostly α-helical. In contrast, CAP160 has fewer potential intrahelical salt bridge linkages which is in keeping with a lower α-helical content.

The functional roles that CAPs 85 and 160 play either during freezing or water stress remains unknown. Speculations about the function of repeating motifs based on computer simulations of structure for a variety of desiccation responsive proteins (Dure, 1993) may also apply for the same repeating elements in CAP85. However, for CAP160, its primary sequence provides less obvious clues to what its function might be during times of stress. Answers to the basic questions of function and mode of action will not only require 3-dimensional structural analyses of the purified protein, but detailed biochemical and genetic analyses.

The protein initially designated as CAP79 (Guy and Haskell, 1987), is now known to be a group of closely migrating proteins all of which belong to the heat shock 70 kDa class of proteins (Neven et al., 1992). Here we present RNA blot data for two members of the HS70 family. Clones for at least four HS70 genes have now been isolated from spinach. RNA blot analyses indicate several different expression patterns in response to low temperature (Neven et al., 1992; Anderson et al., 1993). The ER-luminal HSC70 shows a moderate upregulation (about 2 fold), while another HS70 member is not expressed at 20°C but is induced at 5°C (Anderson et al., 1993). The pattern of expression for BiP in this study is quantitatively different from that found previously (Anderson et al., 1993). The plants used for the RNA blot studies shown in Fig. 1A were grown at 100% relative humidity. For some unknown reason, this resulted in a very high level of expression of BiP at 20°C.

Lowering the temperature to 15°C resulted in a much lower expression level. This would suggest a relationship with temperature. Growing the plants at 100% relative humidity in a closed atmosphere may have caused a greenhouse effect elevating the ambient temperature by a few degrees during the photoperiod. However, if the high expression level is due to a heat stress, then it is quantitatively different than that elicited by a heat shock where the expression is dramatically lower (Fig. 1B). Actually, the expression level over the duration of this experiment was quite variable for BiP. We suspect that growing the plants under a constant 100% relative humidity has a complex influence on the normal physiology of the tissue that may in some way affect the secretory process and the function of the ER.

At least two cytosolic HSC70s, HSC701 and HSC703 are also moderately upregulated in response to lower temperatures (Fig. 1A, only HSC701 is shown). Interestingly, all three HSC70s, the ER-luminal form (Anderson et al., 1993) and the two cytosolic forms show increased mRNA levels upon return to 20° from 5°C at 100% relative humidity. Unlike CAPs 85 and 160, none of the HS70 family members we have tested to date show increased expression during water stress (Fig. 1B).

The synthesis of HSC70s in spinach leaf tissue is clearly increased during exposure to 5°C (Guy and Haskell, 1987) and parallels an increase in the mRNA steady-state level (Neven et al., 1992). However, the steady-state protein pool does not dramatically change (Fig. 3A; Neven et al., 1992). Increased synthesis with relatively constant steady-state levels implies that a certain homeostatic standing pool size is maintained possibly by an enhanced turnover mechanism. The response of HS70s observed with spinach is quite different in chilling sensitive species like tomato and bean. The level of HS70s in these species is dramatically increased (Neven et al., 1992) suggesting that the influence of low temperature is much more drastic. Perhaps there is a greater need for the functions that HS70s perform in these species at low temperature. One possibility is that protein folding is somehow affected (Neven et al., 1992). If this were the case, then it would be expected that abundance of another important molecular chaperone associated with protein folding and assembly, GroEL, would also be increased in chilling sensitive plants. This is exactly what is observed with tomato and bean (Fig. 3B). Like the response with HS70s, the GroEL homolog pool in spinach is not altered by low temperature exposure (Fig. 3B).

It is now clear that the conformational stability of globular proteins is surprisingly low. Defined as the difference in free energy between the folded and unfolded state at physiological conditions $[\Delta G(H_2O)]$, the conformational stability of most globular proteins ranges between 5-15 kcal/mol (Pace, 1990). Plots of ΔG vs temperature are known as protein stability curves and not unexpectedly, conformational stability can be described by the Gibbs-Helmholtz equation (Pace, 1990). What is interesting about this relationship is that maximal conformational stability of a given protein occurs at a specific temperature and above or below that temperature stability declines. Cold denaturation therefore, appears to be a general property of globular proteins (Privalov, 1990). Some 30 different enzymes have been shown to undergo cold inactivation. Most are multimeric and cold inactivation appears to involve dissociation into subunits (Privalov, 1990). Hydrophobic interactions are largely entropy driven with little change in enthalpy (Kauzmann, 1959). At low temperatures, contributions to stability by these interactions are diminished and this may cause either unfolding or dissociation of oligomeric complexes (Bock and Frieden, 1978).

Cold inactivation of many plant enzymes is known, and most are from plants adapted to warm environments. That certain proteins could be cold labile suggests an obvious parallel with the well-known influence of heat on proteins, and the potential for an association with certain heat shock proteins (Pelham, 1986). In fact, heat shock treatment protects *Drosophila* larvae from cold stress (Burton et al., 1988) and germinating *Neurospora* conidiospores from freeze stress (Guy et al., 1986). Furthermore, cold shock causes the accumulation of heat shock proteins in insects (Joplin et al., 1990; Petersen et al., 1990) and cold acclimation of spinach causes increased synthesis of HSC70s (Guy and Haskell, 1987; Neven et al., 1992). Similarly, heat or cold shock induce the synthesis of a developmental regulated membrane protein (Maniak and Nellen, 1988). That both cold and heat shock could induce specific as well as common proteins could be related to a common sensing mechanism in *E. coli* (VanBogelen and Neidhardt, 1990).

Why does spinach show only modest changes in the abundance of HS70s and GroEL in response to cold while chilling sensitive bean and tomato accumulate both proteins? At least two mechanisms could be proposed. The first would involve the dissociation and/or cold denaturation of already existing proteins (Privalov, 1990). The appearance of improperly folded or denatured proteins at low temperature would signal the need for increased production of molecular chaperones in an attempt to refold or prevent the aggregation of the cold

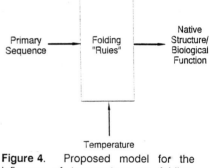

Figure 4. Proposed model for the influence of temperature on folding. Adapted from Nilsson et al., 1990.

denatured proteins. In bean and tomato where the function of the cells have been evolutionarily optimized for warm temperature conditions, the conformational stability may also be equally optimized for warm temperatures (Hardies and Garvin, 1991). Reducing the growth temperature some 20° to 25°C may lower the conformational stability of a number of critically important proteins to the point where the equilibrium between the disassembled ⇌ assembled or unfolded ⇌ folded state is shifted to the left (Privalov, 1990). On the other hand, spinach is a cool-season plant that overwinters as a rosette and resumes growth when springtime temperatures are still in the 0 - 10°C range. Consequently, spinach has evolved to function at lower temperatures, and part of this may have included selection for enhanced conformational stability at lower temperatures (Alber, 1989). Thus, at 5°C the potential for cold denaturation processes may not be as great in spinach as in tomato and bean. The second mechanism that may be operable has to do with the actual folding and assembly of nascent chains into a functional native state. To proceed from a totally random coil to a native conformation, folding appears to occur only along a few productive pathways out of a very large number of possible pathways (Creighton, 1990). That proteins fold on a biologically relevant time scale indicates that finding the correct folding pathway is not a random process (Creighton, 1990). Studies with phage tailspike proteins have shown that temperature-sensitive-folding mutants alter an intermediate state along the folding pathway (King et al., 1990) so as to render folding to a native state temperature sensitive. In these mutants, only folding is blocked at the restrictive temperature with all folded mutants having normal functionality. Virtually all of these mutants are restrictive at higher temperatures, but there is no known reason why low temperature restrictive folding mutations could not also occur. Surprisingly, GroEL in *E. coli* is one example of a folding/assembly blockage at low temperature. GroEL monomer from *E. coli* will not self assemble to an oligomeric form, but appears to undergo a cold denaturation. However, once assembled, oligomeric GroEL is quite stable at low temperature (Lissin et al., 1990). This leads us to propose that like evolutionary optimization of biological function for a prevailing set of conditions, the actual folding process might also be similarly optimized. In plants one of the major factors that would influence folding processes would be temperature (Fig. 4). Thus, one reason but not the only one, why tomato, bean, and most mesophilic plants might be chilling sensitive is that some proteins would be unable to fold into a biologically-functional native form.

Members of the 70 kDa heat shock family belong to a diverse and growing assortment of proteins known as molecular chaperones (Ellis and van der Vies, 1991). A defining characteristic for all molecular chaperones is an ability to bind other peptides. As originally conceived, a central role for molecular chaperones was to prevent or minimize incorrect or inappropriate interactions between proteins during biogenesis or stressful conditions that favored protein denaturation (Ellis and van der Vies, 1991). However, it is now clear that molecular chaperones may serve many diverse functions (Gething and Sambrook, 1992). In the case of the HS70 family, they have been connected with stabilization of folded (Minton et al., 1982) and unfolded proteins (Pelham, 1986), binding and stabilization of nascent peptide chains as they emerge from the ribosome (Beckmann et al., 1990), maintaining precursors in a translocation competent state (prevent folding) for import into organelles (Deshaies et al., 1988; Chirico et al., 1988), asymmetrically facilitating import of precursors both outside and inside the ER (Brodsky et al., 1993), assisting in assembly of oligomeric complexes (Knittler and Haas, 1992), disassembly of oligomeric complexes (Schmid et al., 1985), renaturation of stress induced unfolded peptide chains (Gaitanaris et al., 1990; Skowyra et al., 1990; Buchner et al., 1992), and targeting proteins for degradation (Chiang et al., 1989). How such varied functions are achieved is not yet understood, but some aspects have been delineated. Many of the functions listed above are associated with ATP binding and/or hydrolysis (Zylicz et al., 1983; Braell et al., 1984; Schmid et al., 1985; Lewis and Pelham, 1985; Skowyra et al., 1990). Binding ATP causes a conformation change in the HS70 protein (Kassenbrock and Kelly, 1989; Liberek et al., 1991b; Palleros et al., 1991), and hydrolysis appears to be necessary for bound peptide release (Munro and Pelham, 1986). In *E. coli*, the ATPase activity of DnaK is stimulated by the presence of DnaJ and GrpE (Liberek et al., 1991a). DnaJ appears to interact with DnaK to stabilize a bound peptide in an unfolded state, then GrpE and ATP hydrolysis facilitate transfer of the peptide to GroEL for further folding (Langer et al., 1992). The relationship between the ATP induced conformational change and various functions of HS70s remains clouded with uncertainty. The N-terminal domain is responsible for the ATPase activity, and X-ray crystallography has revealed the topology of the hydrolytic nucleotide binding site (Flaherty et al., 1990), and its similarity with actin (Flaherty et al., 1991) and sugar kinases (Bork et al., 1992). The peptide binding domain is located in the C-terminal half of the protein.

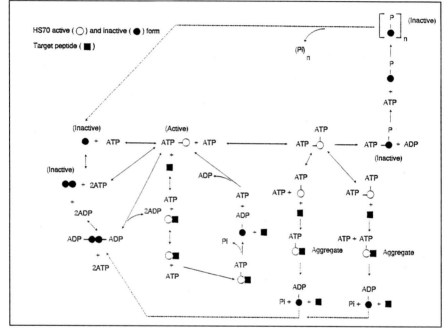

Figure 5. Working model on the interactions of ATP with the monomeric, dimeric, and oligomeric forms of HS70s.

Understanding the relationship of HS70s with ATP is further complicated by three features of their interaction. DnaK and HS70s, in general, will undergo autophosphorylation (Zylicz et al., 1983) in a Ca^{2+} (Leustek et al., 1989) and temperature dependent fashion (Sherman and Goldberg, 1992). In the case of the DnaK, *in vitro* phosphorylation is correlated with increased ATPase activity (McCarty and Walker, 1991). On the other hand, BiP shows reduced phosphorylation during stress, and only the dephosphorylated forms appear to be bound to other proteins (Freiden et al., 1992). Similarly, plant HSC70s that are phosphorylated are less stimulatory in membrane translocation than the unphosphorylated form (Miernyk et al., 1992). The phosphorylated form of BiP is also unable to bind ATP (Gaut and Hendershot, 1992). Is it possible that BiP, HSC70 and DnaK act in completely opposite ways? A second complicating factor is the apparent existence of a second ATP binding site on HSC70s and BiP (Schmid et al., 1985; Gaut and Hendershot, 1992). The second ATP binding site of HSC70 (clathrin uncoating enzyme) is implicated in triskelion sequestration and dimer dissociation, but the ATP bound at this site is not hydrolyzed (Schmid et al., 1985). The function of the second site is not at all clear, but it appears to have a regulatory nature. Can this second site play a role in determining which of the various activities of HSC70s or BiP are expressed? The third action of ATP is to influence the degree of oligomerization of HS70s. Phosphorylated forms of BiP exist either as a monomer, dimer, or as an aggregate (Freiden et al., 1992; Gaut and Hendershot, 1992), while the dephosphorylated form is found primarily as a monomer (Freiden et al., 1992). Purified recombinant BiP, presumably unmodified, formed both monomer and dimers, but when ATP was present much of the dimeric form was converted to monomer (Carlino et al., 1992) in a manner very similar to the clathrin uncoating enzyme (Schmid et al., 1985). Similarly, HSC70 was monomeric with ATP present, and would form dimers and oligomers in its absence (Kim et al., 1992). Unexpectedly, HSP70 oligomerizes in the presence of ATP or ADP at elevated temperatures, binds ADP more tightly, and binds target peptides more rapidly than free HSP70 (Palleros et al., 1991). Thus, the picture of which forms are active (phosphorylated, dephosphorylated, monomeric, dimeric, or oligomeric) in peptide binding, ATP hydrolysis, and other important functions is still not clear. Based on our interpretation of the collective experimental body of evidence, we have attempted to formulate a model to use as a guide in dissecting some of the complexity of the interactions with ATP and functions of HS70s (Fig. 5). When this model is considered in the context of spinach responses to low temperature, direct relationships between mRNA levels, rates of HS70 synthesis, and total steady-state protein pools may not always coincide. Given that the total pool is composed of several forms, some active and others not, large changes in the total pool size may not be required to compensate for stress situations, only conversion from inactive to an active form. At the present time, how the low temperature stress conditions influence the turnover of the inactive and active forms remains unknown, but will need to be factored into the overall picture.

Characterization of the HS70 gene family and the functions of the constituent members has largely been confined to yeast, *Drosophila* and *Caenorhabditis* (Craig, 1989, Rubin et al., 1993; Snutch et al., 1988). Until very recently very few HS70 genes had been characterized from plants, and no more than three members of the family had been isolated from a single species. Consequently, very little is known about the developmental and environmental expression patterns in plants. The few studies that have been reported have only begun to examine developmental (Duck et al., 1989; Jones and Bush, 1991; Hopf et al., 1992) and/or environmental regulation patterns. This situation is compounded by the possibility that plants may have more HS70 genes than the organisms listed above (Vierling, 1991). Our earlier studies on protein synthesis and mRNA abundance (Guy and Haskell, 1987; Neven et al., 1992) assessed the total pool of HS70s in spinach leaf tissue during cold acclimation and did not distinguish the responses of individual members. More recent investigations have shown that a number of different expression patterns occur in response to low temperature (Anderson et al., 1993). Our goal is to characterize the mRNA and protein expression pattern for all members of the HS70 family in an effort to identify which ones are responsive to low temperature. Once this is known, we will be in a better position to understand the role of this molecular chaperone in low temperature tolerance mechanisms.

LITERATURE CITED

Alber, T (1989) Mutational effects on protein stability. Ann Rev Biochem **58**: 765-798

Anderson, JV, Li, Q-B, Haskell, DW, Guy, CL (1993) Structural organization of the spinach ER-lumenal HSC70 gene and its expression during cold acclimation. In review.

Ausubel, FM, Brent, R, Kingston, RE, Moore, DD, Seidmon, JG, Smith, JA, Struhl, K (1989) *Current Protocols in Molecular Biology*, Wiley Interscience, New York.

Beckmann, RP, Mizzen, LA, Welch, WJ (1990) Interaction of Hsp 70 with newly synthesized proteins: Implications for protein folding and assembly. Science **248**: 850-854

Bock, PE, Frieden, C (1978) Another look at the cold lability of enzymes. Trends Biochem Sci **3**: 100-103

Bork, P, Sander, C, Valencia, A (1992) An ATPase domain common to prokaryotic cell cycle proteins, sugar kinases, actin and hsp70 heat shock proteins. Proc Natl Acad Sci USA **89**: 7290-7294

Braell, WA, Schlossman, DM, Schmid, SL, Rothman, JE (1984) Dissociation of clathrin coats coupled to the hydrolysis of ATP: Role of an uncoating ATPase. J Cell Biol **99**: 734-741

Brodsky, JL, Hamamoto, S, Feldheim, D, Schekman, R (1993) Reconstitution of protein translocation from solubilized yeast membranes reveals topologically distinct roles for BiP and cytosolic Hsc70. J Cell Biol **120**: 95-102

Burton, V, Mitchell, HK, Young, P, Petersen, NS (1988) Heat shock protection against cold stress of *Drosophila melanogaster*. Mol Cell Biol **8**: 3550-3552

Carlino, A, Toledo, H, Skaleris, D, DeLisio, R, Weissbach, H, Brot, N (1992) Interactions of liver Grp78 and *Escherichia coli* recombinant Grp78 with ATP: multiple species and disaggregation. Proc Natl Acad Sci USA **89**: 2081-2085

Chiang, H-L, Terlecky, SR, Plant, CP, Dice, JF (1989) A role for a 70-kilodalton heat shock protein in lysosomal degradation of intracellular proteins. Science **246:** 382-385

Chirico, WJ, Waters, MG, Blobel, G (1988) 70K heat shock related proteins stimulate protein translocation into microsomes. Nature **332:** 805-810

Close, TJ, Kortt, AA, Chandler, PM (1989) A cDNA-based comparison of dehydration-induced proteins (dehydrins) in barley and corn. Plant Mol Biol **13:** 95-108

Craig, EA (1989) Essential roles of 70kDa heat inducible proteins. BioEssays **11:** 48-52

Creighton, TE (1990) Understanding protein folding pathways and mechanisms. *In* LM Gierasch and J King ed, Protein Folding: Deciphering the Second Half of the Genetic Code. AAAS, Washington, DC, pp 157-170

Deshaies, RJ, Koch, BD, Werner-Washburne, M, Craig, EA, Schekman, R (1988) A subfamily of stress proteins facilitates translocation of secretory and mitochondrial precursor polypeptides. Nature **332:** 800-805

Dickens, BF, Thompson, GA (1981) Rapid membrane response during low temperature acclimation. Correlation of early changes in the physical properties and lipid composition of *Tetrahymena* microsomal membranes. Biochim Biophys Acta **664:** 211-218

Duck, N, McCormick, S, Winter, J (1989) Heat shock protein hsp70 cognate gene expression in vegetative and reproductive organs of *Lycopersicon esculentum*. Proc Natl Acad Sci USA **86:** 3674-3678

Dure III, L, (1993) A repeating 11-mer amino acid motif and plant desiccation. Plant J **3:** 363-369

Ellis, RJ, van der Vies, SM (1991) Molecular chaperones. Ann Rev Biochem **60:** 321-347

Flaherty, KM, DeLuca-Flaherty, C, McKay, DB (1990) Three-dimensional structure of the ATPase fragment of a 70k heat-shock cognate protein. Nature **346:** 623-628

Flaherty, KM, McKay, DB, Kabsch, W, Holmes, KC (1991) Similarity of the three-dimensional structures of actin and the ATPase fragment of a 70-kDa heat shock cognate protein. Proc Natl Acad Sci USA **88:** 5041-5045

Freiden, PJ, Gaut, JR, Hendershot, LM (1992) Interconversion of three differentially modified and assembled forms of BiP. EMBO J **11:** 63-70

Gaitanaris, GA, Papavassiliou, AG, Rubock, P, Silverstein, SJ, Gottesman, ME (1990) Renaturation of denatured λ repressor requires heat shock proteins. Cell **61:** 1013-1020

Gaut, JR, Hendershot, LM (1992) The *in vivo* and *in vitro* phosphorylation of BiP occur at different sites but have similar consequences on function. Abstract #3, Molecular chaperones: Functions in protein folding and cellular metabolism.

Gething, M-J, Sambrook, J (1992) Protein folding in the cell. Nature **355:** 33-45

Gilmour, SJ, Artus, NN, Thomashow, MF (1992) cDNA sequence analysis and expression of two cold-regulated genes of *Arabidopsis thaliana*. Plant Mol Biol **18:** 13-21

Goldstein, J, Pollitt, NS, Inouye, M (1990) Major cold shock protein of *Escherichia coli*. Proc Natl Acad Sci USA **87:** 283-287

Guo, W, Ward, RW, Thomashow, MF (1992) Characterization of a cold-regulated wheat gene related to *Arabidopsis cor47*. Plant Physiol **100:** 915-922

Guy, CL (1990) Cold acclimation and freezing tolerance: Role of protein metabolism. Ann Rev Plant Physiol Plant Mol Biol **41:** 187-223

Guy, CL, Yelenosky, G, Sweet, HC (1981) Distribution of ^{14}C photosynthetic assimilates in Valencia orange seedlings at 10° and 25°C. J Am Soc Hort Sci **106:** 433-437

Guy, CL, Plesofsky-Vig, N, Brambl, R (1986) Heat shock protection of germinating *Neurospora crassa* conidiospores against intracellular freezing stress. J Bacteriol **167:** 124-129

Guy, CL, Haskell, D (1987) Induction of freezing tolerance in spinach is associated with the synthesis of cold acclimation induced proteins. Plant Physiol **84:** 872-878

Guy, CL, Huber, JLA, and Huber, SC (1992a) Sucrose phosphate synthase and sucrose accumulation at low temperature. Plant Physiol **100:** 502-508

Guy, CL, Haskell, DW, Neven, LN, Klein, P, and Smelser, C (1992b) Hydration-state-responsive proteins link cold and drought stress in spinach. Planta **188:** 265-270

Hahn, M, Walbot, V (1989) Effects of cold-treatment on protein synthesis and mRNA levels in rice leaves. Plant Physiol **91:** 930-938

Hardies, SC, Garvin, LD (1991) Can molecular evolution provide clues to the folding code? *In* BT Nall, KA Dill, ed, Conformations and Forces in Protein Folding. AAAS, Washington DC, pp 69-76

Hajela, RK, Horvath, DP, Gilmour, SJ, Thomashow, MF (1990) Molecular cloning and expression of *cor* (cold-regulated) genes in *Arabidopsis thaliana*. Plant Physiol **93:** 1246-1252

Hofig, A, Neven, LG, Li, Q-B, Haskell, D, and Guy, CL (1993) Expression of a plant cold stress protein in *Escherichia coli* imparts enhanced freezing tolerance. Submitted.

Hopf, N, Plesofsky-Vig, N, Brambl, R (1992) The heat shock response of pollen and other tissues of maize. Plant Mol Biol **19:** 623-630

Houde, M, Danyluk, J, Laibert, J-F, Rassart, E, Dhindsa, RS, Sarhan, F (1992) Cloning, characterization, and expression of a cDNA encoding a 50-kilodalton protein specifically induced by cold acclimation in wheat. Plant Physiol **99:** 1381-1387

Jones, RL, Bush, DS (1991) Gibberellic acid regulates the level of a BiP cognate in the endoplasmic reticulum of barley aleurone cells. Plant Physiol **97:** 456-459

Joplin, KH, Yocum, GD, Denlinger, DL (1990) Cold shock elicits expression of heat shock proteins in the flesh fly, *Sarcophaga crassipalpis*. J Insect Physiol **36:** 825-834

Kassenbrock, CK, Kelly, RB (1989) Interaction of heavy chain binding protein (BiP/GRP78) with adenine nucleotides. EMBO J **8:** 141-1467

Kauzmann, W (1959) Some factors in the interpretation of protein denaturation. *In* CB Afinsen, ML Anson, K Bailey, JT Edsall eds, Advances in Protein Chemistry. Academic Press, New York, pp 1-63

Kim, PS, Bole, D, Arvan, P (1992) Transient aggregation of nascent thyroglobulin in the endoplasmic reticulum: Relationship to the molecular chaperone, BiP J Cell Biol **118:** 541-549

King, J, Fane, B, Haase-Pettingell, C, Mitraki, A, Villafane, R, Yu, M-H (1990) Identification of amino acid sequences influencing intracellular folding pathways using temperature-sensitive folding mutations. *In* LM Gierasch, J King ed, Protein Folding: Deciphering the Second Half of the Genetic Code. AAAS, Washington, pp 225-240

Knittler, MR, Haas, IG (1992) Interaction of BiP with newly synthesized immunoglobulin light chain molecules: cycles of sequential binding and release. EMBO J **11:** 1573-1581

Kondo, K, Inouye, M (1991) *TIP1*, a cold shock-inducible gene of *Saccharomyces cerevisiae*. J Biol Chem **266:** 17537-17544

Kondo, K, Kowalski, RZ, Inouye, M (1992) Cold shock induction of yeast NSR1 protein and its role in pre-RNA processing. J Biol Chem **267:** 16259-16265

Kurkela, S, Franck, M (1990) Cloning and characterization of a cold- and ABA-inducible *Arabidopsis*. Plant Mol Biol **15:** 137-144

Lång, V, Palva, ET (1992) The expression of a *rab*-related gene, *rab18*, is induced by abscisic acid during the cold acclimation process of *Arabidopsis thaliana* (L.) Heynh. Plant Mol Biol **20:** 951-962

Langer, T, Lu, C, Echols, H, Flanagan, J, Hayer, MK, Hartl, FU (1992) Successive action of DnaK, DnaJ, and GroEL along the pathway of chaperone-mediated protein folding. Nature **356:** 683-689

Levitt, J (1962) A sulfhydryl-disulfide hypothesis of frost injury and resistance in plants. J Theoret Biol **3:** 355-391

Levitt, J (1980) *Responses of Plants to Environmental Stresses.* New York: Academic Press. 697 pp

Leustek, T, Amir-Shapira, D, Toledo, H, Brot, N, Weissbach, H (1992) Autophosphorylation of 70 kDa heat shock proteins Cell Mol Biol **38:** 1-10

Lewis, MJ, Pelham, HRB (1985) Involvement of ATP in the nuclear and nucleolar functions of the 70 kd heat shock protein. EMBO J **4:** 3137-3143

Liberek, K, Marszalek, J, Ang, D, Georgopoulos, C (1991) *Escherichia coli* DnaJ and GrpE heat shock proteins jointly stimulate ATPase activity of DnaK. Proc Natl Acad Sci USA **88:** 2874-2878

Lin, C, Thomashow, MF (1992) DNA sequence analysis of a complementary DNA for cold-regulated *Arabidopsis* gene *cor15* and characterization of the COR15 polypeptide. Plant Physiol **99:** 519-525

Lin, Y, Gross, JK (1981) Molecular cloning and characterization of winter flounder antifreeze cDNA. Proc Natl Acad Sci USA **78:** 2825-2829

Lissin, NM, Venyaminov, SY, Girshovich, AS (1990) (Mg-ATP)-dependent self-assembly of molecular chaperone GroEL. Nature **348:** 339-342

Luo, M, Liu, J-H, Mohapatra, S, Hill, RD, Mohapatra, SS (1992) Characterization of a gene family encoding abscisic acid- and environmental stress-inducible proteins of alfalfa. J Biol Chem **267:** 15367-15374

Maniak, M, Nellen, W (1988) A developmentally regulated membrane protein gene in *Dictyostelium discoideum* is also induced by heat shock and cold shock. Mol Cell Biol **8:** 153-159

McCarty, JS, Walker, GC (1991) DnaK as a thermometer: Threonine-199 is site of autophosphorylation and is critical for ATPase activity. Proc Natl Acad Sci USA **88:** 9513-9517

Miernyk, JA, Duck, NB, Shatters, RG, Folk, WR (1992) The 70-kilodalton heat shock cognate can act as a molecular chaperone during the membrane translocation of a plant secretory protein precursor. Plant Cell **4:** 821-829

Minton, KW, Karmin, P, Hahn, GM, Minton, AP (1982) Nonspecific stabilization of stress-susceptible proteins by stress-resistant proteins: A model for the biological role of heat shock proteins. Proc Natl Acad Sci USA **79:** 7107-7111

Mundy, J, Chua, NH (1988) Abscisic acid and water-stress induce the expression of a novel rice gene. EMBO J **8:** 2279-2286

Munro, S, Pelham, HRB (1986) An Hsp70-like protein in the ER: Identity with the 78 kd glucose-regulated protein and immunoglobulin heavy chain binding protein. Cell **46:** 291-300

Neven LG, Haskell DW, Guy CL, Denslow N, Klein PA, Green LG, Silverman A (1992) Association of 70 kDa heat shock cognate proteins with acclimation to cold. Plant Physiol **99:** 1362-1369

Neven, LG, Haskell, DW, Hofig, A, Li, Q-B, Guy, CL (1993) Characterization of a spinach gene responsive to low temperature and water stress. Plant Mol Biol **21:** 291-305

Nilsson, B, Kuntz, ID, Anderson, S (1990) Expression and stabilization: Bovine pancreatic trypsin inhibitor folding mutants in *Escherichia coli*. *In* LM Gierasch, J King ed, Protein Folding: Deciphering the Second Half of the Genetic Code. AAAS, Washington, pp 117-122

Nordin, K, Heino, P, Palva, ET (1991) Separate signal pathways regulate the expression of a low-temperature-induced gene in *Arabidopsis thaliana* (L.) Heynh. Plant Mol Biol **16**: 1061-1071

Ouellet, F, Houde, M, Sarhan, F (1993) Purification, characterization and cDNA cloning of the 200 kDa protein induced by cold acclimation in wheat Plant Cell Physiol **34**: 59-65

Orr, W, Lu, B, White, TC, Robert, LS, Singh, J (1992) Complementary DNA sequence of a low temperature-induced *Brassica napus* gene with homology to the *Arabidopsis thaliana kin1* gene. Plant Physiol **98**: 1532

Pace, CN (1990) Conformational stability of globular proteins. Trends Biochem Sci **15**: 14-17

Palleros, DR, Welch, WJ, Fink, AL (1991) Interaction of hsp70 with unfolded proteins: Effects of temperature and nucleotides on the kinetics of binding. Proc Natl Acad Sci USA **88**: 5719-5723

Pearce, RS (1988) Extracellular ice and cell shape in frost-stressed cereal leaves: A low-temperature scanning-electron-microscopy study. Planta **175**: 313-324

Pelham, HRB (1986) Speculations on the functions of the major heat shock and glucose-regulated proteins. Cell **46**: 959-961

Petersen, NS, Young, P, Burton, V (1990) Heat shock mRNA accumulation during recovery from cold shock in *Drosophila melanogaster*. Insect Biochem **20**: 679-684

Privalov, PL (1990) Cold denaturation of proteins. Crit Rev Biochem Mol Biol **25**: 281-305

Rubin, DM, Mehta, AD, Zhu, J, Shoham, S, Chen, X, Wells, QR, Palter, KB (1993) Genomic structure and sequence analysis of *Drosophila melanogaster* HSC70 genes. Gene (in press).

Sakai, A, Larcher, W (1987) Frost Survival of Plants: Responses and Adaptation to Freezing Stress. Springer-Verlag, Berlin. pp 321

Sato, N (1992) Cloning of a low-temperature-induced gene *lti2* from the cyanobacterium *Anabaena variabilis* M3 that is homologous to α-amylases. Plant Mol Biol **18**: 165-170

Schmid, SL, Braell, WA, Rothman, JF (1985) ATP catalyzes the sequestration of clathrin during enzymatic uncoating. J Biol Chem **260**: 10057-10062

Sherman, MY, Goldberg, AL (1992) Heat shock in *Escherichia coli* alters the protein-binding properties of the chaperonin groEL by inducing its phosphorylation. Nature **357:** 167-169

Skowyra, D, Georgopoulos, C, Zylicz, M (1990) The *E. coli dnaK* gene product, the hsp70 homolog, can reactivate heat-inactivated RNA polymerase in an ATP hydrolysis-dependent manner. Cell **62:** 939-944

Snutch, TP, Heschl, MFP, Baillie, DL (1988) The *Caenorhabditis elegans hsp70* gene family: a molecular genetic characterization. Gene **64:** 241-255

Steponkus, PL (1984) Role of the plasma membrane in freezing injury and cold acclimation. Ann Rev Plant Physiol **35:** 543-584

Sutton, F, Ding, X, Kenefick, DG (1992) Group 3 *LEA* gene *HVA1* regulation by cold acclimation and deacclimation in two barley cultivars with varying freeze resistance. Plant Physiol **99:** 338-340

VanBogelen, RA, Neidhardt, FC (1990) Ribosomes as sensors of heat and cold shock in *Escherichia coli*. Proc Natl Acad Sci **87:** 5589-5593

Vierling, E (1991) The roles of heat shock proteins in plants. Annu Rev Plant Physiol Plant Mol Biol **42:** 579-620

Weiser CJ (1970) Cold resistance and injury in woody plants. Science **169:** 1269-1278

Willimsky, G, Bang, H, Fischer, G, Marahiel, MA (1992) Characterization of *cspB*, a *Bacillus subtilis* inducible cold shock gene affecting cell viability at low temperature. J Bacteriol **174:** 6326-6335

Zylicz, M, LeBowitz, JH, McMacken, R, Georgopolous, C (1983) The dnaK protein of *Escherichia coli* possesses an ATPase and autophosphorylating activity and is essential in an *in vitro* DNA replication system. Proc Natl Acad Sci USA **80:** 6431-6435

Low Temperature Signal Transduction, Gene Expression, And Cold Acclimation: Multiple Roles of Low Temperature

Rajinder S. Dhindsa and Antonio F. Monroy
Department of Biology, McGill University
1205 Doctor Penfield Avenue
Montreal, Quebec H3A 1B1, Canada

Introduction

Many plants develop resistance to freezing temperatures during exposure to a period of low but nonfreezing temperature (cold acclimation). During the past 5 years cold-induced genes have been cloned and characterized from several plants including alfalfa (Mohapatra et al., 1988, 1989; Monroy et al., 1993a; Wolfraim et al., 1993), Arabidopsis (Kurkela and Franck, 1990), barley (Cattivelli and Bartels, 1990; Dunn et al., 1991), Brassica (Orr et al., 1992), maize (Hahn and Walbot, 1990) and wheat (Houde et al., 1992). Some of the cold-induced genes isolated from alfalfa and wheat are specifically induced by cold, i.e. they are not induced by other environmental stresses or by the plant stress hormone ABA. It is clear from the sequence analysis of the cold-induced genes that, in agreement with the multigenic nature of freezing tolerance, cold acclimation is associated with a diversity of genes. Thus a rapid progress has been made towards understanding the molecular genetic basis of freezing tolerance. However, in order to cold acclimate a plant must be able to sense the low temperature signals and transduce them into specific biochemical processes leading to cold acclimation and development of freezing tolerance. We have been examining the role of calcium and protein phosphorylation in low temperature signal transduction during cold acclimation of alfalfa cells. Here we summarise evidence for the interlinking of calcium influx, protein phosphorylation, gene expression and the development of freezing tolerance during cold acclimation. For details of procedures and methods used in these studies see Mohapatra et al. (1989), Monroy and Dhindsa (1993), Monroy et al. (1993a,b) and Wolfraim et al., 1993).

NATO ASI Series, Vol. H 86
Biochemical and Cellular Mechanisms
of Stress Tolerance in Plants
Edited by J. H. Cherry
© Springer-Verlag Berlin Heidelberg 1994

Calcium and Cold Acclimation

Calcium is known to modulate metabolic and developmental processes and to act as a second messenger in a variety of organisms including plants (Veigle et al., 19984; Trewavas and Gilroy, 1991). Plant cell walls and vacuoles are believed to be the principal sites of calcium storage (Schroeder and Thuleau, 1991; Trewavas and Gilroy, 1991). In plants, a rapid increase in cytosolic calcium is triggered by several environmental stimuli such as touch, wind and cold shock (Knight et al., 1991, 1992). Thus we considered it quite likely that cold acclimation may involve calcium as a second messenger. We reasoned that if influx of cell wall calcium is essential for triggering the process of cold acclimation, then prevention of this influx should inhibit the cold-induced development of freezing tolerance. We have employed two ways to prevent such calcium influx. Cells were treated with either calcium chelator EGTA (50 mM), or with one of the calcium channel blockers La3+ (1 mM), verapamil (0.1 mM) or

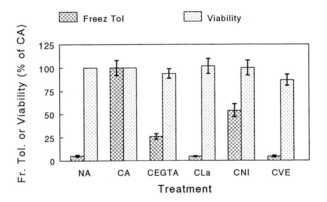

Figure 1. Effects on cell viability and cold-induced freezing tolerance in alfalfa cells of the calcium chelator EGTA (50 mM, CEGTA), and of calcium channel blockers Lanthanum (2 mM, CLa), nitrendipine (0.1 mM, CNI) and verapamil (0.1 mM, CVE). NA, nonacclimated; CA, cold acclimated. Cold acclimation was for 8 days.

nitrendipine (1 mM). The chemicals were added to the cells one hour before they were transferred to 4°C for 8 days of cold acclimation. The chemicals were added a second time at 4 days of cold acclimation to compensate for their possible degradation during a prolonged period of experimentation. Freezing tolerance was measured at the end of 8 days by the TTC reduction assay (Towill and Mazur,1974; Wolfraim et al., 1993). The effects of these chemicals on cell viability were also determined by carrying out TTC reduction assay on the unfrozen cells. It can be seen (Figure 1) that compared to nonacclimated cells (NA), cells cold acclimated for 8 days (CA) develop nearly 20 times greater freezing tolerance. In cells cold acclimated in the presence of EGTA (CEGTA) the development of freezing tolerance is only about 25 % of that developed by cells cold-acclimated in the absence of EGTA. The cells cold acclimated in the presence of the calcium channel blockers La3+ (CLa) and verapamil (CVE) completely lose the capacity to develop freezing tolerance. Interestingly, the channel blocker nitrendipine (CNI) inhibits cold-induced development of freezing tolerance by only about 40 percent. Nitrendipine has been previously shown to be a less effective calcium channel blocker (Graziana et al., 1988). It is noteworthy that none of the above chemicals has appreciable effects on the viability of cells. The differences between the different calcium channel blockers on cold acclimation may be due to a diversity of calcium channel populations present on the plasma membrane.

Calcium and Protein Phosphorylation

Calcium-dependent protein kinases have been well-documented in plants (Roberts and Harmon, 1992). However, clear evidence for the existence of protein kinase C and calcium /calmodulin-dependent protein kinases in plants is still lacking. Nevertheless, there is considerable evidence for the involvement of protein phosphorylation in signal transduction pathways in plants such as transduction of light stimulus (Datta and Cashmore, 1989) and pathogen elicitor signal (Felix et al., 1991). We have attempted to explore the role of protein kinases

504

regulated during cold acclimation of alfalfa cells by using differential inhibitors (Hidaka et al., 1981, 1984) of various types of kinases. For example H7 inhibits protein kinases C but not cAMP-dependent protein kinases. On the other hand, H8 preferentially inhibits cAMP-dependent protein kinases but is relatively ineffective against protein kinase C. Similarly, W7 acts as a calmodulin antagonist and, therefore, inhibits calcium/calmodulin-regulated protein kinases. It also inhibits calcium- dependent protein kinases (CDPKs). W5 is an ineffective analogue of W7. The effects of these chemicals on cold acclimation (8 days) and cell viability are shown in Figure 2. The development of freezing tolerance with 8 days of cold acclimation is affected by H7 and W7, but not by their ineffective analogues, H8 and W5 respectively. Whereas H7 inhibits the development of freezing tolerance by only about 50 %, the inhibition due to W7 is 100 percent. There is little effect of these inhibitors on cell viability. Even in the cells cold-acclimated in the presence of W7, where the cellular capacity to cold acclimate is completely abolished, the cellular viability is nearly 85 % of

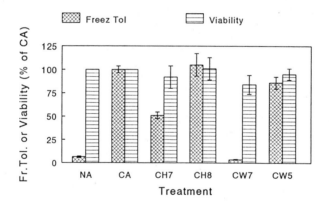

Figure 2. Effects of protein kinase C inhibitor H7 (CH7) and its ineffective analogue H8 (CH8), calmodulin antagonist and CDPK inhibitor W7 (CW7) and its ineffective analogue W5 (CW5), on cell viability and on cold-induced freezing tolerance. NA, nonacclimated; CA, cold-acclimated. Each chemical was used at 100 μM.

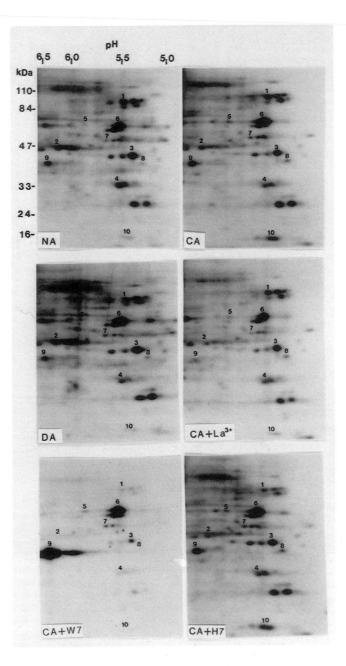

Figure 3. Effects of calcium channel blocker La3+ (CA+La)),
and of protein kinase inhibitors H7 (CA+H7) and W7 (CA+W7)
on cold-induced changes in phosphoprotein profile. Chemicals
were added 1 h before the start of cold acclimation. La3+
was added at 2 mM, while H7 and W7 were added at 100 μM
each. NA, nonacclimated; CA, cold acclimated; DA,
deacclimated. Cold acclimation was for 3 h.

the untreated cold acclimated cells (CA). H8 and W5, the
ineffective analogues of H7 and W7 respectively, have no effect
on either cell viability or on the development of freezing
tolerance. These results show that protein kinases are likely to
be involved in the processes of cold acclimation. Thus cold-
induced changes in phosphoprotein profile of alfalfa cells were
examined.

When alfalfa cells are transferred from 25° to 4° C, rapid
changes in their phosphoprotein profile take place (Fig. 3). Many
proteins of the nonacclimated cells (NA) are in the phosphorylated state. With 3 h of cold acclimation (CA), the relative phosphorylation level of several proteins decreases (e.g. proteins
1-4) while that of several others increases (e.g. proteins 5-10).
Protein-10 shows a dramatic increase in its phosphorylation level
with cold. When cells are returned to 25°C for 1 h of deacclimation (DA), the phosphoprotein profile returns to that of
nonacclimated cells. Presence of the calcium channel blocker
La3+ (CA+La3+) prevents the cold- induced changes in the
phosphoprotein profile which becomes similar to that of
nonacclimated cells. The protein kinase C inhibitor H7 inhibits
cold-induced changes in phosphoprotein profile only partially
(CA+H7).However, the effects of W7, an antagonist of calmodulin
and inhibitor of calcium-dependent protein kinases, are dramatic.
Except proteins 6 and 9, phosphorylation of which shows a large
increase, all proteins show a drastic decrease in their
phosphorylation. Thus W7 affects a large number of proteins
including those which show no relationship with cold acclimation.
This may be due to the multifunctional nature of calmodulin. It
is pertinent to recall that W7 does not have much effect on
viability of cells.

When cold acclimation is carried out in the presence of 150
μM cycloheximide (CA+CH), the phosphoprotein profile is similar
to that of cells cold acclimated in the absence of cycloheximide
(Monroy et al., 1993b). Cycloheximide at this concentration
inhibited protein synthesis by more than 90 %. This suggests that
the cold-induced changes in phosphoprotein profile occur on
preexisting proteins.

The specificity and magnitude of the cold-induced
phosphorylation of protein-10 (of about 15 kDa molecular mass) is
particularly noteworthy. Its phosphorylation level in the
nonacclimated cells is hardly detectable. Furthermore, all
treatments which inhibit cold-induced development of freezing
tolerance (Figures 1 and 2) also inhibit the phosphorylation of
protein-10. A semiquantitative measure of the relative

Table 1. Effects of cold acclimation, deacclimation, ABA,
cycloheximide, protein kinase C inhibitor H7, antagonist of
calmodulin and inhibitor of CDPK W7, and Ca2+ channel
blocker La3+ on the relative phosphorylation level of
protein-10 in alfalfa cells.

Treatment	Relative phosphorylation level
NA	0.59
CA (3 h)	6.91
DA (1 h)	0.56
NA + ABA (75 μM)	0.46
CA + CH (150 μM)	8.28
CA + H7 (100 μM)	7.49
CA + W7 (100 μM)	0.27
CA + La3+ (1 mM)	0.72

phosphorylation level of this protein under different treatments
was obtained with video densitometry and the results are shown in
Table 1. The relative phosphorylation level is shown in arbitrary
units. It can be seen that cold acclimation increases the
phosphorylation of this protein to greater than 10-fold and
subsequent deacclimation (DA) for just 1 h brings the level of
this phosphorylation down to the level seen in nonacclimated (NA)
cells. Relative phosphorylation level of this protein in cells
cold-acclimated in the presence of W7 (CA + W7) or La3+ (CA +
La3+) is similar to that in the nonacclimated cells. It is
noteworthy that phosphorylation of protein-10 is little affected
by H7 although the latter inhibits cold acclimation by 50 %. It
is possible that H7 affects events further downstream from the

site of involvement of protein-10 during cold acclimation.

Treatment of nonacclimated cells with 75 μM ABA (Monroy et et., 1993b), osmotic stress or drought did not reproduce the effects of cold on protein phosphorylation (Monroy and Dhindsa, unpublished data). This is significant because we have previously shown that ABA is much less effective than cold acclimation in inducing cold acclimation-specific gene expression and development of freezing tolerance (Mohapatra et al., 1988).

Calcium and Cold Acclimation-Specific Gene Expression

The expression of cold acclimation-specific (cas) genes cloned from alfalfa seedlings shows a high positive correlation (r $\geq$ 0.965) with the degree of freezing tolerance developed during cold acclimation of several cultivars of alfalfa (Mohapatra et al., 1989). The cell cultures we have been using as experimental material in our recent studies have been developed from the most freezing-tolerant cultivar of alfalfa (Medicago sativa spp. falcata, cv Anik). These cell cultures express the same genes as the intact seedlings in response to cold (Monroy et al., 1993a; Wolfraim et al., 1993). In order to study possible relationship of cold-induced calcium influx and protein phosphorylation to the cold acclimation-specific gene expression, we have determined the effects of chemicals which inhibit cold-induced freezing tolerance and protein phosphorylation, on cold-induced expression of cas15 as marker. Cas15 is rapidly induced by cold and encodes a putative nuclear-targetted protein of about 15 kDa molecular mass (Monroy et al., 1993a). Of all cold-induced genes of alfalfa so far examined, cas15 expression is detected the earliest. It is possible that cas15 plays a role in linking low temperature signal transduction and cold acclimation. The effects of chemicals which inhibit calcium influx or protein kinases on the cold-induced expression of cas15 are shown in Table 2. The transcript level is expressed as percent of that in the cold-acclimated, untreated cells (CA). The level is undetectable in the nonacclimated cells (NA). The calcium chelators EGTA and BAPTA reduce the expression of this gene drastically. The results are shown only for BAPTA which is a

modified EGTA molecule with pH-independent activity and is
effective at lower concentrations than EGTA. All calcium channel
blockers, La^{3+}, nitrendipine and verapamil, inhibit inhibit the
cold-induced expression of cas15. Inhibition due to nitrendipine
is the lowest. It should be mentioned that nitrendipine has been
shown to be a much less effective calcium channel blocker in
other plant systems (Graziana et al., 1988). The calmodulin
antagonist and inhibitor of calcium-dependent protein kinases,
W7, strongly inhibits the cold-induced accumulation of cas15
transcripts. The effects of the protein kinase C inhibitor H7 are
less pronounced.

Table 2. Effects of the calcium chelator BAPTA (2 mM),
calcium channel blockers La^{3+} (1 mM), nitrendipine (0.1 mM)
and verapamil (0.1 mM), and of protein kinase inhibitors H7
and W7 (0.1 mM each), on the cold-induced expression ofcas15
gene. The chemicals were added 1 h before the start of cold
acclimation. Total RNA was extracted and the level of
transcripts corresponding to cas15 was determined by
dot-blot hybridization (Monroy et al., 1993b) and expressed
as percent of that in cold acclimated cells (CA).

Treatment	Transcript Level (% CA)
NA	0
CA	100
CA + BAPTA	10
CA + La3+	45
CA + Nitrendipine	80
CA + Verapamil	35
CA + H7	45
CA + W7	10

It is clear from the foregoing that calcium chelators,
inhibitors of calcium transport and protein kinases interfere
with the process of cold acclimation. It suggests the requirement
for calcium influx during cold acclimation. Since the calcium
chelators and calcium channel blockers are unlikely to enter the
cell, they are probably acting through preventing the entry of
the cell wall calcium into the cytosol. These data suggest that

intracellular calcium is unlikely to play a major role in cold acclimation.

If calcium influx is needed to initiate the process of cold acclimation, is this need for calcium influx permanent or transitory? We have attempted to answer this question by applying the inhibitors of calcium influx at different times in relation to the start of cold acclimation and measuring the level of cas15 transcripts. La^{3+} (La) and Ruthenium Red (RR) were each separately added to the cell cultures at 2 h before, at or at different times during cold acclimation. Whereas La3+ was used as calcium channel blocker, RR is believed to inhibit the release of intracellular calcium. La3+ inhibits cas15 expression only if added either before or within the first 2 h of cold acclimation (Figure 4). Addition of La3+ at 4 h or later during cold acclimation had little effect on the transcript level. Addition of RR did not much affect the transcript accumulation. These data suggest that the need for calcium influx is transitory and that intracellular calcium does not play a major role during low temperature signal transduction.

Relative Roles of Calcium and Cold

Do calcium and cold play specific roles during cold acclimation? It is quite unlikely that the role of low temperature is to merely trigger the calcium influx because cold-induced calcium influx also occurs in a freezing-sensitive plant (Knight et al., 1992). We have attempted to separate the effects of calcium and cold by causing the calcium influx at room temperature and measuring the accumulation of cas gene transcripts. Results of our preliminary experiments show that calcium alone can induce the cold acclimation-specific genes but their transcripts become unstable after about 6 h. Thus low temperature, in some unknown manner, promotes the stabilization of cas gene expression. Since cold-induced calcium influx occurs irrespective of the freezing tolerance of the plant species, the ability to cold acclimate is likely to reside in a specific mechanism, absent in freezing-sensitive plants, which is probably modified by high calcium in order to be regulated by cold. Only

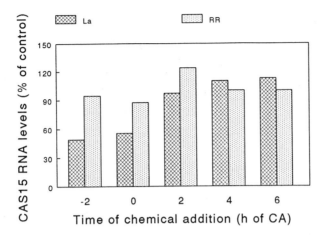

Figure 4. Calcium influx as a transient need for cold acclimation. La3+ (La) as a calcium channel blocker, and Ruthenium Red (RR) as a inhibitor of intracellular calcium release were used at 2 mM each.

when both calcium and cold are present do the processes lead to the sustained cas gene expression and development of freezing tolerance. We suggest the possibility that there are cold-regulated protein kinases which are predisposed to such regulation by high calcium.

Practical Applications of Signal Transduction Studies

Possible reasons for the lack of ability of the cold-sensitive plants to cold acclimate are many. The absence of cold acclimation-specific genes, their defective organization in the genome, and deficient or defective transacting factors have been suggested (Dhindsa and Mohapatra, 1988). The level of expression of cold-induced genes in alfalfa (Mohapatra *et al.*, 1989) and wheat (Houde *et al.*, 1992) shows a strong positive correlation with the degree of freezing tolerance developed during cold acclimation by different varieties of these respective crops. It has also been shown that these genes are present in the freezing-sensitive varieties but they are not sufficiently expressed. One of the possible reasons for this lack of cold-induced expression is that the genes may have defective organization in the genome. Altered genomic organization in sensitive cultivars as compared to tolerant ones has been demonstrated for cas15 gene of alfalfa (Monroy et al., 1993a). In wheat, however, the organization of the cold acclimation-specific genes is similar in freezing-tolerant and freezing-sensitive cultivars (Houde et al., 1992). In such situations, it is possible that the sensitive cultivars have defective components(s) of signal transduction pathway and the cell is unable to take advantage of the cold-induced, transient high intracellular calcium. Thus an understanding of the signal transduction mechanisms underlying the phenomenon of cold acclimation is likely to open additional facets for genetic engineering of freezing tolerance.

Acknowledgements: The research reported here has been supported by Natural Sciences and Engineering Research Council of Canada Grant A2724, and by F.C.A.R. (Quebec) Grants AS-2740 and ER-0575 to R.S.D.

REFERENCES

Cattivelli L, Bartels D (1990) Molecular cloning and characterization of cold-regulated genes in barley. Plant Physiol. **93**, 1504-1510

Datta N, Cashmore AR (1989) Binding of a pea nuclear protein to promoters of certain photoregulated genes is modulated by phosphorylation. Plant Cell **1**, 1069-1077

Dhindsa RS, Mohapatra SS (1988) cDNA cloning and expression of genes associated with freezing tolerance of alfalfa. In: SK Sinha, PV Sane, SC Bhargava, PK Agrawal, eds, Proc Internat Cong Plant Physiol, February 15-18, 1988, Society of Plant Physiology and Biochemistry, New Delhi, India, Pp 908-915

Felix G, Grosskopf DG, Regenass M, Boller T (1991) Rapid changes of protein phosphorylation are involved in transduction of the elicitor signal in plant cells. Proc. Natl. Acad. Sci. U.S.A. **88**, 8831-8834

Graziana A, Fosset M, Ranjeva R, Hetherington AM, Lazdunski M (1988) Ca2+ channel inhibitors that bind to plant cell membranes block Ca2+ entry into protoplasts. Biochemistry **27**, 764-768

Guy CL (1990) Cold acclimation and freezing stress tolerance: role of protein metabolism. Annu.Rev.Plant Physiol.Plant Mol.Biol. **41**, 187-223

Hanson PI, Schulman H (1992) Neuronal Ca2+/ calmodulin-dependent protein kinases. Annu. Rev. Biochem **61**, 559-601

Hidaka H, Asano M, Tanaka T (1981) Activity-structure relationship of calmodulin antagonists naphthalene-sulfonamide derivatives. Mol. Pharmacol. **20**, 571-578

Hidaka H, Inagaki M, Kawamoto S, Sasaki Y (1984) Isoquinoline-sulfonamides, novel and potent inhibitors of cyclic nucleotide dependent protein kinase and protein kinase C. Biochemistry **23**, 5036-5041

Johannes E, Brosnan JM, Sanders D (1991) Calcium channels and signal transduction in plant cells. BioEssays **13**, 331-334

Kikkawa U, Nishizuka Y (1986) The role of protein kinase C in transmembrane signaling. Ann. Rev. Cell Biol. **2**, 149-178

Knight MR, Campbell AK, Smith SM, Trewavas AJ (1991) Transgenic plant aequorin reports the effects of touch and cold-shock and elicitors on cytoplasmic calcium. Nature **352**, 524-526

Knight MR, Smith SM, Trewavas, AJ (1992) Wind-induced plant motion immediately increases cytosolic calcium. Proc Natl Acad Sci USA **89**, 4967-4971

Kurkela S, Franck M (1990) Cloning and characterization of a cold- and ABA-inducible Arabidopsis gene. Plant Mol. Biol. **15**, 137-144

Levitt J (1980) Responses of Plants to Environmental Stresses. Vol. I. Chilling, freezing, and high temperature stresses. Academic Press, New York

Mohapatra SS, Poole RJ, Dhindsa RS (1988) Abscisic acid-induced gene expression in relation to freezing tolerance in alfalfa. Plant Physiol. **87**, 468-473

Mohapatra SS, Wolfraim L, Poole RJ, Dhindsa RS (1989) Molecular cloning and relationship to freezing tolerance of cold-acclimation-specific genes of alfalfa. Plant Physiol. **89**, 375-380

Monroy AF, Dhindsa RS (1993) Induction of cold acclimation-specific genes at 25° C by calcium. (Submitted)

Monroy AF, Castonguay Y, Laberge S, Sarhan F, Vezina LP, Dhindsa RS (1993a) A new cold-induced alfalfa gene is associated with enhanced hardening at subzero temperature. Plant Physiol **102**: 873-879

Monroy AF, Sarhan F, Dhindsa RS (1993b) Cold-induced changes in freezing tolerance, protein phosphorylation, and gene expression. Evidence for a role of calcium. Plant Physiol **102**: 1227-1235

Roberts DM, Harmon AC (1992) Calcium-modulated proteins: Targets of intracellular calcium signals in higher plants. Annu. Rev. Plant Physiol. Plant Mol. Biol. **43**, 375-414

Sarokin LP, Chua N-H (1992) Binding sites for two novel phosphoproteins, 3AF5 and 3AF3, are required for rbcS-3A expression. Plant Cell **4**, 473-483

Schroeder JI, Thuleau P (1991) Calcium channels in higher plant cells. The Plant Cell **3**, 555-559

Towill LE, Mazur P (1974) Studies on the reduction of 2,3,5-triphenyltetra-zolium chloride as a viability assay for plant tissue cultures. Can. J. Bot. **53**, 1097-1102

Trewavas A, Gilroy S (1991) Signal transduction in plant cells. Trends Genet. **7**, 356-361

Veigl ML, Vanaman TC, Sedwick WD (1984) Calcium and calmodulin in cell growth and transformation. Biochim.Biophys.Acta **738**, 21-48

Veluthambi K, Poovaiah BW (1984) Calcium-promoted protein phosphorylation in plants. Science **223**, 167-169

Wolfraim LA, Langis R, Tyson H, Dhindsa RS (1993) Complementary DNA sequence, expression and transcript stability of a cold acclimation-specific gene, cas18 of alfalfa cells. Plant Physiol. **101**, 1275-1282

MOLECULAR ANALYSIS OF COLD-HARDENING IN BARLEY

Cattivelli L.*, C. Crosatti*, M. Grossi*, A. Portesi*, F. Rizza*, A.M. Stanca*, A. Furini**, D. Bartels**

*Istituto sperimentale per la Cerealicoltura, Via S.Protaso 302, 29017 Fiorenzuola d'Arda (PC) - Italy
**Max-Planck-Institut für Züchtungsforschung, Carl-von-Linné-Weg 10 - D-5000 Köln 30 - Germany

INTRODUCTION

The optimum development of barley over its life cycle depends on a number of environmental stress factors that can prevent the plant's expressing its maximum genetic potential. Severe grain losses are often caused by high or low temperatures, drought, anaerobiosis, and such soil anomalies as excess salt. The responses elicited from the plant by these stresses, when not lethal, include alterations in its processes of photosynthesis, respiration, and hormonal regulation through the development of specific, adaptive defense systems and mechanisms that are molecularly controlled. The duration of the stress and the plant's growth stage at the former's onset in turn affect yield. One can also find differing reactions as to plant susceptibility to adverse conditions. Thus, genetic variability plays a primary role in determining positive adaptation to environmental stresses and, hence, in supporting the spread of various barley genotypes to extreme climatic conditions (Stanca et al., 1992).

The ability of plants to survive low winter temperatures

NATO ASI Series, Vol. H 86
Biochemical and Cellular Mechanisms
of Stress Tolerance in Plants
Edited by J. H. Cherry
© Springer-Verlag Berlin Heidelberg 1994

is a phenomenon entailing a number of factors, including frost duration and severity, alternation of frost and thaw periods, synthesis of toxic substances affecting recovery capacity, plant growth stage, and the duration of the hardening period (Olien, 1979). Hardening, or cold acclimation, a physiological process comprising a series of biochemical reactions that enable tissues to enhance frost resistance (Levitt, 1980; Cattivelli and Bartels, 1992), is already found in barley after 2 weeks at 4°C at the coleoptile first-leaf stage. Once this acclimation period is over, the plant can resist intense cold. For example, test ratings of six nonhardened barley varieties, three spring and three winter types, at the first-leaf stage showed and LT_{50} (lethal temperature) of -6°C; when acclimated at 1°C for 3 weeks, the same genotypes recorded LT_{50}s of -9 and -15°C for the spring and winter types, respectively (Kolar et al., 1991).

If hardening is important, frost resistance in barley is even more so, because barley growing has been shifting for a number of years from spring to autumn sowing as winter varieties are higher yielding than spring cultivars. It is, however, a known fact that the winter barley varieties are less hardy than winter wheat, rye, and triticale.

Studies of the physiological mechanisms controlling cold acclimation have focused on three main areas: 1) membrane alteration, 2) changes in enzyme activities and content of soluble components, and 3) modification of gene expression and cold-regulated (COR) genes. Tests performed with rye have shown slight increases in total lipids and phospholipids of membranes during hardening (Cloutier, 1987; Lynch and Steponkus, 1987). An increase in soluble sugars in response to temperature decrease is common to all cereals (Levitt, 1980). Carbohydrate accumulation was found to occur during hardening in winter cereals because the demand for assimilates is reduced due to slower growth at lower temperatures (Livingston et al., 1989). It has also been found that plants exposured to cold (1-5°C) induce new mRNAs, which lead to the production of a specific set of cold-regulated proteins (Marmiroli et al.,

1986; Cattivelli and Bartels, 1989). Their synthesis and accumulation in plant tissues have been correlated with plant acclimation to cold (Hughes and Pearce, 1988). The function of some cold induced proteins in <u>Bacillus</u> <u>subtilis</u> has been recently described (Schindelin et al., 1993; Schnuchel et al., 1993) while in plants with the exception of some polypeptides with antifreezing properties (Kurkela and Franck, 1990; Griffith et al., 1992), the function of most cold induced proteins is at present unknown.

It is also well known that ABA is involved in low temperature response. Many physiological studies established a correlation between cold acclimation and increasing ABA levels in plant tissues (Chen et al., 1983; Heino et al., 1990). Furthermore, exogenous ABA application can induce in plant cell cultures cold-hardening to the same degree as a low temperature treatment does (Chen e Gusta, 1983).

To describe the barley molecular response to low temperature we have first analyzed the changes in gene expression and then cloned several genes positively regulated by cold treatment. We have also investigated the effect of an ABA treatment on the expression of the cold regulated genes.

IDENTIFICATION OF COLD-INDUCED mRNAs

Polyadenylated RNA was extracted from cold-stressed (4 d + 1°C) barley shoots and the <u>in</u> <u>vitro</u> translation products were separated by 2-D electrophoresis. A comparison with the <u>in</u> <u>vitro</u> synthesized proteins from unstressed shoots revealed three kind of modifications: a) several <u>de-novo</u> synthesized <u>in</u> <u>vitro</u> proteins were induced; b) a group of few <u>in</u> <u>vitro</u> translated products increased several fold as response to cold; c) several <u>in</u> <u>vitro</u> products decreased or disappeared during cold treatment.

In open field trials barley shows a wide range of frost

resistance as measured by physiological parameters. However when the poly(A)RNAs induced by low temperature in spring and winter cultivars have been compared, only few additional or missing in vitro-synthetized proteins were detected. Therefore a clear relationship between specific changes in gene expression and the level of frost resistance could not be demonstrated.

Example of protein products obtained from cold-induced mRNAs are presented in figure 1 (Cattivelli and Bartels, 1989).

ISOLATION OF cDNA CLONES ENCODING COLD-REGULATED TRANSCRIPTS

Poly(A) RNA isolated from barley shoots treated at low temperature as in the previous experiment, was used to construct a cDNA library. By differential hybridization several different cDNA clones were selected. In hybridization experiments with RNA from cold-treated shoots it was confirmed that these clones represent different genes (Cattivelli and Bartels, 1990).

All the selected clones detected homologous transcripts in mRNA population already after 24° of low temperatures, however the levels of the transcripts increased during longer cold treatment (up to 3 days). When after 4 days at 1°C the temperature was raised to 22°C, the levels of the transcripts induced in low temperature conditions dropped appreciably after 2 h. After 8 h at 22°C, no cold-induced mRNAs were measured (figure 2).

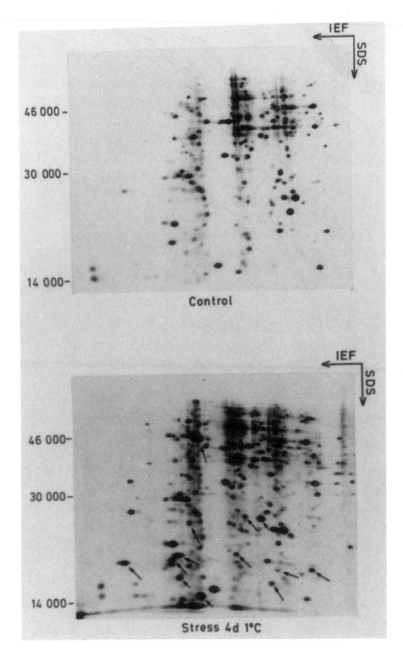

Figure 1. Two dimensional separation of labelled _in_ _vitro_ translation products obtained from poly(A)RNAs isolated from control and cold treated barley shoots. The arrows on the bottom panel indicate the proteins increased or induced by low temperatures.

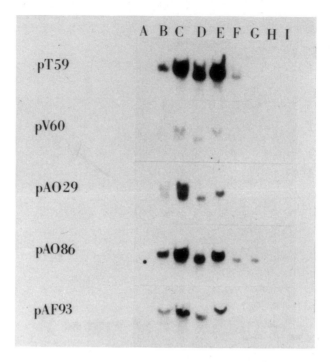

Figure 2. Northern hybridization of the cold-regulated cDNA clones to total RNAs (80 µg per lane loaded) isolated from barley shoots after different temperature treatments. **A**, Control 22°C; **B**, 1 d 1°C; **C**, 4 d 1°C; **D**, 4 d 1°C then shift to 1 h at 22°C; **E**, 4 d 1°C then shift to 2 h at 22°C; **F**, 4 d 1°C then shift to 4 h at 22°C; **G**, 4 d 1°C then shift to 6 h at 22°C; **H**, 4 d 1°C then shift to 8 h at 22°C; **I**, heat shock (2 h 37°C).

The expression of cold-regulated transcripts in cold-stressed tissues other than young shoots was also investigated, RNAs were isolated from roots and mature leaves. Roots were derived from seedlings (one leaf stage) and mature leaves were harvested from two different developmental stages (late tillering and preanthesis). With a single exception all the isolated cDNA clones detected homologous transcripts in the mRNA population of leaves of mature plants exposed for 4 days at low temperatures.

Several cold-regulated clones found homologous mRNAs also

in root tissues after exposure to low temperatures, however cDNA clone representing a mRNA expressed only in leaf tissues has also been isolated. Transcripts homologous to barley cDNA clones were also found inducible by low temperature in others cereal species (wheat and rye).

The most relevant characteristics of some barley cold-related cDNA clones are summarized in table 1.

Table 1. Characteristics of some barley cold-regulated cDNA clones

Clone	Size of corresponding mRNA	Tissue specific expression after cold treatment[a]		leaves[b]		Homologous mRNAs in other cereals[a]			Expression after other stress condition[a]	
		Shoots	Roots	A	B	Wheat	Rye	Oats	ABA	Drought
	kb									
pT59	0.9	XXX		XXX	XXX	X	X			
pV60	0.9	XXX				X	X			
pA029	0.9	XXX	XXXX	XX	XX	X				
pA086	0.9	XX	X	XX	XX	X	X			
pAF93	1.5	XX	XX	X	X					XX

[a]The number of the crosses indicates the intensity of the hybridization signal. [b]A, Tillering stage; B, preanthesis stage

RELATIONSHIP BETWEEN MOLECULAR RESPONSES TO LOW TEMPERATURE AND DROUGHT STRESS

It is well known that ABA is a key hormone involved in the response to different environmental stresses such as drought (Skriver and Mundy, 1990) or cold (Heino et al., 1990). Based on this knowledge we have therefore performed several experiments in order to establish the relationship between the modifications in gene expression induced by low temperature, and by drought stress (Grossi et al., 1992).

Poly(A)RNAs isolated from barley shoots after low temperature treatment, after drought stress and after an exogenous ABA treatment were used for _in_ _vitro_ translation experiments. The 2-dimensional analysis of _in_ _vitro_ translation products obtained from mRNAs isolated after different treatments (drought, ABA and cold) showed that there is also a single stress protein induced in all three conditions. The level of induction of the corresponding mRNAs is similar in drought and cold stress, while it is reduced when the shoots are sprayed with ABA.

In order to identify other molecular similarities between the responses to low temperatures and to water deficit the cold-regulated cDNA clones were tested for their ABA inducibility and for their expression in response to drought. A clear result was obtained with the clone pAF93. Northern analysis performed with pAF93 showed a hybridization signal corresponding to a mRNA of 1.5 kb, characterized by cold and drought induction. No signal was detected in mRNAs extracted from ABA treated shoots. Therefore the mRNA corresponding to pAF93 represents a gene induced by drought and cold stress but not by ABA treatment (figure 3).

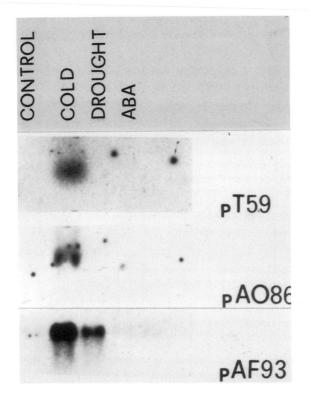

Figure 3. Northern hybridization of the cold-regulated cDNA clones pT59; pA086 and pAF93 to poly(A)RNA isolated from barley shoots in different stress condition: cold treatment (3 d at +2°C); drought stress (water potential = -1.2 MPa); ABA treatment (1 mM ABA). pT59 and pA086 represent mRNAs induced only at low temperatures, while the mRNA corresponding to pAF93 can be induced both by cold and drought stress.

DISCUSSION

The results summarized in this paper and several other works (Dunn et al., 1990 e 1991) suggest that the barley molecular response to low temperatures is complex process involving many genes. Indeed also several genetic experiments suggest that the inheritance of the capacity for cold acclimation-induced freezing tolerance is a quantitative character controlled by a number of additive genes (Guy, 1990; Thomashow, 1990). However, in wheat genetic analysis indicates that some genes with major effects on cold tolerance are located on chromosome 5A, closely linked with the locus Vrn1 controlling vernalisation requirements (Sutka and Snape, 1989).

The cloning and the characterization of genes regulated by low temperature represent the molecular biology approach to study the acclimation process. As expected most of the genes isolated in this way encode for unknown products and therefore the next step will be to prove the direct role of the identified genes in the stress phenomenon.

The availability of many different cDNA clones however it allows to establish same relationship between the acclimation process of different species. Many of the cDNA clones isolated in barley up to date are not ABA regulated, this is in contrast with what has been found in other species. In Arabidopsis, for instance, almost all the low temperature induced cDNA clones are inducible also by ABA (Hajela et al., 1990; Kurkela and Franch, 1990). These evidences may suggest that the cold acclimation process has been evolved in differently in cereals and in Arabidopsis.

The fact that a large number of the cold-regulated cDNA clones from barley have homologous transcripts in other cereal species indicates that the molecular pathways leading to acclimation are conserved among closely related species.

ACKNOWLEDGEMENTS: This work was supported by Italian Ministry of Agriculture Project: "Resistenze genetiche delle piante agrarie agli stress biotici e abiotici".

REFERENCES

Cattivelli L, Bartels D (1989) Cold-induced mRNAs accumulate with different kinetics in barley coleoptiles. Planta 178: 184-188.

Cattivelli L, Bartels D (1990) Molecular cloning and characterization of cold-regulated genes in barley. Plant Physiol 93: 1504-1510.

Cattivelli L, Crosatti C, Grossi M, Pecchioni N, Portesi A, Rizza F, Stanca AM, Terzi V (1991) Molecular response to abiotic stresses in barley. Vortr Pflanzenzüchtg 20: 168-172/5.

Cattivelli L, Bartels D (1992) Biochemistry and molecular biology of cold-inducible enzymes and proteins in higher plants. In: "Society for Experimental Biology Seminar Series 49: Inducible Plant Proteins" (JL Wray ed.), Cambridge University Press, pp. 267-288.

Chen THH, Gusta LV (1983) Abscisic acid-induced freezing resistance in cultured plant cells. Plant Physiol 73: 71-75.

Chen H-H, Li PH, Brenner ML (1983) Involvement of abscisic acid in potato cold acclimation. Plant Physiol 71: 362-365.

Cloutier Y (1987) Lipid and protein changes in cold- and drought-hardened cereals. Phytoprotection 68: 87-96.

Dunn MA, Hughes MA, Pearce R.S, Jack PL (1990) Molecular characterization of a barley gene induced by cold treatment, J Exp Bot 41: 1405-1413.

Dunn MA, Hughes MA, Zhang L, Pearce RS, Quigley AS, Jack PL (1991) Nucleotide sequence and molecular characterisation of the low temperature induced cereal gene, BLT4. Mol Gen Genet 229: 389-394.

Griffith M, Ala P, Yang DSC, Hon W-C, Moffatt B (1992) Antifreeze protein produced endogenously in winter rye leaves. Plant Physiol 100: 593-596.

Grossi M, Cattivelli L, Terzi V, Stanca AM (1992) Modification of gene expression induced by ABA, in relation to drought and cold stress in barley shoots. Plant Physiol Biochem 30(1): 97-103.

Guy CL (1990) Cold acclimation and freezing stress tolerance: role of protein metabolism. Ann Rev Plant Physiol Plant Mol Biol 41: 187-223.

Hajela RK, Horvath DP, Gilmour SJ, Thomashow MF (1990) Molecular cloning and expression of cor (cold-regulated) genes in Arabidopsis thaliana. Plant Physiol 93: 1246-1252.

Heino P, Sandman G, Lång V, Nordin K, Palva ET (1990) Abscisic acid deficiency prevents development of freezing tolerance in Arabidopsis thaliana (L.) Heynh. Theor Appl Genet 79: 801-806.

Hughes MA, Pearce RS (1988) Low temperature treatment of barley plants causes altered gene expression in shoot meristems. J Exp Bot 39: 1461-1467.

Kolar SC, Hayes PM, Chen THH, Linderman RG (1991) Genotypic variation for cold tolerance in winter and facultative barley. Crop Sci. 31: 1149-1152.

526

Kurkela S, Franck M (1990) Cloning and characterization of a cold- and ABA-inducible _Arabidospis_ gene. Plant Mol Biol 15: 137-144.

Levitt J (ed) (1980) Responses of Plants to Environmental Stresses, Vol. I, Chilling, Freezing, and High Temperature Stresses. Academic Press, Orlando FL.

Livingston III DP, Olien CR, Freed RD (1989) Sugar composition and freezing tolerance in barley crowns at varying carbohydrate levels. Crop Sci. 29: 1266-1270.

Lynch DV, Steponkus PL (1987) Plasma membrane lipid alterations associated with cold acclimation of winter rye seedlings (_Secale cereale_ L. cv Puma). Plant Physiol. 83: 761-767.

Marmiroli N, Terzi V, Odoardi Stanca M, Lorenzoni C, Stanca AM (1986) Protein synthesis during cold shock in barley tissues. Theor Appl Genet 73: 190-196.

Olien CR (1979) Physiology of winter hardiness in barley, Barley. USDA Agric. Handb. 338, Washington D.C., pp. 147-154.

Schindelin H, Marahiel MA, Heinemann U (1993) Universal nucelic acid-binding domain revealed by crystal structure of the _B. subtilis_ major cold-shock protein. Nature 364: 164-168.

Schnuchel A, Wiltscheck R, Czisch M, Herrier M, Willimsky G, Graumann P, Marahiel MA, Holak TA (1993) Structure in solution of the major cold-shock protein from _Bacillus subtilis_. Nature 364: 169-171.

Skriver K, Mundy J (1990) Gene expression in response to abscisic acid and osmotic stress. The Plant Cell 2: 503-512.

Stanca AM, Terzi V, Cattivelli L (1992) Biochemical and molecular studies of stress tolerance in barley. In: "Barley: genetics, biochemistry, molecular biology and biotechnology" (P.R. Shewry ed.), C.A.B. International, Wallingford UK, pp. 277-288.

Sutka J, Snape JW (1989) Location of a gene for frost resistance on chromosome 5A of wheat. Euphytica 42: 41-44.

Thomashow MF (1990) Molecular genetics of cold acclimation in higher plants. Adv in Genet 28: 99-131.

REGULATION OF LOW TEMPERATURE-INDUCED GENES DURING COLD ACCLIMATION OF *ARABIDOPSIS THALIANA*

E. Tapio Palva, Björn Welin, Tiina Vahala, Åke Olson, Kerstin Nordin-Henriksson, Einar Mäntylä and Viola Lång

Department of Molecular Genetics

Uppsala Genetic Center

Swedish University of Agricultural Sciences

Box 7003

S-750 07 Uppsala

Sweden

ABSTRACT

Arabidopsis thaliana provides an ideal model system for molecular analysis of plant cold acclimation. This small Cruciferae can readily cold acclimate and the acclimation process is accompanied by expression of a specific set of low temperature-responsive genes. Structural analysis of such *lti* (*low temperature induced*) genes has demonstrated that many of them code for polypeptides related to members of the RAB/LEA/DHN family of water stress responsive proteins. This structural similarity is indicative of common function for these proteins and suggests overlapping responses to freezing and desiccation stress. The other *lti* genes characterized seem to code for novel, often very hydrophilic proteins. Enhanced freezing tolerance of *A. thaliana* can be induced by exposure to low temperature, mild desiccation or exogenous abscisic acid (ABA). Mutant studies have suggested that ABA-controlled processes appear to be required for

NATO ASI Series, Vol. H 86
Biochemical and Cellular Mechanisms
of Stress Tolerance in Plants
Edited by J. H. Cherry
© Springer-Verlag Berlin Heidelberg 1994

a normal acclimation response. In accordance with the observed pattern of freezing tolerance induction, expression of the *lti* genes is responsive to the same three stimuli. However, there are marked differences in the expression patterns of the *lti* genes during the different modes of induction. Furthermore, the genes seem to fall into three different categories with respect to the signal pathways employed for their expression. Existence of separate response pathways to the different stimuli appears to be the most common mechanism for induction of these genes. Accordingly, stimulus specific DNA elements appear to be present in the *lti* promoters.

INTRODUCTION

Cold acclimation is an adaptive response in many temperate plant species that results in enhanced tolerance to extracellular freezing (Levitt, 1980; Sakai and Larcher, 1987). The capability to cold acclimate appears to be of complex polygenic nature and the acclimation process is triggered by several environmental stimuli. In herbaceous plants the main trigger appears to be exposure to low nonfreezing temperatures (LT), but many plant species seem to respond to other stimuli as well i.e. mild desiccation or exogenously added ABA (Cloutier and Siminovitch, 1982; Chen et al., 1979; Chen and Gusta, 1983; Orr et al., 1986; Lång et al., 1989).

Several physiological and biochemical changes have been associated with cold acclimation (Levitt, 1980; Sakai and Larcher, 1987), but only more recently it has been shown that alterations in gene expression can be correlated to the acclimation process (Guy et al., 1985; Mohapatra et al., 1987; Gilmour et al., 1988; Kurkela et al., 1988) as originally proposed by Weiser (1970). Based on such differential expression of acclimation related genes, LT-responsive genes and cDNAs have recently been isolated from several plant species including alfalfa (Mohapatra et al., 1989), *A. thaliana* (Hajela et al., 1990; Kurkela and Franck, 1990; Nordin et al., 1991) barley (Cattivelli and Bartels, 1990; Dunn et al., 1990), *Brassica napus* (Orr et al., 1992), wheat (Houde et al., 1992; Guo et al., 1992) and spinach (Neven et al., 1993). Transcripts corresponding to these genes/cDNAs appear rapidly in plants subjected to LT and often even in response to other stimuli that lead to enhanced freezing tolerance such as desiccation and exogenous application of

ABA. Structural analysis of the genes has demonstrated that many of the polypeptides share conserved amino acid motifs (Gilmour et al., 1992; Guo et al., 1992; Lång and Palva, 1992; Houde et al., 1992; Wolfrain et al., 1993) with the RAB/LEA/DHN family of proteins (Skriver and Mundy, 1990), suggesting a role in tolerance to the freeze-induced cellular dehydration that is a consequence of extracellular ice formation. The protein encoded by the *cor15* gene has been reported to have cryoprotective properties in vitro (Lin and Thomashow, 1992a) but the importance of this or other LT-responsive genes in the acclimation process or enhanced freezing tolerance needs yet to be demonstrated.

To address these questions we have isolated several *lti* genes of *A. thaliana* and are characterizing their regulation and structural properties (Nordin et al., 1991; 1993a, b; Lång and Palva, 1992; Welin et al, in preparation). *A. thaliana* provides an ideal model system for molecular analysis of the mechanism of plant cold acclimation. It is well established as the model for plant molecular biology (Meyerowitz, 1989), in particular the availability of mutants and the advanced molecular genetic techniques makes this plant well suited for such studies. *A. thaliana* has the ability to cold acclimate both by LT exposure (Kurkela et al., 1988; Gilmour et al., 1988) and by exogenous ABA (Lång et al., 1989), and it can rapidly increase its freezing tolerance to the level (LT_{50}-10°C - -20°C) typical for many cold acclimation-proficient herbaceous plant species.

Here we review what has been learned about the structure and regulatory properties of *A. thaliana* LT-responsive genes and what this has told us about the cold acclimation process.

RESULTS AND DISCUSSION

Low temperature-induced genes of *Arabidopsis thaliana*

Isolation of DNA sequences corresponding to the LT-induced transcripts has been mainly accomplished by differential screening of cDNA libraries. These cDNAs in turn have served as probes for obtaining the corresponding genomic clones. Additional genes have been isolated by screening with heterologous probes from other plant species. About 10 LT-responsive genes of *A. thaliana* have been characterized sofar and their salient features are summarized in Table 1. Structural characterization of the isolated genes and

Table 1. *Low temperature induced genes of* Arabidopsis thaliana

Gene/cDNA[a]	Protein[b]		Regulation[c]				Reference
	M_r	Similarity	LT	ABA	D	Class	
lti78 (= lti140, c cor160, rd29A)	78	-/LTI65	+++	+	+	I	Nordin et al., 1993b
lti65 (=rd29B)	65	-/LTI78	+	++	++	III	Nordin et al., 1993b
c cor47	45	RAB/LTI8	+	+	+	I	Gilmour et al., 1992
c lti8	40	RAB/COR47	+++	+	+	I	Welin et al., (in preparation)
c lti13	30	RAB	+++	+	+	I	Welin et al., (in preparation)
rab18	18	RAB	+	+++	+++	II	Lång and Palva, 1992
dhn	13	RAB					Rouse et al., 1992
c cor15	15	-/cp protein	+	+	+	I	Lin and Thomashow, 1992b
kin1	7	-/KIN2/ COR6.6	+	+	(-)		Kurkela and Franck, 1990
c cor6.6 (= kin2)	7	-/KIN1	+	+	+	I	Gilmour et al., 1992

[a]Only genes or cDNAs (indicated by c before name of the gene) where sequence information is available have been included. The reference is to the paper describing the sequence.
[b]The molecular weight shown is in kDa, the similarity indicated by RAB is to the RAB/LEA/DHN family of proteins, cp = chloroplast.
[c]Induction of gene expression by LT, ABA or desiccation (D) is indicated by +. If the relative level of response has been characterized it is indicated by a scale from - (no response) to +++ (strongly induced).

cDNAs has not been particularly informative regarding the putative function of the corresponding polypeptides. However, certain interesting features are starting to emerge from these structural comparisons that may be relevant for understanding the function of these polypeptides. The most common structural feature of the LT-induced proteins is a conserved lysine-rich motif which is present in different permutations in about half of the polypeptides (Fig. 1). Similar motifs have been originally characterized in RAB/LEA/DHN proteins that accumulate during late embryogenesis (LEA; Baker et al., 1988; Dure et al.,

1989) and in vegetative tissues in response to dehydration stress (DHN; Close et al., 1989) or abscisid acid (RAB; Mundy and Chua, 1989; Skriver and Mundy, 1990). Of the LT-induced proteins in *A. thaliana*, RAB18 (Table 1; Fig. 1) most closely resembles group 2 LEA proteins (Dure et al., 1989) in its structural organization. RAB18 contains in addition to the two copies of the lysine-rich repeat also the other structural motifs typical for this class of proteins (Lång and Palva, 1992). In contrast, the other related LT-responsive polypeptides (e.g. COR47, LTI8, LTI13) have a variable number of copies of the lysine-rich repeat (Fig. 1). Most of them also seem to contain the serine rich sequence present in the RAB family (Fig.1). It has been postulated that the RAB/LEA/DHN family of polypeptides protects cells from dehydration stress by stabilization of cellular structures and solvation of structural surfaces (Baker et al., 1988; Dure et al., 1989; Close et al., 1989; Skriver and Mundy, 1990)). Consequently, the structural similarity of the LT-induced proteins to the RAB/LEA/DHN group is indicative of a similar protective role, as would be required during freeze-induced cellular dehydration.

A common feature of the LT-induced gene products characterized so far from *A. thaliana* is their hydrophilic nature. This may be manifested in the "boiling solubility" of many of these polypeptides (Lin et al., 1990), a feature which was originally described for LEA proteins (Baker et al., 1988; Dure et al., 1989). Most of the *lti* genes seem to encode soluble cytoplasmic polypeptides (see Table 1 and references therein), only one of the proteins, COR15, has been found in an extracytoplasmic location and appears to be targeted to the chloroplast (Lin and Thomashow, 1992). Structurally, apart from the RAB-related proteins, the other *lti* genes seem to form a more heterogeneous group that encode different types of novel polypeptides. Counterparts of many of the *A. thaliana* genes can also be found in *Brassica napus*. We have recently characterized two closely related genes *lti78* and *lti65* (Nordin et al., 1993b) that encode novel hydrophilic polypeptides. Using these genes as probes we could detect related transcripts in *B. napus* where their accumulation appeared to correlate with the level of freezing tolerance attained by different rapeseed cultivars (Vahala et al., in preparation). Weretilnyk et al. (1993) have characterized the *B. napus* counterpart of *cor15* and Orr et al. (1992) the counterpart of *kin* (or *cor6.6*) genes. The polypeptides encoded by the *kin*-like genes show same similarity to type I fish antifreeze proteins (Picket et al., 1984) and led Kurkela and Franck

(1990) to propose that the KIN polypeptide may have a similar function. As the observed similarity is mainly due to the high content of alanine residues in these proteins, the functional meaning of this is open to question (Gilmour et al., 1992).

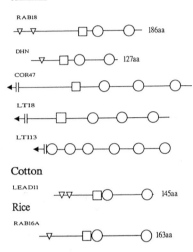

Figure 1. A comparison of the conserved structural motifs present on the polypeptides encoded by LT-responsive genes of *A. thaliana* and some members of the RAB/LEA/DHN family of proteins. The linear arrangement of these elements: -DEYNGP- (inverted triangles), multiple serine residues (open boxes) and the lysine-rich motifs (open circles) is indicated on the left. Alignment of the lysine-rich repeats is shown on the right.

An intriguing feature of the *lti* genes of *A. thaliana* is that they mainly seem to come in pairs of closely related sequences. We have in detail characterized one such gene pair, *lti78* and *lti65* (Nordin et al., 1993b) that clearly seem to have arisen by tandem duplication. Recent sequence analysis of a genomic clone suggests that *cor47* and *lti8* form a similarly organized pair (Welin et al., in preparation) as *lti78* and *lti65* (Fig. 2). Similar organization has also been observed for *kin1* and *kin2* (Kurkela, 1991) and may also exist for the other *A. thaliana* genes (Table 1). The functional significance of these duplications is not clear but the often very different expression patterns of the duplicate genes (see below) may indicate somewhat diverged functions in stress tolerance or adaptation.

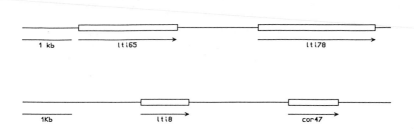

Figure 2. Structural organization of two pairs of LT-responsive genes in the genome of *A. thaliana*. The data presented in the diagram are based on sequence analysis of genomic clones (Nordin et al., 1993b; Welin et al., in preparation). The arrows indicate direction of transcription of the depicted genes.

Induction of freezing tolerance in *A. thaliana*

Exposure of *A. thaliana* to low nonfreezing temperatures results in rapid increase in freezing tolerance (Kurkela et al., 1988; Gilmour et al., 1988). Greenhouse plants acclimated for a few days survive readily at -10°C but axenically grown seedlings are somewhat less tolerant (about -7°C) (Nordin et al., 1993a). In addition to LT, also other external stimuli such as exogenous ABA (Lång et al., 1989) or mild desiccation (Mäntylä et al., in preparation) result in enhanced freezing tolerance of *A. thaliana* seedlings (Table 2). It has been proposed that ABA may be the endogenous signal of the cold acclimation response and it is conceivable that ABA could trigger both LT and desiccation induced cold acclimation. Accumulating evidence implies that ABA could have a central role in plant cold acclimation. In addition to results showing that ABA can substitute for the LT stimulus in induction of freezing tolerance in *A. thaliana* (Lång et al., 1989), the endogenous level of ABA is transiently increased during LT acclimation in both *A. thaliana* (Lång et al., submitted) and other plant species (Chen et al., 1983). More direct proof of the involvement of ABA in cold acclimation comes from mutant studies. Both ABA-deficient (*aba*) and ABA-insensitive (*abi*) mutants of *A. thaliana* have been characterized (Koornneef et al., 1982; 1984; 1989). We have earlier shown that the ABA-deficient mutant (*aba-1*) was impaired in cold acclimation (Heino et al., 1990) and our more recent data indicate that the ABA-insensitive *abi1* mutant appears to exhibit a delayed acclimation phenotype (Mäntylä et al., in preparation). In conclusion, several types of experiments imply that ABA controlled processes are essential for normal cold acclimation in *A. thaliana*.

Table 2. *Cold acclimation of* Arabidopsis thaliana[a]

Acclimation treatment	LT$_{50}$ (^{0}C)			
	0	1	3	days of acclimation
Low temperature	-3	-4	-7	
Desiccation	-3	-5	-7	
ABA	-3	-5	-8	

[a]Axenically grown seedlings were exposed to the treatments for the times indicated and the LT$_{50}$ values were estimated by the electrolyte leakage test and confirmed by visual examination (Mäntylä *et al.*, in preparation).

Expression of *lti* genes

As discussed low temperature is not the only stimulus which induces enhanced freezing tolerance in *A. thaliana* but similar response is also achieved by desiccation or ABA treatment. Therefore, it is conceivable that genes required for cold acclimation would also be responsive to the latter two stimuli. Characterization of the expression patterns of genes isolated from *A. thaliana* (Table 1) clearly indicates that the majority of them are indeed induced by LT, desiccation and exogenously applied ABA (Hajela et al., 1990; Kurkela and Franck, 1990; 1992; Nordin et al., 1991; 1993b; Lång and Palva, 1992; Welin et al., in preparation). However, there are marked differences in responsiveness to different stimuli between the genes (Fig. 3). It is particularly noteworthy that often members of the duplicate gene pairs such as *lti78* and *lti65* (Nordin et al., 1991; 1993b) or *kin1* and *kin2* (Kurkela and Franck, 1990, 1992) exhibit differential patterns of expression. The most extreme case so far is provided by the *lti78/lti65* pair (Nordin et al., 1993b) where *lti78* is strongly induced by low temperature but to a clearly lesser degree by desiccation or ABA (Table 1, Fig. 3). In contrast, *lti65* is hardly at all responsive to LT but is strongly induced by both ABA and desiccation (Nordin et al., 1993b). A similar expression pattern can be seen for *rab18* where only a weak and delayed LT response is observed as compared to the strong and rapid induction by ABA or desiccation (Lång and Palva, 1992). Such differential expression of related genes might be indicative of corresponding differences in function. Thus, the LTI78 polypeptide might be more involved in freezing stress tolerance while the related but diverged LTI65 protein could be required for adaptation to desiccation stress. The duplicate nature of the related but differentially expressed genes could also reflect their evolutionary origin. It has been proposed that the original adaptation to withstand drought (Stebbins, 1972) could have been utilized in evolution of the freezing tolerance mechanisms (Guy et al., 1992).

The response to low temperature, ABA and desiccation stress is not a general stress response but seems specific to the above stimuli. A clear indication of the specificity of the response is that *lti* genes are not induced by another type of temperature stress, the heat shock (Hajela et al., 1990; Kurkela and Franck, 1990; Nordin et al., 1991; 1993b; Lång and Palva, 1992). A characteristic feature of the LT-response in *A. thaliana* is that the transcripts start to accumulate within hours of LT-exposure and seem to persist through the acclimation period (Fig. 3) as shown for several genes of *A. thaliana* (Hajela et al., 1990; Nordin et al., 1991; Yamaguchi-Shinozaki and Shinozaki, 1993). However, when plants are returned to normal growth temperatures the transcripts usually disappear within a few hours of deacclimation (Hajela et al., 1990; Nordin et al., 1991). No tissue specificity has been observed with any of the *lti* genes characterized sofar but they seem to be expressed throughout the plant. However, more careful analysis e.g. by in situ hybridization of the cell-type and tissue specific expression patterns of these genes is required before this conclusion can be verified. The expression of the Pti-genes can be correlated to the freezing tolerance of the plant. The temporal pattern of the gene expression during acclimation and deacclimation demonstrates that transcript accumulation precedes tolerance development and transcript decline precedes loss of tolerance. The appearance and disappearance of the corresponding gene products e.g. LTI78 and RAB18 can be more directly correlated to the level of freezing tolerance (Mäntylä et al., in preparation). Furthermore, differential accumulation of LTI78 appears to correlate to different degrees of tolerance displayed by northern and southern ecotypes of *A. thaliana* (Mäntylä et al., in preparation).

536

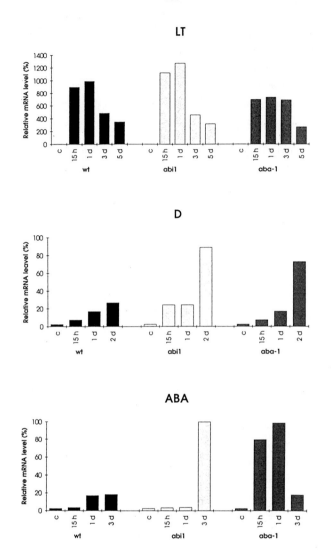

Figure 3. Expression of the *lti78* gene in *A. thaliana*. Accumulation of *lti78* mRNA was detected by Northern hybridization and quantified by phosphor imaging. The top panel shows the response to low temperature (LT), the middle pane to mild desiccation (D) and the lower panel to exogenous ABA. Note that the scale of the y-axis of the top panel is different from the two others. wt denotes wild-type *A. thaliana* and *abi1* and *aba-1* ABA-insensitive and ABA-deficient mutants, respectively.

Signal transduction

Elucidation of the signal pathways involved in the perception of the different stimuli that induce cold acclimation and transduction of the resulting signal to altered gene expression is one of the key tasks required for understanding the mechanism of plant cold acclimation. As discussed, three different stimuli; low temperature, desiccation and ABA trigger the acclimation process in *A. thaliana*, and also induce the expression of *lti* genes albeit to varying degrees. Are these signals perceived and transduced independently or do they share the same signal transduction cascade? If there is a common signal pathway is ABA the mediator of these signals as originally proposed by Chen et al. (1983)? To address these questions we have employed both ABA-deficient (*aba*) and ABA-insensitive (*abi*) mutants of *A. thaliana* (Koornneef et al., 1982; 1984; 1989) and assessed LT, desiccation and ABA-induced gene expression in these mutant backgrounds (Fig. 3). The results from this analysis suggest that ABA is not the common mediator for altered *lti*-gene expression (Nordin et al., 1991; 1993a, b). The majority of the LT-induced genes, designated as class I genes (Table 1), appear to respond independently to the different environmental signals (Table 1, Fig. 3): Reduction of the ABA level or insensitivity to ABA did not markedly affect the LT- or desiccation-induced expression of class I genes (Fig. 3) (Nordin et al., 1991; 1993a, b; Gilmour and Thomashow, 1991). Only two (*rab18*, *lti65*) of the ten *A. thaliana lti*-type genes characterized so far exhibit a different expression pattern that clearly involves ABA. Perception of both the low temperature and desiccation induced signals by the *rab18* gene (class II, Table 1) seems to be by an ABA-mediated signal pathway, as this was blocked both in the ABA-deficient and ABA-insensitive mutants (Lång and Palva, 1992; Nordin et al., 1993a). A similar type of ABA mediated signal pathway appears to control expression of desiccation stress responsive genes in other plants (Bray 1988; Cohen and Bray, 1990). Yet another type of expression was observed with *lti65* (Nordin et al., 1993b). Similar mutant studies as performed with class I and class II genes demonstrated that the desiccation induced expression of *lti65* is by an ABA-mediated signal pathway similar to that of *rab18* while the LT response is not (Table 1). Thus the signal pathways employed by the class III gene, *lti65*, appear as a hybrid between those of class I and class II genes.

Consequently, ABA appears not to be the central mediator of the altered gene expression during cold acclimation but provides one alternative signal pathway. Interestingly, a mutation in the *abi1* locus seems to block this pathway, and suggests that this gene should define a component in this signal transduction cascade. A related gene, *abi3*, which seems to control responsiveness to ABA in seeds was recently cloned and the gene product characterized as a likely transcription factor (Giraudat et al., 1992). By analogy, the *abi1* gene might have a similar function in vegetative tissues.

The existence of separate signal pathways that control the expression of the *lti* genes suggests that the promoters of the genes might contain distinct response elements for the different stimuli. All of the LT-induced genes are responsive to ABA and might therefore contain DNA elements related to the well characterized ABRE-box found in many ABA-responsive genes (Guiltinan et al., 1990; Mundy et al., 1990). Related DNA sequences can indeed be found in all of the LT-induced genes studied (Fig. 4), *rab18* (Lång and Palva, 1992), *kin1* and *kin2* (Kurkela, 1991) as well as *lti78* and *lti65* (Nordin et al., 1993b). However, the location and number of such elements is variable and it remains to be shown whether these sequences indeed function as ABREs or represent binding sites for more general transcription factors.

Comparison of the *lti65* and *lti78* promoters suggested the presence of a specific LT-response element (LTRE; Nordin et al., 1993b). Four tandem copies of the putative LTRE sequence are found in the strongly LT-responsive *lti78* promoter while only one imperfect copy is present in the *lti65* promoter which is only very weakly induced by LT (Fig. 4). Similarly only one related sequence is found in the

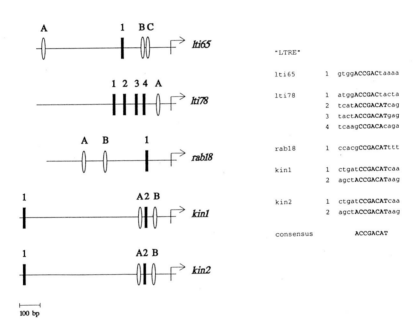

Figure 4. Putative signal response elements in the promoters of LT-responsive genes of *A. thaliana*. The linear organization of the "LTRE" (vertical bars) and ABRE-like (ovals) sequences in the promoters indicated is depicted on the left panel. The indicated "LTRE" sequences are shown on the right.

mainly ABA-responsive *rab18* gene, whereas the LT-responsive *kin1* and *kin2* genes harbor two copies of the LTRE sequence (Fig. 4). Interestingly, a related sequence has been found in the *Brassica napus cor15* homologue,
the *BN115* gene (Weretilnyk et al., 1993) where it has been shown to confer LT-inducibility (J. Singh, personal communication). No desiccation specific DNA-elements have been reported yet, but the ongoing promoter analysis of several of the *lti* genes can be expected to lead to identification of these putative structures in the near future.

In conclusion, the results indicating the presence of distinct response elements (Fig. 4) support the idea of separate response pathways for the different stimuli and suggest that the expression of these stress response genes is under a complex environmental control. Elucidation of the molecular mechanism of this control is one of the most important future challenges in cold acclimation research.

ACKNOWLEDGEMENTS

We are indebted to Dr T.K. Palva for the invaluable help in preparing this manuscript.
This work has been supported by the Swedish Natural Sciences Research Council, the Swedish Council for Forestry and Agricultural Research and the Swedish National Board for Technical Development.

LITERATURE CITED

Baker J, Steele C, Dure L (1988) Sequence and characterization of 6 *Lea* proteins and their genes from cotton. Plant Mol Biol **11**: 277-291

Bray E (1988) Drought-and ABA-induced changes in polypeptide and mRNA accumulation i tomato leaves. Plant Physiol **88**: 1210-1214

Cattivelli L, Bartels D (1990) Molecular cloning and characterization of cold-regulated genes in barley. Plant Physiol **93**: 1504-1510

Chen HH, Gavinlertvatana P, Li PH (1979) Cold acclimation of stem-cultured plants and leaf callus of solanum species. Bot Gaz **140**: 142-147

Chen HH, Li PH, Brenner ML (1983) Involvement of abscisic acid in potato cold acclimation. Plant Physiol **71**: 362-365

Chen HH, Gusta LV (1983) Abscisic acid-induced freezing resistance in cultured plant cells. Plant Physiol **73**: 71-75

Close TJ, Kortt AA, Chandler PM (1989) A cDNA-based comparison of dehydration-induced proteins (dehydrins) in barley and corn. Plant Mol Biol **13**: 95-108

Cloutier Y, Siminovitch D (1982) Correlation between cold- and drought-induced frost hardiness in winter wheat and rye varieties. Plant Physiol **69**: 256-258

Cohen A, Bray EA (1990) Characterization of three mRNAs that accumulate in wilted tomato leaves in response to elevated levels of endogenous abscisic acid. Planta 182: 27-33

Dure IL, Crouch M, Harada J, Ho THD, Mundy J, Quatrano R, Thomas T, Sung ZR (1989) Common amino acid sequence domains among the LEA proteins of higher plants. Plant Mol Biol 12: 475-486

Dunn MA, Hughes MA, Pearce RS, Jack PL (1990) Molecular characterization of a barley gene induced by cold treatment. J Exp Bot 41: 1405-1413

Gilmour SJ, Hajela RK, Thomashow MF (1988) Cold acclimation in Arabidopsis thaliana. Plant Physiol 87: 745-750

Gilmour SJ, Thomashow MF (1991) Cold acclimation and cold-regulated gene expression in ABA mutants of Arabidopsis thaliana. Plant Mol Biol 17: 1233-1240

Gilmour SJ, Artus NN, Thomashow MF (1992) cDNA sequence analysis and expression of two cold-regulated genes of Arabidopsis thaliana. Plant Mol Biol 18: 13-21

Giraudat J, Hauge BM, Valon C, Smalle J, Parcy F, Goodman H (1992) Isolation of the Arabidopsis abi3 gene by positional cloning. Plant Cell 4: 1251-1261

Guiltinan MJ, Marcotte WR Jr, Quatrano RS (1990) A plant leucine zipper protein that recognizes an abscisic acid response element. Science 250: 267-271

Guo W, Ward RW, Thomashow MF (1992) Characterization of a cold-regulated wheat gene related to Arabidopsis cor47. Plant Physiol 100: 915-922

Guy CL, Niemi KJ. Brambl R (1985) Altered gene expression during cold acclimation of spinach. Proc Natl Acad Sci USA 82: 3673-3677

Guy C, Haskell D, Neven L, Klein P, Smelser C (1992) Hydration-state-responsive proteins link cold and drought stress in spinach. Planta 188: 265-270

Hajela RK, Horvath DP, Gilmour SJ, Thomashow MF (1990) Molecular cloning and expression of cor (cold-regulated) genes in Arabidopsis thaliana. Plant Physiol 93: 1246-1252

Heino P, Sandman G, Lång V, Nordin K, Palva ET (1990) Abscisid acid deficiency prevents development of freezing tolerance in Arabidopsis thaliana (L.) Heynh. Theor Appl Genet 79: 801-806

Houde M, Danyluk J, Laliberté J, Rassart E, Dhindsa RS, Sarhan F (1992) Cloning, characterization and expression of a cDNA encoding a 50-kilodalton protein specifically induced by cold acclimation in wheat. Plant Physiol 99: 1381-1387

Koornneef M, Hanhart CJ, Hilhorst HWM, Karssen CM (1989) In vivo inhibition of seed development and reserve protein accumulation in recombinants of abscisic acid biosynthesis and responsiveness mutants in Arabidopsis thaliana. Plant Physiol 90: 463-469

Koornneef M, Jorna ML, Brinkhorst-van der Swan DLC, Karssen CM (1982) The isolation of abscisic acid (ABA) deficient mutants by selection of induced revertants in non-germinating gibberillin sensitive lines of *Arabidopsis thaliana* (L.) Heynh. Theor Appl Genet **61**: 385-393

Koornneef M, Reuling G, Karssen CM (1984) The isolation and characterization of abscisic acid-insensitive mutants of *Arabidopsis thaliana*. Physiol Plant **61**: 377-383

Kurkela S, Franck M, Heino P, Lång V, Palva ET (1988) Cold induced gene expression in *Arabidopsis thaliana* L. Plant Cell Reports **7**: 495-498

Kurkela S, Franck M (1990) Cloning and characterization of a cold- and ABA-inducible *Arabidopsis* gene. Plant Mol Biol **15**: 137-144

Kurkela S (1991) Structure and expression analysis of an *Arabidopsis thaliana* gene family of two cold induced genes. PhD Thesis, University of Helsinki, Helsinki, Finland

Kurkela S, Borg-Franck M (1992) Structure and expression of *kin2*, one of two cold- and ABA-induced genes of *Arabidopsis thaliana*. Plant Mol Biol **19**: 689-692

Levitt J (1980) Responses of plants to environmental stresses: Vol 1. Chilling, freezing and high temperature stresses. Academic Press, New York

Lin C, Guo WW, Everson E, Thomashow MF (1990) Cold acclimation in *Arabidopsis* and wheat: a response associated with expression of related genes encoding 'boiling-stable' polypeptides. Plant Physiol **94**: 1078-1083

Lin C, Thomashow MF (1992a) A cold-regulated *Arabidopsis* gene encodes a polypeptide having potent cryoprotective activity. Biochem Biophys Res Commun **183**: 1103-1108

Lin C, Thomashow MF (1992b) DNA sequence analysis of a complementary DNA for cold-regulated *Arabidopsis* gene *cor15* and characterization of the COR15 polypeptide. Plant Physiol **99**: 519-525

Lång V, Heino P, Palva ET (1989) Low temperature acclimation and treatment with exogenous abscisic acid induce common polypeptides in *Arabidopsis thaliana* (L.) Heynh. Theor Appl Genet **77**: 729-734

Lång V, Palva ET (1992) The expression of a *rab*-related gene, *rab18*, is induced by abscisid acid during the cold acclimation process of *Arabidopsis thaliana* (L.) Heynh. Plant Mol Biol **20**: 951-962

Meyerowitz EM (1989) *Arabidopsis*, a useful weed. Cell **56**: 263-269

Mohapatra SS, Poole RJ, Dhindsa RS (1987) Changes in protein patterns and translatable mRNA populations during cold acclimation of alfalfa. Plant Physiol **84**: 1172-1176

Mohapatra SS, Wolfraim L, Poole RJ, Dhindsa RS (1989) Molecular cloning and relationship to freezing tolerance of cold-acclimation-specific genes in alfalfa. Plant Physiol **89**: 375-380

Mundy J, Chua N-H (1988) Abscisic acid and water-stress induce the expression of a novel rice gene. EMBO J **7**: 2279-2286

Mundy J, Yamaguchi-Shinozaki K, Chua N-H (1990) Nuclear proteins bind conserved elements in the abscisic acid-responsive promoter of a rice *rab* gene. Proc Natl Acad Sci USA **87**: 1406-1410

Neven LG, Haskell DW, Hofig A, Li Q-B, Guy CL (1993) Characterization of a spinach gene responsive to low temperature and water stress. Plant Mol Biol 21: 291-305

Nordin K, Heino P, Palva ET (1991) Separate signal pathways regulate the expression of a low-temperature-induced gene in *Arabidopsis thaliana* (L.) Heynh. Plant Mol Biol 16: 1061-1071

Nordin K, Lång V, Mäntylä E, Heino P, Welin B, Baudo M, Holmström K-O, Palva ET (1993a) Role of ABA in regulation of low temperature-induced genes in *Arabidopsis thaliana*. In Advances in Plant Cold Hardiness, edited by P.H. Li and L. Christersson, pp. 45-56. CRC Press, Boca Raton, Florida

Nordin K, Vahala T, Palva ET (1993b) Differential expression of two related, low-temperature-induced genes in *Arabidopsis thaliana* (L.) Heynh. Plant Mol Biol 21: 641-653

Orr W, Keller WA, Singh J (1986) Induction of freezing tolerance in an embryogenic cell suspension culture of *Brassica napus* by abscisic acid at room temperature. J Plant Physiol 126: 23-32

Orr W, Iu B, White TC, Robert LS, Singh J (1992) Complementary DNA sequence of a low temperature-induced *Brassica napus* gene with homology to the *Arabidopsis thaliana kin1* gene. Plant Physiol 98: 1532-1534

Picket M, Scott G, Davies P, Wang N, Joshi S, Few C (1984) Sequence of antifreeze protein precursor. Eur J Biochem 143: 35-38

Rouse D, Gehring CA, Parish RW (1992) Structure and sequence of a dehydrin-like gene in *Arabidopsis thaliana*. Plant Mol Biol 19: 531-532

Sakai A, Larcher W (1987) Frost survival of plants. Responses and adaptations to freezing stress. Springer-Verlag, Berlin

Skriver K, Mundy J (1990) Gene expression in response to abscisic acid in osmotic stress. Plant Cell 2: 503-512

Stebbins GL (1972) Flowering plants: Evolution above the species level. Belknap, Harvard Boston

Weiser CJ (1970) Cold resistance and injury in wooded plants. Science 169, 1269-1278

Weretilnyk E, Orr W, White TC, Iu B, Singh J (1993) Characterization of three related low-temperature-regulated cDNAs from winter *Brassica napus*. Plant Physiol 101:171-177

Wolfrain LA, Langis R, Tyson H, Dhindsa RS (1993) cDNA sequence, expression and transcript stability of a cold acclimation-specific gene, *cas18*, of alfalfa (*Medicago falcata*). Plant Physiol 101:1275-1282

Yamaguchi-Shinozaki K, Shinozaki K (1993) Characterization of the expression of a desiccation-responsive *rd29* gene of *Arabidopsis thaliana* and analysis of its promoter in transgenic plants. Mol Gen Genet 236: 331-340

GENETICALLY ENGINEERED MODIFICATION OF PLANT CHILLING SENSITIVITY AND CHARACTERIZATION OF CYANOBACTERIAL HEAT SHOCK PROTEINS

H. Hayashi, I. Nishida, O. Ishizaki-Nishizawa, Y. Nishiyama and
N. Murata

National Institute for Basic Biology
Myodaiji
Okazaki, Aichi 444
Japan

Abstract

We conducted the following research in order to understand
molecular mechanism of tolerance and adaptation in higher plants
and microbiral plants to high and low temperature. (1) Glycerol-
3-phosphate acyltransferase from *Arabidopsis* and squash was
introduced into tobacco, resulting in a modified chilling
tolerance in the leaves of the transgenic tobacco. (2) Cells of
a cyanobacterium, *Synechococcus* PCC7002, were grown under high
temperature, and aquired thermotolerance in the photosynthetic
oxygen-evolving complex. The factors responsible for the
enhanced thermal stability appeared to be associated with the
thylakoid membrane. (3) Two *groEL*-homologous genes were cloned
from *Synechococcus* PCC7002 and were then sequenced. Heat shock
treatment markedly increased the mRNAs of both *groEL*-homologous
genes. One gene was accompanied by an upstream *groES* gene, while
the other was not.

NATO ASI Series, Vol. H 86
Biochemical and Cellular Mechanisms
of Stress Tolerance in Plants
Edited by J. H. Cherry
© Springer-Verlag Berlin Heidelberg 1994

1. Transformation with acyltransferase genes alters the chilling tolerance of tobacco plants

1.1. Unsaturated fatty acids and chilling sensitivity

It has been proposed that the primary event in chilling injury is the formation of a lipid gel phase in cellular membranes as temperature decreases [Lyons, 1973]. When a model membrane goes into the phase-separated state, in which gel and liquid crystalline phases coexist, the membrane becomes permeable to small electrolytes, thereby diminishing membrane ion gradients essential for maintenance of cellular activities [Murata and Nishida, 1990]. Thus, the formation of a lipid gel phase is followed by a series of processes leading to cell death.

Higher-plant cells contain monogalactosyldiacylglycerol (MGDG), digalactosyldiacylglycerol (DGDG), sulfoquinovosyl-diacylglycerol (SQDG), and phosphatidylglycerol (PG) in plastid membranes; phosphatidylcholine (PC), phosphatidylethanolanine (PE), diphosphatidylglycerol (DPG), and PG in mitochondrial membranes; and PC, PE, and phosphatidylinositol (PI) in endoplasmic reticulum and plasma membranes [Murata and Nishida, 1990]. Transition temperatures between the gel and liquid crystalline phases of glycerolipid vary markedly with the saturation level of the fatty acyl chains. Molecular species containing only saturated fatty acids, such as 16:0 and 18:0, reveal phase transition temperatures above 40°C, while species containing a *cis*-unsaturated bond reveal phase transition near 0°C [phillips et al, 1972]. The introduction of the second *cis*-unsaturated bond decreases the phase transition temperature to about -20°C [phillips et al, 1972]. The substitution of 16:0 by *trans*-monounsaturated fatty acid, 16:1t, at the C-2 position of PG, shifts the phase transition temperature by 10°C [Kenrick and Bishop, 1986]. These findings suggest that if certain lipid molecular species undergo membrane phase transition above 0°C, they are most likely to be the fully saturated or *trans*-monounsaturated species. Among leaf-cell glycerolipids, only PG contains high levels of saturated and *trans*-monounsaturated species [Murata et al, 1982; Murata, 1983; Raison and Wright, 1983; Murata and Yamaya, 1984].

We demonstrated that chilling-sensitive plants contain much higher proportions of the 16:0/16:0 plus 16:0/16:1t species of PG than do chilling-resistant plants [Murata et al, 1982]. We tested approximately 20 plants and found that 16:0/16:0 plus 16:0/16:1t comprise from 3 to 19% of the total PG in the chilling-resistant plants, and from 26 to 65% in the chilling-sensitive plants [Murata, 1983]. In addition, Roughan [Roughan, 1985] surveyed the fatty-acid composition of PG in 74 plants, and found a correlation between chilling sensitivity and levels of saturated plus *trans*-monounsaturated PG molecular species. Taken together, these findings strongly suggest that saturated plus *trans*-monounsaturated species are closely associated with chilling sensitivity in plants.

1.2. *Affinity of acyltransferase to unsaturated fatty acids.*

On the basis of the research concerning the biosynthesis of PG in plastids [Andrews and Mudd, 1985], we have suggested that levels of saturated and *trans*-unsaturated PG depends on substrate selectivity during the first step of PG synthesis. In this step, the acyl group from acyl-ACP (acyl-carrier protein) is transferred to the C1 position of glycerol 3-phosphate [Murata, 1983] by an enzyme designated acyl-ACP: glycerol-3-phosphate acyltransferase (hereinafter, acyltransferase), located in the stroma of chloroplasts [Frentzen et al, 1983]. Previous research has revealed differences in acyltransferases substrate selectivity between chilling-resistant plants and chilling-sensitive plants. Acyltransferases from chilling-resistant plants prefer *cis*-unsaturated fatty acids, whereas acyltransferases from chilling-sensitive plants cannot distinguish saturated fatty acid from *cis*-unsaturated fatty acids [Frentzen et al, 1987; Nishida et al, 1987]. Thus, genetic engineering of acyltransferase should enable us to convert plants from chilling-sensitive to chilling-resistant, and *vice versa*.

1.3. *Transformation with acyltransferase genes*

cDNAs for acyltransferase have been cloned from a chilling-sensitive plant, squash [Ishizaki et al, 1988], and a chilling-resistant plant, *Arabidopsis thaliana* [Nishida et al, 1993]. Ti-plasmids for over-expressing squash and *Arabidopsis* acyltransferases cDNA in tobacco plants were constructed in pBI-121 under the control of the cauliflower mosaic virus 35S constitutive promotor. The GUS region of pBI-121 was replaced by a cDNA insert containing the full-length coding sequence for a precursor to *Arabidopsis* acyltransferase and its 5'- and 3'-noncoding regions [Nishida et al, 1993]. This plasmid was designated pARA. In another plasmid, the GUS region of pBI-121 was replaced by a fusion construct in which the mature-protein region of squash acyltransferase cDNA [Ishizaki et al, 1988] was ligated to the transit peptide region of pea Rubisco small subunit cDNA [Schreier et al, 1985]. This plasmid was designated pSQ. The Ti-plasmids were introduced into *Agrobacterium tumefaciens* by electroporation, and the resultant transformants were selected by kanamycin resistance and by DNA-DNA hybridization analysis. *Nicotiana tabacum* var. Samsum was transformed using the leaf-disk method.

Fifteen independent clones were arbitrarily chosen from each group of transgenic tobacco plants with the same construct, to examine the level of unsaturation in PG from the leaves. Most of the transformants showed modified levels of PG unsaturation. From the 15 transformants of each group, five clones with large changes in unsaturation, were selected for further analysis.

DNA-DNA hybridization analysis of genomic DNA from these transgenic tobacco plants confirmed the stable genomic integration of introduced cDNA. RNA-DNA hybridization analysis of poly(A)+ RNA from leaves of the transformants indicated that mRNA from the cDNAs was expressed at high levels. Immunoblot analysis using antibodies against acyltransferases from *Arabidopsis* and squash revealed that, in transformants containing *Arabidopsis* and squash cDNAs, mature proteins were detected in both the total protein extracts from leaves and in the soluble protein extracts from isolated chloroplasts. Levels of the over-expressed proteins were estimated to range from 0.1% to 1% of total leaf protein. This is much higher than the endogenous level of acyltransferase in wild-type tobacco plants. These observations demonstrate that the over-expressed proteins are transported into chloroplasts where they are then processed to mature proteins.

1.4. *Chilling tolerance of transgenic tobacco plants*

Examination of lipids from leaves of the transgenic tobacco plants revealed no discernible change in overall fatty-acid compositions between transformed and non-transformed plants. Analysis of individual lipid classes indicated that PG was the only lipid class whose fatty-acid unsaturation level was significantly altered by transformation with acyltransferase cDNAs (Table I). When pSQ (containing squash cDNA) was introduced, the level of saturated and *trans*-unsaturated PG increased from 36% to 76%. In contrast, the introduction of pARA (containing *Arabidopsis* cDNA) caused a small but significant decrease in saturated and *trans*-unsaturated species, from 36% to 28%.

The transgenic plants were then analyzed for chilling sensitivity. The pBI-121 transformant and wild-type tobacco showed some chilling sensitivity, but introduction of squash acyltransferase cDNA dramatically increased chilling sensitivity. In contrast, introduction of cDNA for *Arabidopsis* acyltransferase decreased chilling sensitivity [Murata et al, 1992].

Exposure of whole plants to a temperature of 1°C for 10 days produced a marked difference in leaf appearance among the transformants with different constructs. These changes included varying degrees of chlorosis and leaf deterioration. Differences in the extent of damage became more distinct when the plants were kept at 25°C for 2 days after the chilling treatment. Leaves of wild-type tobacco and pBI-121 transformants showed only partial chlorosis, while leaves of pSQ transformants were severely damaged or completely dead. Furthermore, pARA transformants were much more resistant to chilling than wild-type tobacco or pBI-121 transformants. These differences in chilling sensitivity correlated well with levels of PG fatty-acid unsaturation in these plants.

Table I. *Changes in unsaturation of tobacco PG due to transformation with squash and* Arabidopsis *acyltransferase cDNA (calculated from data in [Ishizaki et al, 1988])*

Transgenic plant	Molecular species				
	16:0 16:0/16:1t PG	18:0 16:0/16:1t PG	18:1 16:0/16:1t PG	18:2 16:0/16:1t PG	18:3 16:0/16:1t PG
	(% of molecular species)				
Wild type	32	4	8	22	34
Transgenic with					
pBI121 (control)	34	2	12	22	30
pSQ	62	14	4	6	14
pARA	24	4	14	24	34

Abbreviations: 16:0, hexadecanoic acid (palmitic acid); 16:1t, Δ3-trans hexadecenoic acid; 18:0, octadecanoic acid (stearic acid); 18:1, Δ9-octadecenoic acid (oleic acid); 18:2, Δ9,12-octadecadienoic acid (linoleic acid); 18:3, Δ9,12,15-octadecatrienoic acid (α-linolenic acid).

2. The thermotolerance of photosynthetic oxygen evolution in cyanobacteria

2.1. Acclimation of cyanobacteria to high temperature

When plants are exposed to heat, their photosynthesis is irreversibly inactivated. Some plants, however, have the ability to adapt to high-temperature environments and to enhance the thermal stability of their photosynthetic apparatus [Berry and Björkman 1980]. These phenomena are observed in several species of plants, though the mechanism of the adaptation is unclear. Among photosynthetic activities, PS2 activity (oxygen-evolution) appears to be the most heat sensitive [Katoh and San Pietro 1967, Yamashita and Butler 1968, Nash et al, 1985]. Thus, the protection of the PS2 oxygen-evolving complex against heat inactivation is regarded as the major mechanism for photosynthetic adaptation to high temperature.

Cells from *Synechococcus* PCC7002 were grown at temperatures ranging from 25°C to 45°C and their thermal stability was examined [Nishiyama et al, 1993]. Figure 1 shows profiles of heat inactivation of photosynthesis measured by oxygen evolution in cells grown at 25°C and 40°C. Growth temperature had a distinct effect on the thermal stability of photosynthesis. Temperature for 50% inactivation were 46°C and 50°C in cells grown at 25°C and 40°C, respectively. These observations indicate that cells grown at 40°C acquired enhanced thermal stability of photosynthesis compared with cells grown at 25°C.

A similar change in thermal stability was observed in the PS2-mediated electron transport from H_2O to *p*-benzoquinone, as measured by oxygen evolution in the presence of *p*-benzoquinone as the electron acceptor. In cells grown at 25°C and 40°C, 50% inactivation temperatures occurred at 44°C and 48°C, respectively. These results suggest that PS2, which is the most sensitive of photosynthetic activities can adapt to high temperature.

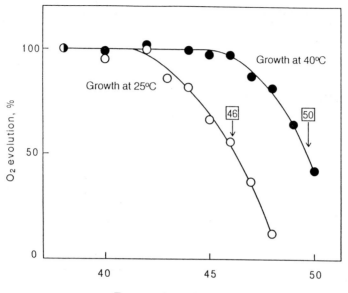

2.2. A possible thermoprotecting protein of cyanobacteria

Thylakoid membranes were prepared using glass beads to rupture the cyanobacterial cells. We then examined the variation in 50% inactivation temperatures for PS2 activity in response to heat treatment [Nishiyama et al, 1993]. Temperature for 50% inactivation were at 38°C and 42°C in thylakoid membranes isolated from cells grown at 25°C and 40°C, respectively. The profiles of heat inactivation in thylakoid membranes were shifted toward lower temperature than those in intact cells. Nevertheless, an increase in thermal stability acquired by cells grown at 40°C was retained in thylakoid membranes. This result indicates that factors responsible for enhanced thermal stability in PS2 are associated with thylakoid membranes, and that thylakoid membranes retain these factors even after isolation.

3. *Two groEL-homologous genes in* Synechococcus *PCC7002*

3.1. *Cloning of* groESL *genes*

There are a wide variety of heat shock proteins, most of which are highly conserved in both prokaryotes and eukaryotes. Cellular levels of these proteins increase dramatically upon heat shock treatment. Some heat shock proteins act as molecular chaperones which assist in the correct folding of other polypeptides and their assembly into oligomeric protein structures [Ellis and Hemingsen, 1989]. To examine the role of heat shock proteins in adaptation to high temperature stress in cyanobacteria, we attempted to isolate the genes for the heat shock protein GroEL from one of the transformable cyanobacteria, *Synechococcus* PCC7002.

Mixed oligonucleotides were synthesized based on conserved amino acid residues in GroEL of *E. coli* and *Synechococcus* PCC7942 [Hemmingsen et al, 1988; Webb et al, 1990] as well as HSP60 of *S. cerevisiae*, *A. thaliana*, and *Z. mays* [Reading et al, 1989; Prasad and Stwart, 1992], to be used as PCR (polymerase chain reaction) primers. A 250 bp product with a nucleotide sequence highly homologous to a region of the *groEL* gene from *Synechococcus* PCC7942 [Webb et al, 1990] was amplified from *Synechococcus* PCC7002 genomic DNA. The 250 bp PCR product was used to probe a genomic library of *Synechococcus* PCC7002 constructed using phage vector, λDASH II. The probe hybridized strongly to a 12 kb fragment of *Synechococcus* PCC7002 chromosomal DNA, which was isolated and subcloned into plasmid vector, pBluescript. Nucleotide sequence analysis of the fragment revealed the presence of the *groEL* gene accompanied by an upstream *groES* gene. The amino acid sequence was predicted from the *groEL* nucleotide sequence and was approximately 85% identical to *groEL* amino acid sequence from *Synechococcus* PCC7942 [Webb et al, 1990] and *Synechocystis* PCC6803 [Lehel et al, 1993]. RNA-DNA hybridization analysis of total RNA, using both *groES* and *groEL* specific probes, revealed a marked increase in 2.4 kb mRNA in response to a heat shock treatment of 46°C for 1 h, regardless of which probe was used. These results indicate that *Synechococcus* PCC7002 also contains the *groESL* operon. The 12 kb fragment of *Synechococcus*

PCC7002 chromosomal DNA also contained a 9-bp inverted repeat, commonly observed in the promotor regions of several bacteria [Narberhaus and Bahl, 1992, Schmidt et al, 1992], with a sequence identical to that found in the *groESL* operon of *Synechocystis* PCC6803 [Lehel et al 1993].

3.2. *Another* groEL *homologous gene*

The *Synechococcus* PCC7002 genomic library was also screened with a DNA probe corresponding to an internal part of the *Synechocystis* PCC6803 *cpn60* gene [Chitnis and Nelson, 1991]. Among 10 positive clones, 5 clones contained a DNA fragment identical to the *groESL* operon of *Synechococcus* PCC7002. The other 5 clones contained a DNA fragment similar to the *groEL* gene. The deduced amino acid sequence from the fragment resembled the predicted amino acid sequence of *groEL* and *cpn60* from *Synechocystis* PCC6803 [Lehel et al, 1993; Chitnis and Nelson, 1991] with the amino acid identities of 64% and 78 %, respectively. RNA-DNA hybridization analysis, using a probe specific to this gene, revealed a marked increase in 1.9 kb transcript upon heat shock treatment of 46°C for 1 hr. Since the 1.9 kb transcript corresponds to the expected size of the *groEL*-homologous gene coding region, this gene must not be accompanied by the *groES* gene. Thus, it is likely that this *groEL*-homologous gene corresponds to *Synechocystis* PCC6803 *cpn60*, but not to *Synechocystis* PCC6803 *groEL*.

The present study demonstrates the existence of two *groEL*-homologous genes in a single species of cyanobacteria. Heat shock treatment increased the levels of both *groEL*-homologous gene mRNAs, however, there are distinct differences between the two genes. For example, one *groEL*-homologous gene is accompanied by the *groES* gene as in *E. coli groEL*. The other lacks the *groES* gene but retains the carboxyl terminal repeat of Gly-Gly-Met observed in *E. coli* GroEL and yeast HSP60 [Hemmingsen et al, 1988; Reading et at, 1989]. Further molecular biological studies will elucidate the necessity of two *groEL* genes and the role each gene plays in the response to stress.

REFERENCES

Andrews J, Mudd JB (1985) Phosphatidylglycerol synthesis in pea chloroplasts. Plant Physiol 79: 259-265

Berry J, Björkman O (1980) Photosynthetic response and adaptation to temperature in higher plants. Annu Rev Plant Physiol 31: 491-543

Chitnis PR, Nelson N (1991) Molecular cloning of the genes encoding two chaperone proteins of the cyanobacterium *Synechocystis* sp. PCC6803. J Biol Chem 266: 58-65

Ellis RJ, Hemmingsen SM (1989) Molecular chaperones: Proteins essential for the biogenesis of some macromolecular structures. Trends Biochem Sci 14: 339-342

Frentzen M, Heinz E, Mckeon TA, Stumpf PK (1983) Specificities and selectivities of glycerol-3-phosphate acyltransferase and monoacylglycerol-3-phosphate acyltransferase from pea and spinach chloroplasts. Eur J Biochem 129: 629-636

Frentzen M, Nishida I, Murata N (1987) Properties of the plastidial acyl-(acyl-carrier protein): glycerol-3-phosphate acyltransferase from the chilling-sensitive plant squash (*Cucurbita moschata*). Plant Cell Physiol 28: 1195-1201

Hemmingsen SM, Woolford C, van der Vies SM, Tilly K, Dennis DT, Georgopoulos CP, Hendrix RW, Ellis RJ (1988) Homologous plant and bacterial protein chaperone oligomeric assembly. Nature 333: 330-334

Ishizaki O, Nishida I, Agata K, Eguchi G, Murata N (1988) Cloning and nucleotide sequence of cDNA for the plastid glycerol-3-phosphate acyltransferase from squash. FEBS Lett 238: 424-430

Katoh S, San Pietro A (1967) Ascorbate-supported NADP photoreduction by heated *Euglena* chloroplasts. Arch Biochem Biophys 122: 144-152

Kenrick JR, Bishop DG (1986) Phosphatidylglycerol and sulphoquinovosyldiacylglycerol in leaves and fruits of chilling-sensitive plants. Phytochem 25: 1293-1295

Lehel C, Los D, Wada H, Gyorgyei J, Horvath I, Kovacs E, Murata N, Vigh L (1993) A second *groEL*-like gene, organized in a *groESL* operon is present in the genome of *Synechocystis* sp. PCC6803. J Biol Chem 288: 1799-1804

Lyons JK (1973) Chilling injury of plants. Annu Rev Plant Physiol 24: 445-466

Murata N, Sato N, Takahashi N, Hamazaki Y (1982) Compositions and positional distributions of fatty acids in phospholipids from leaves of chilling-sensitive and chilling-resistant plants. Plant Cell Physiol 23: 1072-1079

Murata N (1983) Molecular species composition of phosphatidylglycerols from chilling-sensitive and chilling-resistant plants. Plant Cell Physiol 24: 81-86

Murata N, Yamaya J (1984) Temperature-dependent phase behavior of phosphatidylglycerols from chilling-sensitive and chilling-resistant plants. Plant Physiol 74: 1016-1024

Murata N, Nishida I (1990) Lipids in relation to chilling sensitivity of plants. In: Wang CY (ed) Chilling Injury of Horticultural Crops, pp. 181-199. CRC Press, Boca Raton, FL

Murata N, Ishizaki-Nishizawa O, Higashi S, Hayashi H, Tasaka Y, Nishida I (1992) Genetically engineered alteration in the chilling sensitivity of plants. Nature 356: 710-713

Narberhous F, Bahl H (1992) Cloning, sequencing, and molecular analysis of the *groESL* operon of *Chlostridium acetobutylicum*. J Bacteriol 174: 3282-3289

Nash D, Miyao M, Murata N (1985) Heat inactivation of oxygen evolution in photosystem II particles and its acceleration by chloride depletion and exogenous manganese. Biochim Biophys Acta 807: 127-133

Nishida I, Frentzen M, Ishizaki O, Murata N (1987) Purification of isomeric forms of acyl-[acyl-carrier-protein]: glycerol-3-phosphate acyltransferase from greening squash cotyledons. Plant Cell Physiol 28: 1071-1079

Nishida I, Tasaka Y, Shiraishi H, Murata N (1993) The gene and the RNA for the precursor to the plastid-located glycerol-3-phosphate acyltransferase of *Arabidopsis thaliana*. Plant Mol Biol 21: 267-277

Nishiyama Y, Kovács E, Lee CB, Hayashi H, Watanabe T, Murata N (1993) Photosynthetic adaptation to high temperature associated with thylakoid membranes of *Synechococcus* PCC7002. Plant Cell Physiol 34: 337-343

Phillips MC, Hauser H, Paltauf F (1972) The inter- and intramolecular mixing of hydrocarbon chains in lecithin/water systems. Chem Phys Lipids 8: 127-133

Prasad TK, Stewart CR (1992) cDNA clones encoding *Arabidopsis thaliana* and *Zea mays* mitochondrial chaperonin HSP60 and gene expression during seed germination and heat shock. Plant Mol Biol 18: 873-885

Raison JK, Wright LC (1983) Thermal phase transitions in the polar lipids of plant membranes; their induction by disaturated phospholipids and their possible relation to chilling injury. Biochim Biophys Acta 731: 69-78

Reading DS, Hallberg RL, Myers AM (1989) Characterization of the yeast hsp60 gene coding for a mitochondrial assembly factor. Nature 337: 655-659

Roughan PG (1985) Phosphatidylglycerol and chilling sensitivity in plants. Plant Physiol 77: 740-746

Schmidt A, Schiesswohl M, Volker U, Hecher M, Schumann W (1992) Cloning, sequencing, mapping and transcriptional analysis of the *groESL* operon from *Bacillus subtilis*. J Bacteriol 174: 3993-3999

Schreier PH, Seftor EA, Schell J, Bohnert HJ (1985) The use of nuclear-encoded sequences to direct the light-regulated synthesis and transport of a foreign protein into plant chloroplasts. EMBO J 4: 25-32

Webb R, Reddy KJ, Sherman LA (1990) Regulation and sequence of the *Synechococcus* sp. strain PCC7942 groESL operon, encoding a cyanobacterial chaperonin. J Bacteriol 172: 5079-5088

Yamashita T, Butler WL (1968) Inhibition of chloroplasts by UV-irradiation and heat-treatment. Plant Physiol 43: 2037-2040

Low Temperature Regulation of Gene Expression in winter *Brassica napus*.

Jas Singh and Theresa White
Plant Research Centre
Agriculture Canada
Ottawa, Ontario, K1A 0C6
Canada

INTRODUCTION

The oilseed crucifer *Brassica napus* exists as both spring and winter types. The winter type, which requires vernalization, is planted in the fall, cold acclimates and overwinters. Although spring *Brassica napus* is also capable of cold hardening, the winter type develops a higher degree of freezing tolerance after acclimation (Laroche et al., 1992). The high level of freezing tolerated by cold-acclimated winter *Brassica napus* (Laroche et al., 1992) and the ability of the species to be transformed (Moloney et al., 1989) suggests that it is potentially a suitable species for the study of gene expression and regulation during cold acclimation. Recently, there has been considerable interest in the identification of low temperature-induced genes in higher plants capable of developing frost tolerance (Guy, 1990; Thomashow, 1990). As low temperature is the main trigger in this process, it is hoped that at least some of these genes will be related to the development of cellular

NATO ASI Series, Vol. H 86
Biochemical and Cellular Mechanisms
of Stress Tolerance in Plants
Edited by J. H. Cherry
© Springer-Verlag Berlin Heidelberg 1994

frost tolerance. Therefore, knowledge of the regulation of
expression of these genes and elucidation of the function of
their products will provide a better understanding of the
mechanisms of cold acclimation and cellular frost tolerance, in
addition to the possibility of genetic engineering for hardier
plants.

RESULTS AND DISCUSSION

Using differential screening of a cDNA library prepared from
poly A$^+$ mRNA isolated from leaves of cold-acclimated winter
Brassica napus cv Jet neuf seedlings, we have isolated cDNAs
corresponding to three distinct transcripts that are induced by
low temperature (Orr et al., 1992; 1993; Weretilnyk et al.,
1993). The properties and expression characteristics of the
cDNAs corresponding to the low temperature-induced transcripts
are summarized in Table 1. Cold-induced genes BN115, BN26 and
BN19 may be members of a small gene family (Weretilnyk et al.,
1993). Accumulation of transcripts hybridizing to all three
genes occurs very rapidly on exposure of the plants or leaf-discs
to low temperature, remains at an elevated level even up to 10
weeks of cold acclimation and totally disappears within 24 hours
of return of the plants to room temperature. Transcripts
hybridizing to BN115 and its homologues are detectible only in
leaves and are the most abundant (0.1-0.15% of total mRNA) of the
cold-induced mRNA in *Brassica napus*.

Deduced amino acid sequences of all three genes suggest that
they encode very similar polypeptides (Fig. 1) and are similar to

Table 1: Properties and Expression Characteristics of cDNAs
          corresponding to low temperature-induced transcripts

Clone	Induction	Homolog	Function
BN115	Cold	COR 15 (Arabidopsis leaves)	Cryoprotectant? (COR 15)
BN79	Cold	"	
BN26	Cold	"	
BN59	Cold, ABA	Tonoplast ATPase 70kD subunit (Carrot roots)	H^+-ATPase
BN28	Cold, ABA	Kin 1 (Arabidopsis)	Antinucleator? Cryoprotectant?

```
BN115   MAMSLSGSAVLIGIGSSFSSGIAK------QSGVGAVGFGRKTELVVVAQ    44
BN19    MAMSLSGSAVLSGIGSSFSSGAAK------QSGVGAVGFGRKTEFVVVAQ    44
BN26    MAMSFSG-AVLSGINSSFPSGVAKKSGVAKQSGVGAVRFGRKTELVVVAQ    49
COR15   MAMSFSG-AVLTGMASSFHSG-AK------QSSFGAVRVGQKTQFVVVSQ    42
        ****.** *** *..*** ** **      **. ***  *.**..***.*

BN115   RKKSLIYADKGDGNILDDLNEATKRASDYATEKTKEALKNGEKAKDYVVD    94
BN19    RKKSLIYADKGDGNILDDLNEATKRASDYATEKTNEALKHGEEAKDYVVD    94
BN26    RKKSLIYAEKGDGNILDDINEATKRASDYVTDKTKEALKDGEKAKDYVDE    99
COR15   RKKSLIYAALGDGNILDDLNEATKKASDFVTDKTKGALADGEKAKDYVVE    92
        ********. ********.*****.***..*.**..** .**.******* .

BN115   KNVEAKDTAVDEAQKALDYVKAKGNEAGNKVAEFVEGKAGEAKDATKA   142
BN19    KNVEDKDTAVDEAQKALDYVKAKGNEAGNKVAEFVEGKAGEAKDATKA   142
BN26    KNVEAKDTALDEAQKVLDYVKEKGNEAG-----------EDKDTTKA   135
COR15   KNSGTADTLGKEAEKAAAYVEEKGKEAANKAAEFAEGKAGEAKDATK-   139
        ** .. **  .**.*. .**..**.**.**.***.******.**.***
```

Fig. 1: Comparisons of deduced amino acid sequences of BN115, 26, 19 and COR15. Asterisks indicate homology and dots indicate conservative substitutions.

that of a cold-induced gene, COR15, isolated from *Arabidopsis thaliana* (Lin and Thomashow, 1992). Unlike COR15, however, transcripts hybridizing to the *Brassica* genes were not inducible by ABA (Weretilnyk et al., 1993; White et al., 1992). When both *Arabidopsis* and *Brassica* leaves were incubated with 100 uM ABA in liquid media at room temperature and the isolated RNA hybridized against BN115, a strong hybridization signal (same intensity as cold induction) was observed with the *Arabidopsis* but not the *Brassica* RNA (unpublished). It has been demonstrated that the product of COR15 provides cryoprotection to lactic dehydrogenase in in *vitro* freezing and that the polypeptide is targeted into the chloroplast stroma with the first 50 amino acids removed to generate the mature peptide (Lin ad Thomashow, 1992). Comparisons of the deduced amino acid sequences (Fig. 2) of all three *Brassica* genes (BN115, 19, 26) suggest that they possess similar although not identical

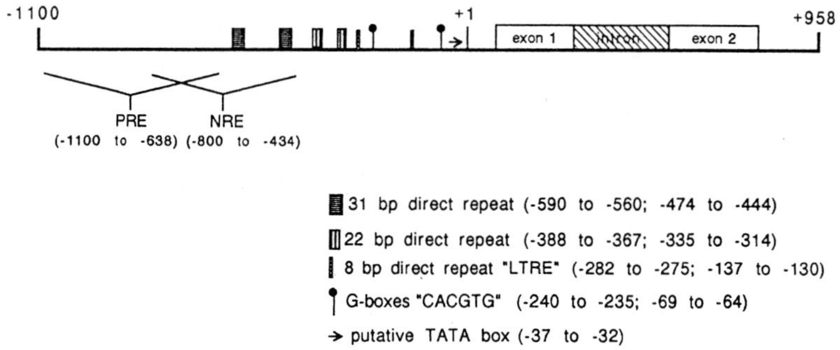

Fig. 2: Linear map of the genomic clone of BN115.

targeting sequences as COR15. While the sequence of the putative transit peptides are similar between BN115 and BN19, there is significant divergence between that of BN26 and BN115, BN19. These differences may reflect the targeting of the peptides to different cellular organelles or may simply be a reflection of the *rapa* and *oleracea* composition of the *napus* genome. A search of the nucleic and amino acid databases did not produce any known sequences with homology to BN115, 19 and 26. Similarly, no transcripts hybridizing to BN115 was observed in RNA isolated from leaves of cold-acclimated spinach or wheat. Transgenic *Brassica napus* have been made with sense and antisense constructs of BN115 (in collaboration with Wilf Keller, PBI, Saskatoon) and analyses of these plants are in progress.

Transcripts hybridizing to BN59 accumulated to between five and ten-fold above room temperature levels after 7 days of acclimation at 2°C. Nucleic acid sequence analyses (unpublished) of BN59 suggested it is ninety-six percent identical to the 70 kD subunit of carrot root tonoplast or endomembrane H^+-ATPase (Zimniak et al., 1988), with most of the divergence in deduced amino acid sequence from the carrot gene occurring within the first twenty amino acids of the N-terminus. Transcripts hybridizing to BN59 are inducible by both cold and ABA application. Transcript levels peak at day 7 of cold acclimation (Orr et al., 1993) and seemed to coincide with the peak of transient low temperature-induced accumulation of ABA in leaves of *Brassica napus* during cold acclimation (unpublished). Northern blot hybridization of other cruciferous species including flixweed (*Descurania sophia*) and *Arabidopsis* also showed an accumulation of transcripts corresponding to BN59 upon cold acclimation (Orr et al., 1993).

Hydropathy comparisons using the Intelligenetics SOAP program (Kyte and Doolittle, 1982) between the deduced amino acid sequences of BN59 and the carrot ATPase 70 kDa or A subunit indicated that BN59 is more hydrophillic at the N-terminus. The A and B subunits of the tonoplast ATPase are peripheral membrane proteins which are cold-labile, released from the complex in the cold and has been implicated in chilling injury in mung bean (Matsuura-Endo et al., 1992). It is therefore possible that decreased hydrophobicity may be required to maintain structure-functional integrity during cold acclimation in *Brassica*. Interestingly, analyses of mRNA isolated from leaves of tobacco and soybeans, two species not capable of cold acclimation, but grown in the cold for one week, did not show significant accumulation of transcripts corresponding to BN59 (unpublished). Observations of the transcript accumulation encoding the 70 kD subunit of endomembrane ATPase was consistent with the observations of osmotic adjustments (Table 2) and proliferation of endomembranes (Singh et al., 1987) during cold acclimation in *Brassica napus*.

Table 2:   Osmotic adjustment during cold acclimation

Days at 2°C	π MPa
0	0.9
1	1.1
4	1.3
7	1.4
14	1.7

BN28 (Orr et al., 1992) is similar to the *kin1* gene isolated from *Arabidopsis thaliana* (Kurkela and Franck, 1990). Based on limited deduced amino acid similarity to the flounder antifreeze protein, an antinucleating function has been proposed for the gene product. In this laboratory, transgenic tobacco plants expressing BN28 (transcript and protein) at room temperature did not show increases in frost tolerance over the wild type (unpublished). Again, *Brassica napus* have been transformed with sense and antisense constructs of BN28 and analyses of the development of frost tolerance at room temperature are in progress.

As BN115, 19 and 26 represent the most abundant of the low temperature-induced transcripts and the only group of low temperature-induced genes from winter *Brassica napus* that are not also induced by ABA, the genomic clone of BN115 was isolated in order to identify the elements involved in low temperature-regulated gene expression. Figure 2 shows a 2kb *SalI/XbaI* fragment containing the entire coding region of BN115, interspersed by a 0.26 kb intron, and over 1 kb of the 5' regulatory region. The 5' region upstream of the putative TATA box (Fig.2) contains large direct repeats of 31 bp and 22 bp and several smaller direct repeats including an 8 bp direct repeat (TGGCCGAC) staggered between 2 G-boxes (Guiliano et al., 1988). The 5' regulatory region from -1100 to +102 was fused in frame to the GUS reporter gene. Particle bombardment of the leaves with the BN115 promoter/GUS construct showed that transient activity was 5-fold higher after post-bombardment incubation at 2°C than at 22°C (Fig. 3). The level of cold induction by the promoter was in the order of 5 to 10% of expression by a 35S/GUS construct obtained by particle bombardment followed by incubation at room temperature. In contrast, post-bombardment

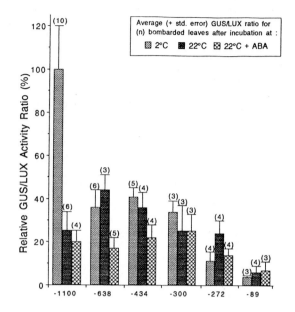

Fig. 3: Relative fluorescent GUS and Luciferase activity ratios of Jet neuf leaves co-bombarded with the BN115 promoter/GUS and pEPLUX plasmid. Young expanding sterilized leaves in semi-solid MS media were incubated for 24 to 48 hours at 2° or 22°C after bombardment, extracted and assayed for fluorescent GUS and LUX activities. Values were obtained from the mean (+ S.E.) net GUS/net LUX activities (nmol MU min^{-1} mg^{-1} / mV.sec ug^{-1}) of three separate sets of experiments with 4-6 leaves bombarded per condition per experiment, and expressed as a percentage of the 2°C activity of the undeleted promoter.

incubation with 100 uM ABA at 22°C did not show any increase in transient activity.

Deletion of the promoter down to -638 resulted in loss of over 60% of the GUS activity

at 2°C. No further significant changes in cold inducibility were observed with

truncating the promoter to -300. Deletion of only another 28 bp to -272 resulted in

reduction of cold inducible activity to approximately only 15% of the original.

Interestingly, this deletion removes the 5'-most member of the 8 bp repeat suggested

to be a low temperature regulatory element (Nordin et al., 1993).

A small, but significant rise in room temperature activity was consistently observed with the promoter truncated from -638 to -300. Analyses of leaves from 11 independent tobacco transgenic plants transformed with the BN115 promoter/GUS construct did not show significant differences in GUS activity between 2°C and 25°C induction. In addition, higher levels of transient GUS expression from the BN115 promoter/GUS construct was observed after incubation at 25°C than at 2°C in particle-bombarded soybean leaves. Thus, in both of these non-acclimating plants, the BN115 promoter was not induced by cold. Taken together with the deletion analyses, these data suggest the possibility that either the non-acclimating plants lack the regulatory factors required for cold-induced expression from the BN115 promoter or the regulation of BN115 expression in *Brassica napus* involves, in part, repression at higher temperatures.

The promoter analyses also confirmed earlier observations that transcripts hybridizing to BN115 was not inducible by ABA application at room temperature (Weretilnyk et al., 1993). It is not known why the BN115 family of genes from winter *Brassica napus* differ from COR15 of *Arabidopsis* in its inability to be induced by ABA at room temperature. There are recent reports of other cold-induced genes which are not induced by ABA in alfalfa (Wolfraim et al., 1993) and wheat (Houde et al., 1992). Interestingly, these observations are also from overwintering species which require extended periods of cold acclimation and vernalization in the fall. Winter *Brassica napus* requires a long period of cold acclimation to develop a high level of frost tolerance (Laroche et al., 1992) and therefore sustained induction by cold may be essential. As the induction of endogenous ABA accumulation by the onset of low temperature is a rapid transient phenomena, its role in sustained cold acclimation required in a winter species may not be as critical as in *Arabidopsis thaliana*.

Acknowledgement

Contribution No.   of the Plant Research Centre

References

Guiliano G, Pichersky E, Malike VS, Timko MP, Scolnik PA, Cashmore AR (1988) An evolutionary conserved protein binding sequence upstream of a plant light-regulated gene. Proc Natl Acad Sci USA **85**: 7089-7093

Guy CL (1990) Cold acclimation and freezing stress tolerance: Role of protein metabolism. Ann Rev Pl Physiol Pl Mol Biol **41**:787-796

Houde M, Danyluk J, Laliberte J-F, Rassart E, Dhindsa R, Sarhan F (1992) Cloning, characterization and expression of a cDNA encoding a 50-kilodalton protein specifically induced by cold acclimation from wheat. Plant Physiol **99**: 1381-1387

Kyte J, Doolittle RF (1982) A simple method for displaying the hydropathic character of a protein. J Mol Biol **157**: 105-132

Kurkela S, Franck M (1990) Cloning and characterization of a cold- and ABA-inducible *Arabidopsis* gene. Plant Mol Biol **15**: 137-144

Laroche A, Geng X-M, Singh J (1992) Differentiation of freezing tolerance and vernalization responses in *Cruciferae* exposed to low temperatures. Plant Cell Environ: *15: 439-445*

Lin C, Thomashow MF (1992) DNA sequence analysis of a cDNA for cold regulated *Arabidopsis* Gene *cor15* and characterization of the COR15 polypeptide. Plant Physiol **99**: 519-525

Matsuura-Endo C, Maeshima M, Yoshida S (1992) Mechanism of the decline in vacuolar H^+-ATPase activity in mung bean hypocotyls during chilling. Plant Physiol **100**: 718-722

Moloney M, Walker J, Sharma K (1989) High efficiency transformation of Brassica napus using Agrobacterium vectors. Plant Cell Rep **8**: 238-242

Nordin K, Vahala T, Palva ET (1993) Differential expression of two related low temperature-induced genes in *Arabidopsis thaliana*. Plant Mol Biol **21**: 641-653

Orr W, Iu B, White TC, Robert LS, Singh J (1992) cDNA sequence of a low temperature induced *B. napus* gene with homology to the *A. thaliana kin1* gene. Plant Physiol **98**: 1532-1534

Orr W, White TC, Iu B, Singh J (1993) cDNA sequence and characterization of a low temperature-induced *B. napus* gene with homology to the carrot vacuolar H^+-ATPase. Plant Physiol **102(suppl)**: 82

Singh J, Iu B, Johnson-Flanagan AM (1987) Membrane alterations in winter rye and Brassica napus cells during lethal freezing and plasmolysis. Plant Cell Environ **10**: 163-168

Thomashow MF (1990) Molecular genetics of cold acclimation in higher plants. Adv Genet **28**:99-131

Weretilnyk E, Orr W, White T, Iu B, Singh J (1993) Characterization of three related low-temperature regulated cDNAs from winter Brassica napus. Plant Physiol **101**: 171-177

Wolfraim LA, Langis R, Tyson H, Dhindsa RS (1993) cDNA sequence, expression and transcript stability of a cold acclimation-specific gene, *cas18*, of alfalfa cells. plant Physiol **101**: 1275-1282

Sorting Genes Controlling Freezing Stress Resistance[1]

Strategy for Moving Desired Traits by Merging Physiological and Genetic Approaches

Jiwan P. Palta
Department of Horticulture, University of Wisconsin, Madison, WI 53706, USA

Abstract

Factors that contribute to a plant's ability to survive winter or a frost episode include: a) freezing tolerance, i.e. tolerance of stresses associated with extracellular ice; b) freezing avoidance, i.e. ability to prevent ice formation; c) capacity to cold acclimate, i.e. ability to increase freezing tolerance and/or avoidance upon exposure to low (usually non-freezing) temperatures; and d) speeds of acclimation and deacclimation. We have demonstrated that freezing tolerance in the nonacclimated state and capacity to cold acclimate are under independent genetic control. In support of these genetic studies we have found that distinct plasma membrane lipids are associated with freezing tolerance in the nonacclimated state, and these lipids are different than those that are associated with increase in freezing tolerance during cold acclimation. We have also found evidence for genetic variations for the speeds of cold acclimation and deacclimation. In addition we have found that it is possible to independently select for freezing tolerance and freezing avoidance and then recombine them to get desired plants. Our results not only provide insight into molecular mechanisms of freezing stress resistance, but also provide avenues for improving frost/winter survival of crop plants.

Introduction

Freezing stress resistance is regarded as a very complex trait with polygenic inheritance (Marshall, 1982; Stushnoff et al., 1984; Thomashow, 1990). Lack of understanding of this complexity, in part, is due to heavy reliance on field selection for improving frost or winter survival. Field selection has many inherent problems. Ideally one hopes for a 'test frost episode' or a 'test winter' that is severe enough to kill the most sensitive genotypes, cause various degrees of injury to intermediate genotypes, and cause no injury to most resistant genotypes. However, such test frost episodes or test winters are rare.

---

[1] Portions of research summarized here were supported by the USDA National Research Initiative Grant (Agreement No 91-37100-6636) and by the College of Agriculture and Life Sciences, University of Wisconsin, Madison.

NATO ASI Series, Vol. H 86
Biochemical and Cellular Mechanisms
of Stress Tolerance in Plants
Edited by J. H. Cherry
© Springer-Verlag Berlin Heidelberg 1994

Many factors contribute to frost or winter injury in plants. For example injury to plants from a frost episode is influenced by ice nucleation temperature, freezing (cooling) rate, lowest temperature reached, duration of exposure to ice, thawing rate and post-thaw conditions (Palta and Weiss, 1993). Similarly speed of acclimation in fall, variation in snow cover, mid-winter thaw periods followed by very cold temperatures and speed of deacclimation in spring are major factors that contribute to winter survival. To develop a comprehensive approach for improving frost or winter survival we must address the following questions: Are there different physiological/biochemical/molecular parameters associated with different components that contribute to frost or winter survival? Do different components of freezing stress resistance have separate genetic control? How can we individually select for various components of freezing stress resistance?

We have addressed, to some degree, these fundamental questions in our studies. (Stone et al., 1993; Teutonico et al., 1993; Palta and Weiss, 1993; Palta and Simon, 1993). Our results show that three major components of freezing stress resistance namely non-acclimated freezing tolerance, ability to cold acclimate and thereby increase in freezing tolerance, freezing avoidance are under separate genetic control. In addition we have found evidence for the existence of genetic variability for the speeds of acclimation and deacclimation (Sutinen et al., 1992). These results suggest that study of individual components not only provide insight into molecular mechanisms of freezing stress resistance, but also provide avenues for improving frost/winter survival of crop plants. A brief account of our recent research is given here.

Independent Genetic Control for a Non-Acclimated Freezing Tolerance and Capacity to Cold Acclimate

Two major components of freezing stress resistance are (a) freezing tolerance in the non-acclimated state (i.e. normal growing conditions) and (b) capacity to cold acclimate (i.e. increase in freezing tolerance upon exposure to chilling temperatures for few days to few weeks period). Both of these traits are important for frost or winter survival depending on the climatic conditions preceding the frost episode and on the time of season.

The plant will be expected to rely on its nonacclimated tolerance if it is subjected to a sudden frost episode during the growing season or during the late spring and early fall periods. This is because of relatively long chilling period (few days to few weeks) are required for cold acclimation. During early spring or late fall a frost episode will be preceded by a cool period allowing plant to cold acclimate. Thus under these conditions ability to cold acclimate will be expected to play an important role in plant survival.

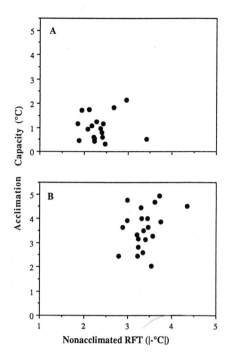

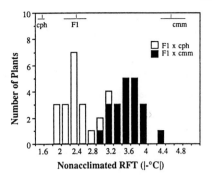

Figure 2. Distribution of nonacclimated RFTs of the backcross progenies (see Figure 1 legend for details). The range and mean RFT values for both parents (cmm and cph) and F₁ (cmm x cph) are indicated at the top of the graph. Source: Stone et al. (1993).

Figure 1. Correlation between nonacclimated freezing tolerance (RFT) and acclimation capacity (increase in RFT following cold acclimation) of individuals in segregating backcross populations of two diploid potato species *S. commersonii* (cmm) and *S. cardiophyllum* (cph). (*A*) F₁ x cph backcross progeny (*r* = -0.066, *P.* > 0.5). (*B*) F₁ x cmm backcross progeny (*r* = -0.317, *P* > 0.2). Source: Stone et al.(1993).

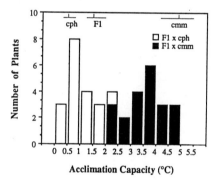

Figure 3. Distribution of acclimation capacity of backcross progenies (see Figure 1 legend for details). The range and mean acclimation capacity values for both parents (cmm and cph) and the F₁ (cmm x cph) are indicated at the top of the graph. Source: Stone et al. (1993).

This trait will also be important in winter survival since sufficient time is usually available for cold acclimation before plants are subjected to very cold winter temperatures.

We have recently demonstrated that nonacclimated tolerance and capacity to cold acclimate have distinct genetic control (Stone et al., 1993). In these studies we utilized two wild diploid potato species, *Solanum commersonii* and *Solanum cardiophyllum* as parents. These two species represent the extremes of non-acclimated freezing tolerance and acclimation capacity. *Solanum commersonii* is a frost tolerant species (nonacclimated tolerance -4·5°C) which is able to acclimate (acclimated freezing tolerance -9.6°C). *Solanum cardiophyllum* on the other hand is a frost sensitive species (nonacclimated tolerance -1.6°C) and is unable to cold acclimate. The F_1 of these parents was backcrossed to the parents to develop *Solanum* populations segregating for both the traits. Precise assessment of these two traits allowed small but significant differences among genotypes. Relative freezing tolerance (RFT) of these populations were assessed, using electrolyte leakage method, before and after cold acclimation. No correlation was found between nonacclimated RFT and acclimation capacity (Figure 1) suggesting that the two traits have independent genetic control. These results from the *Solanum* populations have been confirmed in rapeseed cultivars where no relationship between nonacclimated RFT and acclimation capacity was found (Teutonico et al., 1993).

Analyses of generation means indicated that all the variance for acclimation capacity and a major proportion of the variance for nonacclimated freezing tolerance can be best explained by an additive dominance model with both traits being partially recessive (Stone et al., 1993). The partial recessiveness of these two traits in *Solanum* populations is also evident from the fact that the F_1 was closer to the sensitive parent for both traits (Figures 2 and 3).

Only Limited Number of Genes Confer Cold Acclimation Ability in Potato

The fact that we were able to recover both parental phenotypes for acclimation ability in a limited population (Figure 3) suggests that cold acclimation ability is controlled by relatively few genes. The generation mean analyses also showed that cold acclimation ability can be explained by a simple additive-dominance model (Stone et al., 1993). Thus our studies suggest that the cold acclimation trait is not as genetically complex as has been thought.

Implications for Improving Frost Tolerance of Cultivated Potatoes

Frost impacts potato production in several ways. In the temperate regions such as Northern USA and Canada, frost damage in spring and fall to the potato crop can limit the growing season. In the Andean highlands of South America where potato is the staple food, frost can damage the potato crop at any time during the growing season, thus limiting the cultivation of high yielding sensitive varieties. For example, low temperature of -2 to -4°C can reduce yield by about 30% (Estrada, 1987). Our results (Stone et al., 1993) show that we can improve frost survival of potato by independently selecting for nonacclimated tolerance and for capacity to cold acclimate. Once selected the desired genotypes could be intermated for moving both these traits to high yielding varieties. Having high acclimation capacity will ensure frost survival during spring and fall periods whereas a high degree of nonacclimated tolerance will help protect potato plants from frost episodes during the summer period.

Independent selection for freezing tolerance and freezing avoidance and combining these traits to alleviate frost problems in carrots

In addition to tolerating extracellular ice, plants are capable of surviving freezing temperatures by simply avoiding the formation of ice. Freezing avoidance has been found to be important for the survival of overwintering seeds, buds and xylem ray parenchyma cells of many woody species (see reviews of Burke et al., 1976; Sakai and Larcher, 1987; Li and Sakai, 1978, 1982). These tissues avoid ice formation by supercooling and preventing nucleation. There is also evidence that some tissues in herbaceous plants are capable of surviving by freezing avoidance (Sakai and Larcher, 1987).

Different parts of the same plant can survive freezing temperatures by separate mechanisms. For example, in overwintering woody plants, the wood and buds are known to survive by freezing avoidance (supercooling) whereas bark tissue actually freezes and survives by tolerance of ice and tolerance of other stresses associated with freeze-thaw process (Sakai and Larcher, 1987). Such separation of different survival mechanisms for various plant parts within the same plant have not been distinguished in herbaceous plants. It is possible that some of the crown tissue in alfalfa (*Medicago sativa L.*) and winter cereals have potential to survive by freezing avoidance whereas the roots of these plants invariably freeze and thus must survive by freezing tolerance. It is important to know wheather root and crown survival during winter in these plants, is controlled by separate genes. Because if they are, these traits should be selected separately and then recombined.

We have recently demonstrated that separate selection for both freezing tolerance and freezing avoidance is needed to improve frost survival of carrots (Palta and Simon, 1993). In various parts of France, Great Britain and the Netherlands, mature carrots are left in the field during fall and the winter months for fresh market. Where they are subjected to periodic frosts. The injury to the plants by freeze-thaw stress is manifested as foliar damage, cracks on the roots (especially on the crown) crown rot and root rot (Tucker, 1974; Tucker and Cox, 1978; Palta and Simon, 1993). In addition, freeze-thaw injured foliage left in the field is subjected to secondary biotic stresses resulting in rot which often enters the root. Carrots are preferably harvested by machines, which pulls on the foliage. A damage to foliage can prevent mechanical harvest. The cracks are rot in the crown and root tissue decreases the yield and quality of carrots.

From a systematic study of 13 breeding lines of carrots we found that freeze-thaw injury observed on the foliage in the field was highly correlated to freezing tolerance of leaf tissue (measured as ion leakage) determined by controlled freeze-thaw test in our laboratory (Table I). Crown rot was significantly correlated to freezing tolerance of the foliage. Considerable variation existed for the freezing tolerance among the lines we tested.

Table I: Spearman's Rank Correlation Coefficient Between Freeze-Thaw Injury in the Field and Various Parameters Important in Freezing Tolerance and Freezing Avoidance of Carrot Plants. Source: Palta and Simon (1993).

	Spearman's rank correlation coefficients		
	Freezing tolerance parameter % ion leakage		Freezing avoidance parameter
Freeze-thaw injury in the field	Leaf tissue	Root tissue	Crown position in relation to soil surface
On foliage	0.88**	0.53*	0.59*
Crown cracks	0.50	0.11	0.68*
Crown rot	0.68**	0.52	0.61*

*   Significant at $\alpha = 0.05$.
** Significant at $\alpha = 0.01$.

In this study we also found that prevention of cracks in the root tissue was related to freezing avoidance of the root. Since carrot root is relatively rigid and dense tissue cracks develop as soon as ice forms in this tissue. This meant that avoidance of ice formation in the root and crown tissue was essential for the prevention of cracks in these tissues. The carrot breeding lines we tested varied considerably for the crown position in relation to soil surface. In some lines, the crown was up to 3.5 cm above the soil surface, whereas in other lines, crown position was about 1 cm below soil surface. Clearly the carrot crowns and roots below the soil surface will be better at avoiding ice in the tissue, thus avoiding cracks. A significant correlation between crown cracks (observed in the field) and crown position in relation to soil surface was found (Table I).

Thus a breeding strategy for obtaining a desired carrot plant (alleviate frost related injuries) should include, separate selection for freezing tolerance of foliage and freezing avoidance of crown and root, and then recombination of these traits to get the desired plant (Figure 4). By combining the characteristics that avoid ice in the crown tissue (crown position below soil surface) with the characteristics that reduce foliage and root injury by ice (freezing tolerance of foliage) we may be able to obtain the desired plant.

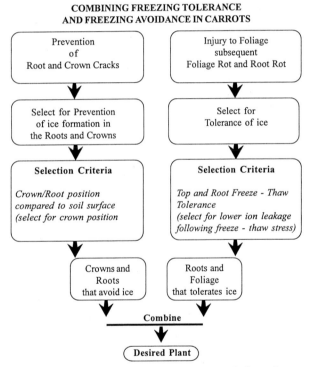

Figure 4. A flow diagram illustrating possible strategy for improving frost survival of carrot plants. Source: Palta and Simon (1993).

Genetic Variation Exists for Speeds of Acclimation and Deacclimation

Speeds of acclimation and deacclimation appear to be two of the several important factors responsible for frost and winter survival of plants. Ability of gain freezing stress resistance rapidly in response to decreasing air temperatures will certainly be advantageous for survival from unexpected very cold spell (freezing temperatures) in fall. On the other hand not able to deacclimate rapidly in response to mid-winter thaw will be of advantageous against late winter cold spells and spring freezes. Thus in addition to acclimation ability and nonacclimated tolerance it will be important to be able to select for rapid acclimation and slow deacclimation abilities. Our recent studies suggest that genetic variation for these traits exist (Sutinen et al., 1992).

We investigated seasonal changes in the freezing stress resistance of needles of red pine (*Pinus resinosa* Ait.) and Austrian pine (*Pinus resinosa* Arnold) trees (Sutinen et al., 1992). Red pine needles attained much higher freezing stress resistance during the mid-winter than Austrian pine (Figure 5). Thus red pine needles have much higher capacity to acclimate than Austrian pine even though they both have similar nonacclimated freezing stress resistance. Red pine needles also acclimated and deacclimated faster than Austrian pine needles (Figures 5 and 6). During fall there was a sharp increase in the freezing stress resistance of red pine needles whereas the freezing stress resistance of Austrian pine needles increased gradually (Figure 5). During the spring red pine needles deacclimated rapidly and its needles were less hardy than Austrian pine needles in late April (Figure 6). These results help explain why red pine is better adapted to very cold winters and shorter growing season in the northern latitudes.

Plasma Membrane Lipids Associated with Genetic Variability in Freezing Tolerance and Cold Acclimation Ability in *Solanum* Species

Plasma membrane is regarded as a key site of injury by freeze-thaw stress (Palta and Li, 1980; Stephonkus, 1984; Arora and Palta, 1988; Iswari and Palta, 1989; Palta, 1989). Since cold acclimation results in increased tolerance to freeze-thaw stress, it follows that this membrane must undergo changes to withstand a higher intensity of stresses imposed by freeze-thaw stress. Over the last 20 years changes in membrane lipids that coincide with increase in freezing tolerance during cold acclimation have been investigated. The results of early studies are highly controversial. These studies were based on the notion that an increase in the fatty acid unsaturation would increase membrane "fluidity" which could help stabilize membrane structure and function at low temperatures. This idea was extended from the purposed association of chilling injury with the lipid phase transition temperatures (Raison et al., 1971). In early studies on membrane lipid changes during cold acclimation some researchers reported an increase in the

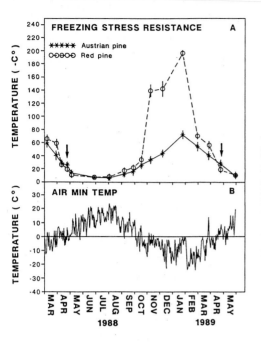

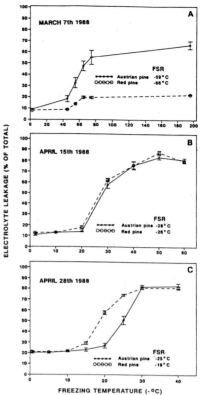

Figure 5 (A) Seasonal changes in freezing stress resistance of Austrian and red pine needles. The sampling was done between March 7, 1988 and May 31, 1989. Current-year needles were used except in June. In July, the newest needles were used when they were fully expanded. Freezing stress resistance was estimated by the electrolyte leakage method except during winter (November-February), when it was estimated by visual observation. The arrows show the period when red pine needles are deacclimating faster than Austrian pine needles. (B) The daily minimum air temperature (2 m above ground level) at the experimental site.
Source: Sutinen et al. (1992).

Figure 6. Changes in the relationship between percent electrolyte leakage and freezing temperatures (laboratory assay for estimating freezing stress resistance, FSR) during deacclimation in spring in Austrian and red pine needles.
Source: Sutinen et al. (1992).

proportion of unsturated fatty acids (de la Roche et al., 1972; Grenier et al., 1972; Willemot, 1975 and 1977) while others failed to detect such changes (de la Roche et al., 1975; de la Roche, 1979; Vigh et al., 1985). These contradictory results can be explained, in part by the fact that these researchers analyzed the lipid composition of whole tissues, organelle preparation or crude membranes rather than purified membrane of interest.

Only in limited studies changes in the lipid composition during cold acclimation of purified plasma membrane has been studied (Lynch and Steponkus, 1987; Uemura and Yoshida, 1984; Yoshida and Uemura, 1984). However, in these studies it is impossible to distinguish lipid changes related to cold acclimation *per se* (i.e. increase in freezing tolerance), from lipid changes that can result from low temperature adjustment of plant metabolism. We have avoided this problem in our recent study (Palta et al., 1993a). We compared plasma membrane lipid changes during cold acclimation in genetically related potato species. In this study the cultivated potato species *Solanum tuberosum* (freezing sensitive and unable to cold acclimate) was compared with the wild potato species *Solanum commersonii* (freezing tolerant and able to cold acclimate). Comparisons were made before and after cold acclimation.

Under normal growing conditions (nonacclimated) the wild *S. commersonii* has a higher degree of freezing tolerance (-4.5°C) compared with the cultivated *S. tuberosum* (-2.5°C). Under these conditions there were some lipid compositional differences between the two species (Table II-IV). These included, a lower sterol/phospholipid ratio (Table II), lower total membrane lipid per mg membrane protein (Table II) and a higher unsaturated/saturated fatty acid ratio (Table IV) in *S. commersonii* as compared with *S. tuberosum*. In addition *S. commersonii* had a higher proportion of acylated steryl glycosides (mol % of total steryl lipid) and a lower proportion of free sterols (mol % of total steryl lipid).

Comparative studies of cold acclimating and nonacclimating species allowed us to differentiate, lipid changes associated with increased freezing tolerance following cold acclimation, from lipid changes that can result from metabolic adjustment to low temperature. Lipid changes detected only in acclimating species (*S. commersoniii*) included an increase in phosphatidylethanolamine (Table III), a decrease in sterol/phospholipid ratio (Table II), an increase in linoleic acid (Table IV) and a decrease in linolenic acid (Table IV). These changes were either opposite or absent in the nonacclimating species *S. tuberosum*, suggesting an association of these plasma membrane lipid changes with cold acclimation.

Table II. **Lipid composition of the plasma membrane isolated from nonacclimated (NA) and cold acclimated (CA)** *Solanum tuberosum* **and** *Solanum commersonii* **leaf tissue.**
Source: Palta et al. (1993a)

Values ± SE of three separate samplings expressed as μmol per mg membrane protein and as mole percent of total lipid.

Solanum tuberosum	NA	CA	NA	CA
	$\mu mol \cdot mg^{-1}$ *protein*		*mol % of total lipid*	
Phospholipid	0.53 ± 0.05	0.55 ± 0.02	46.4 ± 1.5	46.8 ± 1.2
Total Sterols	0.52 ± 0.03	0.55 ± 0.05	45.0 ± 1.1	46.2 ± 1.7
Cerebroside	0.07 ± 0.01	0.06 ± 0.01	6.5 ± 1.3	5.0 ± 0.6
MGDG	0.01 ± 0.00	0.01 ± 0.00	0.9 ± 0.2	0.9 ± 0.3
DGDG	0.01 ± 0.00	0.01 ± 0.00	1.3 ± 0.3	1.1 ± 0.1
Total	1.15 ± 0.08	1.19 ± 0.06		
Sterol/ Phospholid	0.96	0.99	0.97	0.99
Solanum commersonii	NA	CA	NA	CA
	$\mu mol \cdot mg^{-1}$ *protein*		*mol % of total lipid*	
Phospholipid	0.41 ± 0.03	0.50 ± 0.03	48.3 ± 2.0	51.1 ± 1.4
Total Sterols	0.36 ± 0.07	0.42 ± 0.06	41.6 ± 2.8	40.7 ± 2.0
Cerebroside	0.05 ± 0.00	0.05 ± 0.00	6.1 ± 0.5	4.9 ± 0.8
MGDG	0.02 ± 0.00	0.02 ± 0.00	2.2 ± 0.2	1.9 ± 0.1
DGDG	0.02 ± 0.00	0.01 ± 0.00	1.8 ± 0.2	1.4 ± 0.2
Total	0.86 ± 0.11	1.01 ± 0.10		
Sterol/ Phospholipid	0.88	0.81	0.86	0.80

Table III. Phospholipid composition of the plasma membrane isolated from nonacclimated (NA) and cold acclimated (CA) *Solanum tuberosum* **and** *Solanum commersonii* **leaf tissue. Source: Palta et al. (1993a).**

Values ± SE of three separate samplings expressed as μmol per mg membrane protein as mole percent of total lipid, and as mole percent of total.

Solanum tuberosum	NA	CA	NA	CA	NA	CA
	$\mu mol \cdot mg^{-1}$ protein		mol %		mol % Total Lipid	
PC	0.20±0.01	0.20±0.00	38.4 ± 1.4	36.6 ± 0.7	17.7 ±0.1	17.2 ± 0.7
PE	0.20±0.02	0.21±0.01	35.7 ± 0.6	37.6 ± 0.1	16.6 ± 0.3	17.6 ± 0.4
PA	0.06±0.01	0.06 ±0.00	10.2 ± 1.5	10.0 ± 0.4	4.7 ± 0.8	4.7 ± 0.3
LPC	0.02±0.00	0.02±0.00	4.3 ± 0.5	4.4 ± 0.4	2.0 ± 0.3	2.1 ± 0.2
LPE + PI	0.04±0.01	0.04±0.00	7.5 ± 0.6	7.7 ± 0.1	3.5 ± 0.4	3.6 ± 0.1
PG	0.02±0.00	0.02 ±0.00	3.9 ± 0.2	3.6 ± 0.5	1.8 ± 0.1	1.7 ± 0.2
Total	0.53±0.05	0.55±0.02	46.4 ± 1.5	46.8 ± 1.2		
PC/PE	0.93	1.03				

Solanum commersonii	NA	CA	NA	CA	NA	CA
	$\mu mol \cdot mg^{-1}$ protein		mol % total lipid		mol % phospholipid	
PC	0.18 ± 0.01	0.19 ± 0.01	42.6 ± 0.2	36.3 ± 1.5	20.6 ± 0.9	18.5 ± 1.0
PE	0.15 ± 0.01	0.21 ± 0.01	36.2 ± 0.6	41.2 ± 0.8	17.5 ± 0.9	21.1 ± 0.9
PA	0.04 ± 0.00	0.05 ± 0.01	9.5 ± 1.7	9.6 ± 2.0	4.6 ± 1.0	4.9 ± 1.1
LPC	0.01 ± 0.00	0.02 ± 0.01	2.6 ± 0.6	3.4 ± 0.8	1.2 ± 0.2	1.7 ± 0.4
LPE + PI	0.02 ± 0.00	0.03 ± 0.01	5.5 ± 0.5	3.4 ± 0.8	2.6 ± 0.1	3.0 ± 0.3
PG	0.02 ± 0.00	0.02 ± 0.01	3.7 ± 0.9	3.7 ± 0.8	1.8 ± 0.4	1.9 ± 0.4
Total	0.41 ±0.03	0.51 ±0.04			48.3 ± 2.0	51.1 ± 1.4
PC/PE	1.17	0.87				

Table IV. Fatty acid composition of phosphatidylcholine and phosphatidylethanolamine from plasma membrane isolated from nonacclimated (NA) and cold acclimated (CA) *Solanum tuberosum* **and** *Solanum commersonii* **leaf tissue.**

Values expressed as weight percent of total PC or PE fatty acids

	Solanum tuberosum				*Solanum commersonii*			
Fatty Acid	Phosphatidylcholine		Phosphatidyl-ethanolamine		Phosphatylcholine		Phosphatidyl-ethanolamine	
	NA	CA	NA	CA	NA	CA	NA	CA
16:0	36.5	32	39.9	36.8	28.2	24.1	31.7	29.5
16:1	0.6	0.6	0.5	0.4	0.7	0.7	0.5	0.9
18:0	4.3	6.1	3.6	3.8	4.2	3.8	3.2	2.8
18:1	4.7	6.5	2.2	2.7	12.3	12.1	5.2	5
18:2	34.4	33.3	35.9	38.1	33.3	39.8	40.5	43.5
18:3	18.4	19.9	15.1	15.9	19	17.1	17	16.7
20:0	0.7	1.1	2.4	1.9	1.7	1.8	1.5	1.3
18:2/18:3	1.88	1.67	2.38	2.4	1.75	2.33	2.38	2.6
18:2/16:0	0.94	1.04	0.9	1.04	1.18	1.65	1.28	1.47
Unsaturated/ Saturated	1.38	1.54	1.17	1.34	1.92	2.35	1.74	1.97

Clearly, plasma membrane lipids that were associated with the development of freezing tolerance during cold acclimation were different from lipids associated with a higher degree of freezing tolerance in the nonacclimated state. For example, although the hardy species *S. commersonii* had higher proportion of unsaturated fatty acids in the nonacclimated state, yet this ratio increased during cold acclimation in both the acclimating and nonacclimating species (Table IV). Furthermore 18:2 increased and 18:3 decreased with cold acclimation in *S. commersonii* although there were no significant differences in the contents of 18:2 or 18:3 among the two species in the non acclimated state (Table IV). These results suggest that the differences in freezing tolerance in the nonacclimated conditions and the ability to increase freezing tolerance during cold acclimation cannot be explained by the same biochemical and/or genetic mechanisms. These results are in agreement with our recent results (Stone et al., 1993) on separate genetic control for nonacclimated tolerance and capacity to acclimate (Figures 1-3).

Table V. Correlation coefficients of plasma membrane fatty acid contents purified from nonacclimated and cold acclimated leaves (generated from *S. commersonii* and *S. cardiophyllum*, their F$_1$ hybrid, and backcrosses to each parent) with relative freezing tolerance (RFT) in nonacclimated and acclimated states and of changes in the fatty acid contents with change in freezing tolerance following cold acclimation (cold acclimation capacity = acclimated RFT - nonacclimated RFT).

Population	Trait	Correlation coefficients Fatty acid		
		16:0	18:2	18:3
Total	Nonacclimated RFT	0.65**	0.29^{NS}	-0.65**
	Acclimated RFT	0.20^{NS}	0.54**	-0.68**
	Cold acclimation capacity	-0.54**	0.45*	-0.01^{NS}
commersonii	Nonacclimated RFT	0.07^{NS}	0.58I	-0.79II
	Acclimated RFT	-0.20^{NS}	0.83**	-0.86**
	Cold acclimation capacity	-0.24^{NS}	0.65**	-0.06^{NS}
cardiophyllum	Nonacclimated RFT	0.70*	-0.21^{NS}	-0.03^{NS}
	Acclimated RFT	-0.36^{NS}	0.24^{NS}	-0.31^{NS}
	Cold acclimation capacity	-0.64*	0.33^{NS}	-0.17^{NS}

Contents of Specific Plasma Membrane Fatty Acids Co-segregate with Nonacclimated Freezing Tolerance Which are Distinct from the Fatty Acids Contents that Cosegregate with Capacity to Acclimate

As discussed above, we have recently demonstrated that the two major components of freezing tolerance (i.e. nonacclimated tolerance and capacity to acclimate) are under independent genetic control (Stone et al., 1993). We have now analyzed the plasma membrane lipid composition of these segregating *Solanum* populations. We found that relative contents of specific fatty acids correlated with specific traits of freezing stress resistance (Harbage et al., 1992; Table V). The correlation between fatty acid contents and different traits were only highly significant when backcross populations of the two parents were analyzed separately (Table V). For example 18:2 contents were correlated best in the *S. commersonii* backcross population. These results further provide evidence that distinct biochemical and genetic mechanisms are responsible for nonacclimated freezing tolerance and acclimation capacity. Furthermore thesse results demonstrate that separate examination of closely related populations is necessary to establish relationships between fatty acids and components of freezing tolerance.

Plasma Membrane Linoleic Acid (18:2) Appears to Have an Important Role in Freezing Tolerance

The results of our studies suggest an interesting potential role of 18:2 in freezing tolerance. An increase in 18:2 in both the phosphatidylcholine and phosphatidylethanolamine was found only cold acclimating species *S. commersonii* (Palta et al. 1993a; Table IV). Furthermore, 18:2 contents best correlated with acclimated freezing tolerance and cold acclimation capacity (Table V). An increase in 18:2 in plasma membrane enriched fraction following cold acclimation in potato species was reported by us about five years ago (Palta and Meade, 1989). We have also found similar increases in 18:2 with increase in freezing tolerance during cold acclimation in cranberry leaves (Abdallah and Palta, 1989)and pine needles (Sutinen et al., 1989) in fall and winter. It is possible that 18:2 fatty acid has a specific interaction with key membrane enzymes such as ATPases (Palmgren et al., 1988) which are known to be sensitive to freeze-thaw stress (Palta, 1989; Iswari and Palta, 1989). The reason for this could be that 18:2 has larger molecular areas as compared to 18:0 and 18:1 and a smaller length (Cook, 1985). Thus an increase in 18:2 following cold acclimation may be responsible for imparting cryostability to plasma membrane.

Acknowledgements

I thank Ms. Sharon Jaekel for help in typing of this manuscript and in preparing some figures.

Literature Cited

Abdallah AY, Palta JP (1989) Seasonal changes in membrane polar lipid fatty acid composition coincide with initiation of fruit ripening and changes in freezing stress resistance of leaves of cranberry (abstract No. 595). Plant Physiol **89**: S-100

Arora R, Palta JP (1988) *In vivo* perturbation of membrane-associated calcium by freeze-thaw stress in onion bulb cells. Simulation of this perturbation in extracellular KCl and its alleviation by calcium. Plant Physiol **87**: 622-628

Burke MJ, Gusta LV, Quamme HA, Weiser CJ, Li PH (1976) Freezing injury in plants. Annu Rev Plant Physiol **27**: 507-528

584

Cook HW, (1985) Fatty acid desaturation and chain elongation. *In* DE Vanc, JE Vance, eds, Biochemistry of Lipids and Membranes. Benjamin Cummings Pub. Co. California pp 181-212

de la Roche AI (1979) Increase in linolenic acid is not a prerequisite for development of freezing tolerance in wheat. Plant Physiol **63**: 5-8

de la Roche AI, Andrews CJ, Pomeroy MK, Weinberger PA, Kates M (1972) Lipid changes in winter wheat seedlings at temperatures inducing cold hardiness. Can J Bot **50**: 2401-2409

de la Roche AI, Pomeroy MK, Andrews CJ (1975) Changes in fatty acid composition in wheat cultivars of contrasting hardiness. Cryobiology **12**: 506-512

Estrada NR (1987) Utilization of wild and cultivated diploid potato species to transfer frost resistance into tetraploid common potato *Solanum tuberosum L.* in PH Li, ed, Plant Cold Hardiness. Alan R Liss, Inc, New York, pp 339-354

Grenier G, Tremolieres A, Therrien HP, Wiollemot C (1972) Changements dans les lipides de la luzeme en conditions menant a l' endurcissement au froid. Can J

Harbage JF, Weiss LS, Stone JM, Bamberg JB, Palta JP (1992) Increased plasma membrane linoleic acid is correlated with acclimation capacity in a *Solanum* population segregating for freezing tolerance and acclimation capacity. Plant Physiol **99**: 10

Iswari S, Palta JP (1989) Plasma membrane ATPase activity following reversible and irreversible freezing injury. Plant Physiol **90**: 1088-1095

Levitt J (1980) Responses of Plants to Environmental Stresses, Ed 2. Academic Press, New York

Li PH, Sakai A (1978) Plant Cold Hardiness and Freezing Stress: Mechanisms and Crop Implications. Academic Press, NY

Li PH, Sakai A (1982) Plant Cold Hardiness and Freezing Stress: Mechanisms and Crop Implications. Academic Press, NY

Lynch DV, Steponkus PL (1987) Plasma membrane lipid alterations associated with cold acclimation of winter rye seedlings (*Secale cereale* L. cv Puma). Plant Physiol **83**: 761-767

Marshall HG (1982) Breeding for tolerance to heat and cold. *In* MN Christiansen, CJ Lewis, eds, Breeding Plants for Less Favorable Environments. John Wiley and Sons, New York, pp 47-69

Palmgren MG, Sommarin M, Ulvskov P, Jogensen PL (1988) Modulation of plasma membrane H^+- ATPase from oat roots by lysophosphatidylcholine, free fatty acids and phospholipase A_2. Physiol Plant **74**: 11-19

Palta JP (1989) Plasma membrane ATPase as a key site of perturbation in response to freeze-thaw stress. *In* DD Randall, DG Blevins, eds, Current Topics in Plant Biochemistry and Physiology, Vol 8. University of Missouri, Columbia, MO, pp 41-68

Palta JP, Li PH (1980) Alterations in membrane transport properties by freezing injury in herbaceous plants: evidence against the rupture theory. Physiol Plant **50**: 169-175

Palta JP, Mead LS (1988) During cold acclimation of potato species, an increase in 18:2 and decrease in 16:0 in plasma membrane phospholipid coincides with an increase in freezing stress resistance. Plant Physiol **89**: 89

Palta JP, Simon G (1993) Breeding potential for improvement of freezing stress resistance: Genetic separation of freezing tolerance, freezing avoidance, and capacity to cold acclimate. *In* PH Li, L Christersson, eds, Advances in Plant Cold Hardiness. CRC Press, Boca Raton, pp 299-310

Palta JP, Weiss LS (1993) Ice formation and freezing injury: An overview on the survival mechanisms and molecular aspects of injury and cold acclimation in herbaceous plants. *In* PH Li, L Christersson, eds, Advances in Plant Cold Hardiness. CRC Press, Boca Raton, pp 143-176

Palta JP, Whitaker BD, Weiss LS (1993a) Plasma membrane lipids associated with genetic variability in freezing tolerance and cold acclimation of *Solanum* species. Plant Physiol **103**: 793-803

Palta JP, Weiss LS, Harbage JF, Bamberg JB, Stone JM (1993b) Molecular mechanisms of freeze-thaw injury and cold acclimation in rbaceous plants: Merging physiological and genetic approaches. *In* MB Jackson, CR Black, eds, Interacting Stresses on Plants in a Changing Climate, Springer-Verlag, Berlin, pp 659-680

Raison JK, Lyons JM, Keith AD (1971) Ttemperature-induced phase changes in mitochondria membranes detected by spin labeling. J Biol Chem **246**: 4036-4040

Sakai A, Larcher W (1987) Frost Survival in Plants: Responses and Adaptations to Freezing Stress. Springer-Verlag, New York

Steponkus PL (1984) Role of the plasma membrane in freezing injury and cold acclimation. Annu Rev Plant Physiol **35**: 543-584

Stone JM, Palta JM, Bamberg JB, Weiss LS, Harbage JF (1993) Inheritance of freezing resistance in tuber-bearing *Solanum* species: evidence for the independent control of nonacclimated freezing tolerance and cold acclimation capacity. Proc Nat Acad Sci **90**: 7869-7873

Stushnoff C, Fowler DB, Brule-Babel A (1984) Breeding and selection for resistance to low temperature. *In* PB Voss, ed, Plant Breeding - A Contemporary Basis, Pergamon Press, Elmsford, NY, pp 115-136

Sutinen ML, Rybarczyk SJ, Palta JP (1989) Seasonal changes in polar fatty acid composition coincide with changes in freezing stress resistance in pine needles. *In* DD Randall, DG Blevins, eds, Current Topics in Plant Biochemistry and Physiology, Vol 8. University of Missouri, Columbia, MO, p 286

Sutinen ML, Palta JP, Reich PB (1992) Seasonal differences in freezing stress resistance of needles of *Pinus nigra* and *Pinus resinosa*: evaluation of the electrolyte leakage method. Tree Physiol **11**: 241-254

Teutonico RA, Palta JP, Osborn TC (1993) In vitro freezing tolerance in relation to winter survival of rapeseed cultivars. Crop Sci **33**: 103-107

Thomashow MF (1990) Molecular genetics of cold acclimation in higher plants. *In* JG Scandalios, ed, Advances in Genetics, Vol 28, Genomic Responses to Environmental Stress, Academic Press, NY pp 99-125

Tucker WG (1974) Freezing injury in carrots. J Hortic Sci **49**: 29-35

Tucker WG, Cox CJ (1978) Studies on overwinter losses of carrots stored *in situ*. J Hortic Sci **53**: 291-294

Uemura M, Yoshida S (1984) Involvement of plasma membrane alterations in cold acclimation of winter rye seedlings (*Secale cereale* L. cv Puma). Plant Physiol **75**: 818-826

Vigh L, Horvath I, van Hasselt PR, Kuiper PJC (1985) Effect of frost hardening on lipid and fatty acid composition of chloroplast thylakoid membranes in two wheat varieties of contrasting hardiness. Plant Physiol **79**: 756-759

Willemot C (1975) Stimulation of phospholipid biosynthesis during frost hardening of winter wheat. Plant Physiol **55**: 356-359

Willemot C (1977) Simultaneous inhibition of linolenic acid synthesis in winter wheat roots and frost hardening by BASF 13-338, a derivative of pyridazinone. Plant Physiol **60**: 1-4

Yoshida S, Uemura M (1984) Protein and lipid compositions of isolated plasma membranes from orchard grass (*Dactylis glomerata* L.) and changes during cold acclimation. Plant Physiol **75**: 31-37

Chapter 5

ENGINEERING PLANTS FOR STRESS TOLERANCE

ENGINEERING PLANTS FOR STRESS TOLERANCE VIA ORGANELLE GENOMES

Henry Daniell[1], Ananta PoroboDessai[3], Channapatna S. Prakash[3] and William J. Moar[2]

[1]Molecular Genetics Program, Department of Botany & Microbiology,[2]Department of Entomology, Auburn University, Auburn, AL 36849-5407, USA; [3]School of Agriculture, Tuskegee University, Tuskegee, AL 36088, USA.

INTRODUCTION

Most common strategies for gene transfer derive from the fact that Agrobacterium tumefaciens contains Ti plasmids which facilitate transfer of T-DNA from the plasmid genome into plant cells with attendant integration of the T-DNA region into the plant nuclear genome (Binns,1990). Alternatively, DNA may be introduced into protoplasts using various chemical agents (Jenes et al, 1993) or by electroporation (Fromm et al., 1987) or microinjection (Neuhans and Spangenberg, 1990). In all these approaches foreign DNA is introduced into the nuclear genome. Sanford and coworkers (1990) have developed a transformation technique that relies on bombardment of recipient cells or tissues with high velocity tungsten microprojectiles coated with foreign DNA. Using this delivery system several groups have demonstrated stable transformation of foreign genes into nuclear genomes of higher plants, mitochondrial genomes of yeast, and chloroplast genomes of Chlamydomonas and higher plants (Sanford, 1990). Microprojectile bombardment is replacing traditional methods of transformation because of the use of simple vectors and unlimited host range. DNA delivery by the Gene Gun into cultured tissues or the scutellum of immature embryos of several cereal crops is generally followed by selection for transformed tissues. Regenerated transgenic plants have been obtained from maize, rice, wheat, oat and numerous dicots (Guerin and Guerin, 1993), demonstrating the broad applicability of the Gene Gun. The Gene Gun is also routinely used in studies on transient foreign gene expression and is a reliable method to deliver foreign DNA into organelles (Daniell, 1993).

Chloroplast Transformation in Higher Plants:

Till now it has been possible to engineer a single economically important trait via the nuclear genome (eg., resistance to a particular herbicide). However, the presence of gene products that could confer tolerance to herbicides, insects, heat, salt/drought in the chloroplast has opened the door to genetically engineer plants that are resistant or tolerant to several adverse environmental factors. Parallel expression of genes that are important in plant response to multiple environmental factors should provide an extraordinary opportunity to understand interactions between biotic and abiotic stress that affect plant biochemical processes and mechanisms of stress tolerance in this cellular compartment.

Several approaches have been used in the past to manipulate genes in chloroplasts, i.e., generation of chloroplast mutants, protoplast fusion, organelle inactivation and chloroplast recombination. Daniell and McFadden (1987) reported the uptake and expression of bacterial and cyanobacterial genes by isolated chloroplasts. Recent reports of the introduction of chloroplasts into albino protoplasts and the observation of variegated progeny with the transfer of only two chloroplasts (Eigel et al., 1991) opens up the possibility of introduction of transformed chloroplasts into recipient protoplasts. Daniell et al. (1990) demonstrated the first transient expression of foreign genes in plastids of higher plants using the chloramphenicol

NATO ASI Series, Vol. H 86
Biochemical and Cellular Mechanisms
of Stress Tolerance in Plants
Edited by J. H. Cherry
© Springer-Verlag Berlin Heidelberg 1994

acetyltransferase (cat) gene and the biolistic device. Since then several laboratories have reported stable and transient foreign gene expression in chloroplasts of higher plants. Ye et al. (1990) have reported conditions for optimal DNA delivery into plastids of cultured tobacco cells using an improved biolistic device. Daniell et al. (1991) have also reported expression of β-glucuronidase in different cellular compartments following biolistic delivery of chloroplast or nuclear vectors into wheat leaves and calli derived from anther culture or immature embryos. Stable complementation of tobacco chloroplasts using a modified chloroplast 16S rDNA gene conferring antibiotic resistance has also been accomplished using the biolistic device (Svab et al., 1990; Staub and Maliga, 1992). High frequency plastid transformation in tobacco by selection for a chimeric aadA gene has also been reported (Svab and Maliga, 1993).

Biolistic transformation appears to be the only reliable and reproducible method to date for delivering foreign DNA into chloroplasts of higher plants. Agrobacterium mediated chloroplast transformation (de Block et al., 1985) has met with limited success; the published report of PEG-mediated chloroplast transformation (Sporlein et al., 1991) is in disagreement with findings from other laboratories (Ye et al., 1990).

Stable transformation for agriculturally important genes whose products are localized within chloroplasts is currently in progress in our laboratory. This article describes ongoing efforts to engineer chloroplast genomes for conferring stress tolerance in plants.

The Gene Gun:

In the standard gun powder driven PDS-1000 biolistic device (DuPont Company), a gun powder charge is used to drive a plastic plunger (macroprojectile) down a barrel to accelerate DNA-coated tungsten particles (microprojectiles). The macroprojectile hits a stopping plate allowing the microprojectiles to continue to travel into a partially evacuated chamber. Microprojectiles traveling at high speed forcefully enter the target cells or tissues which are placed at a given height inside the bombardment chamber. The gas and debris from the gunpowder charge is vented into a filter housing unit and is trapped in the vacuum oil.

DuPont has recently developed a retrofit kit for the PDS-1000, providing the user with a gas driven particle delivery system, the PDS-1000/He. The PDS- 1000/He uses high pressure helium instead of gunpowder as the driving force for the macrocarriers. In this method, a small chamber is sealed at one end with rupturable membrane(s) and filled with helium to high pressure. A solenoid-driven lance then ruptures the membrane releasing a shock wave, which enters the launch assembly device that accommodates a removable sleeve. There are removable rings inside the sleeve to facilitate launching of microprojectiles on a macrocarrier. A nylon mesh is locked in place across the axis of the sleeve; DNA-coated microprojectiles are loaded directly onto the center of the mesh. The helium shock wave atomizes and accelerates the microprojectiles as it passes through the mesh. In yet another modification called the "flying disc" method, a plastic membrane is loosely held in the same position as the nylon mesh. Particles are dried on its surface. The disc is accelerated down the sleeve 1 cm, upon firing, where it impacts against a stopping screen releasing microprojectiles which hit the target cells. For a detailed description of plant cell and tissue culture media, subculture conditions, generation of tissues for bombardment, preparation of plant material and microprojectiles for bombardment and protocols for bombardment of cells or tissues the reader is referred to a recent article in Methods in Enzymology (Daniell, 1993).

591

Chloroplast Expression Vectors:

Work in our laboratory for the past several years has involved construction of chloroplast expression vectors either for stable integration of foreign genes into the chloroplast genome or for transient/stable expression and autonomous replication of introduced plasmids inside chloroplasts or mitochondria. As a first step towards achieving this, we have constructed a series of chloroplast expression vectors, using the promoter selection vector pKK232-8 (Pharmacia), which is a pBR322 derivative containing a promoterless cat gene. A multiple cloning site (MCS) has been placed 5'-proximal to the cat gene to facilitate insertion and analysis of promoter fragments. Transcription/translation of cat can be used to quantify the strength of promoters inserted into the MCS of pKK232-8. The plasmid contains the ribosomal RNA T1 and T2 terminators distal to the cat gene to allow cloning of strong promoters and three stop codons between MCS and the AUG of the cat gene to prohibit translational read-through into the cat gene. Restriction fragments of chloroplast DNA containing the entire promoter region and 5'-untranslated region of the psbA gene from spinach (pMP450, a gift from Dr. Wilhelm Gruissem) generating pHD306 or pea (pPPBX10218, a gift from Dr. John Mullet) generating pHD312 or alternatively, the rbcL and atpB promoter region from maize (pPBI443, a gift from Dr. Anthony Gatenby) generating pHD103, have been individually inserted into the MCS site; colonies have been screened on LB plates containing chloramphenicol. Plasmids containing chloroplast promoter fragments have been investigated by analyzing transient expression of cat in cucumber etioplasts using the methodology of Daniell and McFadden (1987). The spinach or pea psbA promoter has been found to be the strongest among the promoters tested (Daniell et al., 1990).

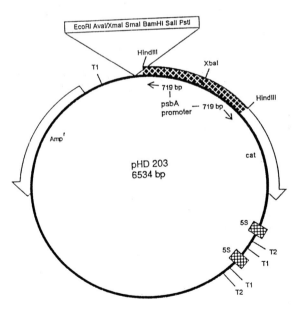

The chloroplast expression vector pHD203 (shown above) contains a double psbA promoter fragment with the promoters in opposite orientation to facilitate insertion of additional genes. While one psbA promoter region would drive the cat gene, the second promoter fragment is placed upstream of a MCS containing sites for AvaI; XmaI; BamHI, BglII, SalI, SmaI; HincII, PstI and HindIII. There is a ribosomal RNA T1 terminator distal to the MCS that facilitates subcloning genes driven by strong promoters. E. coli uidA gene coding for β-glucuronidase (GUS) (Jefferson et al., 1986) has been inserted into the MCS of pHD203 at PstI-SmaI sites as described in detail by Ye et al. (1990).

Autonomously Replicating Chloroplast Expression Vectors:

In order to increase the copy number of the introduced plasmid, origin of replication sequences from plastid genomes may be included in chloroplast vectors. Several pea chloroplast DNA fragments (Meeker et al., 1988) containing one or both replication origins identified as displacement loops (D-loops) have been tested for in vitro DNA replication using a replication fraction isolated from pea chloroplasts containing RNA polymerase, DNA polymerase, DNA primase and topoisomerase I activities. A well-defined chloroplast replicon has been inserted into the chloroplast expression vector pHD312, which contains the pea psbA promoter 5'-proximal to the promoterless cat gene, resulting in the construction of pHD407 (Daniell, 1993). Alkaline agarose gel analysis of in vitro replication products revealed the presence of single stranded DNA molecules of about 9.9 kbp, corresponding to the full length of pHD407. Restriction digests of in vitro replication products showed that fragments of the expected sizes were labeled in pHD407 (Daniell et al, 1990).

Nuclear Expression Vectors:

The nuclear expression vector pPBI121 carries a GUS gene driven by a CaMV-35S promoter and flanked at the 3' end by a polyadenylation signal from the nopaline synthase gene of the Agrobacterium tumefaciens Ti-Plasmid (Jefferson, 1987). For negative controls, pUC19 DNA or appropriate vector DNA should be used in all bombardments. The nuclear expression vector pUC8 CaMV CAT-N is a 4.2 kbp plasmid containing a cat gene driven by a CaMV-35S promoter, flanked by a 3' nos PstI poly A fragment.

Reporter Gene Expression in Chloroplast and Nuclear Compartments:

Cultured tobacco cells collected on filter papers were bombarded with tungsten particles coated with pUC118 (negative control), 35S CAT (nuclear expression vector), pHD312 (repliconless chloroplast expression vector) and pHD407 (chloroplast expression vector with replicon). There was no detectable cat activity in the autoradiograms of sonic extracts of cells bombarded with pUC118 (Daniell et al, 1990). Maximal nuclear expression of cat was observed 72h after bombardment. There was a low level of expression in cells bombarded with chloroplast expression vectors until 48 h incubation and maximal expression at 72 h. The plasmid pHD312 showed about 50% of the maximal activity of pHD407. After 72 h incubation, the expression of nuclear-cat and pHD312 decreased while a high level of cat expression remained in cells bombarded with pHD407 (Daniell et al., 1990).

Various plasmid constructions were introduced into tobacco protoplasts by electroporation to check organelle-specific expression of cat in appropriate compartments. No activity was observed with any vector except the nuclear expression vector, 35S-<u>CAT</u> (Daniell et al, 1990). It is known that electroporation of protoplasts results in DNA delivery into the cytosol and not inside organelles. In the report of Cornelissen and Vandewiele (1989) the <u>psb</u>A promoter was fused to <u>npt</u>II or <u>bar</u> coding sequences and introduced into tobacco by T-DNA mediated transfer; this resulted in transcriptional activity that was too low to produce any detectable mRNA; their attempts to induce the tobacco <u>psb</u>A promoter to function efficiently in the nuclear compartment revealed the absolute need to insert 35S promoter or enhancer elements 5'proximal to the <u>psb</u>A promoter region. Bogorad and coworkers (1988) also observed that when the <u>psb</u>A gene was placed in the nuclei of transgenic plants it was not transcribed from its own promoter.

The continued maintenance of high levels of <u>cat</u> expression in cells bombarded with pHD407 might have been due to several factors. Replication of pHD407 should have resulted in increased copy number and this would maintain a high level of <u>cat</u> expression. In cultured tobacco cells, copy number of plastid DNA per cell has been shown to increase 11-fold within one day after the addition of fresh medium, and replication of plastid DNA was most frequently observed on the second day (Takeda et al., 1992).

We have recently reported an efficient and highly reproducible delivery system, using an improved biolistic device, that facilitates transient expression of GUS in chloroplasts of cultured plant cells (Ye et al., 1990). Tobacco cells bombarded with a nuclear expression vector (pBI505) showed high levels of GUS expression with blue color distributed evenly throughout the cytosol. On the other hand, when the chloroplast expression vector pHD203-GUS was bombarded, the indigo dye was subcellularly localized within the chloroplasts. Compartmentalized expression of GUS by these vectors was further confirmed by introducing them into tobacco protoplasts by PEG mediated transformation, followed by GUS assays. While the nuclear expression vector (pBI505) showed a high level of GUS expression, no GUS activity was observed for pUC118 or pHD203-GUS (Ye et al, 1990). Chloroplast transformation efficiencies increased dramatically (about 200-fold) using an improved helium-driven biolistic device as compared to the more commonly used gun powder charge driven device. Using GUS as a reporter gene and the improved biolistic device, optimal bombardment conditions were established, consistently producing several hundred transformants per petri plate. Chloroplast transformation efficiency was found to be increased further (20-fold) with supplemental osmoticum in the bombardment and incubation media. We have also reported transient expression of GUS in different cellular compartments following biolistic delivery of chloroplast or nuclear expression vectors into wheat leaves or calli derived from anther culture or immature embryos (Daniell et al., 1991). All these observations demonstrate a reproducible foreign DNA delivery system into higher plant chloroplasts.

<u>Engineering Herbicide Resistance via Chloroplast Genomes:</u>

In plants, many economically important gene products (e.g., the protein conferring atrazine sensitivity or resistance) are either encoded by the chloroplast genome (Cheng et al., 1988) or, if they are nuclear encoded, their gene products are functional within the chloroplasts (e.g., 5-enol-pyruvyl shikimate-3-phosphate synthase, EPSPS which confers sensitivity or resistance to glyphosate, Shah et al, 1986 or acetolactate synthase conferring resistance or sensitivity to sulfonylureas or imidazolinones, Mazur and Falco, 1989) or mitochondria (e.g., arylacylamidase, which confers sensitivity or resistance to propanil, Gaynor and Still, 1983; Daniell et al., 1981). Selective herbicides are routinely applied to control weeds that would otherwise compete for available nutrients, space and light, thereby reducing crop yield. Because a single new gene is all that is needed to confer resistance to a particular herbicide, several plant species resistant to a variety of herbicides have been engineered successfully.

The broad spectrum weed killer glyphosate (N-phosphonomethyl glycine) acts by inhibiting the enzyme EPSP synthase, which converts phosphoenol pyruvate and 3-phospho-shikimic acid to 5-enolpyruvyl-3-phosphoshikimic acid in the shikimic acid pathway in bacteria (Comai et al., 1983); EPSP synthase is thus needed for the synthesis of the essential amino acids phenylalanine, tyrosine and tryptophan. Following mutagenesis of Salmonella typhimurium, an altered EPSP synthase enzyme resistant to glyphosate has been identified and the altered aroA gene that encodes it has been cloned and introduced into E. coli to confer resistance to glyphosate (Comai et al., 1983). This mutant gene has also been transferred into tobacco and it has been shown that resultant plants do not grow when sprayed with the herbicide; the glyphosate-insensitive bacterial enzyme was found to be cytoplasmically located since it lacked the transport sequence needed to direct it into the chloroplast. Subsequent to these studies, mutant bacterial genes were fused to plant EPSPS chloroplast transit peptide sequences and then transferred to plants; the chimeric precursor was imported into petunia chloroplasts and conferred glyphosate resistance in transgenic plants (Mazur and Falco, 1989). Using a different approach, glyphosate-tolerant transgenic Petunia plants were engineered by Shah et al (1986) by targeting an overproduced EPSP synthase to the chloroplasts via the 72-residue transit peptide.

The chloroplast gene psbA codes for the photosynthetic quinone-binding membrane protein, QB, which is the target of the herbicide atrazine. The mutant psbA gene from an atrazine resistant biotype of Amaranthus hybridus has been modified by Bogorad and coworkers, by fusing its coding region to transcription regulation and transit-peptide-encoding sequences for SSU of RuBisCO (Cheng et al., 1988). The constructs were introduced into the nuclear genome of tobacco by using the Agrobacterium tumor inducing (Ti) plasmid system, and the protein product of the nuclear psbA gene was identified in the photosynthetic membranes of chloroplasts; though transformed plants survived longer than controls, all of them eventually succumbed in the presence of atrazine. The chloroplast transformants were observed to contain a mixed population of photosystem II complexes that are resistant and sensitive to atrazine. A method for inactivating specific plastid genes in higher plants should be developed in order to overcome this problem. This is the first published report of relocating a chloroplast gene into the nuclear genome, wherein a chloroplast protein has been successfully biosynthesized in the cytosol and delivered across the organellar membranes. Foreign gene products, i.e. β-glucuronidase or nptII have been successfully targeted into chloroplasts; however they do not have interfering endogenous genes or gene products. Thus, the feasibility of this approach is presently limited to genes whose products are not chloroplast encoded.

Since transient and stable expression of foreign genes have already been accomplished in higher plant chloroplasts using the Gene Gun (see Daniell, 1993) it is clear that chloroplast genomes could now be used to engineer foreign genes whose products function within plastids. Several herbicide binding proteins (e.g., the protein conferring atrazine resistance) are either encoded by the chloroplast genome or, if nuclear encoded, their gene products function within chloroplasts (e.g., EPSP synthase conferring glyphosate resistance or acetolactate synthase conferring resistance to sulfonylureas or imidazolinones) or mitochondria (e.g., arylacylamidase conferring resistance to propanil). Therefore, herbicide resistant genes have been used as selectable markers to engineer foreign genes via chloroplast genomes in our laboratory.

Engineering Insect Resistance via Chloroplast Genomes:

Bacillus thuringiensis produces many types of crystalline inclusions which are toxic to insects. The proteins comprising these inclusions have been categorized based on insecticidal host range, and protein homology (Höfte and Whiteley 1989). The CryI and CryII toxins have insecticidal activity against lepidoptera, or lepidoptera and diptera, respectively (Höfte and Whiteley 1989). CryI protoxins are 130-135 kDa in size which are enzymatically cleaved into proteins of 65 kDa for insecticidal activity. CryIIA protoxin is 65kDa in size which probably needs to be cleaved into a protein with a molecular mass of 60-62 kDa for insecticidal activity. Donovan et al. (1988) was the first group to report the cloning, sequencing, expression, and insecticidal activity of a cryIIA gene and its encoded protein. Widner and Whiteley (1989) reported that the cryIIA gene of the HD-1 strain of B. thuringiensis subsp. kurstaki is the distal gene of a three-gene operon. They also reported that expression of the entire operon or the cryIIA gene resulted in CryIIA protein inclusions with similar insecticidal activity against the tobacco hornworm, Manduca sexta (L.) (LC$_{50}$=0.15μg/cm^2). Expression of the first two genes of the operon (orf 1, orf 2) resulted in no insecticidal activity against M. sexta. Crickmore and Ellar (1992) reported that a possible chaperonin, orf 2, is required for an efficient expression of the cryIIA gene in B. thuringiensis. In the absence of a functional ORF 2 polypeptide, the CryIIA toxin does not form crystalline inclusions in B. thuringiensis . They suggest two possible roles for the ORF 2: 1) to protect the CryIIA toxin from proteolytic degradation before it is laid down within the crystalline inclusion; 2) to serve as a matrix to direct the assembly of CryIIA molecules into the regular cuboidal crystal. Results from our laboratory demonstrate the cuboidal structure of CryIIA in native B. thuringiensis P-2 inclusion bodies containing CryIIA, expression of the full-length cryIIA operon in E. coli, and expression of the full length cryIIA operon in a crystal minus strain of B. thuringiensis (Fig. 1 a,b,c). However, expression of a truncated cryIIA operon containing primarily the cryIIA coding sequence results only in an amorphous inclusion body (Fig. 1d). This amorphous inclusion body also was observed to be more soluble under denaturing conditions. These results further suggest that although the orf2 is not required for expression of CryIIA in E. coli (contrary to expression in B. thuringiensis, Crickmore and Ellar, 1992) expression of orf1 and orf2 does affect solubility and crystal shape.

The CryIIA protein has promise for expression in plants to control insects because 1) CryIIA protoxin is only about 1/2 the size as the CryI protoxin (similar in size to activated CryI toxins [ca. 65kDa]) and therefore should result in higher levels of expressed toxin without concerns of instability associated with use of truncated CryI toxins. 2) Many economically important insect pests (especially in the family Pyralidae) are susceptible to CryIIA toxin: European corn borer, Ostrinia nubilalis (Hübner), lesser cornstalk borer, Elasmopalpus lignosellus (Zeller), cowpea pod borer, Maruca testulalis (Geyer), tobacco budworm, Heliothis virescens (F.), tobacco hornworm, Manduca sexta and gypsy moth Lymantria dispar (Dankocsik et al. 1990; Widner and Whiteley, 1989; Moar et al. 1994, WJM unpublished data).

One of the major hurdles in engineering insect resistant plants continues to be the low level of expression of B.t. toxin genes. Fischhof et al. (1987) reported B.t. transgenic tomato plants that were toxic to M. sexta and H. virescens with an expression level of 50-ng of toxin/gram of fresh leaf (0.00001%). Vaeck et al. (1987) reported B.t. transgenic tobacco plants that were toxic to M. sexta with expression levels as high as 3 μg CryIA(b) toxin / gram fresh leaf (0.0003%). Warren et al. (1992) reported that B.t. transgenic tobacco plants expressing cryIA(b) were toxic to H. virescens and M. sexta in a field situation with expression levels as

596

Fig. 1. Shadow cast transmission electron micrographs of CryIIA from four expression systems: (a) Native P-2 inclusion bodies containing CryIIA; (b) full-length cryIIA operon expressed in E. coli; (c) full-length cryIIA operon expressed in a crystal minus strain of Bacillus thuringiensis; (d) truncated cryIIA operon expressed in E. coli. (Electron micrographs courtesy of T. Cottrell, Department of Entomology, Auburn University).

high as 1 μg CryIA(b) toxin / gram fresh leaf (0.0001%). However, it should be noted that H. virescens and M. sexta are very sensitive to most B.t. toxins as compared with other economically important lepidopteran pests. Although the observed levels of expression are toxic to sensitive insects (but not to many other economic pests), the low levels of expression also may lead to generation of insect populations that will ultimately develop resistance to toxins. An increase in B.t. toxin gene expression of up to 500-fold has been achieved through specific modification of the B.t. toxin coding sequence from prokaryotic composition to suit the eukaryotic nature of plant nuclei (Perlak et al., 1991). This is not surprising because chloroplast genes transferred to the nuclear plant genome in the course of evolution have adjusted to nuclear base composition and codon usage (Oliver et al., 1990). Koziel et al. (1993) reported transgenic corn using a synthetic truncated cryIA(b) gene. These transformed plants were resistant to attack by european corn borer, Ostrinia nubilalis. Levels of expression averaged 12.75 μg CryIA(b) toxin/ gram fresh leaf (0.0013%). In yet another study, Wong et al (1992) reported a higher level of expression of the insecticidal protein when the cryIA(c) gene was fused with rbcS leader and transit peptide sequences.

A recent report by Staub & Maliga (1993) shows that the level of expression of the uidA gene (GUS) driven by the psbA promoter in chloroplasts was 2.5% of total cellular protein. Therefore, the problem of insufficient expression of foreign genes, especially if the foreign gene is of bacterial origin, could be overcome by engineering those genes via organelle genomes. Chloroplasts are prokaryotic compartments in eukaryotic cells; genes of bacterial origin have been expressed at high levels in chloroplasts (Daniell, 1993). Furthermore, ribulose bisphosphate carboxylase/oxygenase, the most abundant protein in nature (it constitutes more than 80% of the soluble chloroplast proteins), is expressed in the chloroplast and assembled with the small subunit (which is synthesized in the cytosol and targeted into the chloroplast). Therefore, chloroplasts can indeed support abundant synthesis and assembly of certain proteins without deleterious effects on overall metabolism. In stable chloroplast transformation studies high copy number has been accomplished by introducing one or two copies of the foreign gene into each chloroplast genome of tobacco; each plant cell contains 5,000-10,000 chloroplast genomes per cell and, therefore, it has been possible to express thousands of copies of the introduced foreign gene per cell (Svab et al., 1990).

Yet another problem with expressing foreign genes in the eukaryotic nuclear compartment is the inability to express an operon from a prokaryote, cleave and process the polycistronic mRNA. For example, it is highly desirable to express the entire operon containing all three open reading frames (orf1, orf2, CryIIA) in transgenic plants; this should result in the production of a stable crystalline insecticidal protein. Therefore, the entire operon containing orf1, orf2 and the cryIIA gene will be inserted into the tobacco chloroplast genome and transgenic plants will be examined for the production of insecticidal proteins in our laboratory.

Goldburg & Tjaden (1990) express concern of expressing B. thuringiensis toxins in plant products which will be consumed. Their concern is that mammalian toxin studies that have been conducted do not use the same toxins which are expressed in plants. B. thuringiensis toxins found in formulated materials which have been tested for mammalian safety must be cleaved in an alkaline environment (mammalian G. I. tracts are acidic) and activated by specific digestive enzymes. Many B. thuringiensis toxins expressed in plants already have been "cleaved" and "activated". Therefore, expression of any B. thuringiensis toxin (full-length or truncated) in chloroplasts would allow for tissue-specific and temporal expression. For example, genes transcribed in chloroplasts are not transcribed in chromoplasts or amyloplasts that are found in fruits or storage organs (Kobayashi et al. 1990; Ngernprasirtsiri, 1988 a,b; Ohta et al. 1991). Furthermore, ß-glucuronidase gene under control of the psbA 5' and 3'-regulatory regions integrated into the tobacco plastid genome showed a high level of expression in photosynthetic tissues (e.g., leaves) but not in non-photosynthetic tissues (e.g., roots; Staub and Maliga, 1993). Therefore, the cryIIA gene driven by psbA regulatory signals is not likely to be expressed in non-green fruits or storage organs.

Engineering Osmotolerance via Chloroplast Genomes:

Plants are able to regulate the ionic strength of the interior of their constituent cells in response to fluctuations in the osmotic conditions of their environment. Most organisms accomplish osmotic adjustment by increasing the cellular concentration of inorganic ions and low molecular weight organic compounds, including quaternary ammonium and tertiary sulfonium compounds. Such osmolites have been reported in the cytoplasm of osmotically stressed cells of bacteria (Csonka and Hanson, 1991), algae (Blunden and Gordon, 1986), plants (Rhodes and Hanson, 1993) and animals (Somero, 1986). Genes involved in biosynthetic pathways of these osmolites are potential targets for genetic engineering of drought or salt tolerance in plants (McCue and Hanson, 1990).

For example, glycine betaine can act as osmoprotectant in many organisms and the ability to accumulate glycine betaine in response to drought or salt tolerance is widespread in higher plants, occurring in at least eleven families. In plants, glycine betaine is synthesized by a two step oxidation of choline through the formation of betaine aldehyde. The first step is catalyzed by a stromal enzyme, choline monoxygenase, which introduces an atom of oxygen from 0_2 into choline, resulting in the hydrate form of betaine aldehyde (Lerma et al., 1988). The second step is catalyzed by betaine aldehyde dehydrogenase (BADH) which is also present in the chloroplast stroma. BADH cDNA clones containing the entire protein coding sequence have been isolated from spinach (Weretilnyk and Hanson, 1990) and sugarbeet (McCue and Hanson, 1992). The spinach and sugarbeet BADH cDNA sequences lacking typical transit peptides have been expressed in tobacco where they confer the capacity to oxidize exogenous betaine aldehyde to glycine betaine (Hanson and Burnet, 1993). Experiments are in progress to express BADH genes (kindly provided by Dr. Andrew Hanson) directly in chloroplasts using appropriate chloroplast expression vectors and analyze transgenic plants for osmotolerance.

Among other organic metabolites accumulated in response to water or salt stress, proline is found at high concentrations during osmotic stress in bacteria (Measures, 1975), cultured plant cells (Handa et al., 1983), whole plants (Jeffries, 1980), and even invertebrate animals (Fyhn, 1976). Proline overproducing mutants of Salmonella typhimurium show increased tolerance to osmotic stress (Csonka, 1981). Furthermore, when proline is added to the medium of cultured tobacco cells their growth rate is increased specifically under conditions of water stress (Handa et al., 1986). The presence of proline in extremely high concentrations [1M] under osmotic stress in barley cultivars is suggestive of its key role in the osmotic response (Wyn Jones and Storey, 1978). One of the interesting hypotheses concerning the accumulation of high concentrations of proline is that it confers enzyme protection and increases membrane stability (Schobert and Tschesche, 1978). Proline is also known to confer mitochondrial enzymes heat stability and protect against heat inactivation (Nash et al., 1982).

Proline biosynthesis in E.coli consists of four reactions. First, glutamate is phosphorylated by glutamate kinase to produce γ-glutamyl phosphate; γ-glutamyl phosphate is further reduced by glutamate semialdehyde dehydrogenase using NADPH to make glutamate semialdehyde. Then glutamate semialdehyde is spontaneously cyclized to produce 1-pyrroline 5-carboxylate. Finally, 1-pyrroline 5-carboxylate is reduced by 1-pyrroline 5-carboxylate reductase using NADPH to generate proline. The genes encoding glutamate kinase, glutamate semialdehyde dehydrogenase and 1-pyrroline 5-carboxylate reductase are proB, proA and proC.

A mutation in the most osmotolerant Salmonella typhimurium strain that has been reported is in the gene encoding γ-glutamyl kinase. In the wild-type strain the rate of proline biosynthesis is regulated by feedback inhibition of γ-glutamylkinase by proline. The enzyme in the most osmotolerant mutant has lost sensitivity to feedback inhibition by proline, resulting in increased proline biosynthesis. Expression of proB74, that specifies an enzyme resistant to feedback inhibition by proline in yeast resulted in enhanced resistance to proline analogue L-azetidine-2-carboxylate and elevated intracellular proline levels (Orser et al., 1988). It is anticipated that if proB74 is introduced into the chloroplast genome, it may direct the synthesis of a feedback insensitive γ-glutamyl kinase in transgenic plants. Proline overproducing plants will be analyzed for osmotolerance.

Engineering Thermotolerance via Chloroplast Genomes:

Thermotolerance is believed to be conferred by the induction of heat shock proteins (HSPs). Among the most abundantly induced HSPs, a class of nuclear encoded proteins range in size from 15 to 30 kDa; these are referred to as small HSps. In a number of organisms sHSPs aggregate in oligomeric complexes of 200 to 800 kDa and form heat shock granules (Nover et al. 1989). Although a number of functions may be attributed for the sHSPs, it is clear that they can prevent thermal aggregation of other proteins and facilitate reactivation of denatured enzymes (Jakob et al. 1993). Results showing a correlation between the level of sHSP accumulation and acquired thermotolerance suggest that sHSPs exert a protective function in plants exposed to high temperatures (Vierling, 1991). The sHSPs in plants are encoded by four distinct gene families and the gene products are localized in the cytoplasm, endoplasmic reticulum and chloroplasts (Osteryoung et al. 1993).

Among several sHSPs, HSP21 has been studied in greater detail. HSP21 is localized in the chloroplast compartment and cDNA has been isolated and characterized from several plant species (Osteryoung et al. 1993). HSP21 is synthesized in the cytoplasm as a precursor protein containing a N-terminal transit peptide for targeting to the chloroplast. HSP21 is not detectable under normal growth conditions but is induced several hundred fold in heat stressed plants and is accumulated in the chloroplast stromal fraction as a 200 kDa complex (Osteryoung et al. 1993). The chloroplast localized sHSPs should provide an ideal system to understand the mechanism of thermotolerance in plants. It is clear that chloroplast genomes could be used to express sHSPs, in particular HSP21. Chloroplast transgenic plants that overexpress HSP21 may be comparatively studied with control plants that do not express HSP21 to observe possible changes in chloroplast ultrastructure and photosynthetic characteristics in order to understand the role of HSP21 in conferring thermotolerance. Nuclear transgenic plants of Arabidopsis that either underexpress or overexpress HSP21 have recently been studied to understand its role in thermotolerance (Osteryoung et al. 1993).

EPILOGUE

It is clear that chloroplast genomes could be used to engineer foreign genes whose products are functional within plastids. Chloroplasts are prokaryotic compartments in eukaryotic cells; therefore, genes of bacterial origin could be expressed at high levels in chloroplasts. Extensive modifications of the coding sequence of bacterial genes to suit the eukaryotic nature of nuclei may no longer be necessary.

The sexual transfer of genes from crop plants to weedy species to create more persistent weeds is probably the greatest environmental risk of genetic engineering crops for stress tolerance. Escape of engineered foreign genes via pollen grain has indeed been documented recently (Ellstrand and Hoffman, 1990; Dale, 1992; see also Science 251: 878). Umbeck et al. (1991) have reported 5-6% of pollen dispersal by insects from a field test of genetically engineered cotton. Barring a few exceptions, uniparental-maternal inheritance of cytoplasmic organelles is a widespread phenomenon in the plant kingdom. Therefore an introduced foreign gene is likely to be contained within transgenic plants; engineering foreign genes via organelle genomes may be used as an alternative transformation system for those plants that exhibit a high frequency of outcross.

In addition to photosynthesis, several other biochemical pathways including amino acid biosynthesis, fatty acid biosynthesis and pigment biosynthesis pathways are compartmentalized within plastids. It has been postulated that a unicellular cyanobacterium evolved to cyanelles which were evolutionary precursors of chloroplasts of both green and non-green algae. The latter probably diverged from ancestors of green algae as evidenced by the occurrence of large and small subunit genes for ribulose bisphosphate carboxylase-oxygenase in the chloroplast genomes of chromophytic algae. In contrast, the gene for the small subunit has been moved into the nucleus in the evolution of green algae and higher plants. The evolutionary advantages of this integration are uncertain (McFadden et al, 1986). Interestingly, chloroplast genes transferred to the nuclear plant genome have adjusted to nuclear base composition and codon usage (Oliver et al, 1990). Similarly, fatty acid biosynthesis occurs primarily in chloroplasts but none of the genes encoding enzymes involved in this process has yet been identified in the chloroplast genome of higher plants. In contrast, the gene for acyl carrier protein, required for fatty acid biosynthesis, has been localized in chloroplast genomes of two chromophytic algae (Hwang and Tabita, 1991). More recently, the gene coding for a fatty acid biosynthesis enzyme β-ketoacyl-ACP-synthase III has been localized in the chloroplast genome of the red alga Porphyra umbilicalis (Reith, 1993). It is anticipated that the chloroplast genome may be the target for metabolic genetic engineering, especially for pathways that are compartmentalized in chloroplasts.

BIBLIOGRAPHY

Binns AN (1990) Agrobacterium mediated gene delivery and the biology of host range limitations. Physiol. Plant 79:135-139.
de Block M, Shell J, van Montage M (1985) Chloroplast transformation by Agrobacterium tumefaciens EMBO J. 4:1367-1372.
Blunden G, Gordon SM (1986) Betaines and their sulphonio analogues in marine algae. Progress in Phycological Research 4:39-79.
Cheng A, Bogorad L, Montage M, Schell J (1988) Relocating a gene for herbicide tolerance:

a chloroplast gene is converted into a nuclear gene. Proc. Natl. Acad. Sci. USA 85:391-395.

Comai L, Sen LC, Stalker DM (1983) An altered aroA gene product confers resistance to the herbicide glyphosate. Science 221:370-371.

Cornelissen M, Vandewiele M (1989) Nuclear transcriptional activity of the tobacco plastid psbA promoter. Nucleic Acid Res. 17:19-29.

Crickmore N, Ellar DJ (1992) Involvement of a possible chaperonin in the efficient expression of a cloned CryIIA d-endotoxin gene in Bacillus thuringiensis. Mol. Microbiol. 6: 1533-1537

Csonka LN (1981) Proline over-production results in enhanced osmotolerance in Salmonella typhimurium. Mol. Gen. Genet. 182:82-86.

Csonka LN, Hanson AD (1991) Prokaryotic osmoregulation: genetics and physiology. Annual Review of Microbiology 45:569-606.

Dale PJ (1992) Spread of engineered genes to wild relatives. Plant Physiol. 100:13-15.

Daniell H, Sarojini G, Chinnayan PK, Kulandaivelu G (1981) Action of propanil on in vivo chlorophyll a fluorescence in Echinochloa crus-galli and rice. Weed Research 21:171-177.

Daniell H, McFadden BA (1987) Uptake and expression of bacterial and cyanobacterial genes by isolated cucumber etioplasts. Proc. Natl. Acad. Sci. USA 84:6349-6353.

Daniell H, Vivekananda J, Nielsen B, Ye GN, Tewari KK, Sanford JC (1990) Transient foreign gene expression in chloroplasts of cultured tobacco cells following biolistic delivery of chloroplast vectors. Proc. Natl. Acad. Sci. USA 87:88-92.

Daniell H, Krishnan M, McFadden BA (1991) Expression of ß-glucuronidase gene in different cellular compartments following biolistic delivery of foreign DNA into wheat leaves and calli. Plant Cell Reports 9:615-619.

Daniell H (1993) Foreign gene expression in chloroplasts of higher plants mediated by tungsten particle bombardment. Methods Enzymol. 217:536-556.

Daniell H, Kaliappan SB, Krishnan M, Paszty C (1993) A novel method to study DNA replication in organelles. Nucleic Acids Res. 21:1503-1504.

Dankocsik C, Donovan WP, Jany CS (1990) Activation of a cryptic crystal gene of Bacillus thuringiensis subspecies kurstaki by gene fusion and determination of the crystal protein insecticidal specificity. Molecular Microbiology 4, 2087-2094.

Della-Cioppa G, Bauer SC, Klein BK, Shaw DM, Fraley RT, Kishore GM (1986) Translocation of the precursor of 5-enolpyruvylshikimate-3-phosphate synthase into chloroplasts of higher plants in vitro. Proc. Natl. Acad. Sci. USA 83:6873-6877.

Donovan WP, Dankocsik CC, Gilbert MP, Gawron-Burke MC, Groat RG, Carlton BC (1988) Amino acid sequence and entomocidal activity of the P2 crystal protein. J. Biol. Chem. 263: 561-567.

Eigel L, Oelmuller R, Koop HU (1991) Transfer of defined number of chloroplasts into albino protoplasts using an improved subprotoplast/protoplast microfusion procedure: transfer of only two chloroplasts leads to variegated progeny. Mol. Gen. Genet. 227:446-451.

Ellstrand NC, Hoffman CA (1990) Hybridization as an avenue of escape for engineered genes. BioScience 40:438-442.

Fischhoff DA, Bowdish KS, Perlak FJ, Marrone PG, McCormick SM, Niedermeyer JG, Dean DA, Kusano-Kretzmer K, Mayer EJ, Rochester DE, Rogers SG, Fraley RT (1987) Insect tolerant transgenic tomato plants. Bio/Technology 5: 807-813.

Fromm M, Callis J, Taylor LP Walbot V (1987) Electroporation of DNA and mRNA into plant protoplasts. Methods Enzymol. 153:351-366.

Fyhn HJ (1976) Haloeuryhalinity and its mechanisms in a cirriped crustacean, Balanus improvisus. Comp. Biochem. Physiol. 53A:19-30.

Gaynor JJ, Still CC (1983) Subcellular localization of rice leaf aryl acylamidase activity. Plant Physiol. 72:80-85.

Goldburg, RJ, G Tjaden 1990. Are B.t.k. plants really safe to eat? Bio/Technology 8: 1011-1015.

Guerin TF, Guerin PM (1993) Recent developments in oat molecular biology. Plant Mol. Biol. Rep. 11:65-72.

Handa S, Bressan RA, Handa AK, Carpita NC, Hasegawa PM (1983) Solutes contributing to osmotic adjustment in cultured plant cells adapted to water stress. Plant Physiol. 73:834-843.

Handa S, Handa AK, Hasegawa PM, Bressan RA (1986) Proline accumulation and the adaptation of cultured plant cells to water stress. Plant Physiol. 80:938-945.

Hanson AD, Burnet M (1993) Evolution and metabolic engineering of osmoprotectant accumulation in higher plants. NATO Biology Series, this volume.

Höfte H, Whiteley HR 1989. Insecticidal crystal proteins of Bacillus thuringiensis. Microbiol. Rev. 53: 242-255.

Hwang SR, Tabita FR (1991) Acyl carrier protein derived sequence encoded by the chloroplast genome in the marine diatom Cylindrotheca sp. J. Biol. Chem. 266:13492-13494.

Jakob U, Gaestel M, Engel K, Buchner J (1993) Small heat shock proteins are molecular chaperones. J. Biol. Chem. 268:1517-1520

Jefferson RA, Burgess SM, Hirsh D (1986) ß-Glucuronidase from Escherichia coli as a gene-fusion marker. Proc. Natl. Acad. Sci. USA 83:8447-8451.

Jeffries, RL (1980) In: Genetic Engineering of Osmoregulation, ed. Rains, DW et al., pp. 135-154. Plenum Press: New York.

Jenes B, Moore H, Cao J, Zhang W, Wu R (1993) Techniques for gene transfer in Transgenic Plants: Engineering and Utilization, Kung S, Wu R (eds.), Vol. I, 125-146.

Kobayashi H, Ngernprasirtsiri J and Akazawa T (1990) Transcriptional regulation and DNA methylation in plastids during transitional conversion of chloroplasts to chromoplasts. EMBO J. 9: 307-313.

Koziel MG, Beland GL, Bowman C, Carozzi NB, Crenshaw R, Crossland L, Dawson J, Desai N, Hill M, Kadwell S, Launis K, Lewis K, Maddox D, McPherson K, Meghji MR, Merlin E, Rhoades R, Warren GW, Wright M, Evola SV (1993) Field performance of elite transgenic maize plants expressing an insecticidal protein derived from Bacillus thuringiensis. Bio/Technology 11: 194-200.

Lerma C, Hanson AD, Rhodes D (1988) Oxygen-18 and deuterium labeling studies of choline oxidation by spinach and sugarbeet. Plant Physiology 88:695-702.

Mazur BJ, Falco SC (1989) The development of herbicide resistant crops. Ann. Rev. Plant Physiol. 40:441-470.

McCue KF, Hanson AD (1990) Drought and salt tolerance: towards understanding and application. Trends in Biotechnology 8:358-362.

McCue KF, Hanson AD (1992) Salt inducible betaine aldehyde dehydrogenase from sugarbeet: cDNA cloning and expression. Plant Mol. Biol. 18:1-11.

McFadden BA, Torres-Ruiz J, Daniell H, Sarojini G (1986) Interaction, functional relations and evolution of large and small subunits in Rubisco from prokaryota and Eukaryota. Phil. Trans. R. Soc. Lond. 313:347-358.

Measures JC (1975) Role of amino acids in osmoregulation of nonhalophylic bacteria. Nature 257:398-400.

Meeker R, Nielsen BL, Tewari KK (1988)   Localization of replication origins in pea chloroplast DNA. Mol. Cell. Biol. 8:1216-1223.

Moar WJ, Trumble JT, Hice RH, Backman PA (1994) Insecticidal activity of the CryIIA protein from the NRD-12 isolate of <u>Bacillus</u> <u>thuringiensis</u> subsp. <u>kurstaki</u> expressed in <u>E.coli</u>, <u>B. thuringiensis</u> and in a leaf colonizing strain of <u>Bacillus</u> <u>cereus</u>.   Appl. Environ. Microbiol. In Press.

Nash D, Paleg LG, Wiskich JT (1982) Effect of proline, betaine and some other solutes on the heat stability of mitochondrial enzymes. Aust. J. Plant Physiol. 9:47-57.

Neuhans G, Spangenberg G (1990) Plant transformation by microinjection techniques. Physiol. Plant. 79:213-217.

Ngernprasirtsiri J, Kobayashi H, Sugiyama T, Akazawa T (1988a) DNA methlation as a mechanism of transcriptional regulation in non-photosynthetic plastids in plant cells. Proc. Natl. Acad. Sci. USA 85, 4750-4754.

Ngernprasirtsiri J, Kobayashi H and Akazawa T (1988b) DNA methlation occurred around lowly expressed genes of plastid DNA during tomato fruit development. Plant Physiol. 88, 16-20.

Nover L, Scharf KD, Neumann D (1989) Cytoplasmic heat shock granules are formed from precursor particles and are associated with a specific set of mRNAs Mol. Cell. Biol. 9:1298-1308.

Ohta N, Sato N, Kawano S and Kuroiwa T (1991) Methylation of DNA in the chloroplasts and amyloplasts of the pea <u>Pisum</u> <u>sativum</u>. Plant Science 78, 33-42.

Oliver JL, Marin A, Martinez-Zapter JM (1990) Chloroplast genes transferred to the nuclear plant genome have adjusted to nuclear base composition and codon usage. Nucleic Acids Res. 18:65-73.

Orser CS, Goodner BW, Johnston M, Gelvin SB, Csonka LN (1988) The <u>E. coli</u> proB gene corrects the proline auxotrophy of <u>Saccharomyces cerevisiae</u> pro1 mutants. Mol. Gen. Genet. 212:124-128.

Osteryoung KW, Pipes B, Wehmeyer N, Vierling E (1993) Studies of a chloroplast localized heat   shock protein in <u>Arabidopsis</u>, in NATO Biology series, this volume.

Perlak FJ, Fuchs RL, Dean DA, McPherson, SL, Fischhroff DA (1991) Modification of the coding sequence enhances plant expression of insect control protein genes. Proc. Natl. Acad. Sci. USA 88:3324-3328.

Reith M (1993) A ß-ketoacyl-acyl carrier protein (<u>fab</u>H) is encoded on the chloroplast genome of the red alga <u>Porphyra umbilicalis</u>. Plant Mol. Biol. 21:185-189.

Rhodes D, Hanson AD (1993) Quaternary ammonium and tertiary sulfonium compounds in higher plants. Annual Review of Plant Physiol. and Plant Mol. Biol. 44:357-384.

Sanford JC (1990) Biolistic plant transformation. Physiol. Plant 79:206-209.

Schobert B, Tschesche H (1978) Unusual solution properties of proline and its interaction with proteins. Biochim. Biophys. Acta 541:270-277.

Schulz A, Wengenmayer F, Goodman HM (1990) Genetic engineering of herbicide resistance in higher plants. Critical Rev. in Plant Sciences 9:1-15.

Shah DM, Horsch RB, Klee HJ, Kishore GM, Winter JA, Tuner NE, Hironaka CM, Sanders PR, Gasser CS, Aykent S, Siegel NR, Rogers SG, Fraley RT (1986) Engineering herbicide tolerance in transgenic plants. Science 233:478-481.

Somero GN (1986) Protons, osmolytes, and fitness of internal milieu for protein function. American Journal of Physiology 251:197-213.

Sporlein B, Streubel M, Dahlfeld G, Westhoff P, Koop HU (1991) PEG-mediated transformation: a new system for transient gene expression assays in chloroplasts. Theor. Appl. Genet. 82:717-722.

604

Staub J, Maliga P (1992) Long regions of homologous DNA are incorporated into the tobacco plastid genome by transformation. Plant Cell 4:39-45.

Staub JM, Maliga P (1993) Accumulation of D1 polypeptide in tobacco plastids is regulated via the untranslated region of the psbA mRNA. EMBO J. 12: 601-606.

Svab Z, Hajdukiewitz P, Maliga P (1990) Stable transformation of plastids in higher plants. Proc. Natl. Acad. Sci. USA 87: 8526-8530.

Svab Z, Maliga P (1993) High frequency plastid transformation in tobacco by selection for a chimeric aadA gene. Proc. Natl. Acad. Sci. USA 90:913-917.

Takeda Y, Hirokawa H, Nagata T (1992) The replication origin of proplastid DNA in cultured cells of tobacco. Mol. Gen. Genet. 232:191-198.

Umbeck PF, Barton KA, Nordheim EV, McCarty JC, Parrott WL, Jenkins JN(1991) Degree of pollen dispesal by insects from a field test of genetically engineered cotton. J. Econ. Entomol. 84: 1943-1950.

Vaeck M, Reynaerts A, Höfte H, Jansens S, De Beuckeleer M, Dean C, Zabeau M, Van Montagu M, Leemans J (1987) Transgenic plants protected from insect attack. Nature 328: 33-37.

Vierling E (1991) The roles of heat shock proteins in plants. Annu. Rev. Plant Physiol. Plant Mol. Biol. 42: 579-620.

Warren GW, Carozzi NB, Desai N, Koziel M (1992) Field evaluation of transgenic tobacco containing a Bacillus thuringiensis insecticidal protein gene. J. Econ. Entomol. 85:1651-1659.

Widner WR, Whiteley HR (1989) Two highly related insecticidal crystal proteins of Bacillus thuringiensis subsp. kurstaki possess different host range specificities. J. Bacteriol. 171: 965-974.

Weretilnyk EA, Hanson AD (1990) Molecular cloning of a plant betaine-aldehyde dehydrogenase, an enzyme implicated in adaptation to salinity and drought. Proc. Natl. Acad. Sci. USA 87:2745-2749.

Wong EY, Hironaka CM, Fischoff DA (1992) Arabidopsis thaliana small subunit leader and transit peptide enhance the expression of Bacillus thuringiensis proteins in transgenic plants. Plant Molecular Biology 20: 81-93.

Wyn Jones RG, Storey R (1978) Salt stress and comparative physiology in the Gramnieae. II. Glycine betaine and proline accumulation in two salt and water stressed barley cultivars. Austr. J. Plant Physiol 5:817-829.

Ye GN, Daniell H, Sanford JC (1990) Optimization of delivery of foreign DNA into higher plant chloroplasts. Plant Mol. Biol. 15:809-820.

NATO ASI Series H

NATO ASI Series H

NATO ASI Series H

NATO ASI Series H

NATO ASI Series H

Springer-Verlag
and the Environment

We at Springer-Verlag firmly believe that an international science publisher has a special obligation to the environment, and our corporate policies consistently reflect this conviction.

We also expect our business partners – paper mills, printers, packaging manufacturers, etc. – to commit themselves to using environmentally friendly materials and production processes.

The paper in this book is made from low- or no-chlorine pulp and is acid free, in conformance with international standards for paper permanency.

Printing: Druckerei Zechner, Speyer
Binding: Buchbinderei Schäffer, Grünstadt